Green Energy and Technology

Climate change, environmental impact and the limited natural resources urge scientific research and novel technical solutions. The monograph series Green Energy and Technology serves as a publishing platform for scientific and technological approaches to "green"—i.e. environmentally friendly and sustainable—technologies. While a focus lies on energy and power supply, it also covers "green" solutions in industrial engineering and engineering design. Green Energy and Technology addresses researchers, advanced students, technical consultants as well as decision makers in industries and politics. Hence, the level of presentation spans from instructional to highly technical.

Indexed in Scopus.

More information about this series at http://www.springer.com/series/8059

Mohammad Jawaid · Akil Ahmad ·
Norli Ismail · Mohd Rafatullah
Editors

Environmental Remediation Through Carbon Based Nano Composites

Springer

Editors
Mohammad Jawaid
Laboratory of Biocomposite Technology
Universiti Putra Malaysia
Serdang, Selangor, Malaysia

Akil Ahmad
School of Industrial Technology
University Sains Malaysia
Penang, Pulau Pinang, Malaysia

Norli Ismail
School of Industrial Technology
Universiti Sains Malaysia
Penang, Pulau Pinang, Malaysia

Mohd Rafatullah
School of Industrial Technology
University Sains Malaysia
Penang, Pulau Pinang, Malaysia

ISSN 1865-3529 ISSN 1865-3537 (electronic)
Green Energy and Technology
ISBN 978-981-15-6698-1 ISBN 978-981-15-6699-8 (eBook)
https://doi.org/10.1007/978-981-15-6699-8

This Springer imprint is published by the registered company Springer Nature Singapore Pte Ltd.
The registered company address is: 152 Beach Road, #21-01/04 Gateway East, Singapore 189721, Singapore

The Editors are honored to dedicate this book to Dr. Akil father (Late Jamil Ahmad) and Uncle (Late Taufique Ahmad Khan)

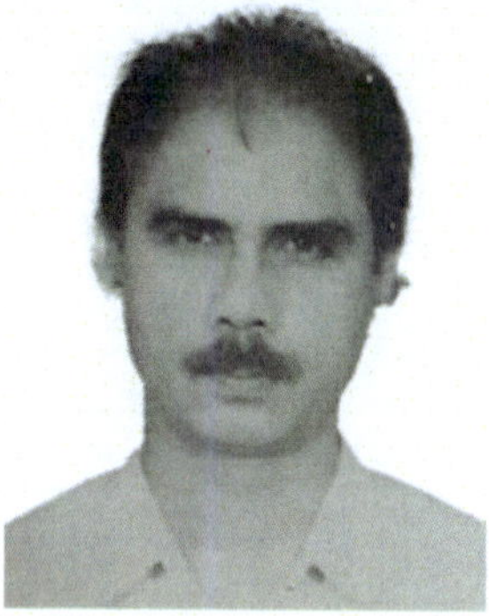

They inspired me for higher studies, and I attribute them whatever I have achieved.

Preface

The specialism in carbon-based nanocomposites are inaugurated by its outstanding properties like high surface area, pore size, mechanical strength and toughness, electrical and thermal conductivity. Unlike other materials, this nanocomposite is easily synthesized and fabricated and doped with various metal oxide nanoparticles. Carbon-based nanocomposites are currently considered as an efficient material for pollutants removal as compared to other available materials. This may be due to its unique characteristics such as low cost, high regenerability, high adsorption capacity, environmental friendly and sustainability. The synthesis and characterization of nanocomposites play a crucial role to find out its potentiality in different real-world applications. It elaborates the basic synthetic route of these nanocomposites to make them highly efficient such as the presence of active sites, high mechanical strength, conductivity and thermal stability properties. Added with significant morphological and structural properties for clear understanding of nanocomposite materials for removal of environmental pollutants have been discussed.

This book gives a sound knowledge of carbon-based nanocomposite to the readers regarding the modern design of nano-sorbents, membrane and photocatalytic degradation materials and manufacture engineering with numerous example illustrations, methods and results for graduate students, researchers and industrialists. Besides that, it also covers the different aspects of carbon-based nanocomposite materials and its application in various environmental fields such as wastewater treatment, air and soil remediation by removal of toxic pollutants. The special features of this book summarize illustration and tables with up-to-date information on research carried out on carbon-based nanocomposite materials and its various applications in different fields.

We are highly thankful to all authors who contributed the chapters and provided their valuable ideas and knowledge in this edited book. We attempt to gather all the scattered information of authors working on carbon-based nanocomposite materials and related research areas from Malaysia, India and Saudi Arabia and finally complete this venture in a fruitful way.

We greatly appreciate the contributor's commitment for their support to compile our ideas. We are highly thankful to Springer Nature team for their generous cooperation at every stage of the book production.

Georgetown, Malaysia — Akil Ahmad
Serdang, Malaysia — Mohammad Jawaid
Georgetown, Malaysia — Norli Ismail
Georgetown, Malaysia — Mohd Rafatullah

Contents

About the Editors

Dr. Mohammad Jawaid is currently working as High Flyer Fellow (Professor) at Biocomposite Technology Laboratory, Institute of Tropical Forestry and Forest Products (INTROP), Universiti Putra Malaysia (UPM), Serdang, Selangor, Malaysia, and also has been Visiting Professor at the Department of Chemical Engineering, College of Engineering, King Saud University, Riyadh, Saudi Arabia since June 2013. He has more than 14 years of experience in teaching, research, and industries. His area of research interests includes hybrid composites, lignocellulosic reinforced/filled polymer composites, advance materials: graphene/nanoclay/fire retardant, modification and treatment of lignocellulosic fibers and solid wood, biopolymers and biopolymers for packaging applications, nanocomposites and nanocellulose fibers, and polymer blends. So far, he has published 37 books, 65 book chapters, more than 350 peer-reviewed international journal papers, and several published review papers under top download/cited articles in science direct during 2013–2020. He also obtained 2 Patents and 6 Copyrights. H-index and citation in Scopus are 49 and 11027 and in Google scholar, H-index and citation are 58 and 15327. He is founding Series Editor of Composite Science and Technology Book Series from Springer-Nature, and also Series Editor-Springer Proceedings in Materials, Springer-Nature. He worked as guest editor of special issues of SN Applied Science, Current Organic Synthesis and Current Analytical Chemistry, International Journal of Polymer Science, IOP Conference Proceeding. He is also Editorial Board Member of Journal of Polymers and The Environment, Journal of Plastics Technology, Applied Science and Engineering Progress Journal, Journal of Asian Science, Technology and Innovation and the Recent Innovations in Chemical Engineering. Besides that, he is also a reviewer of several high-impact international peer-reviewed journals of Elsevier, Springer, Wiley, Saga, ACS, RSC, Frontiers, etc. Presently, he is supervising 20 Ph.D. students (5 Ph.D. as Chairman, and 15 Ph.D. as Member) and 7 Master's students (1 Master as Chairman, and 6 Master as Member) in the fields of hybrid composites, green composites, nanocomposites, natural fiber-reinforced composites, nanocellulose, etc. 24 Ph.D. and 12 Master's students graduated under his supervision in 2014–2020. He has several research grants at university, national, and international levels on polymer composites of around 3 million Malaysian ringgits (USD 700,000). He also delivered plenary and invited talks in international conferences related to composites in India, Turkey, Malaysia, Thailand, the United Kingdom, France, Saudi Arabia, Egypt, and China. Besides that, he is also a member of technical committees of several national and international conferences on composites and material science. Recently Dr. Mohammad Jawaid received Excellent Academic Award in Category of International Grant-Universiti Putra Malaysia-2018 and also Excellent Academic Staff Award in industry High Impact Network (ICAN 2019) Award. Beside that Gold Medal-Community and Industry Network (JINM Showcase) at Universiti Putra Malaysia. He also Received Publons Peer Review Awards 2017, and 2018 (Materials Science), Certified Sentinel of science Award Recipient-2016 (Materials Science) and 2019 (Materials Science and Cross field). He is also Winner of Newton-Ungku Omar Coordination Fund: UK-Malaysia Research and Innovation Bridges Competition 2015.

Dr. Akil Ahmad currently working at Universiti Sains Malaysia as Teaching Fellow and having the experience of five years as postdoc and six months as Visiting Researcher from Chemical Engineering, Centre of Lipid Engineering and Applied Research, University Technology Malaysia (UTM), Malaysia. He has also teaching experience at the college level (Gagan College of Management and Technology) and lectured the graduate and undergraduate students. He has completed Ph.D. in Analytical Chemistry (2011) with the topic “Modification of resin for their use in the separation, preconcentration and determination of metal ions” from Aligarh Muslim University (AMU), India. His research interest in the areas of environmental pollutants and their safe removal, synthesis of nanoparticles and Nano-sorbents (GO, CNT), photo-degradation and antimicrobial effects, preparation of various Chelating sorbents/adsorbent to prevent the environment, water and wastewater treatment, adsorption and ion-exchange, instrumental methods of analysis, thermodynamic and kinetic studies, and preconcentration and method validation with standard reference materials. He has published more than 75 research articles and chapters in the journals and publishers of international repute such as Scientific reports, Talanta, Journal of Chemical Engineering and Data, RSC Advances, Journal of Environmental Sciences, Analytical methods, Journal of Industrial and Engineering Chemistry, Journal of Molecular Liquids etc. H-index and citation in Scopus are 15 and 850 and in Google scholar, H-index and citation are 18 and 1087. He is editorial board member of various journals namely Current World Environment (ISSN: 0973-4929), Sriwijaya Journal of Environment (P-ISSN: 2527-4961) and Journal of Environmental Science, Computer Science and Engineering & Technology (E-ISSN: 2278-17X).

Dr. Norli Ismail holds a Bachelor Degree in Environmental Science from Universiti Putra Malaysia, Masters Degree in Chemical Processes and a Ph.D. in Environmental Technology from Universiti Sains Malaysia. Dr. Norli completed her Ph.D. studies in October 2003 before proceeding to join the School of Industrial

Technology as a lecturer attached to Environmental Technology Division. She has research experienced in various areas of environmental science and technology with particular emphasis on water quality, management issues, and treatability studies in relation to water, wastewater and analytical testing. She is actively involved in the technical aspects of biological and physico-chemical treatability studies, bioremediation research, environmental analytical techniques, sampling and data validation. She was involved in research projects on wastewater treatment and acclimatization studies for a semiconductor industry and as a co-researcher for sewage treatment plant development modification of private company; and also as a co-researcher for the Bioremediation of an Industrial Waste (Latex Effluent) project. She has research experience collaborating on hydrothermal gasification of palm oil mill effluent (POME) with Osaka Gas Ltd., Japan, at the Eco-energy department. Her research focuses on low cost treatment technologies of different types of wastewater which may be used in developing countries.

Her current research fields of interest are an enhancement of biogas production from agricultural biomass through anaerobic fermentation processes and bioflocculant production for water, wastewater and solid waste treatment. Her research interests also expanded to microbiology of wastewater, pharmaceutical waste, heavy metals biosorption, and XOCs bio-degradation. She has supervised more than 10 Ph.D. students and 16 master students. She authored and co-authored more than 88 articles in international peer-reviewed journals with 3930 total citations and 14 H-index as of September 27, 2018. She has authored and co authored 6 of Book chapters.

Dr. Mohd Rafatullah currently is a Senior Lecturer of Environmental Technology in the School of Industrial Technology, Universiti Sains Malaysia (USM), Malaysia. He joined this School in the year 2008 as a Postdoctoral Fellow. He completed his education; Ph.D. in Environmental Chemistry, Master of Science in Analytical Chemistry and Bachelor of Science in Chemistry from Aligarh Muslim University (AMU), India. His research interest is in the areas of environmental pollutants and their safe removal, preparation of various nano-materials to prevent the environment, water and wastewater treatment, adsorption and ion-exchange, thermodynamic and kinetic studies, and activated carbons and its electrochemical properties etc. He has published several reviews articles and regular research papers in the journals of international repute and presented his research work in various national and international conferences. He is a life member of Indian Society for Surface Science and Technology. Dr. Rafatullah won the Hall of fame award in the research publication category from his University. H-index and citation in Scopus are 24 and 3208 and in Google scholar, H-index and citation are 28 and 4466.

Toxicology and Environmental Application of Carbon Nanocomposite

Asim Ali Yaqoob, Mohamad Nasir Mohamad Ibrahim, Akil Ahmad, and A. Vijaya Bhaskar Reddy

Abstract Releasing of toxic metals (Pb, Cd, Hg, Zn, Cr, etc.) and organic compounds (dyes, PAHs, volatile organic compounds, etc.) into the environment has become main resources of environmental pollution which affects the human being and other organism life directly or indirectly. These inorganic and organic compounds have adverse effect on living being even in trace amounts. Various materials like plant-derived biosorbents, resins, metal oxides, natural fibres, carbonaceous materials, etc., have been used for the treatment of organic and inorganic pollutants to clean the environment. From the last decade, carbonaceous materials like graphene have been extensively used for the treatment of environmental pollutants. In this book chapter, we discuss the carbonaceous materials properties, and it is used in the treatment of toxic inorganic and organic compounds removal from the environment.

Keywords Nanocomposites · Toxicity · Metals · Dyes · Environment

1 Introduction

Environmental pollution is one of burning issues that affects the ecosystem, human health and biodiversity globally by polluting the water bodies and natural soil. This urgent issue cannot be addressed by using traditional and casual tools or strategies. The accumulation and collection of toxic organic and inorganic pollutants like toxic metals, dyes, organic compounds, inorganic compounds in water resources, air and soil increase the pollution level to affect the life of living organism [1–4]. Due to

A. A. Yaqoob · M. N. M. Ibrahim (✉)
School of Chemical Sciences, Universiti Sains Malaysia, 11800 Penang, Malaysia
e-mail: mnm@usm.my

A. Ahmad (✉)
School of Industrial Technology, Universiti Sains Malaysia, 11800 Penang, Malaysia
e-mail: akilchem@yahoo.com

A. Vijaya Bhaskar Reddy
Ultra International Limited, Quality Control Division, Sahibabad, Ghaziabad, Uttar Pradesh 201020, India

M. Jawaid et al. (eds.), *Environmental Remediation Through Carbon Based Nano Composites*, Green Energy and Technology,
https://doi.org/10.1007/978-981-15-6699-8_1

growing population and their unplanned anthropogenic activities lead to the degradation of soil quality. Degradation of soil surface may be due to the improper utilization of natural resources. Other important reason is inappropriate dumping of wastes [5]. Furthermore, the progress in scientific world modern technology and increase in industries have led to an rise in discarding the wastes, fluctuating from casual waste to fissionable waste into the environment. This poses a dangerous health threat for existence of humankind on the earth. The wastewater remediation becomes vital for controlling environment contaminants, even with influence on aquatic environment. Generally, different types of wastes such as farming waste, industries waste, local waste, emitting radiation, nuclear waste have been released and enter into the environment. There are different regulations and guidelines to treat the several wastes. Therefore, the sewage shows a main environmental challenge and should be accurately preserved and disposed to regulate the several types of hazardous. Sewage is completely originated through different activities of human being [6]. The valuable materials can be reproduceable by using this sewage as organic sludge or gaseous state like methane or CO_2 which can be collected from it. Similarly, another example, microbial fuel cell is used to treat wastewater, and simultaneously, it helps to produce electricity to overcome the energy issue [7]. The most serious concern of today's time is water pollution which led to dangerous situation for living organism on earth. The water pollution reduces the presence of oxygen for aquatic living organism which causes difficulty in breathing. Occasionally, pollution disturbs the whole food chain. The fishes engage with severe pollutants which are dangerous chemicals for their life. These toxins damage the life cycle. Water pollution from some actions such as oil spills, acid rain entirely destroys the aquatic environments. The water pollutants are mainly organic, inorganic and dyes-based, but the toxic metals including heavy metals and some dyes are very dangerous to aquatic life even in trace amounts. The toxic metals contain an actual heterogeneous cluster of elements broadly diverse in their biochemical and living functional properties. The toxic metals are considered as highly toxicant to environment due to adverse effects on animals, microorganisms, sea life and also badly affect the human being health. The metal pollution in soil is the consequence of natural and anthropogenic activities. Anthropogenic actions include smelting operation, mining and farming have enhanced the toxic metals concentration levels such as Co, Ni, Cr, Cd, Pd and Ni [2, 3, 8, 9]. Metals are persistent in nature, consequently get collected in earths. Nutritional intake of several toxic metals done by plants-based food consumption has long period of harmful effects on the human health. The influence of metals on water bodies is due to pollutants movements from several diffuse sources which provide increase in coincidental combinations to ecosystem. Therefore, it poses danger to water fauna especially to fishes, and it is well known as a main source of protein food for human being. The metals like cadmium, mercury and lead do not carry any biological consequence or valuable use and acknowledged to be enormously toxic [10]. If nickel, copper, zinc, manganese, chromium and tin dispersed in zone of biosphere, then these kinds of metals are not easy to degrade or recovered. So, environmental influence of metal is also called permanent pollution. The metal pollution has damaging effect on life structures, and it not endures the process of biodegradation. Lethal metals like Ni, Au, Zn, Pb, Cr, Cd, Ar, Hg and Cu can be distinguished from other types of environmental pollutants, thus producing several diseases and syndromes at comparatively low concentrations. Hence, there is essential requirement to assess the degree of pollution caused due to heavy or toxic metals and to regulate the monitoring process to keep save the environment. However, like metals pollutants, there are some industrial, domestics and agriculturally based pollutants are also present which cause adverse effect to human being and other living organisms [11–14]. From different sources, the inorganic pollutants are discharged into water resources or natural resources. Inorganic-based pollutants are typically minerals, salts, metals and inorganic compounds like nitrates, sulphate, ammonia, etc. Several studies have shown that inorganic material is found naturally, and it can be modified by different researcher to enhance their application. The inorganic material-based pollutant enters into environment through several anthropogenic activities like mine drainage, chemical processes and smelting metallurgical [15]. These pollutants become toxic due to collection in food chain. Similarly, organic-based pollutant including dyes is known as biodegradable pollutants in environment. These types of pollutants are naturally present and produced through different environmental activities. The most important source is anthropogenic activities. Some common organic pollutants such as polycyclic aromatic hydrocarbons, food waste, petroleum, human waste, polychlorinated biphenyls, diphenyl ether and organochlorine substance are of great concern to eliminate from environment [16–19]. The organic pollutant becomes toxic in environment due to certain properties such as high solubility of lipids, better stability, good lipophilicity and high hydrophobicity [20, 21]. These properties give power to compound to accumulate in diverse sphere zone of environment. From the last decade, carbon-based material emerged as potential material, and it is composite with other material which is used for wastewater treatment in several ways like adsorption, membrane technology and microbial fuel cell. The carbonaceous material is carbon cloth, carbon fibre, activated carbon, graphene oxide, reduced graphene, carbon rode, graphitic material, etc. The most emerging carbon material is graphene and its derivatives due to high-tech properties which make it more prolific in field of wastewater treatment [22]. In this chapter, the major focus is to describe the toxicity limit of several environmental water pollutants (both organic and inorganic) which causing severe diseases. The carbon-based nanocomposite carried out several environmental applications such as adsorption, membrane technology, antimicrobial applications are also summarized in this chapter.

2 Toxicity

There are several types of organic, inorganic and biological pollutant present in the environmental to disturb the ecosystem activities. These types of pollutants enter into the human body in different ways, and some are very toxic in nature even at very low concentration such as mercury, arsenics, lead, cadmium, pesticides and persistent organic-based pollutants [16, 19, 23–25]. The major source of these pollutants is

Fig. 1 Some common water pollutant

wastewater which makes the environment unstable for living organism. Generally, persistent organic pollutants (POPs) are the cluster of chemicals which is globally concern because it carries probable for long-term carriage, perseverance in environment, capacity to bio-magnify, etc., and their substantial harmful effects on humanoid health and ecosystem. The two major classes are found which producing water pollution, i.e. organic pollutant and inorganic pollutant which are summarized below, and a graphical presentation is shown in Fig. 1.

2.1 Inorganic Pollutants

The inorganic elements/compounds or composites seem to be amongst the most common contaminants of ecology. Occasionally, they are existing in such concentrations where these inorganic compounds/composites are able to produce lethal effects on human being and other living organisms. The quantification and identification of inorganic materials by traditional systematic techniques permit the contamination level of environment to be quantified. Furthermore, the actual problem is expecting the influence and their harmful effects of these inorganic materials on living organisms' survivals [26]. For active protection, location of reduction approach and consecutive revitalization of ecosystems is essential to distinguish the environmental superiority of external water. This superiority is assumed mainly by organic influence of water surface effluence. Contaminants come through several kinds of resources such as point resources particularly releases of wastewater, diffuse-based, non-points resources and also considering the atmospheric deposition. Materials contained in contamination are often lethal to water organisms [27]. The concentrations of inorganic substance in external water are lower, but the contaminants are existing in water atmosphere for long term. Under these circumstances, the materials can serve as chronically. Danger of chronic influence of surface water contamination is actual frequently underestimated due to concealed long-term action. The influence of these toxic substances leads to the damage of organs of human being. A disaster of significant dynamic functions, which reduce the organism strength, is further common. From this purpose, a typical stability of ecosystem might be influenced, and lastly, it can lead to destruction of ecosystem. The most dangerous inorganic pollutant is toxic metals which are present in wastewater and toxic metals are main source of inorganic pollutant. The contaminated risk of external water effluence is measured from consequences of assessment of deadly danger of inorganic and organic fragment of external aquatic pollution. Subsequent degree of poisonous risk of entire pollution is assumed by main degree sensed in water pollution. The toxic danger of entire surface water pollution is done through the maximum range of lethal risk measured for water pollution. Table 1 demonstrates contaminated risk degrees and reliable classes of external water class and primary actions as refer by Czech national standard [27].

Table 1 List of toxic risk condition, feasible utilization, primary action and type of water quality are summarized

S. No.	Type of water quality	Toxic risk condition	Priority action	Probable utilization of water
1	High pure water	Unimportant risk	No need for action	Appropriate for all utilization
2	Pure water	Reasonable risk	Supported protection from more rise of pollution	Appropriate for common of utilization, particularly for: drinking water, making water sporting fish breeding industries, landscape value
3	Contaminated water	Maximum tolerable risk	Supported defence from more rise of pollution	Appropriate for industrial supply
4	Highly contaminated water	High risk	Contamination has long-lasting impact, need of long term	Partial possibility of utilization like used in experiment at lab scale
5	Very highly contaminated water	Serious risk	Contamination has acute influence, need of urgent treatment action	Not appropriate for any type of utilization

The heavy metals need to be treated by using various effective methods. Recently, Pan et al. [28] studied the removal method called ion-exchange membrane for removing metals concentration from wastewater because some metals are even more toxic at less concentration. The carbon foam for removal of heavy metal from several industrial wastes plating-based water and harmfulness assessment also discussed [11]. The electroplating surplus holds several kinds of toxic materials, like toxic heavy metals, cleansing agents and solvents. Carbon foam was employed as an adsorbent for remediation of toxic heavy metals from actual manufacturing plating wastewater. The sorption volume was associated with a viable ion-exchange resin and a metal-based adsorbent in batch system. The carbon foam has comprised of high sorption volume for Cu and Cr than viable adsorbents for alkali/acid-based wastewater and cyanide wastewater. Furthermore, the cytotoxicity experiment exposed that an advanced adsorbent has lower toxic effects on human cells. Therefore, the carbon foam was showed the high sorption volume for Cu (14.86 g kg^{-1}), Cr (73.64 g kg^{-1}), Ni (7.74 g kg^{-1}) within 14.5 days of operational time. The oxidation pre-treatments through using UV/H_2O_2 rise the removal rate of metal from plating wastewater which also contain the cyanide compounds. The heavy and some other metal tolerance limit and their adverse effect on human body are listed in Table 2.

2.2 *Organic Pollutants*

Many regions of world are suffering with water deficiencies, almost a billion people lacking with freshwater. Furthermore, 90% infectious-based diseases in several countries are conveyed from contaminated water [40]. The organic pollution is showing the presence of organic compounds in larger amount in water sources. It originally comes from different urban sewage, domestic wastes, agricultural effluents and industrial wastes which contain different organic compounds and dyes in excessive amount. In industrial level, food processing industries, plant treatment industries, paper industries and several other are included [41]. When organic pollutant undergoes decomposition process, the dissolved oxygen consumed into water at high ratio than it might be replenished, which producing depletion of oxygen and causing several water-based diseases. Organic pollutant with wastewater has high amount of solid particles which decrease the supply of light to photosynthetic-based organisms, and it is not good for invertebrates. The organic pollutant is usually phenols, biphenyls, pesticides, oils, proteins, fertilizer, pharmaceuticals substances, carbohydrates, detergents, plasticizers, greases, and several types of dyes which also called organic dyes are included. Therefore, organic pollutants are causing many types of environmental issues, and most commonly named is persistent organic pollutants and abbreviated as POPs [42]. It is a great concern of today's time, due to their high toxicity effect, long-term transport stability and bioaccumulation in living organisms. POPs are basically carbonaceous compounds and mixture of twelve pollutants, i.e. industry-based chemicals like dibenzofurans, polychlorinated biphenyls, organochlorine pesticides, polychlorinated dibenzo-dioxins and dichloro-diphenyl-trichloroethane [43]. When

Table 2 List of inorganic-based pollutant effect on human body along with tolerance limits

S. No.	Inorganic pollutant	Tolerance limit (mg/L)	Effect on human body	References
1	Lead	0.05	It caused high blood pressure, attention deficit hyperactivity disorder, reduced fatal growth, liver damage	[29]
2	Chromium	0.05	Damage nervous system	[29]
3	Cadmium	2.00	Cancer, kidney damage, bone marrow diseases	[30]
4	Copper	0.05	Anamnia, liver damage	[29]
5	Zinc	5.00	Nerve disorder, skin diseases	[31]
6	NH_3-N	1.0 ppm		[29]
7	Nickel	0.01	Bronchitis problems, reduced lung function	[32]
8	Iron	0.15	Iron deficiency anaemia and even death	[33]
9	Mercury	0.01	Protoplasm poising, nervous system damage	[34]
10	Selenium	0.05	Lower selenium levels increased risk of heart disease in human	[35]
11	Sulphides	5	Extremely rapid unconsciousness and death	[36]
12	Organo-phosphor compounds	1	Long-term contact to organophosphates can produce the anxiety, loss of remembrance, damage appetite, disorientation, despair and personality fluctuations	[29]
13	Chlorinated hydrocarbons	0.2	Damage central nervous system, reduce reproductive, damage liver, increase kidney toxicity, and cause carcinogenicity	[37]
14	Fluorides	15	Caused skeletal fluorosis	[38]
15	Residual chlorine	1	Caused pulmonary oedema	[39]

these all enter in excessive amount to environment by different sources, they damage the ecosystem and aquatic life. There are many effective and active techniques are available to remove different types of organic pollutants from wastewater which are quiet successful techniques like microbial fuel cell, adoration process, coagulations, ozonation, advance oxidation method, ion-exchange process, precipitation technique and reverse osmosis. Some methods have some drawbacks such as high operational cost and not easy to handle, but another side, many effective methods are playing vital role like microbial fuel cell (MFC) method, adsorption, ion-exchange method and reverse osmosis. Recently, MFC got much attractive interest from scientific world due to several advantages such as low-cost method, energy production as extra merit and easy to handle. The ion-exchange method and reverse osmosis are mostly used, but it is not suitable economically to use at large scale [44]. The adsorption is also got much attention, and, in this process, solid adsorbents illustrate it most effective technique for removal of pollutants. It is very simple and easy to handle, and also, it can work in less budget with no large space required for this operation. This method entirely depends on the performance of adsorbent and most commonly and useful is carbon-based adsorbent to treat wastewater which has high properties in favour of process. Carbonaceous-based material is available easily and at very low cost, e.g. activated carbon can be obtainable from waste material and employ as adsorbent [45–47]. The adsorption is successfully used to remove different types of dyes and carbonaceous material, and its composites are employed to achieve bright performance as shown in Table 3.

The toxicity of organic compounds and dyes is very dangerous when it is exceeding the tolerance limit into water which is dangerous for human health. Qin et al. [63] described that the toxicity effect of organic compounds in freshwater. According to results, the toxicity of organic-based chemicals was entirely depending upon the hydrophobicity factor. A solo typical for mutually non-polar and polar narcotics was established through polarity inclusion descriptor and also with hydrophobic parameter. The vastly hydrophobic polarity could be preserved as non-polar due to their functional group. The small change in polarity greatly influenced the hydrophobicity factor. To examine the toxic mechanism of exploit for volatile compounds, the response-surface method was utilized to progress model's consequent from simply designed descriptors. Benzoic acids are simply absorbed through unicellular bacteria. Therefore, the presence of these toxic substances in water affect fishes which need to be eliminated from the water bodies using suitable techniques [64]. The organic compound and dyes have several bad effects on human being health and different class of organic pollutant [65].

Table 3 List of carbon-based adsorbents

S. No.	Adsorbent material	Adsorption capacity	Target analyte	References
1	Salix psammophila activated carbon	225.89 mg/g	Methylene blue	[48]
2	Bituminous coal-based activated carbon	580 mg/g	Methylene blue	[49]
3	Hemidesmus Indicus carbon	370 ppm	Phenol	[50]
4	Activated carbon	0.27 mmol/g	Reactive blue 2	[51]
5	Coal-based carbon	234.0 mg/g	Methylene blue	[52]
6	Carbon from Posidonia oceanica	285.7 mg/g	Methylene blue	[53]
7	Activated carbon	0.11 mmol/g	Reactive yellow 2	[51]
8	Carbon from cotton stalk	180.0 mg/g	Methylene blue	[54]
9	Carbon from flamboyant pods	890 mg/g	Methylene blue	[55]
10	Commercial-activated carbon	294 ppm	Phenol	[53]
11	Activated carbon	0.24 mmol/g	Reactive red 4	[51]
12	Granular-activated Carbon	74.07 mg/g	Phenol	[56]
13	Waste tea-activated carbon	203.34 mg/g	Acid blue 25	[57]
14	Mesoporous carbon	428 mg/g	Phenol	[58]
15	Activated carbon from oil palm wood	90.9 mg/g	Methylene blue	[59]
16	Natural clay	15 mg/g	Phenol	[60]
17	Active carbon	257 mg/g	Phenol	[58]
18	Activated carbon from oil palm shell	243.9 mg/g	Methylene blue	[61]
19	Activated carbon from oil palm empty fruit bunch	232.56 mg/g	2,4-dichlorophenol	[62]

3 Environmental Application of Carbon Composite Material

The carbon-based material and its derivatives got much attention in environmental bioremediation because carbon has high-tech properties as well as easily available at low-cost value. The carbon is one of the materials which are employed almost in every wastewater treatment. In environmental application, it is called a potential material to use for keeping stable the environment natural condition. It can serve as photocatalyst, as antimicrobial agent, as adsorbent, and many other uses are reported already. Some common application is summarized in this chapter to enhance its prolific in scientific world.

3.1 Carbon-Based and Its Derivatives Materials as Sorbents

Various sorbents such as activated carbon, clay, biosorbents have been used for the removal of different environmental pollutants. Traditional wastewater treatment depends on physicochemical sorption developments for exclusion of several dyes, organic compounds, organic-based dyes and inorganic pollutants. Eras of research have improved our consideration of sorption procedure and enabled the sorbent optimization properties. The sorption measurements of carbon-based sorbents are inadequate. The huge dimensions of sorbents also bound their conveyance due to lower porosity atmospheres and muddle energies in subsurface treatment. Carbon-based nanosorbents with higher surface area towards bulk ratio, measured pore size circulation and manipulatable surface chemistry stunned several of these essential restrictions [66–70]. Therefore, the significant sorption studies through carbonaceous materials at nanoscale report swift rates of equilibrium, higher adsorption measurements, efficiency over a comprehensive pH series, constancy with BET, etc. The organic pollutants, direct sorption at surface of nanomaterial is determined through similar essential dispersion, hydrophobic and dipolar forces which regulate the energy of sorption in ordinary systems. The high rate of equilibrium in carbon nanosorbents is credited to polarizability of π electron or π-π electron/donor/acceptor relations with aromatic sorbates which condensed adsorption energy of heterogeneity and no pore diffusion mechanism present in the process of adsorption. This outcome is achieved by Yang et al. [68–70] associating a diversity of carbon nanoscale sorbents such as nano C-60(C = carbon) nanoparticles, single wall nanotubes (SWNTs), C-60, and variable MWNTs (multi-walled nanotubes) dimensions saleability. Additional benefit to carbon nanosorbents is the simulated by hysteresis absence between desorption isotherms and adsorption, gasses below distinctive pressure. Improved distinctive pressure, appropriate in hydrogen storing applications, might be re-establish hysteresis in a system through decreasing the barrier of energy to seal non-wetting pores of carbon nanotubes (CNT), and nC-60 aggregates contain intraparticle region. The non-precise van der Walls connections powerful adsorption to SWNTs is improved through increasing the interface abilities in schemes with coiled geometries [71]. The activated carbon conventional applications in term of wastewater remediation are organic pollutants reduction and residual sensitivity. Although carbon-based nanoscale sorbents are active in these zones, their price and conceivable toxicity have prohibited wide research in straight and extensive usage for wastewater handling. Savage and Diallo [72] have planned nanosorbents integration on packed-bed reactors; however, particulars on efficiency of several small-material immobilization approaches have not been accessible. Recently, most investigation on nanosorbents environmental applications has targeted the elimination of precise hazardous pollutants, i.e. polycyclic aromatic hydrocarbons, trihalomethanes and naphthalene. Although quick rate of equilibrium and higher sorbent measurements are influential attributes of nanocarbon sorbents, they are fundamentally developments upon an existing model. The accurate revolutionary nanosorbents come in various pathways for fullerene tailored manipulation and surface chemistry of nanotube [73]. Carbonaceous material like CNTs is considered as a potential sorbents for the treatment of organic and inorganic pollutants.

3.2 Carbonaceous Materials as Antimicrobial Application

The carbonaceous-based material is very suitable for environmental application especially serving as antimicrobial agent. The nano-range size of carbonaceous material or its composite has carried out significant importance in biological/medical field. The exclusive and highly valuable properties especially the nano-range dimensions of carbon allotrope fullerenes and the carbon nanotubes (CNT) have elevated concern between environmental and toxicologists' researchers [74]. Although antimicrobial activities mechanisms are not still fully explored and it is under investigation by researchers, toxicity might be depending on physical, structural and chemical features such as density, functional group, diameter, surface chemistry, length and remaining catalyst pollution. There are many investigators are eager to exploit the importance of antimicrobial properties in term of human health and environmental applications [75]. Definite nanomaterials might be employed as agent for surface antimicrobial coatings, laboratory-based microbiology techniques, water disinfection and medical therapies. The cell membranes are a powerful and unique method in order to control the pathogen. Membrane disorderly agents are wide-ranging spectra of antibiotics, and the toxicity mechanism which depends on physical features cannot stimulate the resistance of antibiotic. Current, Escherichia coli toxicity factor through contact with single-walled carbon nano-range tubes proposes that disruption of membrane is an important source of inactivation. Innovative surface antimicrobial coatings which explored the intrinsic susceptibility of microbes to CNTs may offer well-designed engineering explanations to interesting problem of microbe's colonization and development of biofilm in freshwater classifications, medicinal implantation strategies and further submerged surfaces. Effort on toxicity of CNT in the direction of various microbial groups is continuing [76–81]. Currently, a large number of researchers are working on antiviral and antimicrobial nanoparticles for applications of water remediation and circulation systems. The inactivation and elimination of microbes and viruses depend upon single-walled CNTs hybrid filters as referenced earlier. The carbon-based nanotubes are also known as scaffolding agents for semiconducting-based photocatalysts (TiO_2) and Ag nanoparticles antimicrobial applications [82]. The nano C-60 and fullerol pathogen deactivation also act as agent to remediation purpose for wastewater. Fullerol shows good antiviral activities due to the presence of oxygen molecules which act as superoxide (electron donating molecule) under the

UV light [83]. In conclusion, even nano C-60 interruptions showed strong antibacterial activities to physiologically various microbes concluded as an environmental condition.

3.3 Environmental Sensing Applications

Environmental experts, environmental engineers and ecological experts face a stimulating issue in their work. Pollutants stated in absorptions of parts-per-trillion are connected to ecological inferences stated in millions of litres of water, and many people are affected earlier. The environmental sensing, though, is nanotechnology-based unique application with possible to bond this variety of scale [84]. Currently, activities to monitor environment changes via networked sensor systems will notify prognostic models and form future ecological policy. For example, a cohesive sensor offers an excellent tool for the treatment of wastewater and makes it drinkable. The forecasting and scalability at this scale will entail a set of exact and elastic sensors. The CNT detection devices provide many compensations to present platform sensor [85]. Functionalized CNTs are considered as a potential material in electrode or sensor system due to its superior electric conductivity, large surface area, mechanical toughness, chemical stability and thermal stability. In additional application, collections of aligned multi-walled CNTs have grown up on substrate which is SiO_2 to act as anodes in sensors ionization for gaseous molecule detection. The high-pitched orders of nanotubes enable the higher electric arenas at lower voltages, allowing transferrable, battery-operated and low-scale sensors devices. Exposure happens through electrical decomposition of material tailed by cathode process of an exclusive impression for individually gaseous analyte [86–88]. The introduction of carbon nanowire into sensors system carried an advancement for sensor zones. The charged species adsorption to carbon modifies surface of nano conductance, thus starting a source for association between analyte concentration and current fluctuation. Kong et al. [89] industrialized the gaseous sensors which showed the high electric resistance for semiconducting single-walled CNTs altered through the magnitude after contact to concentrations of NH_3 or NO_2 at room temperature. The wild reply time, lower detection bounds and higher sensitivity factors are a meaning of entirely bare carbon surface area. In order to monitoring or sensing microbial pathogen, biosensor material was employed. Some classifications exploit direct accumulation adsorptive of nucleic acid to carbonaceous material surface as electrode range for electric sensing of hybridization.

3.4 Membrane Technology Application

The comparative permeability of carbonaceous material and its allotrope has been employed for gas as well as for liquid-phase partings through using the membrane technology. The membrane separation is a vigour effective and modest method of manufacturing separation which served as fragment of separation of gas and water purification system. The application of carbon-based composites for separation of gas-phase, e.g. graphene sheets made porous via a higher intensity heat treatment and another chemical etching technology. Both methods have been received much attention in synthesis of graphene membrane accomplished of separation of gases, albeit at solitary gaseous molecules and also it not as gaseous mixtures [90]. Theoretic studies employed for porous graphene membranes for separation of gas which shown the membrane functionalization and pore network modification which can have substantial effect on presentation of materials [91]. In application in term of separations of liquid, numerous investigational studies have been shown, predominantly for water treatment and water desalination. The water highly permeability in structure of graphene, due to lower friction and exclusive molecular water preparation in pores, permits the graphene porous membrane to purpose in higher volume applications like water treatment [92]. The graphene membranes are typically organized in two conformations in water remediation, i.e. the nonporous-based graphene membranes and graphene oxide (GO) stacked membranes. In the previous, the ions separation from water is chiefly done via size elimination and electric contacts with surface of graphene. The GO stacked membranes also function through exclusion of size and electrostatic contacts, but it also considered ion adsorption on membrane inner layers, thus showing high performance in separation developments. In graphene-based membranes and ion rejection considered the several electrostatic contacts through the graphene surface contacts, and performance can be changed and adjusted via functionalized graphene surface [93]. Water flux and rate of rejection of dissolved ions mainly depend on the pore size of graphene-based membrane. The presence of various functional groups such as oxygen, hydrogen, nitrogen and fluorine is responsible for the rejection of dissolved ions and water flux. Graphene membranes in the presence of hydroxyl group functionality recover the water penetrability but unable to find selectivity at higher ionic asset of solution. The pore size control in graphene membranes is actual significant for unvarying performance. The graphene via chemical etching can be used for this purpose, as it increases the faults and aggregation which present in pristine graphene sheet [94]. The carbon allotrope graphene got much attention in field of membrane technology.

4 Conclusion and Future Outlook

Environmental pollution is an emerging issue that disturbs the environment stability and human survival on earth, by contaminating the water and many other natural resources. There are two major types of pollutant, i.e. inorganic and organic pollutants, which contaminate the water resources to make unfit for drinking purpose. In this chapter, both are summarized in detailed, and some significant application of carbonaceous material is discussed. The carbonaceous material is very unique material which has many advantageous factors that enhance its value at industrial level. In terms of practical application, the carbon-based materials and its composites have numerous properties such as greater specific surface area, easy functionalization as well as high mechanical strength which make it potential material for several environment-based applications. Many environmental applications of carbon material like liquid/gaseous phase adsorption, photocatalysis, membrane filtration are discussed and have been considered also hypothetically through using computational chemistry or experimental at nanoscale. It would be valuable assets for environment experts and engineers to develop as a modern technology. The performance of carbon-based materials in term of environmental applications is expressively exaggerated through surface and functionalities features. The upcoming research and progress will likely be directed through magnificently tune to precise the applications. However, carbon-based materials are economical as compared to many other polymeric agents available commercially. Energies in this area are used in emerging ways to yield the nanocomposites and make the production costs low. Research is also ongoing to explore the impact on novel functionalities in terms of biological performance. The carbonaceous material reuse and recovery should make part of investigation which may reduce the expenses. As presentation of materials in ecological applications is expressively exaggerated via functionality and surface properties, upcoming research and growth will depend on these properties for definite applications. While fundamental consideration of the exploit of carbon material and composites for antimicrobial activities is inadequate to few interpretations from investigational and theoretic studies, the technology is growing, particularly in the form of antimicrobial nanocomposite materials. Efforts in environment areas are required in emerging ways to fabricate highly effective nanocomposites at nano range and reduce the price of material by using modern technology.

Acknowledgements All authors are highly thankful to Universiti Sains Malaysia for providing research facilities.

References

1. Islam A, Laskar MA, Ahmad A (2013) Preconcentration of metal ions through chelation on a newly synthesized resin containing O, O donor atoms for quantitative analysis of environmental and biological samples. Environ Monit Assess 185:2691–2704
2. Ahmad A, Siddique JA, Laskar MA, Kumar R, Mohd-Setapar SH, Khatoon A, Shiekh RA (2015) New generation Amberlite XAD resin for the removal of metal ions: a review. J Environ Sci 31:104–123
3. Ahmad A, Mohd Setapar SH, Chuo SC, Khatoon A, Wani WA, Kumar R, Rafatullah M (2015) Recent advances in new generation dye removal technologies: novel search of approaches to reprocess waste water. RSC Adv 5:30801–30818
4. Yaqoob AA, Ibrahim MNM (2019) A review article of nanoparticles; synthetic approaches and wastewater treatment methods. Int Res J Eng Technol 6:1–7
5. Cabrera F, Clemente L, Barrientos ED, López R, Murillo JM (1999) Heavy metal pollution of soils affected by the Guadiamar toxic flood. Sci Total Environ 242:117–129
6. Bolong N, Ismail AF, Salim MR, Matsuura T (2009) A review of the effects of emerging contaminants in wastewater and options for their removal. Desalination 239(1–3):229–246

7. Logan BE, Hamelers B, Rozendal R, Schröder U, Keller J, Freguia S, Aelterman P, Verstraete W, Rabaey K (2006) Microbial fuel cells: methodology and technology. Environ Sci Technol 40(17):5181–5192
8. Yaqoob AA, Parveen T, Umar K, Ibrahim MNM (2020) Role of nanomaterials in the treatment of wastewater: a review. Water 12(2):495
9. Yaqoob AA, Khan RM, Saddique A (2019) Review article on applications and classification of gold nanoparticles. Int J Res 6(3):762–768
10. Ahmad A, Khatoon A, Mohd-Setapar SH, Kumar R, Rafatullah M (2016) Chemically oxidized pineapple fruit peel for the biosorption of heavy metals from aqueous solutions. Desalin Water Treat 57:6432–6442
11. Khan MA, Ahmad A, Umar K, Nabi SA (2015) Synthesis, characterization, and biological applications of nanocomposites for the removal of heavy metals and dyes. Indust Eng Chem Res 54(1):76–82
12. Ahmad H, Ahmad A, Islam SS (2017) Magnetic Fe_3O_4@poly(methacrylic acid) particles for selective preconcentration of trace arsenic species. Microchim Acta 184:2007–2014
13. Umar K, Ibrahim MN, Ahmad A, Rafatullah M (2019) Synthesis of Mn-doped TiO_2 by novel route and photocatalytic mineralization/intermediate studies of organic pollutants. Res Chem Interm 45(5):2927–2945
14. Yaqoob AA, Umar K, Ibrahim MNM (2020) Silver nanoparticles: various methods of synthesis, size affecting factors and their potential applications–a review. App Nanosci 13:1
15. Sivry Y, Riotte J, Sonke JE, Audry S, Schäfer J, Viers J, Blanc G, Freydier R, Dupré B (2008) Zn isotopes as tracers of anthropogenic pollution from Zn-ore smelters The Riou Mort-Lot River system. Chem Geol 255(3–4):295–304
16. Ahmad A, Rafatullah M, Vakili MT, Mohd Setapar SH (2018) Equilibrium and kinetic studies of methyl orange adsorption onto chemically treated oil palm trunk powder. Environ Eng Manage J 17:2933–2943
17. Kohzadi S, Shahmoradi B, Ghaderi E, Loqmani H, Maleki A (2019) Concentration, source, and potential human health risk of heavy metals in the commonly consumed medicinal plants. Biol Trace Ele Res 187(1):41–50
18. Matafonova G, Batoev V (2018) Recent advances in application of UV light-emitting diodes for degrading organic pollutants in water through advanced oxidation processes: a review. Water Res 132:177–189
19. Ahmad A, Lokhat D, Mohd Setapar SH, Khatoon A, Rafatullah M (2018) Nanocarbon composites for detection of volatile organic compounds. In: Khan A, Jawaid M, Asiri AM (eds) Nanocarbon and its composites. Elsevier Woodhead Publishing, pp 401–419
20. Azizi A, Bottaro CS (2020) A critical review of molecularly imprinted polymers for the analysis of organic pollutants in environmental water samples. J Chromatograp A. 1614:460603
21. Yaqoob AA, Ahmad H, Parveen T, Ahmad A, Oves M, Ismail IM, Qari HA, Umar K, Ibrahim MNM (2020) Recent advances in metal decorated nanomaterials and their various biological applications: a review. Front Chem 8:341
22. Yaqoob AA, Ibrahim MN, Rafatullah M, Chua YS, Ahmad A, Umar K (2020) Recent advances in anodes for microbial fuel cells: an overview. Materials 13(9):2078
23. Ahmad A, Khatoon A, Laskar MA, Islam A, Mohammed AW, Yong NL (2013) Use of 2-Hydroxy-3-Methoxybenzaldehyde functionalized Amberlite XAD-16 for preconcentration and determination of trace metal ions by flame atomic absorption spectrometry. Der Pharma Chemica 5:12–23
24. Islam A, Ahmad A, Laskar MA (2015) Flame atomic absorption spectrometric determination of trace metal ions in environmental and biological samples after preconcentration on a new chelating resin containing p-Aminobenzene Sulfonic Acid. J AOAC Int 98:165–175
25. Said M, Ahmad A, Mohammad AW (2013) Removal of phenol during ultrafiltration of palm oil mill effluent (POME), effect of pH, ionic strength, pressure and temperature. Der Pharma Chemica 5:190–196
26. Sauvant MP, Pepin D, Bohatier J, Groliere CA, Guillot J (1997) Toxicity assessment of 16 inorganic environmental pollutants by six bioassays. Ecoto Environ Safe 37(2):131–140

27. Soldán P (2003) Toxic risk of surface water pollution—six years of experience. Environ Int 28(8):677–682
28. Pan ZF, An L (2019) Removal of heavy metal from wastewater using ion exchange membranes. Applications of Ion Exchange Materials in the Environment. Springer, Cham, pp 25–46
29. Gandhi MR, Meenakshi S (2013) Recent advancement in heavy metal removal onto silica-based adsorbents and chitosan composites—a review. Nova Science, pp 201–230
30. Suwazono Y, Sand S, Vahter M, Filipsson AF, Skerfving S, Lidfeldt J, Åkesson A (2006) Benchmark dose for cadmium-induced renal effects in humans. Environ Health Perspect 114(7):1072–1076
31. Valberg L, Flanagan PR, Chamberlain MJ (1984) Effects of iron, tin, and copper on zinc absorption in humans. Am J Clinic Nutrit 40(3):536–541
32. Jemaï R, Lahouli R, Hcini S, Rahmouni H, Khirouni K (2017) Investigation of nickel effects on some physical properties of magnesium based ferrite. J Alloy Compd 705:340–348
33. Griffin IJ, Eckman CB (2019) The effects of iron in a rodent model of alzheimer disease. J Nutrit 149(12):2079–2080
34. Alluri HK, Ronda SR, Settalluri VS, Bondili JS, Suryanarayana V, Venkateshwar P (2007) Biosorption: an eco-friendly alternative for heavy metal removal. African J Biotechnol 6(25):2924–2931
35. Wrobel JK, Power R, Toborek M (2016) Biological activity of selenium: revisited. IUBMB Life 68(2):97–105
36. Martinez-Cutillas M, Gil V, Mañé N, Clavé P, Gallego D, Martin MT, Jimenez M (2015) Potential role of the gaseous mediator hydrogen sulphide (H2S) in inhibition of human colonic contractility. Pharma Res 93:52–63
37. Ruder AM (2006) Potential health effects of occupational chlorinated solvent exposure. Ann New York Acad Sci 1076(1):207–227
38. Kanduti D, Sterbenk P, Artnik B (2016) Fluoride: a review of use and effects on health. Mater Socio-Med 28(2):33
39. Mason JY, Rosenberg JN (2019) Methods for inactivating mosquito larvae using aqueous chlorine dioxide treatment solutions. Google Patents. US 2018/ 0009684 A1
40. Pimentel D, Berger B, Filiberto D, Newton M, Wolfe B, Karabinakis E, Clark S, Poon E, Abbett E, Nandagopal S (2004) Water resources: agricultural and environmental issues. Bioscience 54(10):909–918
41. Cachada A, Pato P, Rocha-Santos T, da Silva EF, Duarte AC (2012) Levels, sources and potential human health risks of organic pollutants in urban soils. Sci Total Environ 430:184–192
42. Bruce-Vanderpuije P, Megson D, Reiner EJ, Bradley L, Adu-Kumi S, Gardella JA Jr (2019) The state of POPs in Ghana-A review on persistent organic pollutants: Environ human exposure. Environ Pollut 245:331–342
43. Burkhard LP, Lukasewycz MT (2008) Toxicity equivalency values for polychlorinated biphenyl mixtures. Environ Toxicol Chem Int J 27(3):529–534
44. Crini G (2005) Recent developments in polysaccharide-based materials used as adsorbents in wastewater treatment. Progn Polym Sci 30(1):38–70
45. Gupta VK (2009) Application of low-cost adsorbents for dye removal—a review. J Environ Manage 90(8):2313–2342
46. Ren X, Chen C, Nagatsu M, Wang X (2011) Carbon nanotubes as adsorbents in environmental pollution management: a review. Chem Eng J 170(2–3):395–410
47. Yong Z, Mata V, Rodrigues AE (2002) Adsorption of carbon dioxide at high temperature—a review. Sep Purif Technol 26(2–3):195–205
48. Bao Y, Zhang G (2012) Study of adsorption characteristics of methylene blue onto activated carbon made by Salix psammophila. Energy Proced 16:1141–1146
49. El Qada EN, Allen SJ, Walker GM (2006) Adsorption of methylene blue onto activated carbon produced from steam activated bituminous coal: a study of equilibrium adsorption isotherm. Chem Eng J 124(1–3):103–110
50. Srihari V, Ashutosh D (2009) Adsorption of phenol from aqueous media by an agro-waste (Hemidesmus indicus) based activated carbon. App Ecol Environ Res 7(1):13–23

51. Al-Degs YS, El-Barghouthi MI, El-Sheikh AH, Walker GM (2008) Effect of solution pH, ionic strength, and temperature on adsorption behavior of reactive dyes on activated carbon. Dyes pig 77(1):16–23
52. Gong GZ, Qiang X, Zheng YF, Ye SF, Chen YF (2009) Regulation of pore size distribution in coal-based activated carbon. New Carbon Mater 24(2):141–146
53. Hashlamon A, Ahmad A, Hong LC (2015) Pre-treatment methods for seawater desalination and industrial wastewater treatment: a brief review. Inter J Sci Res Sci, Eng Technol 1:422–428
54. Girgis BS, Smith E, Louis MM, El-Hendawy AN (2009) Pilot production of activated carbon from cotton stalks using H_3PO_4. J Anal App Pyrol 86(1):180–184
55. Vargas AM, Cazetta AL, Kunita MH, Silva TL, Almeida VC (2011) Adsorption of methylene blue on activated carbon produced from flamboyant pods (Delonix regia): study of adsorption isotherms and kinetic models. Chem Eng J 168(2):722–730
56. Maarof HI, Hameed BH, Ahmad AL (2004) Adsorption isotherms for phenol onto activated carbon. Asian J Chem Eng 4(1):70–76
57. Auta M, Hameed B (2011) Preparation of waste tea activated carbon using potassium acetate as an activating agent for adsorption of Acid Blue 25 dye. Chem Eng J 171(2):502–509
58. Haque E, Khan NA, Talapaneni SN, Vinu A, JeGal JG, Jhung SH (2010) Adsorption of phenol on mesoporous carbon CMK-3: effect of textural properties. Bullet Korean Chem Soc 31(6):1638–1642
59. Ahmad A, Loh M, Aziz J (2007) Preparation and characterization of activated carbon from oil palm wood and its evaluation on methylene blue adsorption. Dyes Pigm 75(2):263–272
60. Djebbar M, Djafri F, Bouchekara M, Djafri A (2012) Adsorption of phenol on natural clay. App Water Sci 2(2):77–86
61. Tan I, Ahmad A, Hameed B (2008) Enhancement of basic dye adsorption uptake from aqueous solutions using chemically modified oil palm shell activated carbon. Coll Surfa A: Physicochem Eng Aspect 318(1–3):88–96
62. Shaarani F, Hameed B (2010) Batch adsorption of 2, 4-dichlorophenol onto activated carbon derived from agricultural waste. Desalination 255(1–3):159–164
63. Qin WC, Su LM, Zhang XJ, Qin HW, Wen Y, Guo Z, Sun FT, Sheng LX, Zhao YH, Abraham MH (2010) Toxicity of organic pollutants to seven aquatic organisms: effect of polarity and ionization. SAR QSAR Environ Res 21(5–6):389–401
64. Mangwani N, Kumari S, Das S (2019) Taxonomy and characterization of biofilm forming polycyclic aromatic hydrocarbon degrading bacteria from marine environments. Polycyclic Aromat Comp, pp 1–14
65. Shukla SK, Mangwani N, Rao TS, Das S (2014) Biofilm-mediated bioremediation of polycyclic aromatic hydrocarbons. Microb biodegrad Bioremed, pp 203–232
66. Allen-King RM, Grathwohl P, Ball WP (2002) New modeling paradigms for the sorption of hydrophobic organic chemicals to heterogeneous carbonaceous matter in soils, sediments, and rocks. Adv Water Resour 25(8–12):985–1016
67. Lu C, Chung YL, Chang KF (2005) Adsorption of trihalomethanes from water with carbon nanotubes. Water Res 39(6):1183–1189
68. Yang K, Wang X, Zhu L, Xing B (2006) Competitive sorption of pyrene, phenanthrene, and naphthalene on multiwalled carbon nanotubes. Environ Sci Technol 40(18):5804–5810
69. Yang K, Zhu L, Xing B (2006) Adsorption of polycyclic aromatic hydrocarbons by carbon nanomaterials. Environ Sci Technol 40(6):1855–1861
70. Yang FH, Lachawiec AJ, Yang RT (2006) Adsorption of spillover hydrogen atoms on single-wall carbon nanotubes. The J Phy Chem B 110(12):6236–6244
71. Cheng X, Kan AT, Tomson MB (2005) Uptake and sequestration of naphthalene and 1, 2-dichlorobenzene by C 60. J Nanoparticle Res 7(4–5):555–567
72. Savage N, Diallo MS (2005) Nanomaterials and water purification: opportunities and challenges. J Nanoparticle Res 7(4–5):331–342
73. Ion AC, Ion I, Culetu A, Gherase D (2010) Carbon-based nanomaterials. Environmental applications. Romania, pp 1–20

74. Wörle-Knirsch J, Pulskamp K, Krug H (2006) Oops they did it again! Carbon nanotubes hoax scientists in viability assays. Nano Lett 6(6):1261–1268
75. Sayes CM, Liang F, Hudson JL, Mendez J, Guo W, Beach JM, Moore VC, Doyle CD, West JL, Billups WE, Ausman KD (2006) Functionalization density dependence of single-walled carbon nanotubes cytotoxicity in vitro. Toxicol Letter 161(2):135–142
76. Czarnecka J, Wiśniewski M, Forbot N, Bolibok P, Terzyk AP, Roszek K (2020) Cytotoxic or not? disclosing the toxic nature of carbonaceous nanomaterials through nano-bio interactions. Mater 13(9):2060
77. Jia G, Wang H, Yan L, Wang X, Pei R, Yan T, Zhao Y, Guo X (2005) Cytotoxicity of carbon nanomaterials: single-wall nanotube, multi-wall nanotube, and fullerene. Environ Sci Technol 39(5):1378–1383
78. Kang S, Herzberg M, Rodrigues DF, Elimelech M (2008) Antibacterial effects of carbon nanotubes: size does matter! Langmuir 24(13):6409–6413
79. Kim JW, Shashkov EV, Galanzha EI, Kotagiri N, Zharov VP (2007) Photothermal antimicrobial nanotherapy and nanodiagnostics with self-assembling carbon nanotube clusters. Lasers Surg Med Official J Am Soc Laser Med Surg 39(7):622–634
80. Liu D, Mao Y, Ding L (2018) Carbon nanotubes as antimicrobial agents for water disinfection and pathogen control. J Water Health 16(2):171–180
81. Tang YJ, Ashcroft JM, Chen D, Min G, Kim CH, Murkhejee B, Larabell C, Keasling JD, Chen FF (2007) Charge-associated effects of fullerene derivatives on microbial structural integrity and central metabolism. Nano Lett 7(3):754–760
82. Mir NA, Khan A, Umar K, Muneer M (2013) Photocatalytic study of a xanthene dye derivative, phloxine B in aqueous suspension of TiO_2: adsorption isotherm and decolourization kinetics. Energ Environ Focus 2(3):208–216
83. Tong Z, Bischoff M, Nies L, Applegate B, Turco RF (2007) Impact of fullerene (C60) on a soil microbial community. Environ Sci Technol 41(8):2985–2991
84. Lyon DY, Brown DA, Alvarez PJ (2008) Implications and potential applications of bactericidal fullerene water suspensions: effect of nC60 concentration, exposure conditions and shelf life. Water Sci Technol 57(10):1533–1538
85. Lin Y, Yantasee Y, Wang J (2005) Carbon nanotubes (CNTs) for the development of electrochemical biosensors. Front Biosci 10(492–505):582
86. Allen BL, Kichambare PD, Star A (2007) Carbon nanotube field-effect-transistor-based biosensors. Adv Mater 19(11):1439–1451
87. Hierold C, Jungen A, Stampfer C, Helbling T (2007) Nano electromechanical sensors based on carbon nanotubes. Sens Actuat A: Phy 136(1):51–61
88. Wagner C, Blaudeck T, Meszmer P, Böttger S, Fuchs F, Hermann S, Schuster J, Wunderle B, Schulz SE (2019) Carbon Nanotubes for Mechanical Sensor Applications. Phys Status Solidi 216(19):1900584
89. Kong J, Franklin NR, Zhou C, Chapline MG, Peng S, Cho K, Dai H (2000) Nanotube molecular wires as chemical sensors. Science 287(5453):622–625
90. Han Y, Xu Z, Gao C (2013) Ultrathin graphene nanofiltration membrane for water purification. Adv Funct Mater 23(29):3693–3700
91. Du H, Li J, Zhang J, Su G, Li X, Zhao Y (2011) Separation of hydrogen and nitrogen gases with porous graphene membrane. J Phy Chem C 115(47):23261–23266
92. Jawaid M, Ahmad A, Lokhat D (2019) Graphene-based nanotechnologies for energy and environmental applications. Elsevier, Amsterdam
93. Gugliuzza A, Politano A, Drioli E (2017) The advent of graphene and other two-dimensional materials in membrane science and technology. Curr Opinion Chem Eng 16:78–85
94. Joshi RK, Alwarappan S, Yoshimura M, Sahajwalla V, Nishina Y (2015) Graphene oxide: the new membrane material. App Mater Today 1(1):1–2

The Role of Carbon Nanocomposite Membranes for Water and Wastewater Treatment

Sapna Raghav, Pallavi Jain, Ritu Painuli, and Dinesh Kumar

Abstract The water pollution is greatest global concerns due to contamination of environmental factors, which is rising day by day. To remove the pollutants from the water and wastewater is a challenging task, and the number of adsorbents and membranes has been synthesized to remove water pollutants. Carbon nanocomposite (CNC) membranes have grabbed worldwide attention in the environmental applications due to its higher adsorption capacity. Advanced CNCs offer certain good characteristics like improved permeation, enhanced rejection, and reduced fouling which is beneficial for water contamination removal. The CNC membranes are improved by physical or chemical modification with various functional groups, which enhanced the removal or desalination capabilities of the membrane from water. The present chapter offers an inclusive review of functionalized CNC membranes and their existing and potential applications for contaminant removal from water and its desalination. The application of the CNC membrane showed various advantages such as antifouling capability, improve water permeability, and high selectivity as well. In this chapter, we are discussing about cellulose, carbon nanotubes (CNTs), and graphene-based CNC membranes for contaminant removal from the wastewater and water.

S. Raghav
Department of Chemistry, Banasthali Vidyapith, Banasthali, Tonk 304022, India
e-mail: sapnaraghav04@gmail.com

P. Jain
Department of Chemistry, SRM Institute of Science & Technology, Delhi-NCR Campus, Modinagar 201204, India
e-mail: pallavij@srmist.edu.in

R. Painuli
Department of Chemistry, Uttaranchal University, Arcadia Grant, Chandanwadi, Premnagar, Dehradun, Uttarakhand 248007, India
e-mail: ritsjune8.h@gmail.com

D. Kumar (✉)
School of Chemical Sciences, Central University of Gujarat, Gandhinagar, India
e-mail: dinesh.kumar@cug.ac.in

M. Jawaid et al. (eds.), *Environmental Remediation Through Carbon Based Nano Composites*, Green Energy and Technology,
https://doi.org/10.1007/978-981-15-6699-8_2

Keywords Membrane · Cellulose · Carbon nanotubes · Graphene · Pollutants · CNCs

Abbreviations

MPD	1-methyl-2-pyrrolidinone
PVF	Polyvinylidene fluoride
PES	Polyethersulfone
PVA	Polyvinyl alcohol
PA	Polyamide
TEMPO	2,2,6,6-tetramethyl-1-piperidinyloxy
BKC	Benzalkonium chloride
BC	Bacterial cellulose
CNC	Carbon nanocomposite
CNF	Cellulose nanofiller
NPs	Nanoparticles
TEMPO	2,2,6,6-tetramethyl-1-piperidinyloxy

1 Introduction

There are a lot of worries for water contaminants. Water contaminants are categorized as inorganic, organic, biological, and chemical, and these are making people aware of the disastrous consequences. Dyes, pesticides, and heavy metals are the major pollutants that come from industrial releases. They get in contact with the water streams and hence deteriorate them. Manually they are thrown away using physical methods. These include leaching of fertilizers and biocides through agricultural treatment. They are leeched into underground water networks through rainwater, gas excretions from industries, and running vehicles.

Water-soluble elements are scattered in parts and are soluble in the stream of water and cannot be seen through naked eyes. Despite having rules by the government regarding this, yet there is only partial enforcement done. Logging is one such activity that plays a major part in water pollution in many countries of the world. To decrease the number of pollutants in water, a plenty of chemical methods have been implied. Chemical compounds that include aluminium sulfate, sodium aluminate, sodium hydroxide, and metals with chlorine are considered best to treat wastewater. This is because of their quick action methods, and the lower price is another reason to use them. These chemicals find their usage in many common treatment methods like water logging, clumping of water, settling of water through gravity, and the most prevalent, which is the separation of pollutants from water using membrane. Despite that, the treatment of water with the help of chemicals might raise another set of problems. Chemical compounds, when treating the contaminants present in the water, can change the architecture of the pollutants. Hence, they are known as chemical intermediates. Reactions that occur among the chemical compounds and contaminants can cause the formation of end products that are very dangerous toward marine animals and humans.

So, to cleanse the wastewater has become essential to save the living species and the biodiversity. Innovations are being done to make new technologies as well as to make drinkable water which can sustain till longer duration [31]. It should also be noted that the methods that should be adopted must be cheap, stabilized, and in harmony with the ecosystem. Many methods have been adopted to treat water. The separation membrane method is considered one of the best techniques to purify water and to remove the salinity. The reason is that it induces high efficiency and has no complex features and uses fewer chemicals. The vital reason to use membrane technology is the fast modifications being done to it [29, 30]. The main point to use these membranes is that it can divide the dirt pollutants with the help of the penetrable pores that it has from the clean water. Also, not all particles get removed, but selective removal takes place with good mechanical strength [19, 25, 60, 61].

Studies on this method are gaining a lot of importance in industrial as well as an academic field [21, 73]. The categories of these techniques involve reverse osmosis (RO), pressure retarded osmosis, gas separation, nanofiltration (NF), ultrafiltration (UF), microfiltration (MF), pervaporation, and separation by liquid membranes. The method used depends on the characteristics of the membrane [4, 43, 52, 83, 92, 107]. Several materials from natural to synthetic bio-/polymers have been utilized to produce membranes for the purification of wastewater and water. For example, polyamide (PA) [74], polyvinyl alcohol (PVA) [68], polyethersulfone (PES) [62], carbon nanotubes [111], polyvinylidene fluoride (PVF), and chitosan (CS) [75].

1.1 Evolution of Membrane

To remove foreign particles or pollutants from the water to make it drinkable, RO is the most prevalent technique known to date. This process can be used with other techniques to treat water and recycle it. The water produced will be free of any contaminants present in it [10, 17]. The procedure of RO was first represented by WF21 in Southern California in 1977. The aim was to recycle the already used water and make it into a drinkable liquid. The usage was to lessen already present micropollutants in the water. The water was earlier subject to traditional methods to clear out lime content, recarbonation, and multimedia filtration. Modern pre-treatments involve a single-step microfiltration (MF) technique. It is denser and more systematic for the dismissal of the pollutants [104].

The highly intensified UV light that is subjected to low pressures with hydrogen peroxide (UV/H_2O_2) has great use in ensuring enough reduction of micropollutants like N-nitrosodimethylamine (NDMA) that are only partially removed [24].

The partial removal is done through the membranes in RO. Another treatment technique MF-RO-UV/H_2O_2, is also used in more prevalence in water recycle plants that are movable. An important improvisation is the straight involvement of membrane bioreactor-107 MBR discharge produced by RO. The MBR technique attains a variety of parts of the bioreactor like biomass separation, and RO pre-treatment. The abatement of ongoing pre-treatment helps to cut down the usage of extra space, energy, and cost using this MBR method. Hence, various plants have adopted this technique [53, 79]. Despite the noticeable improvisations, water recycling is still a major concern as it is undergoing a lot of setbacks. A report suggested the vital need to supply high-quality drinkable water with proper details regarding pathogens and micropollutants. Problems like the utilization of a lot of energy, the deterioration of membrane (smelly), and the collection of disposed of particles are critical and must be checked [11, 90].

1.2 Problems to Overcome

Expenditure of energy

- New kinds of membrane that use a low energy and the processes that are hybrid.
- New kinds of membranes with inflated permeability and peak refinement.

Polluted membrane

- Usage of antifouling membranes.
- The outline sketch that performs in a better way, and the results are antifouling.

The eviction of micropollutants

- New membranes developed to remove small dirt particles.
- New design for the whole process.

Concentrate disposal

- New kinds of processes that are hybrid.

2 Cellulose Nanocomposite Membranes

Recently, more concentration is given to the usage of cellulose nanomaterials (CNMs) as they are organically available and are recyclable for the making of permeable membranes. Cellulose nanofibril (CNF) works out to be a good membrane. The role it plays whether a membrane or filter depends on pore structure. Cellulose-based membranes or adsorbents with the large surface area have a higher efficiency to adsorbed pollutants from the wastewater. This is the most important reason they are used in various recycling processes. Another important issue in these processes is a biofouling and organic fouling. The capability of the micropollutants to foul the membrane is due to their hydrophobic nature alongside their surface functionalization [66].

The longevity of these celluloses made membranes is also a setback with its main implementations on water treatment. The greater strength in these nanomaterials is a contributing factor in the development of highly penetrable structures. The role-play of the thickness/thinness is very important in the continuous change of water or the polluted water. The reason is that as the thickness increases, there is a lessening in the rate flow of particles through the membrane. The cellulose-based membranes can be developed through four different techniques, i.e., development of bacterial cellulose membranes through organic material method, infusion of electro-spun mats with cellulose material, vacuum filtration and coating, and cellulose membranes. The ability to retain pollutants by the membrane is measured through dynamic adsorption methods. This requires the quantitative analysis of the polluted particles adsorbed by the penetrable membrane only under specified flow requirements. The elimination of the adsorbate is assumed to be complete only when the size of the nanoparticle is greater than the pore size in the membrane. The elimination is also possible with the help of repulsive powers.

2.1 *Bacterial Cellulose Membranes*

BC is a pure form of cellulose that can be synthesized by some microorganisms, such as acetobacter xylinum and gluconacetobacter xylinus, which are Gram-negative strains of acetic acid-producing bacteria [42, 51]. During BC synthesis by bacteria, a pellicle forms on top of the static cultured growth medium. The developed composition is of a non-toxic and non-allergen cellulose nanofiber complex. This network has huge tensile strength, along with some other features. The features include elastic nature, resilience, longevity, resilience, retention of shape, and a great capacity to bind water. The penetrable structure of this tissue formed as a coat matches enough conditions for membrane filtration. The pores of BC membranes were tested for the process of filtration of bovine serum albumin (BSA) [103]. These are formed in a duration of two days of cultivation. Time alongside cell density was the factor that affected the hydraulic permeability coefficient throughout the formation of the membrane. When we dry the membrane, a shrinkage is developed, which reduces the porosity. Wet BC membranes with porosities of 95–97% caused the elimination of oil from stabilized or non-stabilized emulsions. The droplet size range was less than a micrometer [36]. The harvesting time increase changes nothing in the membrane's porosity. The thin structure causes the BC membrane to show high pure water flux (845 L/h m^2) in a shorter duration of time in contrast to that harvested in longer durations. The efficiency of elimination of oil from the stable emulsion was 98.3 and 99.3% for non-stable emulsion. The bacterial cellulose membrane could be improvised to increase the performance and to promote its implementation in water treatment. Tetra-aminocobalt(II) phthalocyanine was disabled on a membrane of BC, and $NaIO_4$ oxidized beforehand that to produce-CHO groups to which cobalt(II) phthalocyanine-containing $-NH_2$ groups can be joined in a covalent bond. The covalently bonded membrane was checked for the elimination of a dye (X-3B) which is highly reactive from wastewater. Hydrogen peroxide is added to start the oxidation process that works with a catalyst. In certain standard constraints of a rate of dye solution is 6 mL/min, at 50 °C and hydrogen peroxide concentration of 10 mmoL/L, a fade in color is observed of the catalytic membrane reactor. It was $\approx$50 μmol/min g. Permeable membranes that are derived from bacterial cellulose and GO were made by the diffusion of these membranes into the BC formamide gel [20]. The membrane produced represents comprehensible permeation characteristics for varied inorganic and organic ions. The size of these ions also varies. The membranes, that were not dried ever, were glazed with deacetylated chitin sulfonate and utilized as a membrane for separation. The removal was obtained around 85–90% as output for polyethylene glycol (PEG).

2.2 *Insertion of Electrically Spun Mats*

The above technique had its usage for cellulose nanocrystals that were made by a system that was oxidized in an aqueous state. It was derived on 2,2,6,6-tetramethyl-1-piperidinyloxy)-mediated oxidization (TEMPO)/NaBr/NaClO diffused into poly(acrylonitrile) (PAN) nanofibrous structure which is made by electrical spinning. It is held using support in the form of a substrate which is not woven, ethylene terephthalate [64]. The size of the pore of this many-layered nanofibrous microfiltration system can be customized with the help of cellulose nanocrystals content. A complete retaining ability hostile to bacteria was observed after experiments. It is also noticed that the usage of negatively charged cellulose nanocrystals is not much successful in eliminating viruses with the process of adsorption [96].

After the process of cellulose nanofibril infusion, it was noticed that the pore size was reduced from 0.66 to 0.38 micrometers, and the rate of permeation of pure water was reduced. The membrane has the capacity to eliminate the bacteria *Escherichia coli (E. coli)*. The process involved was the exclusion of the size of the membrane. Because of this, a log reduction value was reached, which was of magnitude. Cellulose nanofibril oxidized by cellulose nanofibril inserted with cysteine to enhance the ability, to adsorb of the metal ions contaminant and then pervaded with a PAN setting electrically spun onto a complex PET holder [108, 109].

In contrast to this membrane, a changed and improvised adsorption of 60 and 115 mg/g for Cr(VI) and Pb(II) was noticed, respectively, because of –SH functionalized cellulose nanofibril. A very close technique was utilized to make membranes based on biochemistry derived from cellulose acetate (CA). These membranes were having a coating of HCl-prepared chitin nanocrystals to get filtration of water membranes with customized surface properties. Membranes that have high hydrophilic nature alongside a very great change in the rate of flow of water were produced. Natural membranes culminated exclusion of size, adsorption, and highly hydrophilic nature were produced by pervading cellulose nanocrystals onto electrically spun CA fiber networks [28]. The change in a flow rate of water through the membranes was of a magnitude of 22,000 L/m^2 h and the nanocrystal network when it became continual, then the magnitude decreased. The contact angle also reduces from 102° to 0° after cellulose nanocrystals were coated onto it, thus showing enhanced hydrophilic nature and antifouling of these membrane entities that are hydrophobic. Coating of spray is also proven to be a good approach to produce a lesser thick and consistent barrier layer on the substrate that is electrically spun derived on the gelation nature of 2,2,6,6-tetramethyl-1-piperidinyloxy-mediated oxidized cellulose nanofibril suspension by low pH value [102]. The direct spinning of the cellulose substitute, electrically, is difficult and needs a lot of chemical demands. The nanofibrous membrane was made by the deacetylation of an electrically spun substrate of cellulose acetate membrane [113]. Good resistance insolvent and high air filtration efficacy were observed.

2.3 *Filtration in Vacuum*

It is an easy, quick, attainable, and simple process used to generate layered structures of membranes derived from nanocellulose and nano-papers lead by the optimized hot pressing. Membranes hydrophilic and oleophobic that were used under waters were made from tunic in cellulose nanocrystals in assistance with vacuum by filtration onto a nylon filter membrane [13]. Developed membrane showed efficiency for the dissociation of oily water, i.e., water-in-oil emulsions or oil in water. For isooctane-in-water nanoemulsion, very high performance in regard to dissociation was obtained, and the rate water flow was changed as well. The efficiency of separation was 100%, and the rate of water flow was greater than 1700 L/m^{2}h bar. Nano-papers that were made using BC, cellulose nanofibril, TEMPO-oxidized cellulose nanofillers, and cellulose nanocrystals that played the role of raw material through a papermaking process were used in nanofiltration implementations for organic solvents in wastewater [70]. The pore size of the nanopapers that were prepared played a very important role in the ultrafiltration efficiency and that the ability to penetrate the particles is determined by its grammage. Solvent stable nanofiltration membranes that were made of TEMPO-oxidized cellulose nanofibril also presented that their permeable capacity depends on the hydrophilic nature of the solvent [69]. Ultrasonicated Cladophora cellulose that was basically aqueous suspensions was separated from the pollutants to prepare membranes that had customizable pore size scattering that was suited for removal of virus [57]. Virus particle removal ability was demonstrated, and it was grounded on the principle of size exclusion with a reduction value higher than 6.3. The proof of excellent functioning was checked regarding the xenotropic murine leukemia virus particles [5]. An efficiency repossession higher than 99% was seen for lysozyme and bovine serum albumin [33]. The pore size distribution of nanopapers retentive of viruses can also be customized in 10–25 nm range by governing water evaporation rate through the hot-press drying step that comes after the vacuum filtration process [34]. A change in the evaporation rate causes broad particle dispersal and large pore size. The properties like wet strength were improvised of the Cladophora cellulose membrane. It was done with the help of crosslinking citric acid, thus helping in the increase of pressure gradient without causing any negative change in its integrity [80].

The cellulose surface chemistry may be customized to govern the sorption conduct. Anion removal membranes were produced with the help of vacuum filtration of cellulose nanofibril cations with a grammage of 30 g/m^2. This insertion of cations further enhanced the permeable behavior of the membrane to unchanged cellulose nanofibril. Phosphorylated cellulose nanofibril nano-papers were also produced for industrial uses via a paper making process [71]. The modified nanopapers presented lesser permeability in contrast to that of unchanged cellulose nanofibril nano-papers but showed the ability to adsorb copper ions. The phosphate groups present on the surface of the nano-paper were contributed higher to the total adsorption compared to the other functional groups present within the nanopaper. Cuterpyridine-modified oxidized cellulose nanofibril membranes were made and used to purify the wastewater of the paper industry [37].

The chemical variation of TEMPO/CNF with the coating of Cu/Tpy with the help of an equimolar ratio was assumed to give a five-coordinate complex. The removal efficiency of TEMPO/CNF was found around 93%, and TEMPO/CNF/Cu/Tpy was around 96%, respectively, for the particles suspended in the wastewater. When it comes to electrically spun membranes, a coat of a helping membrane with cellulose materials can be examined. A good deal of layered cellulose nanocrystals membranes were manufactured by vacuum filtration of cellulose nanofibril suspensions ensued by dip coating with sulfated or carboxylated cellulose nanocrystals [45, 48]. The coated layer of cellulose nanocrystals on the membrane enhanced the thickness and mechanical properties. The treatment of acetone on this membrane was done before drying it up. Drying minimized the inter-chain hydrogen bonding and hence enhanced the pore size up to 194 Å from 74 Å, which caused the high increase in water flux. Removal efficacy of Fe^{3+}/Fe^{2+}, Ag^{+}, and Cu^{2+} ions was observed in both cross-flow as well as static mode, and it was proven that the dip-coated cellulose nanocrystals layer improvised the elimination of the metal ion. The flux values were lower despite the high rejection rates and enhanced flux obtained after acetone treatment for real implementations. Vacuum filtration developed the ultrafiltration membranes of high porosity on aPVDF support layer of 2,3-dicarboxylic acid cellulose nanofibril. These membranes were having rejection efficiencies of 74–80% that were considered high for aqueous dextran [93]. To prepare cellulose nanofibril nanoporous membranes of thickness controlled by the producer, the direct filtration method was used. It was done on a microporous cellulose acetate support. The membranes decolorized the methyl blue present in the aqueous solutions. The vacuum filtration technique was deployed to prepare cellulose nanofibril and its functional layer. This layer is made up of TEMPO-oxidized cellulose nanocrystals and gelatin. There was an effective enhancement in the adsorption efficiencies [46, 85].

2.4 Composite Membranes

The composite membranes were derived from nanomaterials made of cellulose can be manufactured to enhance a good command over the characteristics of filtration. Membranes consisting of 2,2,6,6-tetramethyl-1-piperidinyloxy-oxidized cellulose nanofibril and cellulose triacetate were made by molding from NMP mixtures [22, 50]. Improvised flux and increased performance in antifouling were obtained due to the hydrophilic nature of surface occurring due to TEMPO-oxidized cellulose nanofibril. A combination of manually scattered Cladophora cellulose, cellulose nanofibril, and pyrrole was filtered with the help of $FeCl_3$ to produce the pyrrole polymerization. The diffusion of solute between the composite membranes was seen to be quicker because of the lesser complete porosity and greater occurrence of pores, which were narrow for the latter. The flux was proportional to the pore size, which increased with the increase in pore size, and the ion extraction occurred because of the externally applied electric current. It was also observed that the elimination of minute uremic toxins was better and effective.

Improvised clotting characteristics were seen when putting a coating of the stable. The composite showed better compatibility between biological compounds and inflammable properties. The substantial cleansing was essential to eliminate dirt particles and reactive elements present inside the water. This was done to procure a toxic-free material. The suction force generated to pull out composite suspended particles, including cellulose nanofibril, polyamide-amine-epichlorohydrin (PAE), and SiNPs to create membranes [91]. The membranes obtained had a high amount of flux, but the rejection values obtained were less because of the huge size of the pores. Including silica NPs that played the role of spacers permitted the governance of the pore size of the membrane. PAEs played the role of improvising the adhesion among the negatively charged NPs and the cellulose nanofibril along with the moist strength of the membrane. It is also said that the membranes that are already used can no longer be used again or recycled. But it is known that the traditional recycling process of paper to dispose of it. The membrane comprises a layer of support which is formed of cellulose nanofibers. These layers provide manual stabilization and are having a coating of phosphorylated cellulose nanocrystals-gelatin [45, 48]. Microporous membranes were obtained through dry freezing of sulfated cellulose nanocrystals, and chitosan ensued by pressing. These membranes were stabilized with the method of crosslinking with vapors of glutaraldehyde [47]. The membranes eliminated the dyes. This occurred because of the electrostatic attraction that was present between minus charged cellulose nanocrystals and the plus-charged dyes [106]. The coating of cellulose nanofibril has a structure which is mesoporous. The membrane has hydrophilic and oleophobic properties due to the mesoporous structure.

3 CNT Membranes

Many scientific implementations in which many excellent innovative methods have been used for the elimination of liquid pollutants with the help of CNTs membranes. These are being revised, and additional reasons are added in reference to how they will move for future researches.

3.1 Removal of Inorganic Contaminants

Studies have used functional-CNT (f-CNTs) membranes (for the elimination of inorganic particles, like heavy metals from watery solvents). The adsorption capacity is high because of the substantial specific area, the favorably penetrable, and a pipe structure that is hollow. The presence of functional groups on the surface of f-CNTs and other suitable associations among watery pollutants and f-CNTs [112].

Because of all these characteristics, the culmination of f-CNTs into membranes made of polymers substantially increases the elimination of metal ions that are heavy and arsenic onto CNT membranes with the help of adsorption. This process is undergone at four suitable sites on CNTs that include in internal sites, channels that are interstitial, grooves, and surfaces present outside. In the internal area, adsorption is less in nanotubes [39, 40]. The heavy metals adsorption onto f-CNTs is because of the appearance of a variety of positions available for sorption on the surfaces of f-CNTs for heavy metals to get attached.

Usman et al. suggested that zinc ion adsorption is high in the plasma-functioned CNT membrane because of the greater percentage of oxygen groups obtainable for the attachment of the Zn^{2+}, and thus, the removal of protons in the functional groups on CNT surfaces enhanced the positions for binding for the erasure of the cations because of the complexation of the surface. The capacity of adsorption is high in these CNTs compared to that in pristine CNTs. The high adsorption of f-CNTs is due to the electrostatic relationship among the negatively charged CNT and the divalent metal ions [89]. Owing to the addition of proton/removal of the proton of f-CNTs, solution pH is very crucial in the adsorption of metal ions by f-CNTs membrane.

Generally, the effectiveness of cations is greater toward the pH values because of the high electron-rich densities on the f-CNT layer. However, the efficiency in removing is not effective because of the proton added by the functional groups on the CNT.

Vuković et al. studied on the elimination of CD groups from the watery solvent by multi-walled CNT materials functioning by oxidizing ethylenediamine. The rivalry between positive and negative ions on the surfaces of the CNT highly affects the elimination of metal ions [94]. For example, the lead adsorption on f-CNTs is essential in the governance of sodium dodecylbenzene sulfonate because of the production of complex compounds, whereas the lead adsorption reduces rapidly in the existence of BKC because of the combative adsorption. Further, involvement procedures among f-CNT materials and metal ions that are heavy are varied because of the presence of various positive as well as negative charges [56]. The adsorption of Cr(III) on nitrogen-doped magnetic CNTs was because of chemical adsorption, whereas on the acid-modified CNTs, the adsorption mechanism is because of the associations that are electrostatic between f-CNT materials and Cr(III). It is known that the specific area of the f-CNTs is present to adsorb pollutants that are inorganic enhances with a reduction in the carbon's diameter nanotube. But this effect depends on the heavy metal ions adsorption capacity via f-CNTs is not suggested by any study. The metal ions adsorption on f-CNTs is affected by the interactive functional groups present and not on the size of the nanotubes [76].

3.2 Removal of Organic Contaminants

There are a lot of pollutants inside water are in the form of either particles or dissolved materials. The molecular weight dispensation is very broad because of the decomposition of animal or plants, and the activities of humans [84, 87]. Organic elements in water are categorized as components that are water-loving and water-hating have their basis on their attractive nature toward non-ionic resins or paedogenic and aquagenic.

There are a lot of varieties of organics that are excreted by human processes. These contaminants cause a lot of health problems and thus become major concerns to remove from water using membrane. As we know, there are a lot of membranes using low-pressure methods. The tendency to remove the organics dissolved in water is less [54, 55]. It is very essential to develop highly permeable f-CNT membranes which are customizable to eliminate the micropollutants from water. Many researches have shown that f-CNT membranes use the process of adsorptive filtration for micro-organic materials that are dissolute and depth filtration for organic colloids. Wang suggested the elimination of PPCP by CNT nanocomposite membrane and received 95% of the elimination of particles by enhancing the aromatic rings and SSA [95, 97]. Other studies have shown that f-CNT has an adsorption attraction which is very high toward various organic micropollutants. The elimination process of f-CNT membranes toward the micropollutants is happening because of H-bonding, Van der Waals forces, π-π interactions, and chemical adsorption among the f-CNT materials and organic matters. A very same process occurs when pH values enhance which increases the electrostatic repulsion across CNT and natural organic matter. Therefore, the pi–pi interactions cause the elimination of natural matter through f-CNTs. Just like inorganic pollutants, there is a huge race between varied organic chemicals in water that may be seen on the CNT surfaces which declines the adsorption of natural organic contaminants [18, 99]. Therefore, customizing the surface properties of CNT for selectively adsorbing organic matter is a vital research task for improvised water treatment. Also, the pre-treatment of water is also a way that can remove this challenge.

3.3 Removal of Microorganisms

Pathogens are found in water that we drink and the micropollutants like bacteria, viruses and protozoa are present in wastewater. Earlier studies have suggested that nanotubes can deactivate or eliminate a variety of small organisms like bacteria, protozoa, and viruses. Kang reported that SWCNTs successfully inactivated *E. coli* [44]. This was done by penetrating the nanotubes into the cell walls. It is also seen that more advanced nanotubes have a better capability than pristine CNTs and membranes made of polymer to destruct the cell walls of the microbes. The direct involvement of microbes with f-CNTs critically affects cell wall integrity and the whole architecture of bacteria. The highly inactive efficacy of bacteria by carbon nanotubes is characterized due to the insertion into the cell walls. CNT membranes consisting of AgNPs can increase the capacity of membrane to make the bacteria inactive.

4 Graphene-Based CNC Membranes

MWCNTs are economically cheaper than graphene, though there have been various suggestions of several wet chemical methods that can produce affordable price. Exfoliation using electrochemistry, milling through ball, and high shear mixtures are some wet chemical methods that can reduce the pricing of graphene [1]. An increase in the number of researchers that have suggested techniques that increases the benefits of graphene like mechanical stripping, chemical stripping, epitaxial growth, and hot solvent. Graphene Oxide has a structure which has alternating layers made of different materials, and the thickness is around 1–30 μm , and around its planes and edges, it bears functional groups which have lots of oxygen (carboxyl and hydroxyl). Owing to these oxygen-rich functional groups, the structure of Graphene Oxide has a supremely complex structure. Groups that comprise oxygen are affected by the chemical reactions and are deployed for the modification of surface of graphite oxide. Staudenmaier continued the oxidization process with the help of $KClO_3$, HNO_3, and H_2SO_4 by inserting them in the system. Hummer's technique is utilized the all-out as it is fast and reliable. After competence of oxidation reaction, graphite powder is sent for mining to get brown oxidised graphite. Graphite oxidised can be produced with the help of high shear or ultrasonic fierce turbulent peeling. Various scientists have modified this [8, 15, 67]. Substitute methods are yet in the development phase, such as ionic liquid supplementary electronic peeling [63], potassium ferrate [78], sealed oxidation [7], and other new methods, which provide various improvisations to GO oxidation in terms of rate, reliability, and lesser ecological effects.

4.1 Graphene Film

A team of researchers once engaged atmospheric oxygen at a high temperature for manufacturing GOs having nanopores. They manually peeled graphene and placed them in a pipe of mullite which had gas inlets and outlets, placed pipe in a tubular kiln at a specified temperature and proceeded a mixture of Ar and O_2 over it. Nanopores may occur by governance of the flow rate of Ar at 1.1 L/min and oxygen at 0.86 L/min. This helped in the construction of graphene which is porous. Fischbein developed nanopores by means of the electron beam which is focused and does the irradiation. The dimensions of the pores can vary between 0.2 nm and nanometers ranging in magnitude of tens. When this method creates pores, the structure is not changed during passing of time and the existence of these structures does not create abrupt barriers on the architecture of the material. It is seen that when graphene is made through a way that includes a TEM; it means only a minute amount of graphene can be customised and because of this reason, a lot of setbacks arise for industrial implementations.

Bell inserted He and Ga ion rays to strike Graphene, thus making pores of diameter 20 nm [9]. Russo inserted Ar ions as alternatives to obtain modified results. The diameter of these ions ranged from 0.15 nm to 1.35 nm [81]. There are a different method that involves the usage of hydrogen plasma etching [6, 105], ozone/ultraviolet light irradiation oxidation and catalyzed oxidation of gold [14, 27, 38, 49]. Yang devised the plasma-etching method for etching of graphene [110]. Zhao invented a graphene nano-foam, which had a pored layered structure that involves auto assembling of the graphite oxide hydrothermally etching it in the native positions or sites [114].

4.2 Graphene Oxide Membranes

Graphene oxide separation membranes can be manufactured in a lot of ways. Some of them are like assembly generated through the electric field, self-assembly, evaporation, coating, and filtration. These processes employ the graphene oxide being scattered between a penetrable membrane used for a substrate. It can be under pressure or in a vacuum [12, 88, 95, 97, 98, 100, 101] after that it is dried and filtered and is subjugated to various other procedures which finally produced a membrane that has pores in it. Han et al. produced a multi-walled nanotube-intercalated graphene membrane [35]. The scattered graphene grains and CNT undergo filtration in vacuity. The grains are inside a polyvinylidene fluoride (PVDF) UF, which produced the graphene CNT. This method is used to prepare graphene oxide films. The depth of the film varies from nanometers to microns. The interface between the layers is weak; hence, the stability is not good. The coating can be that of a cloth, spin, or spray. Spin is the most prevalent method for coating. In markets, there is a huge need for the graphene oxide sheets that are self-assembled. A novel nanofiltration membrane is developed by placing a coat of ethanol gel on the surface of poly (acrylonitrile) ultrafiltration membrane [86]. Graphene oxide film's production is done by the evaporation method, and the interface involved is that between the two states, liquid and gas. The more the width of the interface, the easier it is to prepare a transverse-sized membrane.

Yang devised GO in a solution that is more intense, around 2 mg/mL. This was done with the help of evaporation, which eliminated the solvent and hence produced a surface that held the film that is self-assembled [108, 109]. These films were at a peak in the adhesive group, carboxyl, along with hydroxyl groups insert layer-by-layer (LBL) auto-assembly. The surface of the GO is it held carboxyl groups; then, a negative charge was created for the groups that were dispersed onto the water. This showed that the self-assembly film that was produced was workable.

5 Applications

The CNC membranes have a variety of applications to treat wastewater, such as the dye, inorganic ions, organic waste, nanoparticle removal, and in desalination of salted water (Fig. 1).

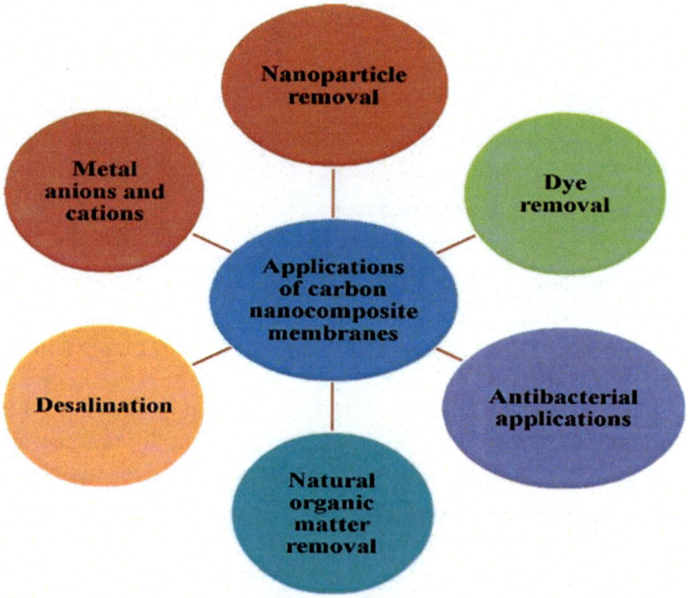

Fig. 1 Application of CNC membranes for wastewater treatment

Table 1 List of CNCs membranes for NPs removal

Membrane	Type of nanoparticle removed	Qe or %Removal (%R) Adsorption capacity (Qe)	Reference
Caron nanofiber NF	Gold nanoparticles having 25 nm		[59]
Porous nanocrystalline silicon (npc) NF	Gold nanoparticles having 15 nm diameter		[26]
PVA NF nanofiller	Gold nanoparticles		[16]
Cellulosic nanofiller with 0.5% wt chitosan	Gold and silver nanoparticles	13.1 and 17.9 mg/g	[65]
PES/polyvinylpyrrolidone NF	nC60 NPs	>99.99%	[23]
Cellulose nanofibre (CNF)/PVDF Modified CNF/PVDF	Fe_2SO_4 NPs	2.498 and 3.984 mg/g	[32]
Gum Karaya/PVA nanofiber	Pt, Ag, Au, CuO, Fe_3O_4	90, 89.4, 84,62, 52.9%	[77]

5.1 Nanoparticle Removal

The CNC membranes have recently been utilized for the removal of toxic NPs from the wastewater and water. The removal of NPs depends upon the pore size of the membranes. The literature survey shows that the membranes process is best suited for the removal of NPs. To enhance the removal efficiency of NPs from water by using a membrane technique, a coupling of other treatment methods helps such as pre- and post-treatment by sedimentation and coagulation. Some CNCs are listed in Table 1.

5.2 Natural Organic Matter Removal

The wastewater comprises natural waste, including natural organic matters which are produced from different sources. These consist of polar and non-polar functional groups. These natural organic matters have negative aesthetic effects on water, such as taste, color, and odor. Due to natural organic matter in the water, the process of water purification is tedious. During filtrations, natural organic matters create fouling of membranes, which cause a severe effect on the water purification. There are several factors responsible for the removal efficiency of natural organic matters by using CNC membranes such its porosity, morphology, pore size, surface chemistry. Some of CNC membranes are shown in Table 2 for the removal of natural organic matters.

Table 2 List of CNCs membranes for organic waste removal

Membrane	Removal of organic contaminant	%R or Qe	Reference
MWCNT/PVB	Humic acid	97.7–94.7%	[95, 97, 98, 100, 101]
f-MWCNT/PANI/PES	River Humic acid	80%	[54, 55]
CNT/PVDF UF	Humic acid	80–100%	[3]
TiO_2 NPs/PES	Humic Acid degradation activity		[57, 58]
Graphene oxide/TiO_2 NPs/PES	10 ppm Humic Acid	99%	
TiO_2 NPs/MWCNTs	2–700 ppm Humic Acid		[19]
ZnO/PES	Humic Acid	97.98%	[2]
DBA/PES GA/PES	Humic Acid	61–81% 61–86%	[72]

5.2.1 Dye Removal

In the wastewater, a number of dyes are present such as congo red, methyl blue, acid black, methyl orange, and direct red, which are harmful to human health. Several CNC membranes were synthesized to eliminate these dyes from the wastewater and water. Some examples are shown in Table 3. The CNC membranes showed good adsorption capability to remove these dyes from the water, and the process is at an

Table 3 List of CNCs membranes for dyes removal

Membrane	Dye	%R or Qe	Reference
Polypropylene/PVA	Brilliant green, crystal violet, victoria blue B	99.8, 99.2, 99.8%	[115]
PES/O-carbomethyl chitosan (0.1, 0.5, 1 wt%)	Direct red	99, 99, 98.5%	[116]
PES, PES/TiO_2, PES/GO/TiO_2	Reactive blue	61.4, 73.5, 81.4%	[82]
HNTs-poly (NASS)/PES	Reactive red 48 and reactive black 5	95, 96%	[41]
Sulfonated HNTs/PES	Reactive red 48 and reactive black 5		[95, 97, 98, 100, 101]
Chitosan-montmorillonite/PES NF	Reactive red 48 and reactive black 5		

optimal cost. The main mechanism of removal of dye is the electrostatic interactions between the functional groups of membranes and the dyes.

5.3 Conclusion

The NMs based desalination and water treatment technology development is an evolving field to provide safe and accessible water around the globe. The CNCs have attracted more, among all, for membrane development owing to its sieving and intrinsic adsorption capabilities, which helps in the removal of contaminants from water. Hence, in these days, the CNC membranes are in demand for wastewater treatment. The NMs used in CNCs have shown a wide range of applications for oil/water emulsion separation, desalination, NOM, and dye removal. Yet, there are lots of GO-based nanocomposite membranes, and CNTs are still in the R and D stage. It is estimated that the constant enhancements in membrane performance, such as ease in the synthesis and functionalization of GO and CNTs, may enhance the commercial market. But, to make it cost-effective and improve their commercial production with long-term stability, further research is desired for the effective production of CNC membranes. Several problems related to CNC membranes are still there, which need to be solved, and despite the lots of attempts made for the functionalization of CNC membranes, there is still a vast space left for the improvement of CNC membranes for desalination and water treatment.

Acknowledgements The authors are thankful to the Banasthali Vidyapith and the Central University of Gujarat. One of the authors, Dr. Pallavi Jain, is thankful to SRM Institute of Science & Technology, Delhi-NCR Campus.

References

1. Abdelkader AM, Kinloch IA, Dryfe RAW (2014) Continuous electrochemical exfoliation of micrometer-sized graphene using synergistic ion intercalations and organic solvents. ACS Appl Mater Interf 6:1632–1639
2. Ahmad A, Abdulkarim A, Ismail S, Ooi B (2015) Preparation and characterisation of PES-ZnO mixed matrix membranes for humic acid removal. Desal Water Treat 54:3257–3268
3. Ajmani GS, Cho HH, Chalew TEA, Schwab KJ, Jacangelo JG, Huang H (2014) Static and dynamic removal of aquatic natural organic matter by carbon nanotubes. Water Res 59:262–270
4. Ajmani GS, Goodwin D, Marsh K, Fairbrother DH, Schwab KJ, Jacangelo JG, Huang H (2012) Modification of low-pressure membranes with carbon nanotube layers for fouling controland characterization of low fouling novel hybrid ultrafiltration membranes based on the blends. Water Res 46:5645–5654
5. Asper M, Hanrieder T, Quellmalz A, Mihranyan A (2015) Removal of xenotropic murine leukemia virus by nanocellulose based filter paper. Biologicals 43:452–456
6. Bai J, Zhong X, Jiang S, Huang Y, Duan X (2010) Graphene nanomesh. Nat Nanotechnol 5:190–194

7. Bao C, Song L, Xing W, Yuan B, Wilkie CA, Huang J, Guo Y, Hu Y (2012) Preparation of graphene by pressurized oxidation and multiplex reduction and its polymer nanocomposites by masterbatch-based melt blending. J Mater Chem 22:6088–6096
8. Becerril HA, Mao J, Liu Z, Stoltenberg RM, Bao Z (2008) Evaluation of solution-processed reduced graphene oxide films as transparent conductors. ACS Nano 2:463–470
9. Bell DC, Lemme MC, Stern LA, Williams JR, Marcus CM (2009) Precision cutting and patterning of graphene with helium ions. Nanotechnology 20:455301
10. Bennett A (2011) Potable water: new technology enables use of alternative water sources. Filtration Separation 48:24–27
11. Cath TY, Childress AE, Elimelech M (2006) Forward osmosis: principles, applications, and recent developments. J Membr Sci 281(12):494–497
12. Chen X, Liu G, Zhang H, Fan Y, Chin (2015) Fabrication of graphene oxide composite membranes and their application for pervaporation dehydration of butanol. J Chem Eng 23:1102
13. Cheng Q, Ye D, Chang C, Zhang L (2017) Facile fabrication of superhydrophilic membranes consisted of fibrous tunicate cellulose nanocrystals for highly efficient oil/water separation. J Membr Sci 525:1–8
14. Cheng YC, Kaloni TP, Zhu ZY, Schwingenschlögl U (2012) Oxidation of graphene in ozone under ultraviolet light. Appl Phys Lett 101:73110
15. Cote LJ, Kim F, Huang J (2009) Flash reduction and patterning of graphite oxide and its polymer composite. J Am Chem Soc 131:1043
16. Dhandayuthapani B, Mallampati R, Sriramulu D, Dsouza RF, Valiyaveettil SF (2014) PVA/gluten hybrid nanofibers for removal of nanoparticles from water. ACS Sustain Chem Eng 2:1014–1021
17. Drewes JE, Reinhard M, Fox P (2003) Comparing microfiltration-reverse osmosis and soil-aquifer treatment for indirect potable reuse of water. Water Res 37(15):3612
18. Engel M, Chefetz B (2016) Adsorption and desorption of dissolved organic matter by carbon nanotubes: effects of solution chemistry. Environ Pollut 213:90–98
19. Esfahani MR, Tyler JL, Stretz HA, Wells MJ (2015) Effects of a dual nanofiller, nano-TiO2 and MWCNT, for polysulfone-based nanocomposite membranes for water purification. Desalination 372:47–56
20. Fang Q, Zhou X, Deng W, Zheng Z, Liu Z (2016) Freestanding bacterial cellulose-graphene oxide composite membranes with high mechanical strength for selective ion permeation. Sci Rep 6:33185
21. Ferraz N, Carlsson DO, Hong J, Larsson R, Fellström B, Nyholm L, Strømme M, Mihranyan A (2012) Haemocompatibility and ion exchange capability of nanocellulose polypyrrole membranes intended for blood purification. J R Soc Interface 9:1943–1955
22. Ferraz N, Leschinskaya A, Toomadj F, Fellström B, Strømme M, Mihranyan A (2013) Membrane characterization and solute diffusion in porous composite nanocellulose membranes for hemodialysis. Cellulose 20:2959–2970
23. Floris R, Moser G, Nijmeijer K, Cornelissen E (2017) Effect of multicomponent fouling during microfiltration of natural surface waters containing nC60 fullerene nanoparticles. Environ Sci Water Res Technol 3:744–756
24. Fujioka T, Khan SJ, Poussade Y, Drewes JE, Nghiem LD (2012) N-nitrosamine removal by reverse osmosis for indirect potable water reuse—a critical review based on observations from laboratory, pilot and full-scale studies. Sep Purif Technol 98:503–515
25. Gaaz T, Sulong A, Akhtar M, Kadhum A, Mohamad A, Al-Amiery A (2015) Properties and applications of polyvinyl alcohol, halloysite nanotubes and their nanocomposites. Molecules 20:22833
26. Gaborski TR, Snyder JL, Striemer CC, Fang DZ, Hoffman M, Fauchet PM, McGrath JL (2010) High-performance separation of nanoparticles with ultrathin porous nanocrystalline silicon membranes. ACS Nano 4:6973–6981
27. Gethers ML, Thomas JC, Jiang S, Weiss NO, Duan X (2015) Holey graphene as a weed barrier for molecules. ACS Nano 9:10909–10915

28. Goetz LA, Naseri N, Nair SS, Karim Z, Mathew AP (2018) All cellulose electrospun water purification membranes nanotextured using cellulose nanocrystals. Cellulose 25:3011–3023
29. Goh K, Karahan HE, Wei L, Bae TH, Fane AG, Wang R, Chen Y (2016) Carbon nanomaterials for advancing separation membranes: a strategic perspective. Carbon 109:694
30. Goh PS, Matsuura T, Ismail AF, Hilal N (2016) Recent trends in membranes and membrane processes for desalination. Desalination 391:43–60
31. Goh PS, Ng BC, Lau WJ, Ismail AF (2014) Inorganic nanomaterials in polymeric ultrafiltration membranes for water treatment. Sep Purif Rev 44(3):216–249
32. Gopakumar DA, Pasquini D, Henrique MA, de MoraisLC GrohensY, Thomas S (2017) Meldrum's acid modified cellulose nanofiber-based polyvinylidene fluoride microfiltration membrane for dye water treatment and nanoparticle removal. ACS Sustain Chem Eng 5:2026–2033
33. Gustafsson S, Manukyan L, Mihranyan A (2017) Protein-nanocellulose interactions in paper filters for advanced separation applications. Langmuir 33:4729–4736
34. Gustafsson S, Mihranyan A (2016) Straegies for tailoring the pore- size distribution of virus retention filter papers. ACS Appl Mater Interfaces 8:13759–13767
35. Han Y, Jiang Y, Gao C (2015) High-flux graphene oxide nanofiltration membrane intercalated by carbon nanotubes. ACS Appl Mater Inter 7:8147
36. Hassan E, Hassan M, Abou-zeid R, Berglund L, Oksman K (2017) Use of bacterial cellulose and crosslinked cellulose nanofibers membranes for removal of oil from oil-in-water emulsions. Polymers 9:388
37. Hassan M, Hassan E, Fadel SM, Abou-Zeid RE, Berglund L, Oksman K (2018) Metallo-terpyridine-modified cellulose nanofiber membranes for papermaking wastewater purification. J Inorg Organomet Polym Mater 28:439–447
38. Huh S, Park J, Kim YS, Kim KS, Hong BH, Nam J (2011) UV/ozone-oxidized large-scale graphene platform with large chemical enhancement in surface-enhanced Raman scattering. ACS Nano 5:9799
39. Ihsanullah, Abbas A, Al-Amer AM, Laoui T, Al-Marri MJ, Nasser MS, Khraisheh M, Atieh MA (2016a) Heavy metal removal from aqueous solution by advanced carbon nanotubes: critical review of adsorption applications. Sep Purif Technol 157:141–161
40. Ihsanullah, Al Amer AM, Laoui T, Abbas A, Al-Aqeeli N, Patel F, Khraisheh M, Atieh MA, Hilal N (2016b) Fabrication and antifouling behaviour of a carbon nanotube membrane. Mater Des 89:549–558
41. Jung D, Cho S, Peck D, Shin D, Kim J (2003) Preparation and performance of a Nafion®/montmorillonite nanocomposite membrane for direct methanol fuel cell. J Power Sources 118:205–211
42. Jung HI, Lee OM, Jeong JH, Jeon YD, Park KH, Kim HS, An WG, Son HJ (2010) Production and characterization of cellulose by acetobacter sp. V6 using a cost-effective molasses-corn steep liquor medium. Appl Biochem Biotechnol 162:486–497
43. Kang GD, Cao YM (2012) Development of antifouling reverse osmosis membranes for water treatment: a review. Water Res 46(3):584–600
44. Kang S, Pinault M, Lisa DP, Elimelech M (2007) Single-walled carbon nanotubes exhibit strong antimicrobial activity. Langmuir 23:8670–8673
45. Karim Z, Claudpierre S, Grahn M, Oksman K, Mathew AP (2016) Nanocellulose based functional membranes for water cleaning: tailoring of mechanical properties, porosity and metal ion capture. J Membr Sci 514:418–428
46. Karim Z, Hakalahti M, Tammelin T, Mathew AP (2017) In Situ TEMPO surface functionalization of nanocellulose membranes for enhanced adsorption of metal ions from aqueous medium. RSC Adv 7:5232–5241
47. Karim Z, Mathew AP, Grahn M, Mouzon J, Oksman K (2014) Nanoporous membranes with cellulose nanocrystals as functional entity in chitosan: removal of dyes from water. Carbohydr Polym 112:668–676
48. Karim Z, Mathew AP, Kokol V, Wei J, Grahn M (2016) High-flux affinity membranes based on cellulose nanocomposites for removal of heavy metal ions from industrial effluents. RSC Adv 6:20644–20653

49. Koenig SP, Wang L, Pellegrino J, Bunch JS (2012) Selective molecular sieving through porous graphene. Nat Nanotechnol 7:728
50. Kong L, Zhang D, Shao Z, Han B, Lv Y, Gao K, Peng X (2014) Superior effect of tempo-oxidized cellulose nanofibrils (tocns) on the performance of cellulose triacetate (cta) ultrafiltration membrane. Desalination 332:117–125
51. Kucińska-Lipka J, Gubanska I, Janik H (2015) Bacterial cellulose in the field of wound healing and regenerative medicine of skin: recent trends and future prospectives. Polym Bull 72:2399–2419
52. Lalia BS, Kochkodan V, Hashaikeh R, Hilal N (2013) A review on membrane fabrication: structure, properties and performance relationship. Desalination 326:77
53. Lay WCL, Lim C, Lee Y, Kwok BH, Tao G, Lee KS, Chua SC, Wah YL, Ghani YA, Seah H (2017) From R&D to application: membrane bioreactor technology for water reclamation. Water Pract Technol 12(1):12–24
54. Lee J, Jeong S, Liu Z (2016) Progress and challenges of carbon nanotube membrane in water treatment. Crit Rev Environ Sci. Technol 46(11–12):999–1046
55. Lee J, Ye Y, Ward AJ, Zhou C, Chen V, Minett AI, Lee S, Liu Z, Chae SR, Shi J (2016) High flux and high selectivity carbon nanotube composite membranes for natural organic matter removal. Sep Purif Technol 163:109–119
56. Li J, Chen S, Sheng G, Hu J, Tan X, Wang X (2011) Effect of surfactants on Pb(II) adsorption from aqueous solutions using oxidized multiwall carbon nanotubes. Chem Eng J 166(2):551–558
57. Li X, Fang X, Pang R, Li J, Sun X, Shen J, Han W, Wang L (2014) Self-assembly of TiO_2 nanoparticles around the pores of PES ultrafiltration membrane for mitigating organic fouling. J Membr Sci 467:226–235
58. Li X, Li J, Fang X, Bakzhan K, Wang L, Van der Bruggen B (2016) A synergetic analysis method for antifouling behavior investigation on PES ultrafiltration membrane with self-assembled TiO_2 nanoparticles. J Colloid Interface Sci 469:164–176
59. Liang HW, Wang L, Chen PY, Lin HT, Chen LF, He D, Yu SH (2010) Carbonaceous nanofiber membranes for selective filtration and separation of nanoparticles. Adv Mater 22:4691–4695
60. Liao Y, Loh CH, Tian M, Wang R, Fane AG (2018) Progress in electrospun polymeric nanofibrous membranes for water treatment: fabrication, modification and applications. Prog Polym Sci 77:69
61. Liu L, Kentish SE (2018) Pervaporation performance of crosslinked PVA membranes in the vicinity of the glass transition temperature. J Membrane Sci 553:63
62. Liu SX, Kim JT (2011) Characterization of surface modification of polyethersulfone membrane. J. Adhes. Sci. Technol. 25:193
63. Lu J, Yang J, Wang J, Lim A, Wang S, Loh KP (2009) One-pot synthesis of fluorescent carbon nanoribbons, nanoparticles, and graphene by the exfoliation of graphite in ionic liquids. ACS Nano 3:2367–2375
64. Ma H, Burger C, Hsiao BS, Chu B (2012) Nanofibrous microfiltration membrane based on cellulose nanowhiskers. Biomacromol 13:180–186
65. Mahanta N, Leong WY, Valiyaveettil S (2012) Isolation and characterization of cellulose based nanofibers for nanoparticle extraction from an aqueous environment. J Mater Chem 22:1985–1993
66. Mansouri J, Harrisson S, Chen V (2010) Strategies for controlling biofouling in membrane filtration systems: challenges and opportunities. J Mater Chem 20:4567–4586
67. Marcano DC, Kosynkin DV, Berlin JM, Sinitskii A, Sun Z, Slesarev A, Alemany LB, Lu W, Tour JM (2010) Improved synthesis of graphene oxide. ACS Nano 4:4806
68. Marin E, Rojas J, Ciro Y (2014) A review of polyvinyl alcohol derivatives: Promising materials for pharmaceutical and biomedical applications. Afr J Pharm Pharmacol 8:674
69. Mautner A, Lee KY, Lahtinen P, Hakalahti M, Tammelin T, Li K, Bismarck A (2014) Nanopapers for organic solvent nanofiltration. Chem Commun 50:5778–5781
70. Mautner A, Lee KY, Tammelin T, Mathew AP, Nedoma AJ, Li K, Bismarck A (2015) Cellulose nanopapers as tight aqueous ultra-filtration membranes. React Funct Polym 86:209–214

71. Mautner A, Maples HA, Kobkeatthawin T, Kokol V, Karim Z, Li K, Bismarck A (2016) Phosphorylated nanocellulose papers for copper adsorption from aqueous solutions. Int J Environ Sci Technol 13:1861–1872
72. Mehrparvar A, Rahimpour A, Jahanshahi M (2014) Modified ultrafiltration membranes for humic acid removal membrane by facile blending with chitosan–montmorillonite nanosheets for dyes purification. Chem. Eng. J 265:184–193
73. Metreveli G, Wågberg L, Emmoth E, Belák S, Strømme M, Mihranyan A (2014) A size-exclusion nanocellulose filter paper for virus removal. Adv Healthcare Mater 3:1546–1550
74. More AS, Pasale SK, Wadgaonkar PP (2010) Synthesis and characterization of polyamides containing pendant pentadecyl chains. Eur Polym J 46:557–567
75. Nataraj D, Sakkara S, Meghwal M, Reddy N (2018) Crosslinked chitosan films with controllable properties for commercial applications. Int. J. Biol. Macromol. 120:1256
76. Ntim SA, Mitra S (2011) Removal of trace arsenic to meet drinking water standards using iron oxide coated multiwall carbon nanotubes. J. Chem. Eng. Data 56:2077–2083
77. Padil VVT, Černík M (2015) Poly (vinyl alcohol)/gum karaya electrospun plasma treated membrane for the removal of nanoparticles (Au, Ag, Pt, CuO and Fe_3O_4) from aqueous solutions. J Hazard Mater 287:102–110
78. Peng L, Xu Z, Liu Z, Wei Y, Sun H (2015) An iron-based green approach to 1-h production of single-layer graphene oxide. Nat Commun 6:5716
79. Qin JJ, Kekre KA, Tao G, Oo MH, Wai MN, Lee TC, Viswanath B, Seah H (2006) New option of MBR-RO process for production of NEWater from domestic sewage. J. Membr. Sci. 272(1):70–77
80. Quellmalz A, Mihranyan A (2015) Citric acid cross-linked nanocellulose-based paper for size-exclusion nanofiltration. ACS Biomater Sci Eng 1:271–276
81. Russo CJ, Golovchenko JA (2012) Atom-by-atom nucleation and growth of graphene nanopores. Proc Natl Acad Sci 109:5953–5957
82. Safarpour M, Vatanpour V, Khataee A (2016) Preparation and characterization of graphene oxide/TiO_2 blended PES nanofiltration membrane with improved antifouling and separation performance. Desalination 393:65–78
83. Shanmuganathan S, Loganathan P, Kazner C, Johir MAH, Vigneswaran S (2017) Submerged membrane filtration adsorption hybrid system for the removal of organic micropollutants from a water reclamation plant reverse osmosis concentrate. Desalination 401:134–141
84. Sillanpaa M 2015, Natural organic matter in water-characterization and treatment methods. Elsevier
85. Soyekwo F, Zhang QG, Lin XC, Wu XM, Zhu AM, Liu QL (2016) Facile preparation and separation performances of cellulose nanofibrous membranes. J Appl Polym Sci 133:43544
86. Sun H, Chen G, Huang R, Gao C (2007) A novel composite nanofiltration (NF) membrane prepared from glycolchitin/poly (acrylonitrile)(PAN) by epichlorohydrin cross-linking. J Membr Sci 297:51–58
87. Tousova Z, Oswald P, Slobodnik J, Blaha L, Muz M, Hu M, Brack W, Krauss M, Di Paolo C, Tarcai Z, Seiler TB, Hollert H, Koprivica S, Ahel M, Schollee JE, Hollender J, Suter MJ, Hidasi AO, Schirmer K, Sonavane M, Ait-Aissa S, Creusot N, Brion F, Froment J, Almeida AC, Thomas K, Tollefsen KE, Tufi S, Ouyang X, Leonards P, Lamoree M, Torrens VO, Kolkman A, Schriks M, Spirhanzlova P, Tindall A, Schulze T (2017) European demonstration program on the effect-based and chemical identification and monitoring of organic pollutants in European surface waters. Sci Total Environ 601–602:1849–1868
88. Tsou C, An Q, Lo S, De Guzman M, Hung W, Hu C, Lee K, Lai J (2015) Effect of microstructure of graphene oxide fabricated through different self-assembly techniques on 1-butanol dehydration. J Membr Sci 477:93–100
89. Usman FM, Luan HY, Wang Y, Huang H, An AK, Jalil KR (2017) Increased adsorption of aqueous zinc species by Ar/O_2 plasma-treated carbon nanotubes immobilized in hollow-fiber ultrafiltration membrane. Chem Eng J 325:239–248
90. Van Houtte E, Verbauwhede J (2013) Long-time membrane experience at Torreele's water re-use facility in Belgium. Desalin Water Treat 51(22–24):4253–4262

91. Varanasi S, Low ZX, Batchelor W (2015) Cellulose nanofibre composite membranes—biodegradable and recyclable UF membranes. Chem Eng J 265:138–146
92. Vatanpour V, Zoqi N (2017) Surface modification of commercial seawater reverse osmosis membranes by grafting of hydrophilic monomer blended with carboxylated multiwalled carbon nanotubes. Appl Surf Sci 396:1478–1489
93. Visanko M, Liimatainen H, Sirviö JA, Haapala A, Sliz R, Niinimäki J, Hormi O (2014) Porous thin membrane barrier layers from 2,3-Dicarboxylic acid cellulose nanofibrils for membrane structures. Carbohydr Polym 102:584–589
94. Vuković GD, Marinković AD, Čolić M, Ristić MĐ, Aleksić R, Perić-Grujić AA, Uskoković PS (2010) Removal of cadmium from aqueous solutions by oxidized and ethylenediamine-functionalized multi-walled carbon nanotubes. Chem Eng J 157(1):238–248
95. Wang J, Lang WZ, Xu HP, Zhang X, Guo YJ (2015) Improved poly (vinyl butyral) hollow fiber membranes by embedding multi-walled carbon nanotube for the ultrafiltrations of bovine serum albumin and humic acid. Chem Eng J 260:90–98
96. Wang R, Guan S, Sato A, Wang X, Wang Z, Yang R, Hsiao BS, Chu B (2013) Nanofibrous microfiltration membranes capable of removing bacteria, viruses and heavy metal ions. J Membr Sci 446:376–382
97. Wang S, Liang S, Liang P, Zhang X, Sun J, Wu S, Huang X (2015) In-situ combined dual-layer CNT/PVDF membrane for electrically enhanced fouling resistance. J Membr Sci 491:37–44
98. Wang X, Xiong Z, Liu Z, Zhang T (2015) Exfoliation at the liquid/air interface to assemble reduced graphene oxide ultrathin films for a flexible noncontact sensing device. Adv Mater 27:1370–1375
99. Wang Y, Huang H, Wei X (2018) Influence of wastewater precoagulation on adsorptive filtration of pharmaceutical and personal care products by carbon nanotube membranes. Chem Eng J 333:66–75
100. Wang Y, Zhu J, Dong G, Zhang Y, Guo N, Liu J (2015) Sulfonated halloysite nanotubes/polyethersulfone nanocomposite membrane for efficient dye purification. Sep Purif Technol 150:243–251
101. Wang Y, Zhu J, Huang H, Cho HH (2015) Carbon nanotube composite membranes for microfiltration of pharmaceuticals and personal care products: capabilities and potential mechanisms. J Membr Sci 479:165–174
102. Wang Z, Ma H, Chu B, Hsiao BS (2017) Fabrication of cellulose nanofiber-based ultrafiltration membranes by spray coating approach. J Appl Polym Sci 134:44583
103. Wanichapichart P, Kaewnopparat S, Buaking K (2002) Characterization of cellulose membranes produced by acetobacter xyllinum. J Sci Technol 24:855–862
104. Wintgens T, Melin T, Schafer A, Khan S, Muston M, Bixio D, Thoeye C (2005) The role of membrane processes in municipal wastewater reclamation and reuse. Desalination 178(1–3 Spec. Iss.):1–11
105. Xie G, Yang R, Chen P, Zhang J, Tian X, Wu S, Zhao J, Cheng M, Yang W, Wang D, He C, Bai X, Shi D, Zhang G (2014) A general route towards defect and pore engineering in graphene. Small 10:2280–2284
106. Xiong Y, Wang C, Wang H, Jin C, Sun Q, Xu X (2018) Nano-cellulose hydrogel coated flexible titanate-bismuth oxide membrane for trinity synergistic treatment of super-intricate anion/cation/oily-water. Chem Eng J 337:143–151
107. Xu L, Shahid S, Shen J, Emanuelsson EAC, Patterson DA (2017) A wide range and high resolution one-filtration molecular weight cut-off method for aqueous based nanofiltration and ultrafiltration membranes. J Membr Sci 525:304–311
108. Yang L, Tang B, Wu P (2014) UF membrane with highly improved flux by hydrophilic network between graphene oxide and brominated poly (2,6-dimethyl-1, 4-phenylene oxide). J Mater Chem A 2:18562–18573
109. Yang R, Aubrecht KB, Ma H, Wang R, Grubbs RB, Hsiao BS, Chu B (2014) Thiol-modified cellulose nanofibrous composite membranes for chromium (VI) and lead (II) adsorption. Polymer 55:1167–1176

110. Yang R, Zhang L, Wang Y, Shi Z, Shi D, Gao H, Wang E, Zhang G (2010) An anisotropic etching effect in the graphene basal plane. Adv Mater 22:4014
111. Yang X, Lee J, Yuan L, Chae SR, Peterson VK, Minett AI, Yin Y, Harris AT (2013) Removal of natural organic matter in water using functionalised carbon nanotube buckypaper. Carbon 59:160–166
112. Yu F, Ma J, Wang J, Zhang M, Zheng J (2016) Magnetic iron oxide nanoparticles functionalized multi-walled carbon nanotubes for toluene, ethylbenzene and xylene removal from aqueous solution. Chemosphere 146:162–172
113. Zhang K, Li Z, Kang W, Deng N, Yan J, Ju J, Liu Y, Cheng B (2018) Preparation and characterization of tree-like cellulose nanofiber membranes via the electrospinning method. Carbohydr Polym 183:62–69
114. Zhao Y, Hu C, Song L, Wang L, Shi G, Dai L, Qu L (2014) Functional graphene nanomesh foam. Energy Environ Sci 7:1913–1918
115. Zheng Y, Yao G, Cheng Q, Yu S, Liu M, Gao C (2013) Positively charged thin-film composite hollow fiber nanofiltration membrane for the removal of cationic dyes through submerged filtration. Desalination 328:42–50
116. Zinadini S, Zinatizadeh A, Rahimi M, Vatanpour V, Zangeneh H, Beygzadeh M (2014) Novel high flux antifouling nanofiltration membranes for dye removal containing carboxymethyl chitosan coated Fe_3O_4 nanoparticles. Desalination 349:145–154

Environmental Monitoring by Removing Air Pollutants Using Nanocomposites Materials

Rekha Sharma, Sapna, and Dinesh Kumar

Abstract This chapter provides an outline of the submission of nanocomposites in ecological monitoring. For the effective removal of biological pollutants and contaminants, nanocomposites propose the potential in ecological remediation. Nanomaterials use for the recognition and elimination of polluted chemicals (heavy metals, manganese, arsenic, nitrate, iron, etc.), organic pollutants (aromatic and aliphatic hydrocarbons), gases (CO, NO_x, SO_2, etc.), and biological substances, for example, antibiotics, parasites, bacteria, and viruses, as catalysts and adsorbents in several morphologies/shapes, i.e., nanotubes, NPs, nanofibers, nanowires, etc. In contrast to other conventional techniques, nanomaterials display improved performance in environmental remediation because of their associated high reactivity, and surface-to-volume ratio (surface area). This chapter focuses on the development of novel nanoscale materials and their current advances and methods for the monitoring of air quality polluted by toxic gases, volatile organic compounds (VOCs), radionuclides, inorganic and organic solutes, viruses, and bacteria. For the handling or monitoring of contaminants and toxins, current advances in the submission of nanocomposite materials are likewise deliberated. Future prospects and research trends are fleetingly deliberated.

Keywords Environmental remediation · Contaminants · Nanomaterials · Nanocomposite · Air pollutants

R. Sharma
Department of Chemistry, Banasthali Vidyapith, Banasthali, Rajasthan 304022, India
e-mail: sharma20rekha@gmail.com

Sapna
Department of Chemistry, Dr. K.N. Modi University, Newai, Rajasthan 304022, India
e-mail: nehrasapna111@gmail.com

D. Kumar (✉)
School of Chemical Sciences, Central University of Gujarat, Gandhinagar 382030, India
e-mail: dinesh.kumar@cug.ac.in

M. Jawaid et al. (eds.), *Environmental Remediation Through Carbon Based Nano Composites*, Green Energy and Technology,
https://doi.org/10.1007/978-981-15-6699-8_3

Abbreviations

SPR	Surface plasma resonance
LOD	Limit of detection
VOCs	Volatile organic compounds
Nano-PM	Nano particulate matter
MOF	Metal–organic framework
LC-MS	Liquid chromatography mass spectrometry
DFT	Density functional theory
MNP-NF	Magnetic nanoparticle-decorated nanofiber
ppb	Parts-per-billion
NPs	Nanoparticles
0-D	Zero-dimensional
2-D	Two-dimensional
3-D	Three-dimensional
CDs/CdS/GCN	Carbon quantum dots/CdS quantum dots/g-C_3N_4

1 Introduction

By producing hazardous wastes and poisonous smoke and gas fumes discharged to the environment, the fast pace of industrial development and the ensuing by-products have affected the environment [42]. Conventional methods such as biological oxidation, adsorption, incineration, and chemical oxidation have been utilized to treat all forms of toxic and organic waste. For the removal of organic toxic waste material, the supercritical water oxidation (SCWO) has been anticipated as a capable method, because of its capability to abolish a huge variation of high-risk wastes ensuing from multifaceted manufacturing biological processing and munitions demilitarization. The high surface area to the mass ratio of nanomaterials can greatly improve the adsorption capacities of sorbent materials. At a similar density, as the thickness shrinks, the surface area of the nanocomposite material grows exponentially because of its reduced size. Because of their minor size, the entire capacity could be rapidly skimmed with trivial quantities of nanomaterials because the movement of nanomaterials is high in solution. Owing to their large radii and condensed size, the surface of nanomaterials is specifically sensitive primarily because of the high concentration at the vortices, edges, and surface of low-coordinated atoms. These inimitable possessions could be functional for scavenging and degrading pollutants in the air [40]. In the treatment of air, the several forms/morphologies/shapes of nanomaterials have an important impression. Many engineered and natural nanomaterials likewise used to consume robust air cleaning properties, counting photocatalytic TiO_2, silver NPs (AgNPs), carbon nanotubes (CNTs), and chitosan [10, 34, 35]. Nanomaterials partake improved photocatalytic and redox possessions [52]. The methods utilized for manufacturing nanomaterials are (1) vacuum evaporation or chemical or physical vapor deposition; (2) gas-phase synthesis techniques, for example, electro explosion, plasma synthesis, flame pyrolysis, and laser ablation; (3) mechanical alloying, milling, and grinding methods (4) sol-gel chemical synthesis methods; and (5) microwave-assisted methods or incineration approaches or delamination of layered materials. With the aim to remove the interaction amongst the biological substances and nanomaterials, to advance optical and surface possessions; and to avoid aggregation, the functionalization process by chemical modification or a coating method applies to nanomaterials. For instance, doping by a suitable dopant that principals to its competence to absorb light in the visible range may cause a red-shift in the band-gap of TiO_2 and advance the photocatalytic activity [21]. The assimilation of NPs principals to an improvement of the optical, electrical, and mechanical properties into polymeric nanocomposites. Through collecting NPs on porous membranes or amalgamation them with inorganic or polymeric membranes [23, 47], NP-based membranes could be made-up [4]. These properties, for example, the hydrophilicity of the surface, porosity, electronegativity, electropositivity, and surface catalytic properties, provide several changes for the developments to membrane surfaces or membranes using nanomaterials. The combined nanoporous materials which could avoid the passageway of various microorganisms and contaminants via the membrane because of the probable size grading. Nanofibers have the competence to trick much smaller contaminants and could likewise deliver an improved filtration with a much smaller porosity. Compared to other conventional filter materials, the inner surface areas of nanofibers are considerably advanced. Additionally, nanofibrous materials could permit high flow rates and have unified open pore structures. This chapter summarizes the submission of nanocomposite materials in the decontamination of air polluted with greenhouse gases, contaminant detection and removal, inorganic and organic solutes, viruses and bacteria, and their recital in ecological treatment, disinfectant manufacture and so on. The removal of air pollutants using mainly the adsorption technique is shown in Fig. 1.

1.1 Air Pollution

The occurrence of undesirable gaseous or solid particles in the air in high quantities causes air pollution toxic towards the environment and human health. It can also be clear that the particulate substance in the air is harmful to living beings known as air pollutants. Pollutants are the particulate matter or gaseous forms produced through several sources by natural events. These are termed as key contaminants, for example, outbreaks of different human and volcanic activities and dust storms, etc. The pollutants like NO_x, VOCs, carbon oxides (CO_2 and CO), suspended particulate matter, and SO_2 are the main contaminants that contribute 90% of global air pollution. To harvest subordinate contaminants, for example, carbonic acid, sulfuric acid, acid rain, and nitric acid, the contaminants react amongst individuals that are fashioned in the atmosphere. Particulate material could be anthropogenic for example cement, mineral dust, fibers, asbestos dust, fly ash smoke particles from fires, metal dust,

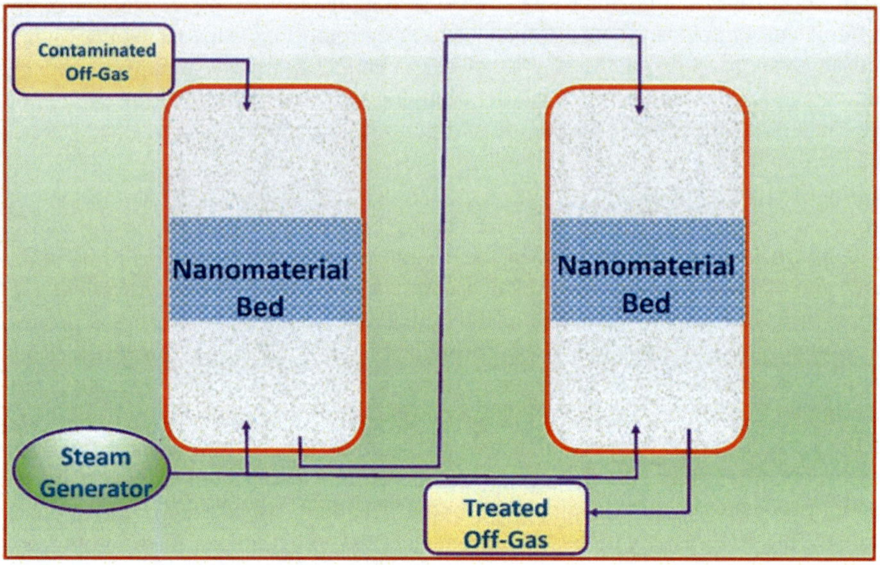

Fig. 1 Adsorption of air pollutants using nanomaterials through adsorption technique

etc., and natural materials, for example, pollen grains, spores, dust, bacteria, viruses, algae and, fungi [5–7].

2 Remediation of Air Pollutants

2.1 *Reduction of NO_x*

After the burning of fossil fuels, the production of NO_x partakes a serious effect on the environment [49]. Approximately 90% of NO_x produced by nitric oxide (NO) via coal incineration in the flue gas. NO attained form nitric acid as soon as combined with a vapor of water in the hazes are the principal causes for acid rain. In the presence of air, because of the thermal incineration method of the O_2 and N_2, NO_x pollutants are formed. NO_x grounds ensuing ecological destruction by the participation in smog production via a hydrocarbon reaction. Staged rich incineration shaped from bound nitrogen could regulate NO_x [44]. Numerous conceivable incineration alterations, together with over-fire, air, low NO_x burners, steam or water injection, and, reburning could considerably lower NO_x production thermally. To eliminate 30–60% NO_x, the direct inoculation of urea or ammonia into the exhaust gas or flue is correspondingly utilized.

For the control of indoor air pollution, the development of a practical and effective approach is vital to the photocatalytic conversion mechanism of NO. Lu et al. [32], hydrothermally synthesized $Bi_2Sn_2O_7$ nanocrystalline material for the photocatalytic removal of NO using stannic chloride pentahydrate and bismuth citrate as precursors. The performance was evaluated underneath the virtual solar light irradiation of the as-prepared $Bi_2Sn_2O_7$ samples in a continuous reactor using photocatalytic degradation of NO. Owing to its enhanced optical absorption capability, smaller particle size, fast diffusion/separation rates of the photogenerated charge carriers, and high specific surface area, the $Bi_2Sn_2O_7$ sample manufactured for BSO-12 (12 h) showed 37% of removal rate for NO which is higher in contrast to $Bi_2Sn_2O_7$ samples manufactured for BSO-24 (24 h) and BSO-36 (36 h) which confirms that $Bi_2Sn_2O_7$ is a promising photocatalyst for indoor air decontamination [32].

To allow charge separation, state, lead-containing, and TiO_2-based perovskite-type photocatalysts display superiority owing to the structural noncentrosymmetric in both ferroelectricity and stability. With the aim of maximizing the efficacy of the ensuing redox reactions, Hailili et al. [16], synthesized $Pb_2Bi_4Ti_5O_{18}$ samples for NO removal, which is an imposing task in photocatalysis which pursues chemically steady photocatalysts having diminished recombination of photoinduced charges. To prepare $Pb_2Bi_4Ti_5O_{18}$ perovskites for removal of NO, Hailili et al. [16], used the molten salt synthesis method to spread their visible light activity with several nanoscale structures and assessed their photocatalytic activity under visible light. The outcomes display contrary to only 15% for commercial P_{25}, perovskite $Pb_2Bi_4Ti_5O_{18}$ samples exhibit outstanding stability, in addition, to show NO removal efficiency over 50% underneath visible light. Leading to the improved photocatalytic activity, they exposed that the photocatalytic $Pb_2Bi_4Ti_5O_{18}$ owns distorted units in which the charge separation ratified because of the dipole-induced internal fields. The Pb-containing perovskite photocatalysts have huge manufacturing benefits because they have a solid–gas reaction by which the lead content is safe in solid-state—the NO_3^- formed by the reaction of O_2^- with NO. The efficient charge separation attained because of enhancing the overall photocatalytic activity by various properties such as high surface area and exclusive structure of layered distorted polyhedral. To avoid secondary Pb pollution, this work offers an applied submission for Pb-based perovskites in a gaseous system as photocatalysts [16].

Chen et al. [9], examined that the photocatalytic performance of Bi nanoparticles could be altered by its structure, morphology, and size. Hence, they utilized a one-step solvothermal technique for the synthesis of Bi@amorphous Bi_2O_3 core-shell nanospheres on NO removal. In this sorbent, the exterior amorphous Bi_2O_3 layer may enable the parting of charge transporters, and under visible light irradiation, the Bi NPs could produce charge carriers by SPR. The Bi_2O_3 layer $1O_2$, $\cdot OH$ radicals, and $\cdot O^{2-}$ are the key responsive substituents intricated in the photocatalysis progressions. Owing to the suited amorphous and suitable size, which can support the competence parting of electrons—holes produced by surface plasma effect of Bi, avoid Bi from oxidation and the $Bi@Bi_2O_3$ sample synthesized for 18 h showed greater photocatalytic activity for degradation of NO under visible light.

On the whole, the solvothermal synthesized nanospheres have a good approach for operative control of air pollution [9].

Zhang et al. [53, 54], used a one-pot solvothermal technique for the synthesis of Ag–$SrTiO_3$ nanocomposites (Ag–STO). In contrast to pristine $SrTiO_3$, because of the improved visible light of Ag NPs, they experience an extensive plasmonic resonance absorption ensuing in ambitious activity for removal of NO. Underneath visible light irradiation, the around 30% of NO was removed in a solitary reaction path, which was higher than the pristine $SrTiO_3$, by the use of 0.5% loading of Ag onto $SrTiO_3$. Primarily, because of the elementary external possessions of strontium sites, the production of NO_2 (destructive intermediate) is mostly inhibited over Ag–STO nanocomposites and $SrTiO_3$. The main reactive species for NO oxidation are ·OH radicals and O^{2-} determined by the ESR spectra. The synthesized Ag–$SrTiO_3$ nanocomposite photocatalyst contains selectivity for NO reduction and high visible light activity via a controllable and facile route. The growth of Ag nanocrystallites and $SrTiO_3$ increased due to the bifunctional role of NaOH. Furthermore, NO_2 production on STO was alleviated compared to P_{25}, which might be helpful for the adsorption of NO_x by the presence of alkaline sites [53, 54]. Gao et al. [13] synthesized Bi/$ZnWO_4$ microspheres via anchoring of $ZnWO_4$ on bismuth (Bi) nanoparticle, as effectual and robust photocatalysts under visible light irradiation for removal of NO at ppb level. In contrast to its single counterparts, i.e., Bi (0.027 min^{-1}) and $ZnWO_4$ (0.004 min^{-1}), the as-synthesized composite with the 50% mass ratio of Bi showed the advanced rate of reaction (0.067 min^{-1}). The Bi/$ZnWO_4$ composites displayed a wide-ranging light absorption in the visible spectrum because of the SPR effect of Bi NPs. In contrast to the pristine materials, the development of the Bi/$ZnWO_4$ heterointerface indorsed the separation of photoexcited electron-hole pairs, which is confirmed by the augmented photocurrent density. The OH radicals were not intricate in the procedure, while to start oxidation of NO, the superoxide radical was the main active species shown by radical scavenger tests. Rather than noble metals, the earth-abundant Bi-material was used to manufacture SPR-enhanced composite photocatalysts due to its workable route for synthesis and economic value viable for the reduction of air pollutants [13].

2.2 *Reduction of SO_2*

Through the incineration in power plants of fossil-derived fuels, factories, automobiles, and houses, sulfur dioxide (SO_2) is often unconfined to the atmosphere. Because of SO_2, acid rain, the corrosion of buildings is a considerate task. For the conversion of SO_2 to sulfur, TiO_2 is the utmost used catalyst by the reaction:

$$SO_2 + 2H_2S /2H_2O + 3S_{solid}$$

Rodriguez [37], examined that the amalgamation of TiO_2 and a gold (Au) system fashioned exceedingly effective desulfurization. Metallic gold consumes a minute catalytic and chemical activity [37, 38]. Though, because of a limited nanoscale size (<10 nm) and charge transfer amongst the gold and oxide, it provided the positive effects of catalytic activity when gold was scattered on MnO_x, TiO_2, MgO, Al_2O_3, Fe_2O_3. With the collective systems of MgO/Au and TiO_2/Au, Rodriguez [37], examined the dissociation effects of SO_2. The major action for the complete dissociation of SO_2 on both oxides supports was detected in systems comprising Au coverages when the size of the Au NPs was beneath 5 nm and that were <1 mL. It was determined that for the altered chemical properties of the functionalized Au NPs and the dissociation of SO_2, TiO_2 showed a straight active role, so the combined system of TiO_2/Au provided further operative dissociation of SO_2 than that of MgO/Au [39]. Catalytic performance tests showed that compared to pure TiO_2 the amalgamation of Au/TiO_2 is a 5–10 times faster because of the following reasons, for instance

(1) by the reaction of SO_2 and CO:

$$SO_2 + 2CO/2CO_2 + S_{solid}$$

(2) by the reduction of SO_2 by H_2S:

$$SO_2 + 2H_2S/2H_2O + 3S_{solid}.$$

In air-based anti-aggregation of gold NPs (AuNPs), Zhang et al. [58], designated a low-cost and straightforward visual technique for on-site recognition of H_2S. The AuNPs are stable, comprising 80 mM NaCl in a Tris buffer solution with the attendance of Tween 80 preserving their red color, resultant because of the bubbling of H_2S corresponds to the development of HS^-, which is functioning as a stabilizing agent for the AuNPs. On the surface of AuNPs, the adsorption of a negatively charged S^{2-} ions also stabilizes the AuNPs. In disparity, the color of AuNPs altered from red to blue and aggregate deprived of the bubbling of H_2S. The expected technique displays outstanding visual sensitivity under optimum circumstances with a naked-eye LOD of 0.5 ppm (v/v), constructing the on-site detection of H_2S probable. By using an unpretentious SO_2 removal device, this technique also owns good selectivity over other gases toward H_2S. The recognition of H_2S concentration indicated the potential submission of this cost-effective technique in local air fruitfully [58].

2.3 Reduction of VOCs

Throughout the photocatalytic air purification, TiO_2 nanotubes (TNT) partakes an exceedingly ordered open structure that indorses the diffusion onto active sites of substrates and dioxygen and displays advance sturdiness in contradiction of deactivation. In this, Weon et al. [48], utilized a new and straightforward technique for the photocatalytic removal of VOCs in commercial and laboratory reactors on {001} facet-exposed TiO_2 nanotubes (001-TNT) and confirmed them as the air cleaner [48]. The 001-TNT's outer surface was favorably associated through {001} facet anatase, while the surface of TNT is habitually collected of {101} facet anatase. TNT did not display any degradation activity for formaldehyde and acetaldehyde, while 001-TNT showed degradation activity under visible light irradiation. The 001-TNT filter was fruitfully installed and scaled up on a commercial air cleaner [48].

For refining the catalytic activity of monotonous transition-metal oxides, Li et al. [28], synthesized MnO_x–CeO_2 crystal facet-dominated surfaces which often deactivate at low temperatures. In this, to adapt the contact of three key surfaces, i.e., active-site behaviors, catalytically active zones, and the quantitative effects and formation of oxygen vacancies, MnO_x–CeO_2 was synthetically managed concerning their distinct surface-active complexes. In contrast to two other low-index facets {110} and {001}, MnO_x–CeO_2 with the exposed {111} facet displayed advanced action for CO_2 selectivity and formaldehyde oxidation. At elevated migration rates, the bulk lattice oxygen can refill the ingesting of surface lattice oxygen, which was related to the stability and activity of oxygen vacancies on the {111} facet. In a household air purifier, the MnO_x–CeO_2-111 catalysts were afterward scaled up to work as filter substrates [28]. To concurrently remove nano-PM and VOCs from manufacturing fumes, Li et al. [27], synthesized a SiC@TiO_2/Pt membrane. Via soaking the SiC membrane, the Pt NPs and TiO_2 transition layer were equipped hooked on the consistent precursor. For refining catalytic action to the operative consignment of Pt NPs, the SiC functionalized with the TiO_2 layer considered being a vital aspirant. In the meantime, this alteration procedure has no consequence on the gas infiltration of the membrane. The PM removal efficiency and mesitylene degradation of SiC@TiO_2/Pt catalytic membrane were assessed. With the inlet concentration of 300, 600, and 800 ppm, the complete alteration could be attained for mesitylene at the reaction temperature of 240, 251, and 263 °C and dwelling time of 1.0 s, correspondingly. The catalytic membrane exhibited 99.98% of adsorption capacity towards Al_2O_3 simulated dust and 100% for the degradation of mesitylene at 262 °C temperature and 1.0 m min^{-1} of gas speed. For the rejection of PM and degradation of mesitylene, a perfect catalytic activity is displayed by the SiC@TiO_2/Pt membrane. With a separation rate of 1 m min^{-1} at 262 °C, almost 100% conversion of mesitylene grasped. Furthermore, 240 mg m^{-3} of inlet dust concentration, a 99.98% removal of dust competence, was attained with 0.3 mg m^{-3} passage concentration of dust [27].

Krishnamurthy et al. [26], synthesized binary mixed-metal oxides (MMOs) with different metal ratios, i.e., TiO_2/SiO_2 and ZrO_2/SiO_2 to remove VOCs, for example, aldehyde compounds. MIL-101(Cr), which is a MOF compound, was manufactured and compared with MMOs as a base adsorbent. Contrary to the titania-based corresponding's item, the zirconia-based materials display a moderately advanced affinity toward formaldehyde. Particularly, at room temperature, the ZrO_2/SiO_2 exhibited a dynamic adsorption capacity of 2.9 mmol g^{-1} with a weight ratio of 25/75 utilizing a HCHO attentiveness of 170 ppm_v. The outcomes described that for the reduction of formaldehyde vapor, the ZrO_2/SiO_2 and TiO_2/SiO_2 could be a potential and efficient solid adsorbent [26].

To remove $PM_{2.5}$ at elevated temperature, Zhang et al. [53, 54], established in height efficacy (>99.5%) polyimide-nanofiber air filters. When temperature extended from 25 to 370 °C, the polyimide nanofibers the $PM_{2.5}$ removal competence was reserved unaffected and exhibited high thermal stability. With a very low-pressure drop, these filters required high air flux. For $PM_{2.5}$ index >300, these polyimide-nanofiber air filters can uninterruptedly work for >120 h. A field test exhibited that at high temperatures, they can efficiently remove >99.5% PM particles from car exhaust [53, 54].

To produce protein functionalized nanostructures, Fan et al. [12], utilized a hierarchically structured all-biomass air filter to advance the manufacturing rate of the electrospinning process, upsurge the percolation capacity, and decrease the pressure drop by applying Pickering emulsions. Precisely, the air filter comprises zein NPs/cellulose nanofibers (CNF) as vigorous fillers equipped from porous structures of microfibers and Pickering emulsions as the frame from wood pulp (WP). To advance the elimination competence of the filters, the CNF/zein, zein-protein-coated NPs donate in several methods. Primarily, counting poisonous particles through interaction mechanisms, the uncovered functional clusters of a zein-protein help to trap air contaminants. Second, the high surface area of NPs is responsible for the high capture competence for small particulate contaminants [11]. In the meantime, the long-micron WP fibers forming a frame decrease the pressure drop with their large pores. For capturing both kinds of contaminants, i.e., chemical gases (CO and HCHO), and particulate matter (PM) via regulating the constituent ratios of Pickering emulsion. Fan et al. [12], reported an enhanced air filter around 1/170 of the zein-based nano air sieve, with the high efficiency and the enormously low normalized pressure drop by electrospinning. Permitting advanced efficacy and a varied range of sizes of NPs, this study provides a cost-effective approach for making a hierarchical micro- and nanostructure of captivating particulate contaminants of additional species. Additionally, by the incorporation of nano- and biotechnology, this is the first report to develop green air filters having high-performance in which Pickering emulsion is useful as a basic method [12, 55].

2.4 Control of Pollutants via H_2 Evolution

Zhu et al. [60], reported a simple approach by the functionalization of 2-D graphite-like carbon nitride ($g\text{-}C_3N_4$) nanosheets (NSs) for the synthesis of 3-D TiO_{2-x}@carbon spheres. Throughout the synthesis procedure, the 0-D tiny TiO_{2-x} NPs were consistently laden with close chemically bonded (Ti–O–C) interfaces onto carbon spheres (CSs). Concurrently, for building exceedingly effective 2D/3D ternary heterostructures and to avoid the oxidation of the directly portrayed Ti^{3+}, the hierarchical sphere like TiO_{2-x}@CSs strongly coated with $g\text{-}C_3N_4$ NSs to form subordinate protecting layer efficiently. In contrast to pristine components or the binary composites, the attained CSs/TiO_{2-x}@g-C_3N_4 heterojunction displayed outstanding improved photocatalytic action in toxic pollutants degradation and hydrogen production. This report overlays an economical, gentle, and green method toward H_2 evolution and CO_2 reduction (solar energy conversion), by the fabrication of further effective TiO_{2-x} novel ternary constituents with multistep electron transfer [60].

Jiang et al. [19], prepared CDs/CdS/GCN photocatalysts. To check the synthesis of CDs/CdS/GCN, systematic characterization, for example, SEM, XRD, UV, TEM, and XPS. The concurrent photocatalytic production of H_2 was proficiently appreciated over the ensuing CDs/CdS/GCN composites attached by the reduction of organic contaminants such as bisphenol A, p-chlorophenol, and called 4-NP, tetracycline, TTC, and BPA, correspondingly. Subsequently, the development of interfaces amongst CdS quantum dots and GCN nanosheets, the as-synthesized CDs/CdS/GCN displays elevated competence of photocatalytic H_2 evolution, and photodegradation rates of biological contaminants of BPA, TTC, and 4- NP underneath visible light illumination corresponds to a useful charge separation competence. Thus, in the concurrent photocatalytic reduction and oxidation system, 4-NP displays advanced photodegradation competence than do BPA and TTC. For a methodical examination directing at manufacture clear the relationship amongst the photocatalytic contaminants and, degradation of the photocatalytic H_2 evolution, the LC-MS, and DFT calculations were utilized [19].

2.5 *Reduction of CO_2*

Yang et al. [50], reported that CO_2 is measured to designate unique of the key greenhouse gases. In 2016, the fast-growing ingestion of fossil fuels (oil, natural, and coal gas) was accountable for the noteworthy upsurge of CO_2 discharges, by surpassing the CO_2 level (400 ppm) in the atmosphere [61]. The increase in atmospheric CO_2 will become a tremendous hazard to human beings and could cause subsequent significant climate changes and global warming. To remove the utmost CO_2, several efforts have been made. The thermochemical, radiochemical, photochemical, biochemical, and electrochemical approaches are the current methods for CO_2 conversion [25, 36]. Amongst these techniques, because of proficient CO_2 reducing and producing various high value-added fuels and chemicals (CH_3OH, CH_4, CO:C1, C_2:C_2H_4, HCOOH; CH_3COOH, C_2H_5OH etc.), CO_2 hydrogenation has established special attention [30]. Depending on the specific pathway, the reduction of CO_2 has a multistep reduction includes the formation of CdH bonds, up to eight protons and electrons, cleavage of CdO bonds, and may also correspond to too many diverse harvests [15]. Though, when using H_2O as an electron donor, several tests remain, for example, the low manufacture evolution rate in μmol h^{-1} and the even subordinate recital. For extenuating CO_2 productions and adapting them hooked on costly fuels and chemicals, electrocatalytic CO_2 reduction is a striking method. In this method, by changing reaction conditions, the harvests could be altered and can partake at atmospheric pressure and moderate temperature. In aqueous electrolytes, the electrochemical CO_2 reduction has been observed as a promising carbon-neutral route [33], in which the H_2 evolution reaction (HER) is a main opposing lateral response. To endorse this conversion reaction in view of the inactive CO_2 molecule reaction kinetics, robust and effective electrocatalysts are essential, categorized in four groups based on elemental composition, i.e., transition-metal oxides, carbon-based materials, chalcogenides, and transition-metal [61]. The morphology, crystal facet, oxidation state, catalyst particle size, grain boundaries, defect, and organic hybrid are some significant features that disturb selectivity and activity. To enable the imminent growth of CO_2, electroreduction, many auspicious approaches, counting chemical modification, surface engineering, composite materials, and nanostructured catalysts are anticipated [50].

Kim et al. [22], established easy-to-fabricate and a novel MNP-NF filter with a low-pressure drop and elevated purification recital. This sieve proficiently eliminates airborne dust, such as metal oxides. The MNP-NF filters were made-up with polymer solutions containing Fe_3O_4 MNPs, which display robust magnetism using a simple electrospinning procedure. To determine that the MNP-NF filter, a field test could be utilized in a real atmosphere [22, 23]. Various other adsorbents and the removed toxic pollutants are summarized in Table 1.

3 Conclusions and Future Perspectives

The request of nanomaterials in the recognition and elimination of contaminants offers lower cost, shorter turn-around times, superior sensitivity, reduced sample sizes, real-time and in-line detection, a higher amount, and movability in ecological remediation. To eliminate metals and organic contaminants by oxidation or the reduction of nanomaterial, metal oxide, and metal nanomaterials could be utilized. By capturing selective target contaminants in air media by functionalization with chemical groups, removal could be improved. This technique is auspicious and operative and could be utilized in the manufacturing of air developments. Nanomembranes partake submissions in water reclamation, drinkable water production, dyes, the elimination of metals, and the removal of NOM and pesticides from polluted water. To selectively eliminate materials, additional developments should be completed in the submission of ecological treatment, such as material should be cost-effective, partake superior steadiness for an extensive period, the chemicals concentrations in polluted water, and greater resistance to changes in pH. Nanofibrous media have a low basis weight, a small pore size, and in height penetrability, that categorize them suitable for broad-ranging filtration applications. Additionally, nanofibers membranes offer exclusive properties, for example, a high specific surface area (contingent on the nanocomposite porosity and diameter of NPs), to incorporate a functionality or active chemistry at a nanoscale, virtuous interconnectivity of the pores. To recognize the properties on the performance of nanocomposites, current investigations are in progress to develop engineered nanomaterials of several morphologies

Table 1 Various adsorbent materials with their removal capacities, particle size

Target pollutants	NPs	Surface area	Filtration capacity (%)	Particle size	References
Aligned carbon nanotube	Aerosols	–	99.98%	0.3 µm	[51]
Nitrogen oxide (NO_x), volatile organic compounds (VOCs), microbial activity	Nanosilver-decorated titanium dioxide (TiO_2) nanofibers	–	21% for NO_x 30% for VOCs	32 nm	[45]
Nitrogen oxide (N_2O)	Ruthenium nanoparticle catalysts (Ru/γ-Al_2O_3)	$199\ m^2\ g^{-1}$	–	1–3 nm	[24]
Sulfur dioxide (SO_2)	Titanium dioxide NPs (TNPs)	–	$60\ L\ s^{-1}$	4 nm	[3]
Carbon disulfide (CS_2)	$Zn_{12}O_{12}$ nanocage	–	–	–	[14]
CO (carbon monoxide)	Ag/SBA-15 nanocomposites		98%	–	[56]
Methane (CH_4), nitrogen (N_2), carbon dioxide (CO_2), carbon monoxide (CO)	Single-walled carbon nanotubes	$483.9\ m^2\ g - 1$	$CO_2 = 10.5$ (w%) $CO = 4.4$ (w%) $CH_4 = 1.5$ (w%) $N_2 = 1.4$ (w%) $H_2 = 0.5$ (w%)	7.14 nm	[29]
VOCs (ethylene)	TiO_2 NPs (TNPs)	$151\ m^2\ g^{-1}$	70%	150–250 nm	[18]
VOCs (aniline and benzene)	Electrospun nanofibers (NF) (hydroxypropyl-beta-cyclodextrin (HPβCD), NF, and hydroxypropyl gamma-cyclodextrin (HPγCD) NF)	HPβCD/DMF-NF = $3.83\ m^2\ g^{-1}$ HPγCD/DMF-NF = $3.48\ m^2\ g^{-1}$	–	(HPβCD) NF = (3–71 µm range) (HPγCD) NF = (1–21 µm range)	[8]

(continued)

Table 1 (continued)

Target pollutants	NPs	Surface area	Filtration capacity (%)	Particle size	References
Sulfur dioxide (SO_2)	Mg ferrite nanospheres ($MgFe_2O_4$)	116 $m^2\ g^{-1}$	1.4 mmol g^{-1}	300–400 nm	[59]
Nitrogen oxide (NO_2), sulfur oxides (SO_2)	Zinc oxide and zirconium hydroxide NPs	–	–	ZnO = 20 nm $Zr(OH)_4$ = 7 μm	[43]
Methane (CH_4), carbon dioxide (CO_2), nitrogen (N_2), carbon monoxide (CO)	Horn-shaped carbon nanotubes	728.6 $m^2\ g^{-1}$	CO_2 (45.1 $cm^3 g^{-1}$), CH_4 (17.0 $cm^3 g^{-1}$), and N_2 (5.1 $cm^3 g^{-1}$)	50 nm	[41]
NO_x (mixture of NO and NO_2)	(SWNTs and MWNTs)	MWNTs = 155 $m^2\ g^{-1}$ SWNTs = 380 $m^2\ g^{-1}$	–	2.5–30 nm	[31, 57]
VOCs	CNTs deposited on quartz filters	150–3000 $m^2\ g^{-1}$	–	–	[1]
CO_2	(CNTs-APTS), modified CNTs using 3aminopropyltriethoxysilane (APTS)	1458 $m^2\ g^{-1}$	75 mg g^{-1}	1.7–100 nm	[46]
CO and CH_3OH gases	Si-doped and Boron-doped SWCNTs	–	–	–	[2, 20]
Isopropyl vapor	SWNTs/NaClO	365 $m^2\ g^{-1}$	103.56 mg g^{-1}	100 nm	[17]

and nanocomposites diameters. The adsorption of contaminant recognition consequence in a greener environment by the ecological submissions of polymer-supported nanocomposites in chemical/photocatalytic catalysis degradation. Though, the study of the collaboration amongst the encapsulated NPs and the host polymers and its result on the dispersal in contaminated air is essential. The significant manufacture of polymer-supported nanocomposites and their additional applied submissions persist open. In ecological remediation, the widespread request of sorbents has shown the competence of adsorbing organic contaminants and metals from contaminated air. TiO_2 nanomaterials, polymeric adsorbents, and iron-based nanomaterials have revealed increased adsorption selectivity and capacities. For process optimization, the surface alteration of adsorbents is deliberate. To decrease the cost in ecological monitoring, the extension of the lifespan of adsorbent, and increasing the recycling ability of adsorbents required to be traveled. Various sensors have been established for the recognition of bacteria, VOCs, chemicals, and various gases. To attain the requirement for trace recognition and the removal of contaminants in the air, further progress is essential in the efficient possession of nanomaterials. Various mechanistic and significant vital studies are essential to discover their actual capacities completely. For the separation of nanocomposite materials, multiple methods such as magnetic and photocatalytic methods have been utilized, though high energy is essential for these techniques. At a large-scale ecological contamination managing, the magnetic composites or metal oxide are auspicious materials, for the advancement of applied requests of these composites, further efforts are essential. For the filtration of air quality, more polymer composites should be addressed, which partake the long-term efficacies as a vital practical aspect.

Acknowledgements The authors gratefully acknowledge the support from the Ministry of Human Resource Development Department of Higher Education, Government of India under the scheme of Establishment of Centre of Excellence for Training and Research in Frontier Areas of Science and Technology (FAST), for providing the financial support to perform this study vide letter No, F. No. 5–5/201 4–TS.Vll. Dinesh Kumar is also thankful DST, New Delhi, for financial support to this work (sanctioned vide project Sanction Order F. No. DST/TM/WTI/WIC/2K17/124(C).

References

1. Amade R, Hussain S, Ocaña IR, Bertran E (2014) Growth and functionalization of carbon nanotubes on quartz filter for environmental applications. J Environ Eng Sci 3:2
2. Azam MA, Alias FM, Tack LW, Seman RNAR, Taib MFM (2017) Electronic properties and gas adsorption behaviour of pristine, silicon-, and boron-doped (8, 0) single-walled carbon nanotube: A first principles study. J Mol Graph Model 75:85–93
3. Baltrusaitis J, Jayaweera PM, Grassian VH (2010) Sulfur dioxide adsorption on TiO_2 nanoparticles: influence of particle size, coadsorbates, sample pretreatment, and light on surface speciation and surface coverage. J Phys Chem C 115:492–500
4. Bottino A, Capannelli G, Comite A, Di Felice R (2002) Polymeric and ceramic membranes in three-phase catalytic membrane reactors for the hydrogenation of methylenecyclohexane. Desalination 144:411–416

5. Brimblecombe P (2003) The effects of air pollution on the built environment, air pollution reviews, vol 2. Imperial College Press, London, UK. ISBN 978-1-86094-291-4
6. Brimblecombe P, Maynard RL (2003) The urban atmosphere and its effects, air pollution reviews, vol 1. Imperial College Press, London. ISBN 978-1-86094-064-4
7. Brimblecombe P (2015) Urban pollution and changes to materials and building surfaces, air pollution reviews, vol 5. Imperial College Press, London. ISBN 978-1-78326-885-6
8. Celebioglu A, Sen HS, Durgun E, Uyar T (2016) Molecular entrapment of volatile organic compounds (VOCs) by electrospun cyclodextrin nanofibers. Chemosphere 144:736–744
9. Chen M, Li Y, Wang Z, Gao Y, Huang Y, Cao J, Lee S (2017) Controllable synthesis of core–shell Bi@amorphous Bi_2O_3 nanospheres with tunable optical and photocatalytic activity for NO removal. Ind Eng Chem Res 56:10251–10258
10. Cho M, Chung H, Choi W, Yoon J (2005) Different inactivation behaviors of MS-2 phage and *Escherichia coli* in TiO_2 photocatalytic disinfection. Appl Environ Microbiol 71:270–275
11. Diallo MS, Savage N (2005) NPs and water quality. J Nanoparticle Res 7:325–330
12. Fan X, Wang Y, Zhong WH, Pan S (2019) Hierarchically structured all-biomass air filters with high filtration efficiency and low air pressure drop based on Pickering emulsion. ACS Appl Mater Interfaces 11:14266–14274
13. Gao Y, Huang Y, Li Y, Zhang Q, Cao JJ, Ho W, Lee SC (2016) Plasmonic Bi/$ZnWO_4$ microspheres with improved photocatalytic activity on NO removal under visible light. ACS Sustain Chem Eng 4:6912–6920
14. Ghenaatian HR, Baei MT, Hashemian S (2013) $Zn_{12}O_{12}$ nano-cage as a promising adsorbent for CS_2 capture. Superlattice Microst 58:198–204
15. Habisreutinger SN, Schmidt-Mende L, Stolarczyk JK (2013) Photocatalytic reduction of CO_2 on TiO_2 and other semiconductors. Angew Chem 52:7372–7408
16. Hailili R, Dong G, Ma Y, Jin S, Wang C, Xu T (2017) Layered perovskite $Pb_2Bi_4Ti_5O_{18}$ for excellent visible light-driven photocatalytic NO removal. Ind Eng Chem Res 56:2908–2916
17. Hsu S, Lu C (2007) Modification of single-walled carbon nanotubes for enhancing isopropyl alcohol vapor adsorption from air streams. Sep Sci Technol 42:2751–2766
18. Hussain M, Ceccarelli R, Marchisio DL, Fino D, Russo N, Geobaldo F (2010) Synthesis, characterization, and photocatalytic application of novel TiO_2 nanoparticles. Chem Eng J 157:45–51
19. Jiang XH, Wang LC, Yu F, Nie YC, Xing QJ, Liu X, Dai WL (2018) Photodegradation of organic pollutants coupled with simultaneous photocatalytic evolution of hydrogen using quantum-dot-modified g-C_3N_4 catalysts under visible-light irradiation. ACS Sustain Chem Eng 6:12695–12705
20. Kang S, Pinault M, Pfefferle LD, Elimelech M (2007) Single-walled carbon nanotubes exhibit strong antimicrobial activity. Langmuir 23:8670–8673
21. Karn B, Masciangioli T, Zhang WX, Masciangioli TM (2004) Nanotechnology and the Environment. Am Chem Soc
22. Kim J, Chan Hong S, Bae GN, Jung JH (2017) Electrospun magnetic nanoparticle-decorated nanofiber filter and its applications to high-efficiency air filtration. Environ Sci Technol 51:11967–11975
23. Kim YK, Park HB, Lee YM (2003) Carbon molecular sieve membranes derived from metal-substituted sulfonated polyimide and their gas separation properties. J Membrane Sci 226:145–158
24. Komvokis VG, Marti M, Delimitis A, Vasalos IA, Triantafyllidis KS (2011) Catalytic decomposition of N_2O over highly active supported Ru nanoparticles ($\leq$3 nm) prepared by chemical reduction with ethylene glycol. Appl Catal B Environ 103:62–71
25. Kondratenko EV, Mul G, Baltrusaitis J, Larrazábal GO, Pérez-Ramírez J (2013) Status and perspectives of CO_2 conversion into fuels and chemicals by catalytic, photocatalytic and electrocatalytic processes. Energ Environ Sci 6:3112–3135
26. Krishnamurthy A, Thakkar H, Rownaghi AA, Rezaei F (2018) Adsorptive removal of formaldehyde from air using mixed-metal oxides. Ind Eng Chem Res 57:12916–12925

27. Li C, Zhang F, Feng S, Wu H, Zhong Z, Xing W (2018) SiC@ TiO_2/Pt catalytic membrane for collaborative removal of VOCs and NPs. Ind Eng Chem Res 57:10564–10571
28. Li H, Ho WK, Cao JJ, Park D, Lee SC, Huang Y (2019) Active complexes on engineered crystal facets of MnOx–CeO_2 and scale-up demonstration on an air cleaner for indoor formaldehyde removal. Environ Sci Technol 53:10906–10916
29. Lithoxoos GP, Labropoulos A, Peristeras LD, Kanellopoulos N, Samios J, Economou IG (2010) Adsorption of N_2, CH_4, CO and CO_2 gases in single walled carbon nanotubes: a combined experimental and Monte Carlo molecular simulation study. J Supercrit Fluid 55:510–523
30. Liu Y, Chen S, Quan X, Yu H (2015) Efficient electrochemical reduction of carbon dioxide to acetate on nitrogen-doped nano diamond. J Am Chem Soc 137:11631–11636
31. Long RQ, Yang RT (2001) Carbon nanotubes as superior sorbent for dioxin removal. J Am Chem Soc 123:2058–2059
32. Lu Y, Huang Y, Cao JJ, Ho W, Zhang Q, Zhu D, Lee SC (2016) Insight into the photocatalytic removal of NO in air over nanocrystalline $Bi_2Sn_2O_7$ under simulated solar light. Ind Eng Chem Res 55:10609–10617
33. Mistry H, Varela AS, Kuehl S, Strasser P, Cuenya BR (2016) Nanostructured electrocatalysts with tunable activity and selectivity. Nat Rev Mater 1:16009
34. Morones JR, Elechiguerra JL, Camacho A, Holt K, Kouri JB, Ramírez JT, Yacaman MJ (2005) The bactericidal effect of silver NPs. Nanotechnol 16:2346
35. Qi L, Xu Z, Jiang X, Hu C, Zou X (2004) Preparation and antibacterial activity of chitosan NPs. Carbohydr Res 339:2693–2700
36. Qiao J, Liu Y, Hong F, Zhang J (2014) A review of catalysts for the electroreduction of carbon dioxide to produce low-carbon fuels. Chem Soc Rev 43:631–675
37. Rodriguez JA (2004) Activation of gold NPs on titania: a novel $DeSO_x$ catalyst
38. Rodriguez JA (2006) The chemical properties of bimetallic surfaces: Importance of ensemble and electronic effects in the adsorption of sulfur and SO_2. Prog Surf Sci 81:141–189
39. Rodriguez JA, Perez M, Jirsak T, Evans J, Hrbek J, Gonzalez L (2003) Activation of Au NPs on oxide surfaces: reaction of SO_2 with Au/MgO (100). Chem Phys Lett 378:526–532
40. Sánchez A, Recillas S, Font X, Casals E, González E, Puntes V (2011) Ecotoxicity of, and remediation with, engineered inorganic NPs in the environment. Trends Anal Chem 30:507–516
41. Sawant SY, Somani RS, Bajaj HC, Sharma SS (2012) A dechlorination pathway for synthesis of horn shaped carbon nanotubes and its adsorption properties for CO_2, CH_4, CO and N_2. J Hazard Mater 227:317–326
42. Sharma R, Raghav S, Nair M, Kumar D (2018) Kinetics and adsorption studies of mercury and lead by ceria nanoparticles entrapped in tamarind powder. ACS omega 3:14606–14619
43. Singh J, Mukherjee A, Sengupta SK, Im J, Peterson GW, Whitten JE (2012) Sulfur dioxide and nitrogen dioxide adsorption on zinc oxide and zirconium hydroxide nanoparticles and the effect on photoluminescence. Appl Surf Sci 258:5778–5785
44. Spengler JD, Sexton K (1983) Indoor air pollution: a public health perspective. Science 221:9–17
45. Srisitthiratkul C, Pongsorrarith V, Intasanta N (2011) The potential use of nanosilver-decorated titanium dioxide nanofibers for toxin decomposition with antimicrobial and self-cleaning properties. Appl Surf Sci 257:8850–8856
46. Su F, Lu C, Cnen W, Bai H, Hwang JF (2009) Capture of CO_2 from flue gas via multiwalled carbon nanotubes. Sci Total Environ 407:3017–3023
47. Taurozzi JS, Arul H, Bosak VZ, Burban AF, Voice TC, Bruening ML, Tarabara VV (2008) Effect of filler incorporation route on the properties of polysulfone–silver nanocomposite membranes of different porosities. J Membrane Sci 325:58–68
48. Weon S, Choi E, Kim H, Kim JY, Park HJ, Kim SM, Choi W (2018) Active 001 facet exposed TiO_2 nanotubes photocatalyst filter for volatile organic compounds removal: from material development to commercial indoor air cleaner application. Environ Sci Technol 52:9330–9340
49. Xia Y, Zhao J, Li M, Zhang S, Li S, Li W (2016) Bioelectrochemical reduction of Fe (II) EDTA–NO in a biofilm electrode reactor: performance, mechanism, and kinetics. Environ Sci Technol 50:3846–3851

50. Yang L, Yang L, Ding L, Deng F, Luo XB, Luo SL (2019) Principles for the application of nanomaterials in environmental pollution control and resource reutilization. In: Nanomaterials for the removal of pollutants and resource reutilization. Elsevier, pp 1–23
51. Yildiz O, Bradford PD (2013) Aligned carbon nanotube sheet high efficiency particulate air filters. Carbon 64:295–304
52. Zhang L, Fang M (2010) Nanomaterials in pollution trace detection and environmental improvement. Nano Today 5:128–142
53. Zhang Q, Huang Y, Xu L, Cao JJ, Ho W, Lee SC (2016a) Visible-light-active plasmonic Ag–$SrTiO_3$ nanocomposites for the degradation of NO in air with high selectivity. ACS Appl Mater Interfaces 8:4165–4174
54. Zhang R, Liu C, Hsu PC, Zhang C, Liu N, Zhang J, Cui Y (2016b) Nanofiber air filters with high-temperature stability for efficient $PM_{2.5}$ removal from the pollution sources. Nano Lett 16:3642–3649
55. Zhang TC, Surampalli RY, Lai KC, Hu Z, Tyagi RD, Lo IM (Eds.) (2009) Nanotechnologies for water environment applications. ASCE
56. Zhang X, Qu Z, Li X, Zhao Q, Wang Y, Quan X (2011) Low temperature CO oxidation over Ag/SBA-15 nanocomposites prepared via in-situ "pH-adjusting" method. Catal Commun 16:11–14
57. Zhang X, Yang B, Dai Z, Luo C (2012) The gas response of hydroxyl modified SWCNTs and carboxyl modified SWCNTs to H_2S and SO_2. Prz Elektrotechniczn 88:311–314
58. Zhang Z, Chen Z, Wang S, Qu C, Chen L (2014) On-site visual detection of hydrogen sulfide in air based on enhancing the stability of gold NPs. ACS Appl Mater Interfaces 6:6300–6307
59. Zhao L, Li X, Zhao Q, Qu Z, Yuan D, Liu S, Chen G (2010) Synthesis, characterization and adsorptive performance of $MgFe_2O_4$ nanospheres for SO_2 removal. J Hazard Mater 184:704–709
60. Zhu C, Chen X, Ma J, Gu C, Xian Q, Gong T, Sun C (2018) Carbon nitride-modified defective TiO_{2-x}@carbon spheres for photocatalytic H_2 evolution and pollutants removal: Synergistic effect and mechanism insight. J Phys Chem C 122:20444–20458
61. Zhu DD, Liu JL, Qiao SZ (2016) Recent advances in inorganic heterogeneous electrocatalysts for reduction of carbon dioxide. Adv Mater 28:3423–3452

Synthesis, Characterization, and Properties of Carbon Nanocomposites and Their Application in Wastewater Treatment

V. Madhavi, A. Vijaya Bhaskar Reddy, and G. Madhavi

Abstract The concern towards the increasing challenges in water treatment technologies made researchers towards significant innovations in recent years. The incorporation of nanotechnology in this field obtained remarkable results due to its exceptional properties. Among them, carbon nanocomposites (CNCs) have proved to be promising materials due to their large surface area, enhanced processibility, stability, synergetic properties, cost-effectiveness, and less impact on the environment. The transformation of CNTs to CNCs by functionalization involves physical/chemical modification of CNTs that improves the capability of CNCs in wastewater treatment technologies. This chapter discusses various synthesis methods, tailored characteristics, spectacular properties, and different functions of CNCs in water treatment. The comprehensive focus is extended on the effective CNC based wastewater treatment technologies of three main classes such as adsorption, desalination, and disinfection. Finally, the concerns of CNCs in environmental health and safety in the direction of future research are discussed.

Keywords Carbon nanocomposites · Functionalization · Tailored characteristics · Synergic properties · Wastewater treatment

V. Madhavi (✉)
Department of Chemistry, BVRIT-H, JNTUH, Hyderabad 500090, India
e-mail: madhuchem9@gmail.com

A. Vijaya Bhaskar Reddy
Centre of Research in Ionic Liquids, Universiti Teknologi PETRONAS, 32610 Seri Iskandar, Perak, Malaysia

G. Madhavi
Department of Chemistry, Sri Venkateswara University, Tirupati 517502, India

M. Jawaid et al. (eds.), *Environmental Remediation Through Carbon Based Nano Composites*, Green Energy and Technology,
https://doi.org/10.1007/978-981-15-6699-8_4

1 Introduction

Water is an inestimable natural reserve for human life and the ecosystem. However, population boom, climate change, rapid development of urbanization, and industrialization present serious challenges and extremely requires cost-effective water treatment technologies for freshwater supply. Water pollutants such as inorganic, organic, and harmful bacteria encompassed as threats and the impact is harming the entire biosphere confronting water-based vulnerability. This issue is hard felt where novel cutting edge technologies are of reliable interest for the removal of detrimental contaminants to achieve water sustainability. Traditional water treatment techniques such as chemical precipitation, coagulation, ion exchange, oxidation, and electrodeposition are available but not entirely satisfactory in terms of cost, efficiency, technical constraints, tedious design, and environmental impact. These conventional water treatment methods eliminate natural organic matter, inorganic sediments, and toxic microbes from the water before the distribution [1]. Due to rapid industrialization, man-made artificial contaminants such as heavy metals, organic dyes, pharmaceuticals, and released by-products of various water treatment methods seriously affect the environmental ecosystems [2]. In this regard, nanotechnology has been proved as a promising prospect to extend advanced materials for effective water purification. The properties of nanomaterials such as large aspect ratio, specific reactivity, both hydrophilic and hydrophobic interactions can be deliberately manipulated at nanolevel to exhibit high performance at an affordable cost. However, agglomeration of nanoparticles due to their high surface area restricts their use and can be reduced by converting nanomaterials to nanocomposites. Nanocomposites are defined as multiphased materials in which at least one of the phases confirms in the nanorange dimensions and maintain an interface between its components with enhanced synergistic characteristics. Of nanomaterials, carbon nanotubes(CNT) due to their high surface area, exceptional chemical inertness, and ease of chemical functionalization have received extraordinary consideration for the removal of organic, heavy metal and microbial impurities in water purification applications. CNTs are considered as 1D allotropes of carbon that are described as graphene sheets rolling a layer into cylinders of nanoscale diameter. In particular, CNTs are classified into single-wall (SWCNTs) and multiwall CNTs (MWCNTs) based on the number of graphene layers. The remarkable properties possessed by cylindrical carbon molecules make them reliable towards numerous applications for electronics, optics, material science, and environmental applications. CNT-based composites are other extensively used carbon-containing nanoparticles owing to their desired synergistic properties such as low density, high aspect ratio, mechanical and thermal stability. In this chapter, we discuss the recent progress in the field over the latest research and explore the novel synthesis methods, characterization, and properties of carbon nanocomposites with their performance evaluation in wastewater treatment. Further, it assesses the current progress and challenges with future perspectives.

2 Carbon Nanotubes in Nanotechnology

CNTs exhibit outstanding adsorption, catalytic, mechanical, magnetic, thermal, and electrochemical properties and hence have diverse applications in industrial as well as technological applications [3]. These properties in CNTs have made them a potential material in the fields of biomedical, sensors, energy storage, solar cells, textiles, environment, and support many areas. CNTs were synthesized by several techniques such as Chemical Vapor Deposition (CVD), electric-arc discharge, Laser ablation, hydrothermal, electrolysis and spray pyrolysis, etc. [4]. Both the laser ablation and arc discharge methods have disadvantages such as scaling up of the process, tangled nature of CNTs, and presence of unwanted carbon impurities that need further purification. CVD method is considered the best technique that is most extensively used due to its comparatively simple, inexpensive, flexible, energy-efficient, and easy operation [5]. This method appears to be the most potential way for the large scale production with a controllable structure having high purity [6]. Extensive efforts are presently underway for diverse applications including biosensors, bioengineering, nanotechnology, and water purification. However, CNTs easily agglomerate and form clusters owing to their high surface energy and high Vander Waals forces between the tubes that tend to produce samples resulting in a combination of assorted diameters and chiralities with metallic and amorphous carbon contamination. Dispersion of CNTs in solvents is the other main factor due to these forces that strongly influence the properties of nanocomposites. The deagglomeration followed by distribution of nanomaterials in the matrices or solvents is known as dispersion. The length, volume fraction, sonication duration, attractive forces, and entanglement density of CNTs determine their dispersion in solvents or matrices [7]. A suitable functionalization of the nanotubes represents the strategy to overcome these limitations and activate the CNTs surface to become an attractive field in nanotechnology. Functionalization improves the dispersibility and processibility of CNTs that develops the interaction with other entities such as organic, inorganic solvents and matrices that allow combined and inimitable properties of CNTs with that of other materials and thus may be utilized for various applications [8–11]

Fundamentally, covalent and non-covalent functionalizations as interactions are used depending on reaction mechanisms. Covalent functionalization employs on the covalent linkage of functional entities to deal with CNTs, thus intends to intact functional groups such as hydroxyl, carboxyl, and aminoacid groups at the open ends and holes in the defect sites and sidewalls of CNTs. Several covalent routes are more likely for functionalization such as amidation, oxidative purification, thiolation, esterification, hydrogenation, halogenations, cycloadditions, and electrochemical functionalizations have been demonstrated by Khan et al. [12]. In contrast, non-covalent functionalization involves physical adsorption, non-destructive utilization of surfactants, and polymers onto the surface of CNTs which involves hydrogen bonding and weak interactions [13]. In general, non-covalent functionalization occurs without any effects on the intrinsic and basic plane structure, properties of CNTs. However, in most cases, the surface of CNTs has to be modified to avoid the adverse damage to the structure of CNTs and to improve carbon-matrix interaction.

3 Carbon Nanocomposites

3.1 Synthesis and Characterization Methods of Carbon Nanocomposites

The tailorable characteristics of CNCs such as appreciable mechanical strength, high specific area, and excellent chemical inertness make them ideal for various applications. The emergence of fascinating advanced carbon nanocomposites is leading to the next-generation sophisticated materials. The adding up of these nanotubes to an array of matrices such as metals, metal oxides, polymers, etc. can improve the electrical, mechanical, thermal, and chemical properties. Carbon nanocomposites can be synthesized in a number of ways among which are impregnation, CVD, Ball milling, Sol-gel, extrusion, etc. [14, 15]. At different stages, more than one method can be combined in the process of formation of nanocomposites. It has been found that the synthesis method has huge influences on the surface morphology of the carbon nanocomposites. Sharififard et al. [16] prepared Iron-activated carbon (IAC) nanocomposite from the evaporation of iron salt solution by anchorage of iron oxide-hydroxide nanoparticles on the activated carbon surface. The synthesis method was facile at low temperature for the formation of IAC nanocomposite that possesses good adsorption properties. SEM micrograph of IAC shows the presence of iron oxide-hydroxide has "silver almonds" shape and iron nanoparticles are effectively dispersed on the AC surface. TiN@C nanocomposites were synthesized by an annealing approach using the oleic acid as a carbon resource for enhancing the electrochemical properties of TiN nanoparticles by Lei et al. [17]. The prepared TiN@C nanocomposites were characterized by EDX, XRD, TEM techniques. The TEM results confirmed that TiN nanoparticles are coated by carbon, and the coated carbon covering has a thickness of about 3 nm. A new rapid, simple, cheap, and effective method for synthesis of magnetic carbon encapsulated Co nanoparticles by catalytic carbonization of cobalt(II) fulvate is presented by Litvin and Galagan [18]. The TEM study of the material shows graphite-like phase that confirms the presence in the composition of the elongated structures of carbon nanotubes with large number of layers or carbon nanofibers. Sovizi et al. [19] obtained magnetic-activated carbon nanocomposite (m-Fe_3O_4@ACCs) for the lead ions removal from wastewater. Experimental results showed that greater than 99% of Pb(II) was removed by m-Fe_3O_4@ACCs at the optimal operational conditions and followed the pseudo-second-order kinetic model for the adsorption of Pb(II). TEM image of the resultant m-Fe3O4@ACCs reveals the presence of iron oxide nanoparticles of diameter 40–80 nm. TGA of the magnetic nanoparticles revealed that 10 wt% of iron oxide present inside the synthesized nanocomposite. Peng et al. [20] prepared CNTs-iron oxide magnetic composites for the adsorption of Pb(II) and Cu(II) from water and recovered above 98% of adsorbent after use. The SEM of the prepared composites shows the entangled networks of CNTs with clusters of iron oxide appended to it and implies the formation of carbon nanotubes/iron oxides composites. The adsorption capacities were 0.51 and 0.71 mmol g^{-1} for Pb(II) and Cu(II) respectively in the concentration range studied at pH 5.0.

A novel attapulgite clay@carbon (ATP@C) nanocomposite adsorbent was synthesized by the hydrothermal carbonization process by Chen et al. [21]. FESEM and TEM images of the as-prepared composite revealed the rod-like nanocomposite with a length of 200–1000 nm and diameter of 40–80 nm similar to that of the original ATP template. The granular nanospheres of several tens of nanometers size were dispersed on the surface of the ATP. Mojoudi et al. [22] synthesized a porous activated carbon/nanoclay/thiolated graphene oxide nanocomposite. The FTIR analysis confirms that carboxylic acid and hydroxylic groups present on the surface of AC/NC/TGO are the main contributors in uptake of contaminant from aqueous solution. Ag@ZnO/MWCNT (Ag-doped ZnO/multiwall carbon nanotubes) nanocomposite was synthesized by Ahmadi Azqhandi et al. [23] SEM images of the prepared nanocomposite demonstrated the uniform distribution of nanoparticles on the MWCNT surface. The composite resulted in the simultaneous removal of high contents of BY28 and MB dyes from aqueous solutions in the presence of ultrasonic power with Ag@ZnO/MWCNT-NC and the process is fast, low-cost, and efficient. The (HAP/TE/GAC) nanocomposite, i.e., granular activated carbon (GAC) was layered with both hydroxyapatite (HAP) nanoflakes and turmeric extract (TE) was obtained by Chathumal Jayaweera et al. [24] to remove the heavy metals and bacterial contaminants that can be utilized as a point-of-use water filter material. The SEM analysis reveals that the turmeric extract is deposited as flakes that are almost in the micrometer range in between the surface of the composite and mesh of nano HAP. The properties of CNTs lead to extraordinary properties when used as a fortification in polymeric materials that have caused a great deal of concern in the researchers' attention. It has been found that considerable variations of thermal, electrical, mechanical, and barrier properties come about with the inclusion of a very low dose of carbon-based nanofillers. Jose et al. [25] prepared PVA/MWCNT nanocomposites by solution casting method that demonstrated the interaction and filler-filler network arrangement of MWCNT in a-MWCNT/PVA nanocomposites by high resolution optical microscopy. Al-Hobaib [26] obtained polyphenylene diamine (PMD) membranes by incorporating carboxylated MWCNT in the polymer that displayed the clean and smooth tube surface with 10–20 nm in diameter in TEM micrographs. Polyvinyl chloride (PVC) membranes containing pristine and modified multiwall carbon nanotube (MWCNT) were prepared by Masoumi et al. [27]. The FESEM images indicated that the number of pores on the membrane surface increased at the presence of pristine and modified MWCNT and pore size distribution curves shifted towards smaller pores. It has been revealed that the antifouling properties of the membranes increased with increasing nanotube concentration, especially COOH-MWCNT. Cellulose acetate (CA)/carbon nanotubes (CNT) membranes have been prepared by using phase inversion method by El-Dein et al. [28] by dispersion of different ratio of CNTs in CA casting solution. Morphology results by SEM showed that porosity of CA membrane decreased with an increase in polymer ratio. The addition of CNTs enhanced the formation of the porous structures and macrovoids that is resulted from instantaneous demixing in the coagulation bath.

The membranes prepared by polyethersulfone as matrix polymer in which acid oxidized multiwalled carbon nanotubes (MWCNTs) were embedded are evaluated by Vatanpour et al. [29] for their efficiency and antifouling properties of mixed matrix nanofiltration membranes. The morphology studies by SEM demonstrated that large macrovoids appeared by the addition of less quantities of functionalized MWCNTs leading to increase in both pure water flux and salt rejection of the membranes. The membrane has lower roughness (0.04 wt% MWCNT/PES) represented the remarkable antifouling property. The raw MWCNT, PAA modified MWCNT, and grafting efficiency of PAA on the characteristics of nanocomposite polyethersulfone (PES) nanofiltration (NF) membranes were investigated by Daraei et al. [30]. The membranes possessing negatively charged surface due to functional groups of modified MWCNTs showed highest salt rejection, superior antifouling properties, and high water flux that reveals the success of simultaneous use of diverse modification methods.

3.2 Properties of Carbon Nanocomposites

carbon nanomaterials with highly ordered zero-, one-, two-, and three-dimensional carbon structures including fullerene, carbon nanotubes (CNTs), graphene, and graphene oxide have attracted increasing interest owing to their unique morphological regularity, chemical inertness, high surface area, biocompatibility, etc. These inimitable functions and properties due to their small or intermediate size make carbon-based materials ideal for reinforcing fillers in nanocomposites that can use and create structures, devices, and systems. In addition, the low-cost and flexibility of carbon raw materials are beneficial aspects for carbon-based applications when considering the economical factors. The carbon nanocomposites also are capable to manipulate on the atomic scale. Utilizing these fascinating aspects of carbon-based nanomaterials and composites, flexible, high performance and reliable materials can be produced for diverse applications. The efficient use of CNT-based composites depends robustly on their ability of homogeneous dispersion throughout the matrix without tearing out the integrity of CNTs. The impact of dispersion, alignment, aspect ratio, and weight fraction of CNTs in matrices are crucial for the capability and applications of carbon-based nanocomposites. Therefore, motivated by technological and scientific potential aspects of CNCs, over two decades, research work has been extensively done on carbon nanocomposites and this field of research is still growing stronger.

Carbon nanocomposites have attracted the interest of researchers owing to their exceptional electrical, optical, mechanical, thermal, and catalytic properties thereby extending their field of applicability. Baik et al. [31] have fabricated carbon nanotube - copper nanocomposites where CNTs are homogeneously dispersed within the copper matrix by mechanical and molecular-level mixing process and these composites showed a significant decrease in electrical resistance. The electrical properties of agarose/DWCNT nanocomposite hydrogels and the effect of DWCNT content on the composite properties were investigated by Guillet et al. [32]. The experiments on AC, DC measurements at different voltage results suggest that these nanocomposite hydrogels can be promising materials as electrode materials in drug delivery by electropermeabilization. MWCNT/epoxy and GNP/epoxy nanocomposites with different filler contents and hybrid epoxy nanocomposites filled with CNTs/GNPs as reinforcement were synthesized and the effects of different individual CNT/GNP contents and combination on electrical properties were evaluated by Kranauskaitė et al. [33]. The electrical conductivity of hybrid nanocomposites containing MWCNTs and GNPs in ratio 5:1 exhibits the highest value of 0.009 S/m, which is more than 4 times higher than that of composites containing only MWCNTs (0.002 S/m). That could be considered as a synergistic effect between GNPs and MWCNTs due to the well distribution of MWCNTs and the tunnelling of electrons between GNPs and MWCNTs. Electronics utilize the applications of conductive polymer nanocomposites that have the potential to be used in electronics, sensors, and actuators [34, 35]. At a critical filler concentration, i.e., percolation threshold, conductance can be observed in nanocomposites due to the formation of conductive networks of nanoparticles in them. [36, 37]. The percolation concentration is experimentally determined by the electrical conductivity at different filler concentrations.

According to Awasthi et al. [38], the conductivity of polyethylene oxide (PEO)-MWCNT composite films resulted in an enhancement of eight orders 6.52 S cm^{-1} of magnitude in conductivity compared to that of bare PEO film. Polyvinyl alcohol (PVA)-vapor growth carbon fiber (VGCF) and PVA-MWCNT were fabricated by Bin et al. [39] using gelation/crystallization methods. The percolation threshold of electrical conductivity for the PVA/MWCNT was <1 wt% MWCNT loading that was much lower than that of PVA/VGCFs composites. The optical properties of silicon incorporated diamond-like carbon (Si-DLC) nanocomposite thin films due to the change in the electronic structure of carbon nanocomposites have been reported by Alam et al. [40]. Si-DLC film showed broad photoluminescence (PL) peak centered at 467 nm, in the range of visible radiation that intensified with an increase in % of Si. The optical, mechanical, electrical properties of combined CNT and metal nanoparticles are interestingly enhanced due to the specific surface area of CNTs [41, 42]. In addition to this, Barberio et al. [43] presented very special electronic and optical properties of MWCNT with metal matrices (Al, Ag, Au, Co, Cu, Fe, Ni, and Ti) than to pristine CNTs. The surface roughness of nanocomposites has been decreased to about 50% that results in strong visible photoluminescence.

Guler et al. [44] synthesized carbon nanotubes hybrid zinc oxide (ZnO-CNTs) nanocomposites using ball mill technique and studied for optical properties. The decrease in reflectance of the composites with CNTs is due to the increase in absorbance of the nanocomposites. Wang et al. [45] used Molecular Dynamics (MD) simulation to evaluate the mechanical properties of CNT reinforced Poly-ether-ether-ketone (PEEK) nanocomposites. The overall mechanical efficiency of CNT/PEEK nanocomposite was improved by introducing H-bonds between CNTs and PEEK matrix. The elastic modulus and tensile strength of the synthesized nanocomposite were envisaged to be 24.5 GPa and 2.47 GPa, respectively, validating the MD model in the evaluation of mechanical properties of CNT/PEEK nanocomposite. Nam et al. [46] investigated on thermal properties of CNT/Al–Cu that was prepared by high energy ball milling followed by spark plasma sintering. The thermal conductivity of these nanocomposites decreased with an increase in the content of CNTs due to the interface thermal resistance between CNTs and Al–Cu matrix. Vahedi et al. [47] reported on the thermal conductivity of CNT/paraffin nanocomposites using multiscale modeling. Molecular dynamics simulations were evaluated for the thermal conductance and their findings reported that the effects of volume fraction and geometric parameters of fillers provide effective thermal conductivity of CNT/paraffin nanocomposites. The increase in the thermal and electrical conductivity of nanocomposites is due to the functionalization of CNTs.

Functionalized MWCNTs performed the best filler materials that simultaneously improve thermal and electrical properties of the composites. The surface functionalization on SWCNTs increases the interaction between the CNTS and matrix but leads to the formation of defects, which obstruct the acoustic phonon transport in SWCNTs [48]. Amrin and Deshpande [49] fabricated Polyvinyl alcohol(PVA) and carboxyl functionalized MWCNTs using a solution cast method to investigate mechanical and dielectric properties of carbon nanocomposites. MWNT-COOH/PVA was found to have higher dielectric constant and AC conductivity due to interfacial polarization effect. Sui et al. [50] used carbon nanofiller in Polypropylene (PP) polymer nanocomposites as two phases, i.e., crystalline and amorphous prepared by melt blending method. Their results showed that PP nanocomposite with CNF (5 wt%) revealed unexpectedly high dielectric constant at wide sweep frequencies with small dielectric loss. According to their reports, there is an improvement of thermal and electrical properties of nanocomposites with an increase in carbon nanofiber content. The carbon nanotube/amino-functionalized poly(arylene ether ketone) composites were prepared by solution blending technique and their dielectric and mechanical properties were evaluated by Zhang et al. [51]. The amino-functionalized MWCNTs dispersed well in polymer that showed a higher dielectric constant of about 130 at 10% volume fraction of CNTs and the nanocomposites had tensile strength and tensile modulus of 69.2 MPa of 3.0 GPa, respectively.

Multiwalled carbon nanotubes/polyaniline/magnetite (MWCNTs/PANI/Fe_3O_4) ternary nanocomposites were successfully fabricated via oxidative polymerization followed by co-precipitation and their optical properties were studied by Ibrahim et al. [52]. The optical absorption showed that MWCNTs/PANI/Fe_3O_4 thin films with 300 nm thickness have both indirect and direct energy band gaps with allowed transitions in the energy range of 2.906–3.41 eV. The dark current-voltage characteristics of the MWCNTs/PANI/Fe_3O_4 thin films were non-linear and exhibited the rectification ratio (RR) of the forward and reverse currents at the same voltages ($V = \pm 3$ V) was found to be 5 at room temperature. Further, Aydin [53] found that electrical conductivity increases with CNT content in CNT and titanium dioxide nanocomposites. The direct current electrical conductivity values σ_{dc} of the nanocomposites were found to be in the range of 5.96×10^{-3}–0.47 S/cm. The obtained band gap (E_g) values were decreased in composites with an increase in CNT contents.

There is a significant enhancement in mechanical properties of carbon nanocomposite scaffold when they are coupled with thermoplastic polyurethane [54]. Carbon nanotubes with highest specific-volume ratio can be functionalized easily are coupled with thermoplastic polyurethane by solution-based fabrication method. These composite scaffolds showed an enhancement in tensile modulus about 200-fold over the pristine polymer at 19 wt% MWCNT loading. These scaffolds were thermally stable above their decomposition temperatures and extended the mechanical reliability by suppressing the mobility of polymer chains.

3.3 Carbon Nanocomposites for Water Purification

The global industrial revolution has led to a drastic increase in effluent discharge causing serious life-threatening problems for environment as well as water contamination. The contamination of water with diverse toxic chemicals and their treatment has turned into a major environmental problem. Water treatment is a comprehensive environmental concern that requires consistent interest in the removal and reduction of hazardous pollutants. The problem is hardly felt in the developing countries where water treatment technologies such as ion exchange, electrochemical treatment, chemical precipitation, membrane filtration, reverse osmosis are not easily accessible at the field scale due to financial constraints. For this, there is a need for technologies that are capable to remove harmful pollutants to a safe level, rapidly, efficiently, and within a reasonable cost framework. CNCs, attributing to high surface area and minimized aggregation has extraordinary consideration for their properties in separation of pollutants from acknowledged water. Carbon nanotubes with tactical combinations of other matrices in the form of composites can facilitate synergistic properties for the facile processing of water treatment. Introduction of specific functional groups to CNTs can be easily incorporated onto composite materials that enhance removal of certain species from water. Various types of carbon nanocomposites (CNCs) with unique and novel properties make them ideal and promising materials for their diverse applications including water treatment. Generally, CNTs are combined with solid support materials like metals, metal oxides, polymers, etc. to facilitate the tailored applications in water purification and filtration. One of the promising parts of CNCs for water treatment is the necessarily less amount of material required to achieve high filtration and contaminant removal capacity.

4 Functions of CNCs in Water Treatment Technologies

4.1 Adsorption

Adsorption process is a surface phenomenon where pollutant concentrations would adsorb onto the layers of solid materials due to the intermolecular (physical/chemical) forces of attractions. It is a prominent process among the most successful techniques for the removal of color, odor and both organic and inorganic pollutants in global effluents. Adsorption is observed as finer method in water treatment at the outline due to its effortlessness activity. CNTs have been frequently used as good adsorbents for capturing ample variety of pollutants owing to their adaptable properties. [55–58]. The scope for fabricating CNT-based composite materials is large as the two main forms of CNTs namely, SWCNTs and MWCNTs allow further flexibility for water purification materials. Both these forms have been demonstrated as efficient adsorbent materials for chemical species. However, the adsorption rate of CNTs is determined by external surface area, pore density, functionalities, purity, and so on. The major adsorption sites such as inner CNT holes, interstitial channel, grooves, and outer surfaces on CNTs play a major role in adsorption of water pollutants. For instance, in open-ended CNTs, an inner hole acts as suitable adsorption site and hence unzipped CNTs have more adsorption sites than pristine CNTs. The open-ended SWCNTs with lower diameter stimulates adsorption and is more suitable for multiple adsorbates than MWCNTs [59, 60]. Small pollutants can entrap into CNTs due to the presence of interstitial channels. SWCNTs can generate more interstitial channels because of their better aggregation than MWCNTs. Grooves and the outermost surfaces of the CNT bundles provide positive impacts for adsorbing various water pollutants. These sites offer accessible spaces for hosting both inorganic and organic contaminants.

CNTs when introduced with specific functionalities like metals, metal oxides, and polymers have the beneficial effect of enhanced adsorption with certain species. These functionalities increase CNT solubility and avoid aggregation in homogeneous solutions. Hence, this helps to enhance the interaction between CNT surfaces and water pollutants. The efficiency of carbon nanocomposites in water treatment is based on the nature of interaction between CNCs and water contaminants. Covalent bonding, hydrogen bonding, hydrophobic interactions, electrostatic interactions, p-p electron coupling, ion exchange, etc. are the general interactions that remarkably perform with CNCs for water treatment. The influencing factors that determine the extent of adsorption are available surface area and functional groups and these parameters have been more emphasized in recent research [61]. Salam et al. [62] fabricated the MWCNT/chitosan nanocomposite with the ratios 25:75 wt%. The nanocomposite of MWCNT and chitosan was utilized for the removal of Zn, Cd, Cu, and Ni ions from aqueous solution. CNTs oxidized with H_2O_2, $KMnO_4$, and HNO_3 were evaluated [63] for Cd(II) adsorption in water. The addition of different functional groups such as –carboxyl (COOH), hydroxyl (–OH), and carbonyl (C=O) made CNTs more soluble in aqueous solution and enhanced the adsorption and ion exchange capacities of functionalized CNTs. H_2O_2, HNO_3, and $KMnO_4$ oxidized CNTs showed the adsorption capacity of 2.6, 5.1, and 11.0 mg/g whereas it is only 1.1 mg/g for pristine CNTs. The performance of CNC for pollutant removal in water by adsorption is presented in Table 1.

4.2 *Desalination*

In view of the fact that the sources of seawater account for almost 98% on the earth, desalination through various technologies imparts a huge impact on water scarcity concerns. CNTs gained more interest in the field of water desalination technologies owing to their diverse and remarkable properties. Desalination utilizes three important kinds of water treatment innovations: (i) Chemical methods, (ii) distillation processes, and (iii) membrane technologies. Tofighy and Mohammadi [75] synthesized CNT sheets with nitric acid oxidizer through CVD method and used as a viable adsorbent for the desalination of salty water. Yang et al. [76] presented that the modified MWCNT with carboxylic and hydroxyl groups for the exclusion of humic acid from water. The results showed that the CNTs with different functionalities enhanced the hydrophilicity and removal capability of humic acid through bucky paper. Adsorption techniques are simple however not validated for desalination of water. Membrane technologies in this regard have received attention because of their fascinating inherent features.

Membrane separation technology does not require chemicals and there can be no regeneration of secondary pollutants and hence it is considered as viable and acceptable route to offer more sustainable process in water treatment. At present, a number of membrane separation techniques such as RO, NF, UF, MF, distillation, dialysis, and electrodialysis are available. Polymers such as polysulfone, polyamides, cellulose nitrate, polyethersulfone are the most favored for membrane technology because of their cost-effective, facile synthesis, good thermal stability, high mechanical strength, and biocompatibility. However, bacteriological contamination and pore blockings by adsorption of inorganic/organic impurities; low throughput and fouling limit those for desalination process through membrane technology. CNT-based membrane technology utilizes the synergistic effects and has been recognized as a viable and effective approach for wastewater treatment. Polymer supported CNT membranes offer the benefit of desirable properties with the ability to upscale production. Suitable filler material, organic and inorganic modification of CNT exterior surfaces decreases agglomeration and increases miscibility in aqueous solutions. Functionalization also controls pore size and diameter which are suitable for fabricating uniform CNT membranes for optimum water desalination. Wang et al. [77] reported the incorporation of CNTs in PES membranes that exhibited higher flux and salt rejection than the individual PES membranes. The study revealed that the highest water flux (38.91 L m^{-2} h^{-1}) and Na_2SO_4 rejection (87.25%) at 4 bar were obtained at 0.1 wt% of CNT concentration. Shawky et al. [78] have fabricated MWCNT/polyamide nanocomposite membrane that demonstrated to have excellent mechanical strength

Table 1 Summary of performance of CNC for pollutant removal in water

CNC	Method	Target species	Performance	Reference
COOH-MWCNT	Batch Adsorption	1,8-Dichlorooctane, quinolone, alkylphenoletoxilate	Adsorption of target species affected by CNT functional groups and relative hydrophobicity of each	Patiño et al. [64]
Acid activated MWCNTs + TiO2 nanoparticles	Batch Photocatalyzed degradation	4-Chlorophenol	Two-fold increase in photocatalyzed degradation with MWCNTs	Zouzelka et al. [65]
Carrageenan modified acid-treated MWCNTs + Fe3O4 nanoparticles	Batch Adsorption	Methylene blue	Carrageenan modified MWCNT composites outperformed non-carrageenan Composite could be magnetically separated	Duman et al. [66]
MWCNTs + Graphene + Fe_3O_4 nanoparticles	Batch Flow Adsorption	Ar(III) and (V)	Flow outperformed batch adsorption for removal of Ar(III) and Ar(V), 100% and 74% removal respectively	Park et al. [67]
PVDF membrane + (i) Native MWCNTs (ii) Hydroxylated MWCNTs (iii) Animated MWCNTs (iv) Large inner diameter MWCNT	Flow Adsorption	Triclosan, prometryn, 4-acetylamino-antipyrine, carbendazim, caffeine, ibuprofen, acetaminophen	Highest adsorption performance observed for hydroxylated MWCNTs high specific surface area improved performance	Wang et al. [68]
Carbon nanofiber of polyacrylonitrile + MWCNTs	Flow Adsorption	Atrazine, sulfmethoxazole	Comparable adsorption	Peter et al. [69]
Acid-treated MWCNTs + chitosan	Electrosorption	Aniline	26.4 mg/g adsorption	Ma et al. [9]
Hydroxylated MWCNTs + PANI + PES	Flow Size exclusion Adsorption	Humic acid	80% removal of humic acid high water flux	Lee et al. [70]
SWCNTs + parylene membrane	Flow Size exclusion	Direct Blue 71, AuNP, Dengue virus	High water flux Rejection of $\geq$5 nm species	Bui et al. [71]

(continued)

Table 1 (continued)

CNC	Method	Target species	Performance	Reference
SWCNTs + Kevlar nanofibers	Batch Adsorption	Phenylic compounds, heavy metals, dyes	Composite material displayed higher adsorption capacity for all compounds when compared to polyethersulfone and polysulfone adsorbents	Nie et al. [72]
CNT/silica nanoparticle sponges	Batch Adsorption	Oils, organic solvents	High adsorption capacity to oils and solvents, up to 1885 weight %	Siddiqa et al. [73]
MWCNT/TiO2/polysulfone	Flow Adsorption Exclusion	Humic acid	Composite outperformed native polysulfone material in terms of both adsorption/rejection of humic acid and water permeability Composite was flexible for different concentrations of humic acid	Esfahani et al. [74]

and fabulous salt rejection ability (76.1%) with high permeability (0.71 L/m^2/h bar). Farahbakhsh et al. [79] prepared a thin film nanocomposite (TFN) membrane with raw and oxidized MWCNTs. The water flux of 25.9 and 28.9 L/m^2 h was observed for raw and oxidized MWCNTs based TFN respectively. The prepared TFN membrane showed a NaCl rejection of 98.1% with 0.002 wt% of raw MWCNTs and 97.8% with 0.002 wt% of oxidized MWCNTs. It is reported that the decrement of salt rejection was due to agglomeration of the MWCNTs (Fig. 1). A mixed matrix membrane of Carboxylated CNTs/polyethersulfone (PES) was reported by Wang et al. [77]. The lowest contact angle of 54.95° was observed for MWCNTs having a diameter of 20 nm. Chan et al. [80] simulated and synthesized zwitterion functionalized CNT/polyamide nanocomposite membranes (diameter 1.5 nm) and achieved 100% ion rejection. The increased ion rejection with the zwitterion functionalized CNTs is attributed to a steric hindrance from the functional groups that partially blocks the tube ends and electrostatic repulsion between functional groups and ions. Corry [81] has effectively removed 100% Na++ and Cl− with functionalized CNTs containing –COOH, –NH_3, and –OH groups.

MWCNT-PA nanocomposite membranes exhibited high chlorine resistance when used in aqueous solutions of NaClO [82]. The experimental results specify that the presence of MWCNT within the PA matrix in membranes significantly modifies both the surface shape and the molecular topology. The separation efficiency of these membranes after chlorine exposure (4800 ppm h) remained unchanged (99.9%) but was considerably reduced to 82% in the absence of MWCNT. Ratto et al. [83] have patented a CNT membrane with greater than 99% of ion rejection efficiency which indicates remarkable potentiality of CNT membranes in water desalination. Yang et al. [76, 84] have tailored CNTs by plasma treatment that showed adsorption capacity of exceeded 400 wt% of salt. In a recent study by El Badawi et al. [85] multi-walled carbon nanotube/cellulose acetate (CNT/CA) nanocomposite membranes

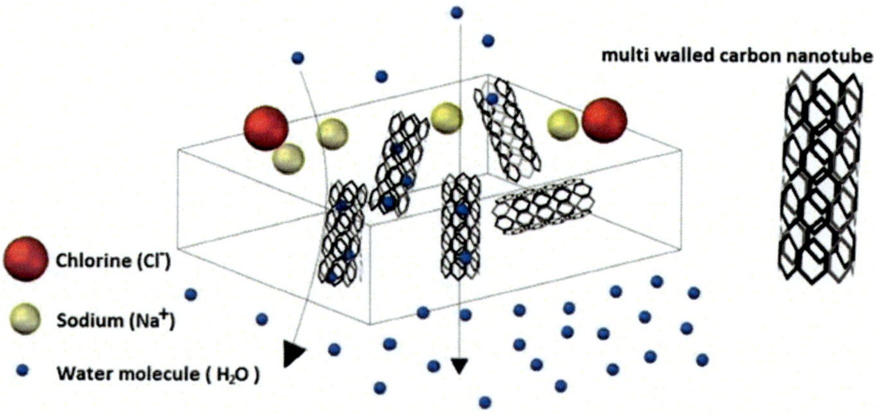

Fig. 1 Schematic representation of MWCNT TFN membrane [79]. Copyright 2017. Reproduced with permission from Elsevier

were prepared successfully and investigated on nanocomposite membrane permeation and salt retention rates with 1000 ppm NaCl solution. Permeation and salt retention rates were found to 54% and −6% respectively for the membranes with the lowest CNT content. Table 2 presents the performance of carbon-based nanocomposites by desalination for water treatment (Fig. 2).

Table 2 Carbon-based nanocomposites for desalination

CNT	Polymer	Salt solution	Performance of the membrane	Reference
CNT	CA	NaCl, 5 g L21	• Improved water flux and NaCl separation • Increased hydrophilicity	El-Din et al. [86]
MWCNT	Polyamide	NaCl	• Increased salt rejection • Improved Youngs modulus, toughness and tensile strength	Shawky et al. [78]
Carboxylated MWCNT	PES	NaCl, 1000 mg L21	• Improvement in salt rejection and water permeability • Decreases average pore radius of the substrate surface	Wang et al. [68]
Zwitterion functionalized CNT	PA		• Increased both flux and salt rejection ratio • High flow rate and better selectivity	Chan et al. [80]
Acid oxidized MWCNTs	PES	Na_2SO_4, $MgSO_4$, NaCl	• Improved antifouling ability • Increased salt rejection and hydrophilicity	Vatanpour et al. [29]
MWCNT	PA	NaClO	• Improved degradation resistance • Increased both flux and salt rejection	Ortiz-Medina et al. [82]
plasma-modified ultralong carbon nanotubes	MCE	NaCl	• Ultrahigh adsorption capability	Yang et al. [84]

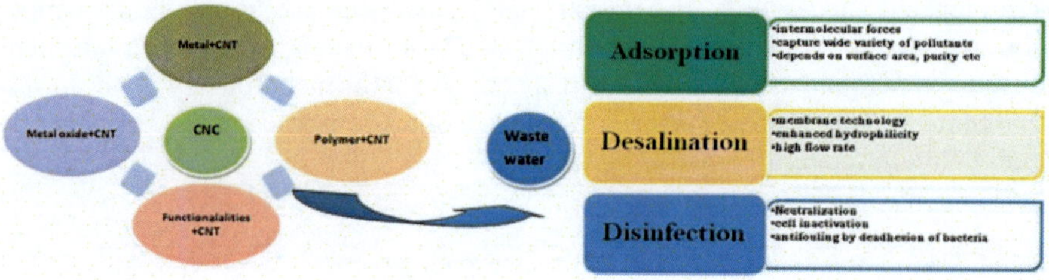

Fig. 2 Representation of CNCs from addition of CNTs to various matrices and possible functions of CNCs in water treatment

4.3 Disinfection

Microorganisms are accountable for many waterborne diseases that deliberately affect human health and lead to high mortality. Therefore, ensuring microbe-free water is of prime importance. The treatment methods for removal of harmful biological organisms from water, in general, include ozonisation, peroxidation, chlorination, etc. However, these methods have drawbacks such as dispersion of undesirable chemical by-products that can have many effects and hence required to be treated that needs additional efforts. Therefore, there is a need to find improved ways for efficient microbial disinfection. Advances in nanoscience can resolve current problems involving water quality using nanosorbents, nanoparticle enhanced membranes, nanocatalysts, etc. CNTs and their composites have also received significant consideration from many researchers because of their strong antimicrobial properties with their diverse mechanisms of action. The promising applications of CNTs and their composites are not only neutralization or exclusion of bacteria but also antifouling property i.e., prevention of adhesion of bacteria in water purification. Pristine CNTs exhibit strong antimicrobial activity towards Gram +ve and Gram -ve bacteria, as well as bacterial spores that are attributed to interference and destruction of pathogenic cell wall. Kang et al. [87] reported that direct contact of *E. coli* cell with SWCNTs leads to rigorous membrane damage followed by cell inactivation.

The other study reported that using N-carbazole-SWNT nanocomposite at 3 wt% of SWNT more than 90% of bacteria have been inactivated [88]. Brady-Estévez and Elimelech [89] used PVDF microporous membrane coated with a thin layer of SWNTs and recorded up to 5–7 log removal of MS2 bacteriophages. Al-Hakami et al. [90] reported that carbon-18 functionalized CNTs with microwave radiation showed the 100% removal of *E. coli* bacteria. Vecitis et al. [91] demonstrated that electrochemical MWCNT filter inactivated and removed *E. coli* bacteria and virus (MS2) in the sewage below the detection limit. Carbon nanocomposite made of silver nanoparticles coated on multiwalled carbon nanotubes/β-cyclodextrin (1 wt%Ag-MWCNTs/βCD) was investigated for the removal of (*E. coli*), ATCC 25922 microbes from water samples. A 100% antibacterial activity was reported (on 1 wt%Ag-MWCNTs/β-CD) within 10 min of interaction that is attributed to smaller Ag crystallites on MWCNTs/β-CD that played a specific role on bacterial contaminants [92].

In another study, Mostafavi et al. [93] a controllable nanoscale porosity CNT-based filter by using a spray pyrolysis method was fabricated and observed that at pressure of 8-11 bar, maximum removal efficiency of MS2 virus occurred. Cu_2O/MWCNT nanocomposite membranes were explored for the removal of MS2 bacteriophages from infected water. Experiments revealed that the special surface properties of MWCNTs present higher adsorption capacity and noteworthy virus retention capability i.e., retention of up to 4-Log (99.99%) [94]. A hybrid polyaniline/graphene nanosheets/carbon nanotube nanocomposites were fabricated using a well-known in situ polymerization technique and ultrasonic assistance and utilized for the exclusion of two types of bacteria (*E. coli* and *S. aureus*) from infected water by a column method [95]. It has been reported that the removal percentages of *E. coli* and *S. aureus* were 99.2% and 99.5 respectively, and approximately the equivalent adsorption percentage was observed when this material is reused for up to four cycles with negligible adsorption losses

5 Challenges and Perspectives of Carbon-Based Nanocomposites (CNCs)

Carbon-based nanocomposites present a range of promising applications for their use in various technology developments due to their substantial performance. Carbon nanotubes are used in nanocomposites owing to their distinctive properties like lightweight, high conductivity, excellent mechanical, tensile, and thermal properties. The amelioration of carbon nanotubes is possible with various functional groups, metals, metal oxides, polymers to improve their potential for the various environmental applications. CNCs have received significance worldwide due to their exceptional and potential physical, chemical, and mechanical properties. However, such modifications may adapt the hazardous profile from fabrication to end-use and disposal in their life cycle. These modifications can as well resist biodegradation increasing cellular uptake and toxicity to terrestrial and aquatic ecosystems. The extensive usage of CNCs cause concerns regarding potential exposure, environmental safety and health as they resemble pathological effects of asbestos. [96, 97]

The study on biological effects of CNCs has been in research and needs to be explored much in the field. Only a few investigations have demonstrated the biological effects of CNCs. The composites have masked the potentially toxic effects of free CNTs because majority of the CNTs remain in the composites. Unlike carbon nanotubes, the nanocomposites have the least toxic effects as the CNT has small amounts in the composites. Mills and Le Hunte [98] reported that CNT metal composite catalysts, in the process of photodegradation of persistent organic pollutants have produced various degradation products that are more toxic than their parent compounds that cause detrimental health effects. Wohlleben et al.

[99] found that the released fragments from multiwallcarbon nanotube (MWCNT)-based nanocomposites have drastic different properties and their subsequent in-vivo hazards.

Despite enormous progress being made for producing nanocomposites, the degree of dispersion, stability, compatibility, and matrix interaction are some of the key features that must be upturned for feasible applications. CNC membranes, apart from their great improvement, the challenges such as permeability, selectivity, and salt rejection performance need further studies to be explored on the different effects of nanomaterials on the morphological and structural characteristics. Considering the future demands with exceptional properties of materials, further studies are necessary to provide practical usefulness in large scale applications of CNCs. The environmental issues concerned to leaching and toxicity of these composites should be seriously considered.

6 Conclusion

The accessibility of clean and safe water is mainly concerned with ecosystem management, agriculture, and industry. The considerable attention has been paid to CNCs for water and wastewater purification in recent research. CNCs, attributing to their large aspect ratio, have promising applications in the fields of adsorption, desalination, and disinfection of water pollutants. Undoubtedly, CNCs substantially outperformed in the water treatment and the technologies are rapidly expanding. However, certain factors like environmental transformation and degradation of the composite materials should be extensively evaluated by the environmental experts to regulate the risk factors associated. The combination of emerging CNCs with suitable matrices with well-defined properties will be highly applicable in developing the future generation of water treatment technologies.

References

1. Vijaya Bhaskar Reddy A, Moniruzzaman M, Veera Manohara Reddy Y, Madhavi G (2019) Graphene-based nanomaterials for the removal of pharmaceuticals in drinking water sources. Micro Nano Technol 329–358
2. Vijaya Bhaskar Reddy A, Jaafar J, Abdul MZ, Aris A, Umar K, Talib J, Madhavi G (2015) Relative efficiency comparison of carboxymethyl cellulose (CMC) stabilized Fe^0 and Fe^0/Ag nanoparticles for rapid degradation of chlorpyrifos in aqueous solutions. Dig J Nanomater Biostruct 10:331–340
3. Michael FL, De V, Tawfick SH, Baughman RH, Hart AJ (2013) Carbon nanotubes Present and future commercial applications. Science 339:535–539
4. Prasek J, Drbohlavova J, Chomoucka J, Hubalek J, Jasek O, Adam V, Kizek R (2011) Methods for carbon nanotubes synthesis—review. J Mater Chem 21:15872–15884
5. Kumar M, Ando Y (2010) Chemical vapor deposition of carbon nanotubes: a review on growth mechanism and mass production. J Nanosci Nanotechnol 10:3739–3758 [CrossRef] [PubMed]

6. Pradhan BK, Harutyunyan AR, Kim UJ, Chen G, Eklund PC (2002) Fuel Chem Div Preprints 47:431
7. Sobolkina A, Mechtcherine V, Khavrus V, Maier D, Mende M, Ritschel M et al (2012) Dispersion of carbon nanotubes and its influence on the mechanical properties of the cement matrix. Cement Concr Compos 34:1104–1113
8. Bahr JL, Mickelson ET, Bronikowski MJ, Smalley RE, Tour JM (2001) Dissolution of small diameter single-wall carbon nanotubes in organic solvents. Chem Commun 193–194. https://doi.org/10.1039/b008042j
9. Ma C-Y, Huang S-C, Chou P-H, Den W, Hou C-H (2016) Application of a multiwalled carbon nanotube-chitosan composite as an electrode in the electrosorption process for water purification. Chemosphere 146:113–120 [CrossRef] [PubMed]
10. Moniruzzaman M, Winey KI (2010) Polymer nanocomposites containing carbon on nanotubes for polymer-based nanocomposites: a review. Compos Part Appl Sci Manuf 41:1345–1367
11. Thostenson E, Li C, Chou T (2005) Nanocomposites in context. Compos Sci Technol 65:491–516
12. Khan W, Sharma R, Saini P (2016) Carbon nanotube-based polymer composites: synthesis, properties and applications carbon nanotubes. In: Current progress of their polymer composites. https://doi.org/10.5772/62497
13. Georgakilas V, Otyepka M, Bourlinos AB, Chandra V, Kim N, Kemp KC et al (2012) Functionalization of graphene: covalent and non-covalent approaches, derivatives and applications. Chem Rev 112:6156–6214. https://doi.org/10.1021/cr3000412
14. Carreno NLV, Garcia ITS, Raubach CW, Krolow M, Santos CCG, Probst LFD, Fajardo HV (2009) J Power Sources 188(2):527–531
15. Rahim A, Barros SBA, Arenas LT, Gushikem Y (2011) Electrochim Acta 56(3):1256–1261
16. Sharififard H, Pepe F, Soleimani M, Aprea P, Caputo D (2016) RSC Adv. https://doi.org/10.1039/c5ra27923b
17. Lei D, Yang T, Qu B, Ma J, Li Q, Chen L, Wang T (2014) Synthesis of TiN@C nanocomposites for enhanced electrochemical properties. Sustain Energy 2(1):1–4. https://doi.org/10.12691/rse-2-1-1
18. Litvin VA, Galagan RL (2017) Synthesis and properties of Co-carbon nanocomposites using synthetic fulvic acids. Mater Chem Phys 201(1):207–213. https://doi.org/10.1016/j.matchemphys.2017.08.055
19. Sovizi MR, Eskandarpour M, Afshari M (2016) Synthesis, characterization, and application of magnetic-activated carbon nanocomposite (m-Fe_3O_4@ACCs) as a new low-cost magnetic adsorbent for removal of Pb(II) from industrial wastewaters. Desalin Water Treat 57(59):28887–28899. https://doi.org/10.1080/19443994.2016.1193062
20. Peng X, Luan Z, Di Z, Zhang Z, Zhu C (2005) Carbon nanotubes-iron oxides magnetic composites as adsorbent for removal of Pb(II) and Cu(II) from water. Carbon 43:855–894. https://doi.org/10.1016/j.carbon.2004.11.009
21. Chen L-F, Liang H-W, Lu Y, Cui C-H, Yu S-H (2011) Synthesis of an attapulgite clay@carbon nanocomposite adsorbent by a hydrothermal carbonization process and their application in the removal of toxic metal ions from water. Langmuir 27(14):8998–9004. https://doi.org/10.1021/la2017165
22. Mojoudi F, Hamidian AH, Zhang Y, Yang M (2019) Synthesis and evaluation of activated carbon/nanoclay/thiolated graphene oxide nanocomposite for lead(II) removal from aqueous solution. Water Sci Technol 79(3):466–479. https://doi.org/10.2166/wst.2019.071
23. Ahmadi Azqhandi MH, Shekari M, Ghalami-Choobar B (2018) Synthesis of carbon nanotube-based nanocomposite and application for wastewater treatment by ultrasonicated adsorption process. Appl Organometal Chem 32(8):e4410. https://doi.org/10.1002/aoc.4410
24. Chathumal Jayaweera HDA, Siriwardane I, de Silva KN, de Silva RM (2018) Synthesis of multifunctional activated carbon nanocomposite comprising biocompatible flake nano hydroxyapatite and natural turmeric extract for the removal of bacteria and lead ions from aqueous solution. Chem Cent 12:18

25. Jose T, George SC, Maya MG, Thomas S (2015) Functionalized MWCNT and PVA nanocomposite membranes for dielectric and pervaporation applications. J Chem Eng Process Technol 6:3. https://doi.org/10.4172/2157-7048.1000233
26. Al-Hobaib AS, Al-Sheetan K, Shaik MR, Al-Suhybani MS (2017) Modification of thin-film polyamide membrane with multi-walled carbon nanotubes by interfacial polymerization. Appl Water Sci 7:4341. https://doi.org/10.1007/s13201-017-0578-5
27. Masoumi S, Miroliaei AR, Jafarzadeh Y (2018) Preparation and characterization of MWCNT-COOH/PVC ultrafiltration membranes to use in water treatment. Adv Environ Technol 2:95–105
28. El-Dein LAN, El-Gendi A, Ismail N, Abed KA, Ahmed AI (2014) Evaluation of cellulose acetate membrane with carbon nanotubes additives. J Ind Eng Chem. https://doi.org/10.1016/j.jiec.2014.11.037
29. Vatanpour V, Madaeni SS, Moradian R, Zinadini S, Astincha B (2011) Fabrication and characterization of novel antifouling nanofiltration membrane prepared from oxidized multiwalled carbon nanotube/polyethersulfone nanocomposite. J Membr Sci 375:284–294. https://doi.org/10.1016/j.memsci.2011.03.055
30. Daraei P, Madaeni SS, Ghaemi N, Ahmadi Monfared H, Khadivi MA (2013) Fabrication of PES nanofiltration membrane by simultaneous use of multi-walled carbon nanotube and surface graft polymerization method: comparison of MWCNT and PAA modified MWCNT. Sep Purif Technol 104:3244
31. Baik S, Lim B, Ryu S, Choi D, Kim B, Oh S, Sung B, Choi J, Kim C (2007) Mechanical and electrical properties of carbon nanotubes in copper-matrix nanocomposite. Solid State Phenom 120:285–288
32. Guillet J-F, Valdez Nava Z, Golzio M, Flahaut E (2019) Electrical properties of double-wall carbon nanotubes nanocomposite hydrogels. Carbon 146:542–548. ISSN 0008-6223
33. Kranauskaitė I, Macutkevič J, Borisova A, Martone A, Zarrelli M, Selskis A, Aniskevich A, Banys J (2017) Enhancing electrical conductivity of multiwalled carbon nanotube/epoxy composites by graphene nanoplatelets. Lith J Phys 57(4):232–242
34. Zare Y, Rhee KY (2017) Prediction of tensile modulus in polymer nanocomposites containing carbon nanotubes (CNT) above percolation threshold by modification of conventional model. Curr Appl Phys 17:873–879. CrossRef
35. Zhu J-M, Zare Y, Rhee KY (2018) Analysis of the roles of interphase, waviness and agglomeration of CNT in the electrical conductivity and tensile modulus of polymer/CNT nanocomposites by theoretical approaches. Colloids Surf. A 539:29–36
36. Al-Saleh MH (2015) Influence of conductive network structure on the EMI shielding and electrical percolation of carbon nanotube/polymer nanocomposites. Synth Met 205:78–84 CrossRef
37. Logakis E, Pissis P, Pospiech D, Korwitz A, Krause B, Reuter U, Pötschke P (2010) Low electrical percolation threshold in poly(ethylene terephthalate)/multi-walled carbon nanotube nanocomposites. Eur Polym J 46:928–936
38. Awasthi K, Awasthi S, Srivastava A, Kamalakaran R, Talapatra S, Ajayan PM, Srivastava ON (2006) Synthesis and characterization of carbon nanotube-polythylene oxide composites. Nanotechnology 17:5417–5422
39. Bin Y, Mine M, Koganemaru A, Jiang X, Matsuo M (2006) Morphology and mechanical and electrical properties of oriented PVA–VGCF and PVA–MWNT composites. Polymer 47:1308–1317
40. Alam MS, Mukherjee N, Ahmed SF (2018) Optical properties of diamond like carbon nanocomposite thin films. In: AIP conference proceedings, vol 1953, 090018. https://doi.org/10.1063/1.5032865
41. Xin F, Li L (2011) Carbon nanotubes/metal nanoparticle based nanocomposites: improvements in visible photoluminescence emission and hydrophobicity. Compos Part A 42:961
42. Sepahvand R, Adeli M, Astinchap B, Kabiri R (2008) New nanocomposites containing metal nanoparticles, carbon nanotube and polymer. J Nanopart Res 10(8):1309–1318. https://doi.org/10.1007/s11051-008-9411-2

43. Barberio M, Barone P, Stranges F, Romano A, Xu F, Bonanno A (2013) Carbon nanotubes/metal nanoparticle based nanocomposites: improvements in visible photoluminescence emission and hydrophobicity. Opt Photonics J 3:34–40. https://doi.org/10.4236/opj.2013.36A007
44. Guler Ö, Guler SH, Yo F, Aydin H, Aydin C, El-Tantawy F, Duraia E-SM, Fouda AN (2015) Electrical and Optical Properties of Carbon Nanotube Hybrid Zinc Oxide Nanocomposites Prepared by Ball Mill Technique. Fullerenes Nanotubes Carbon Nanostruct 23(10):865–869. https://doi.org/10.1080/1536383x.2015.1022256
45. Wang B, Zhang K, Zhou C, Ren M, Yuantong G, Li T (2019) Engineering the mechanical properties of CNT/ PEEK nanocomposites. RSC Adv 9:12836–12845
46. Nam DH, Cha SI, Lee KM, Jang JH, Park HM, Lee JK, Hong SH (2016) Thermal properties of carbon nanotubes reinforced aluminum-copper matrix nanocomposites. J Nanosci Nanotechnol 16(11): 12013–12016(4). https://doi.org/10.1166/jnn.2016.13635
47. Vahedi A, Lahidjani MHS, Shakhesi S (2018) Multiscale modelling of thermal conductivity of carbon nanotube paraffin nanocomposites. Mater Res Express 5(11):115026. https://doi.org/10.1088/2053-1591/aade72
48. Gulotty R, Castellino M, Jagdale P, Tagliaferro A, Alexander A (2013) Balandin effects of functionalization on thermal properties of single-wall and multi-wall carbon nanotube-polymer nanocomposites. ACS Nano 7(6):5114–5121. https://doi.org/10.1021/nn400726g
49. Amrin S, Deshpande VD (2016) Mechanical and dielectric properties of carbon nanotubes/poly (vinyl alcohol) nanocomposites. In: AIP conference proceedings, vol 1728, p. 020641. https://doi.org/10.1063/1.4946692
50. Sui G, Jana S, Zhong WH, Fuqua MA, Ulven CA (2008) Dielectric properties and conductivity of carbon nanofiber/semi-crystalline polymer composites. Acta Mater 56(10):2381–2388. https://doi.org/10.1016/j.actamat.2008.01.034
51. Zhang Y, Liu X, Zhu M, Rong C, Wang G (2012) Dielectric properties of carbon-nanotube/amino-functionalized poly(arylene ether ketone) composites. High Perform Polym 1–7. https://doi.org/10.1177/0954008311432773
52. Ibrahim A, Abdel-Aziz MH, Zoromba M, Al-Hossainy AF (2018) Structural, optical, and electrical properties of multi-walled carbon nanotubes/polyaniline/Fe_3O_4 ternary nanocomposites thin film. Synth Met 238:1–13. https://doi.org/10.1016/j.synthmet.2018.02.006
53. Aydin H (2014) Electrical and optical properties of titanium dioxide-carbon nanotube nanocomposites. J Nanoelectron Optoelectron 9(5):608–613(6)
54. Kalakonda P, Banne S, Kalakonda PB (2019) Enhanced mechanical properties of multiwalled carbon nanotubes/thermoplastic polyurethane nanocomposites. Nanomater Nanotechnol 9:1–7. https://doi.org/10.1177/1847980419840858
55. Hossain F, Perales-Perez OJ, Hwang S, Román F (2014) Antimicrobial nanomaterials as water disinfectant: applications, limitations and future perspectives. Sci Total Environ 466:1047–1059
56. Mubarak N, Sahu J, Abdullah E, Jayakumar N (2014) Removal of heavy metals from wastewater using carbon nanotubes. Sep Purif Rev 43:311–338
57. Ren X, Chen C, Nagatsu M, Wang X (2011) Carbon nanotubes as adsorbents in environmental pollution management: a review. Chem Eng J 170:395–410
58. Yu J-G, Zhao X-H, Yu L-Y, Jiao F-P, Jiang J-H, Chen X-Q (2014) Removal, recovery and enrichment of metals from aqueous solutions using carbon nanotubes. J Radioanal Nucl Chem 299:1155–1163
59. Burde JT, Calbi MM (2007) Physisorption kinetics in carbon nanotube bundles. J Phys Chem C 111:5057–5063
60. Rawat DS, Calbi MM, Migone AD (2007) Equilibration time: kinetics of gas adsorption on closed-and open-ended single-walled carbon nanotubes. J Phys Chem C 111:12980–12986
61. Gupta VK, Moradi O, Tyagi I, Agarwal S, Sadegh H, Shahryari-Ghoshekandi R, Makhlouf ASH, Goodarzi M, Garshasbi A (2016) Study on the removal of heavy metal ions from industry waste by carbon nanotubes: effect of the surface modification: a review. Crit Rev Environ Sci Technol 46:93–118 [CrossRef]
62. Salam MA, Makki Magdy MSI, Abdelaal YAJ (2011) Alloys Compd 509:2582–2587

63. Li Y-H, Hung T-H, Chen C-W (2009) A first-principles study of nitrogen-and boronassisted platinum adsorption on carbon nanotubes. Carbon 47:850–855
64. Patiño Y, Díaz E, Ordóñez S, Gallegos-Suarez E, Guerrero-Ruiz A, Rodríguez-Ramos I (2015) Adsorption of emerging pollutants on functionalized multiwall carbon nanotubes. Chemosphere 136:174–180 [CrossRef] [PubMed]
65. Zouzelka R, Kusumawati Y, Remzova M, Rathousky J, Pauporté T (2016) Photocatalytic activity of porous multiwalled carbon nanotube-TiO_2 composite layers for pollutant degradation. J Hazard Mater 317:52–59 [CrossRef] [PubMed]
66. Duman O, Tunç S, Polat TG, Bozoğlan BK (2016) Synthesis of magnetic oxidized multiwalled carbon nanotube-κ-carrageenan-Fe_3O_4 nanocomposite adsorbent and its application in cationic methylene blue dye adsorption. Carbohydr Polym 147:79–88 [CrossRef] [PubMed]
67. Park WK, Yoon Y, Kim S, Yoo S, Do Y, Kang J-W, Yoon DH, Yang WS (2016) Feasible water flow filter with facilely functionalized Fe3O4 -non-oxidative graphene/CNT composites for arsenic removal. J Environ Chem Eng 4:3246–3252 [CrossRef]
68. Wang Y, Ma J, Zhu J, Ye N, Zhang X, Huang H (2016) Multi-walled carbon nanotubes with selected properties for dynamic filtration of pharmaceuticals and personal care products. Water Res 92:104–112 [CrossRef] [PubMed]
69. Peter KT, Vargo JD, Rupasinghe TP, De Jesus A, Tivanski AV, Sander EA, Myung NV, Cwiertny DM (2016) Synthesis, optimization, and performance demonstration of electrospun carbon nanofiber–carbon nanotube composite sorbents for point-of-use water treatment. ACS Appl Mater Interfaces 8:11431–11440 [CrossRef] [PubMed]
70. Lee J, Ye Y, Ward AJ, Zhou C, Chen V, Minett AI, Lee S, Liu Z, Chae S-R, Shi J (2016) High flux and high selectivity carbon nanotube composite membranes for natural organic matter removal. Sep Purif Technol 163:109–119 [CrossRef]
71. Bui N, Meshot ER, Kim S, Peña J, Gibson PW, Wu KJ, Fornasiero F (2016) Ultrabreathable and protective membranes with sub-5 nm carbon nanotube pores. Adv Mater 28:5871–5877 [CrossRef] [PubMed]
72. Nie C, Peng Z, Yang Y, Cheng C, Ma L, Zhao C (2016) Kevlar based nanofibrous particles as robust, effective and recyclable absorbents for water purification. J Hazard Mater 318:255–265 [CrossRef] [PubMed]
73. Siddiqa A, Shahid A, Gill R (2015) Silica decorated CNTs sponge for selective removal of toxic contaminants and oil spills from water. J Environ Chem Eng 3:892–897 [CrossRef]
74. Esfahani MR, Tyler JL, Stretz HA, Wells MJM (2015) Effects of a dual nanofiller, nano-TiO_2 and MWCNT, for polysulfone-based nanocomposite membranes for water purification. Desalination 372:47–56 [CrossRef]
75. Tofighy MA, Mohammadi T (2010) Desalination 258:182–186
76. Yang X, Lee J, Yuan L, Chae SR, Peterson VK, Minett AI, Yin Y, Harris AT (2013) Removal of natural organic matter in water using functionalised carbon nanotube buckypaper. Carbon 59:160–166
77. Wang L, Song X, Wang T, Wang S, Wang Z, Gao C (2015) Fabrication and characterization of polyethersulfone/carbon nanotubes (PES/CNTs) based mixed matrix membranes (MMMs) for nanofiltration application. Appl Surf Sci 330:118–125
78. Shawky HA, Chae S-R, Lin S, Wiesner MR (2011) Synthesis and characterization of a carbon nanotube/polymer nanocomposite membrane for water treatment. Desalination 272:46–50
79. Farahbakhsh J, Delnavaz M, Vatanpour V (2017) Investigation of raw and oxidized multiwalled carbon nanotubes in fabrication of reverse osmosis polyamide membranes for improvement in desalination and antifouling properties. Desalination 410:1–9
80. Chan W-F, Chen H-Y, Surapathi A, Taylor MG, Shao X, Marand E, Johnson JK (2013) Zwitterion functionalized carbon nanotube/polyamide nanocomposite membranes for water desalination. ACS Nano 7:5308–5319
81. Corry B (2008) Designing carbon nanotube membranes for efficient water desalination. J Phys Chem B 112(5):1427–1434
82. Ortiz-Medina J, Inukai S, Araki T, Morelos-Gomez A, Cruz-Silva R, Takeuchi K, Noguchi T, Kawaguchi T, Terrones M, Endo M (2018) Robust water desalination membranes against

degradation using high loads of carbon nanotubes. J Sci Rep 8:2748. https://doi.org/10.1038/s41598-018-21192-5
83. Ratto TV, Holt JK, Szmodis AW (2011) Membranes with embedded nanotubes for selective permeability. Google Patents
84. Yang HY, Han ZJ, Yu SF, Pey KL, Ostrikov K, Karnik R (2013) Carbon nanotube membranes with ultrahigh specific adsorption capacity for water desalination and purification. Nat Commun 4. http://doi.org/10.1038/ncomms3220 (Article number: 2220)
85. El Badawi N, Ramadan AR, Esawi Amal MK, El-Morsi M (2014) Novel carbon nanotube–cellulose acetate nanocompositemembranes for water filtration applications. Desalination 344:79–85. https://doi.org/10.1016/j.desal.2014.03.005
86. El-Din LAN, El-Gendi A, Ismail N, Abed KA, Ahmed AI (2015) Evaluation of cellulose acetate membrane with carbon nanotubes additives. J Ind Eng Chem 26:259–264
87. Kang S, Mauter MS, Elimelech M (2008) Physicochemical determinants of multiwalled carbon nanotube bacterial cytotoxicity. Environ Sci Technol 42:7528–7534
88. Ahmed F, Santos CM, Vergara R, Tria MCR, Advincula R, Rodrigues DF (2012) Antimicrobial applications of electroactive PVK-SWNT nanocomposites. Environ Sci Technol 46(3):1804–1810
89. Brady-Estévez AS, Elimelech SM (2008) A single-walled carbon-nanotube filters for removal of viral and bacterial pathogens. Small 4(4):481–484
90. Al-Hakami SM, Khalil AB, Laoui T, Atieh MA (2013) Fast disinfection of *Escherichia coli* bacteria using carbon nanotubes interaction with microwave radiation. Bioinorg Chem Appl. https://doi.org/10.1155/2013/458943
91. Vecitis CD, Schnoor MH, Saifur Rahaman M, Schiffman JD, Elimelech M (2011) Electrochemical multiwalled carbon nanotube filter for viral and bacterial removal and inactivation. Environ Sci Technol 45(8):3672–3679. https://doi.org/10.1021/es2000062
92. Rananga LE, Magadzu T (2015) comparative studies of silver doped carbon nanotubes and β-cyclodextrin for water disinfection. Dig J Nanomater Biostruct 10(3):831–836
93. Mostafavi ST, Mehrnia MR, Rashidi AM (2009) Preparation of nanofilter from carbon nanotubes for application in virus removal from water. Desalination 238(1–3):271
94. Németh Z, Szekeres GP, Schabikowski M, Schrantz K, Traber J, Pronk W, Hernádi K, Graule T (2019) Enhanced virus filtration in hybrid membranes with MWCNT nanocomposite. R Soc Open Sci 6(1). https://doi.org/10.1098/rsos.181294
95. Hussein MA, El-Shishtawy RM, Alamrya KA, Asiria AM, Mohamed SA (2019) Efficient water disinfection using hybrid polyaniline/graphene/carbon nanotube nanocomposites. Environ Technol 40(21):2813–2824. https://doi.org/10.1080/09593330.2018.1466921
96. Poland CA, Duffin R, Kinloch I, Maynard A, Wallace WAH, Seaton A (2008) Carbon nanotubes introduced into the abdominal cavity of mice show asbestos-like pathogenicity in a pilot study. Nat Nanotechnol 3:423–428
97. Ryman-Rasmussen JP, Cesta MF, Brody AR, Shipley-Phillips JK, Everitt JI, Tewksbury EW (2009) Inhaled carbon nanotubes reach the subpleural tissue in mice. Nat Nanotechnol 4:747–751
98. Mills A, Le Hunte S (1997) An overview of semiconductor photocatalysis. J Photochem Photobiol A Chem 108(1):1–35
99. Wohlleben W, Brill S, Meier MW, Mertler M, Cox G, Hirth S et al (2011) On the lifecycle of nanocomposites: Comparing released fragments and their in-vivo hazards from three release mechanisms and four nanocomposites. Small 7(16):2384–2395

Magnetite Carbon Nanomaterials for Environmental Remediation

Reena Saxena, Amit Lochab, and Megha Saxena

Abstract Environmental concerns like water and soil pollution have affected the health of entire ecosystem. This situation is getting even worse as the burden on limited resources is increasing by leaps and bounds. Continuous rise in population is forcing more pollutants in the environment to meet their demands. Removal of these pollutants through eco-friendly methods have become the centre of attention now. Nanomaterials have been applied to remediate polluted water and soil as they possess high surface area for adsorption and sensing of the various toxic pollutants. This chapter gives a precise review of the research work for environmental remediation using magnetite carbon nanomaterials. Magnetite carbon nanomaterials have represented themselves as an efficient alternative for the treatment of both inorganic and organic pollutants. It also includes information about the various techniques being employed for the remediation process of various toxic metal ions and dyes. Further optimized parameters such as pH, temperature, contact time and capacity are also discussed which is essential for effective treatment of pollutants, and their applications in different real samples are briefly discussed.

Keywords Carbon nanomaterials · Environmental remediation · Magnetite · Metals · Dyes

1 Introduction

Our environment is getting polluted day by day due to rapid industrialization. The proliferation of industries and commercial activities is the demand of increasing population. The effluents from these industries are the main cause of pollution. The by-products of industries such as dyes effluents, toxic chemicals, heavy metal, and traces of pesticides have profound impact on environment. Toxic heavy metal ions like Hg, As, Cd, Pb, etc., have ability to either replace the essential metal ions

R. Saxena (✉) · A. Lochab · M. Saxena
Department of Chemistry, Kirori Mal College, University of Delhi,
New Delhi, Delhi 110007, India
e-mail: reenasax@hotmail.com; rsaxena@kmc.du.ac.in

M. Jawaid et al. (eds.), *Environmental Remediation Through Carbon Based Nano Composites*, Green Energy and Technology,
https://doi.org/10.1007/978-981-15-6699-8_5

from our body or change the active confirmation of enzymes and disturb their function. Their presence in water can cause disease like Minamata, Wilson, etc. They are highly contagious, and their prolong exposure can lead to death [75]. The other organic pollutants such as dyes, pesticides and other toxic chemicals which enter into water bodies from various sources like industrial effluents, domestic waste, mining, research laboratories, etc., are also hazardous and carcinogenic. Dyes are used as colouring agent at large scale in textile, paper, food, cosmetics and pharmaceutical industries. The effluents of used dyes are discharged directly into nearby rivers, ponds and other water bodies. These toxic waste products are harmful for both aquatic and terrestrial animals. Other organic pollutants such as pesticides, insecticides, herbicides and fungicides are used at large scale to kill the insects, pests or inhibit the growth of weeds in order to save the crop and increase productivity. These pesticides are highly poisonous as they are target specific and meant for the purpose of killing insects. When they enter in the food chain even in small amount can lead to death. The insecticide like atrazine has tendency to change the gender of frogs, etc. So they are not only harmful for present generation but can also affect the future generation. Various other toxic chemicals used in the laboratories are also dangerous for environment and need to be removed for sustainability of ecosystem [85].

Recently, development of nanomaterial is considered boon in research and showed a new path in research because at nanolevel, the properties of material changed drastically. Nanomaterials are particles having at least one dimension in nanometre range, which can be made up of organic, carbon, metal, metal oxide or other inorganic compounds. They possess different chemical and physical properties at nanolevel in comparison to bulk. The reason for difference in properties is due to the increase in surface area or in other words, higher ratio of surface to volume and the quantum effects. The increased surface area provides higher reactivity and adsorption, whereas quantum realm restricts the flow of electron to a small region that produces magnetic moment in nanomaterials. So, nanomaterials display a lot of new phenomenon that can be applied in solving various environmental and biological issues [110]. Nanomaterials have lots of advantages over other adsorbents and can also be modified easily to enhance their selectivity for particular pollutant. Nanomaterials like carbon nanotubes, graphenes, magnetic nanoparticles, zinc oxide nanoparticles, titanium oxide nanoparticles, etc., have been used for this purpose [99]. Among numerous developed nanomaterials, magnetic nanomaterials are playing an important role in this field due to their easy separation compared to conventional methods like filtration and centrifugation. The biggest achievement of magnetic nanomaterials is in the treatment of water pollution. In developing countries, access to safe water for whole population is very difficult. In addition to that various industrial and agricultural practices are continuously discharging harmful organic and inorganic toxicants in the water streams that end up in drinking water supplies. Due to this, all living creatures get affected, and exposure of these pollutants for long duration can cause serious health problems that can result in death also. There is crucial need to develop methods for removal of the toxic pollutants from the environment that are cheap and eco-friendly.

This chapter focuses on development of research over last five years for the removal of toxic metal ions and dyes from environment using magnetite carbon nanomaterials.

2 Magnetite Nanomaterials

Magnetite nanomaterials are highly used in the removal of various contaminants due to their greater stability and reusability. Apart from their inherent magnetic properties, another advantage is their easy functionalization and high adsorption capacity which is due to small the size of nanoparticle and high adsorption surface area. Studies have revealed that decrease in size from micro- to nanolevel have increased the adsorption capacity approximately 62 times. However, too small size can create problem in magnetic separation because of their Brownian motion and so 10–50 nm size of iron oxide is generally employed for the remediation purpose. Shape of magnetic nanomaterials decides their application in both environmental and medicinal fields. Spherically shaped are efficiently used in environmental remediation due to ease in production at large scale, whereas cuboid shaped are playing vital role in the catalysis, storage and magnetic resonance imaging (MRI) area [81].

2.1 Synthesis of Magnetite Nanomaterials

There are four different methods for the synthesis of magnetite nanomaterials which are shown in Fig. 1 [76]. Each route has their own advantages and disadvantages in terms of factors like yield, reaction time, variation in size distribution, control over shape and crystallinity. All the methods employed for production use an iron precursor which is either thermally degraded or reduced in their solution.

Co-precipitation Method: It is the most conventional way for synthesis of magnetic nanoparticles at commercial level. In this method, precipitation of magnetite is done by using strong basic conditions in an aqueous solution containing

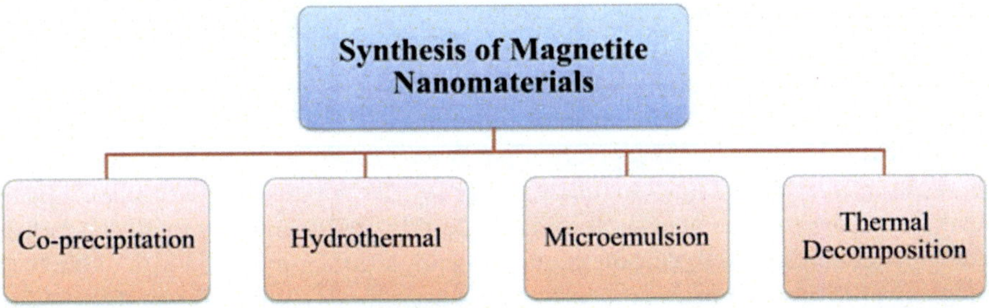

Fig. 1 Synthetic methods for magnetite nanomaterials

Fe (III) and Fe (II) ions. The shape and size of the nanoparticles can be controlled by adjusting ratios of Fe ions, salt anions and pH. Generally, iron oxide nanoparticle diameter size can be varied from 2 to 17 nm using this method. Further functionalization can be performed immediately after their formation. The main advantage of this method is high yield in less time.

Hydrothermal Method: This method is employed for the synthesis of highly crystalline iron oxide nanomaterials. Here, iron salts with stabilizing surfactants are placed in autoclave and heated for certain period of time depending on the required size of the nanoparticles. It has the advantage of narrow range of size distribution with good control over the shape. Its only disadvantage is long reaction periods.

Microemulsion Method: This technique utilizes water, surfactant (dodecylbenzenesulphonate) and oil (xylene) for the formation of emulsion that helps in synthesis of magnetic nanoparticles. Here, similar-sized droplets are produced that gives iron oxide of almost same size stabilized by the surfactant. Reverse emulsion (water in oil) is mostly used for the synthesis of nanoparticles. Its advantage is that it gives reproducible characteristics to the particles if same conditions are repeated. The only disadvantage is that this method requires larger amount of solvent and yield is low.

Thermal Decomposition Method: In this method, iron salt ($Fe(CO)_5$, Fe(III) acetyl acetone, and $Fe(acac)_3$) are thermally degraded with suitable surfactant in a high boiling solvent using an inert atmosphere. Various properties like size distribution, crystalline nature, shape and magnetism of the material can be controlled by using this technique. This method has additional advantage of forming mixed metal oxide nanoparticles with good control over shape and size.

2.2 *Modification of Magnetite Nanomaterials*

Magnetic nanomaterials produced from above methods encounter a common problem of aggregation. These small nanoparticles have inherent magnetism which is much stronger than normal van der Waals forces of attraction that result in aggregation. This decreases the surface area for adsorption and reduces its efficiency. Furthermore, aggregation results in uneven distribution of adsorbent in the solution that also affects the adsorption capacity. To enhance the stability and selectivity of these nanomaterials, further modification is required, which helps in giving selective response to a particular pollutant. The modification or encapsulation can be done with both organic and inorganic nanomaterials as shown in Fig. 2 [108]. Generally, coating with inorganic material solves the above problem efficiently. As organic material stabilized iron oxide can leach out during desorption processes and can cause aggregation due to large size. So it is beneficial to functionalize it with inorganic materials like silica, metal and carbon nanomaterial, as these materials are chemically inert that helps in dispersion and solves the problem of aggregation. Silica-modified iron oxide has drawback of losing magnetic properties to a drastic level that is not desirable, while the composite formed with metal nanoparticles is not economically appreciated.

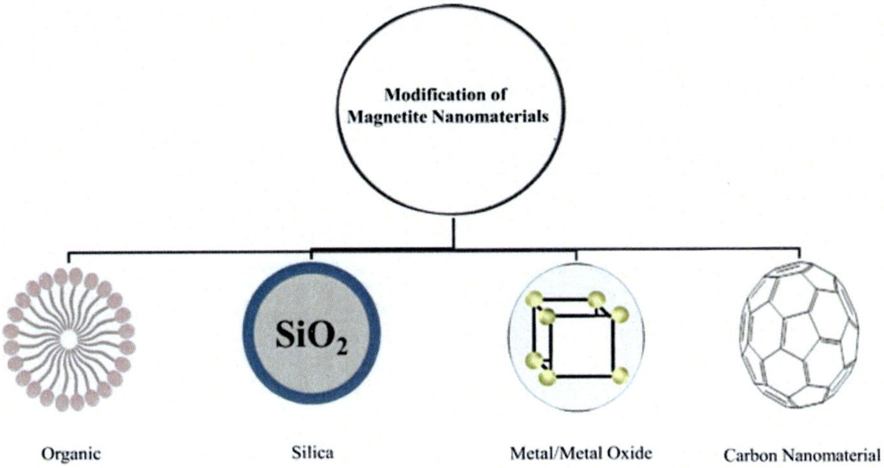

Fig. 2 Modification of magnetite nanomaterials

Table 1 Comparison of various modified magnetite nanomaterials

Modification	Advantages	Disadvantages
• Organic	• Biocompatibility, dispersibility	• Aggregation, leach out
• Silica	• Easy modification, good dispersibility	• Huge loss of magnetism
• Metal/metal oxide	• Low loss of magnetism, chemical inertness	• Costly
• Carbon nanomaterials	• Low loss of magnetism, biocompatible, thermal and chemical stability	• Less development in modification procedure

Magnetite carbon nanomaterials show high potential in both biological and environmental applications. They are biocompatible with high thermal and chemical stability. Research has shown that carbon-encapsulated Fe_3O_4 nanoparticles gave good adsorption capacity for toxic metal ions and organic pollutants due to incorporation of additional terminal groups. Mostly, magnetite carbon nanomaterials are synthesized through hydrothermal method by using suitable surfactants for stabilization [127]. The advantages and disadvantages of various modified magnetite nanomaterial are given in Table 1.

3 Magnetite-Modified Carbon-Based Nanomaterials

Carbon nanomaterial has different morphologies which include spherical, tubular and sheet like structures, e.g. fullerenes, carbon nanotubes and graphene as shown in Fig. 3 [31]. From last decade, there is a huge focus on these materials for their excellent sensing and conductive properties. They are being efficiently used in making

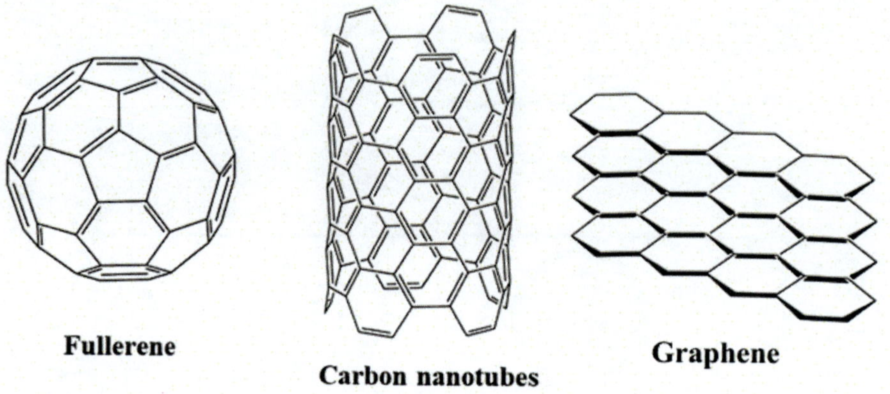

Fig. 3 Carbon-based nanomaterials

portable devices for analytical purposes. Magnetite-modified carbon-based nanomaterials are being efficiently used in environmental remediation due to easy separation and improved adsorption capability. The unique physical and chemical characteristics of different carbon nanomaterials are discussed below individually [28].

Fullerenes: Fullerene consists of 12 pentagons and 20 hexagons where each carbon is attached to three other carbons which are also known as buckyball due to its spherical arrangement. After the discovery of C_{60}, other fullerenes were synthesised C_{70}, C_{78} and smaller ones like C_{36}, C_{28}, etc. in 1990s. They show good mechanical strength and can reshape even after applying high pressure. Apart from this, they show excellent optical properties due to delocalisation of free electron in their p-orbital. So they are opted as an alternative in solar cells for the production of electricity. Fullerene is scarcely used for the purpose of treating environmental pollutants.

Graphene: It is a two-dimensional sheet formed by connected hexagonal structure of sp^2-hybridised carbon atoms. Due to delocalization of electron, they show high conductivity and good mechanical strength. Its wide application is due to its low cost and used in various applications such as fabrication of electrodes, dye-sensitized solar cells (DSSCs) and batteries. Graphene and functionalized graphene (e.g. graphene oxide (GO), reduced graphene oxide, etc.) with high surface area are also used for environmental monitoring using different techniques.

Carbon Nanotubes: CNTs are considered as most popular form of carbon nanomaterials in recent decades as they show high conductivity, rigidity and elasticity. They were firstly introduced in 1991 by Iijima. In the beginning, multiwalled carbon nanotubes (MWCNTs) were synthesized consisting of many rolled graphene sheets in a concentric manner. They possess tube like cylindrical shape with very small diameter compared to its length. Then, Iijima and his co-worker, in 1993 were able to observe single-walled carbon nanotubes (SWCNTs) with smaller diameter compared to MWCNTs. CNTs can show properties of both semiconductors and metals based on the chirality, diameter and electronic density states [124].

4 Environmental Remediation

Environmental remediation is a way or method to remove pollutants from the environment which are present in the environmental sources like water, soil, air, etc. The pollutants are divided into two categories, organic and inorganic pollutants. The different types of environmental pollutants are shown in Fig. 4 [24].

4.1 *Methods for Treatment of Pollutants*

Various methods had been used for environmental remediation like coagulation–flocculation, photocatalytic degradation, oxidation, electrochemical treatment, adsorption, etc. as shown in Fig. 5 [100]. These methods are highly active for removal of pollutants like heavy metal ions, dyes, pesticides and other toxic chemicals.

Coagulation–flocculation: It is a very old method for removal of pollutants. In this method, the pollutant is treated with a coagulating agent, which later can be collected easily. The main drawback of the method is sludge formation [119].

Photocatalytic degradation: It is the method in which the pollutant is treated by a catalyst which degrades the pollutant with large molecular structure like dye, pesticides into less harmful products and by this reduces the impact of the pollutant. The limitation of the method is the production of by-products.

Oxidation: It is another method of treating pollutants, as in this method the pollutants are treated with oxidizing agents like H_2O_2, nitric acid etc. through which the pollutants get oxidised or degrade to a less toxic compound. The major drawbacks

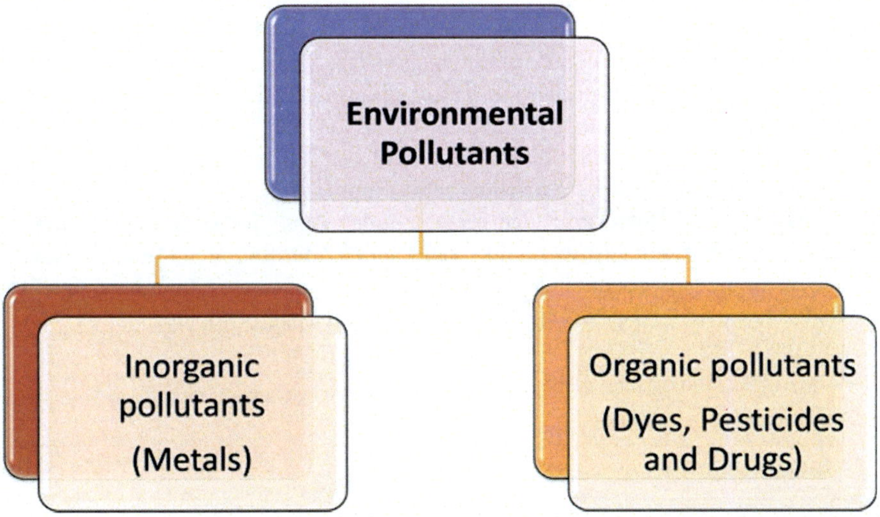

Fig. 4 Types of environmental pollutants

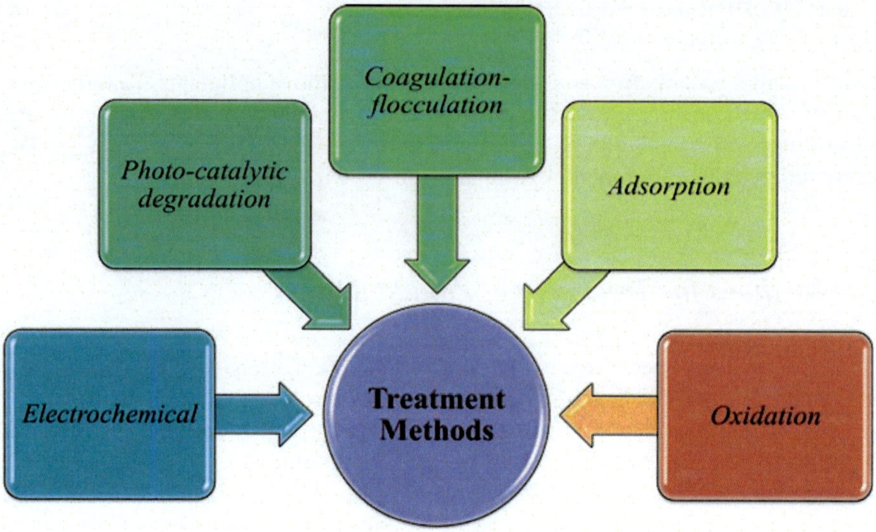

Fig. 5 Treatment methods for environmental remediation

of this method are the requirement of large amount of oxidizing agent which makes the process costly, and it is also not environmentally friendly [90].

Electrochemical: This method is based on the phenomenon in which organic pollutants which are redox active at electrode interface can be converted into less toxic forms. Appearance of the particular peak at certain potential is used for analysis. The method has several advantages like high accuracy, low detection limit and reproducibly. The only drawback of the method is high cost of electricity [66].

Adsorption: It is the most widely used method to remove inorganic pollutants like toxic metal ions as well as organic pollutants like dyes, pesticides and other organic chemicals. The adsorption process is based on surface phenomenon in which an adsorbent provides a surface on which an adsorbate, i.e. pollutant get adsorbed. The attraction between adsorbate and adsorbent can be due to physical or chemical bonding like electrostatic attraction between positively charged adsorbate and negatively charged adsorbent or vice versa, van der Waals forces, H-bonding, π-π interaction, etc. Various adsorbents like sand, silica, alumina metal dust, activated carbon, chitosan, biosorbents like tea waste, coconut shell, etc., have been used to attracts the toxic metal ions or organic pollutants like dyes, pesticides, etc. [114].

4.2 Detection Techniques Used for Monitoring Environmental Pollutants

The detection techniques used for the analysis of organic and inorganic pollutants are different. Detection of inorganic pollutants is generally done by using atomic absorption spectroscopy (AAS) like flame atomic absorption spectrophotometry (FAAS), graphite furnace atomic absorption spectrophotometry (GFAAS), electrothermal atomic absorption spectrometry (ETAAS), inductively coupled plasma atomic optical spectrometry (ICP-OES), inductively coupled plasma mass spectrometry (ICP-MS) and electrochemical. AAS is an optical method of analysis of pollutants like metal ions present in the environmental samples even at trace level. It includes various types depending on the heating source like FAAS which has flame as heating source, GFAAS which has a graphite furnace as heat source and ETAAS which has electrical heating source. ICP-OES is another technique for the detection of chemical elements in which the inductively coupled plasma is used to generate excited ions or atoms that emit the electrochemical radiations which corresponds to the wavelength of particular element. The intensity of emission gives the concentration of element in the sample. ICP-MS is a hyphenated technique in which separation is done by inductively coupled plasma and detection of analyte occurs with the help of mass spectrometry. The two techniques together give highly accurate results even with low quantity of analyte. Organic pollutants due to large size and high conjugation absorb in visible region. So, their determination is possible using UV-visible spectrophotometry, chromatography-based hyphenated techniques like gas-chromatography-mass-spectrometry (GC-MS), liquid-chromatography-mass-spectrometry (LC-MS), high-pressure liquid chromatography (HPLC), etc. as shown in Fig. 6 [117]. These techniques have their own advantages and disadvantages according to pollutant and material used for removal. UV-Visible spectrophotometry is the simple, inexpensive and widely used detection technique for various pollutants. This technique is based on absorption of particular wavelength when passed through the analyte. The change in absorbance corresponds to change in concentration [48]. Gas chromatography is

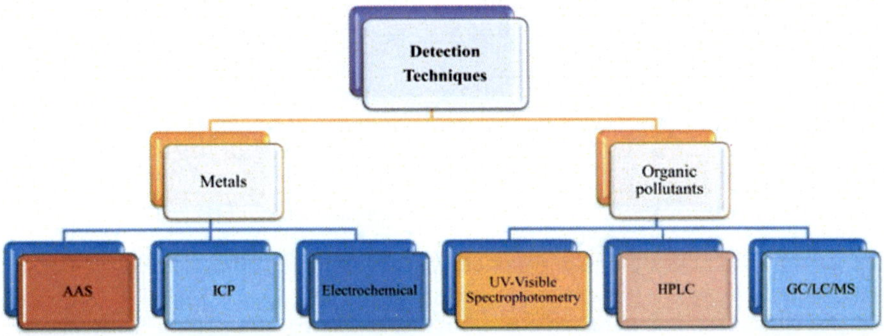

Fig. 6 Detection techniques used in monitoring environmental pollutants

highly compatible with mass spectroscopy and gives highly accurate and precise results. This technique can be used to separate volatile and non-volatile components in the mixture, and mass spectrometry confirms the molecular structure. LC-MS is another hyphenated technique in which separation occurs via liquid chromatography and mass spectrometry which helps in recognition of separated products [145]. HPLC is the chromatographic technique in which liquid is used as mobile phase and separation of analyte from an environmental sample takes place on a column. The detection is done by either UV light source (HPLC-UV) or mass spectrometry (HPLC-MS). This technique has advantages like high separation tendency and precise results. The limitation of this technique is the cost and use of organic solvents [87]. Hyphenated flow-injection technique is an advance technique which gives much better results at low cost. It is used mainly for preconcentration of analyte even at ppb level. This technique can be hyphenated to any detection technique like with FAAS, ICP-OES, UV-Visible spectrophotometer, etc. for metal and other organic pollutants detection [117].

5 Role of Magnetite Carbon Nanomaterial in Removal of Toxic Metal Ions and Organic Pollutants

Excessive production and mining activities are continuously raising the concentration of toxic metals to alarming level in the environment. These toxic metals are non-biodegradable and get accumulated in nature which finally enters our food chain. These toxic metals like lead, cadmium, mercury, chromium and arsenic have high tendency to interfere with biological processes due to their high complex forming tendency. This complex-forming tendency renders various active biomolecules like proteins, enzymes and amino acids inactive leading to fatal diseases. Hence, there is a crucial need for developing fast, easy to handle and environmentally friendly methods to deal with environment pollution caused by these toxic metals [41]. The general mechanism for the removal of above pollutants using carbon nanomaterials is shown in Fig. 7 [39]. The removal of toxic metal ions using magnetic carbon-based nanomaterials is discussed below.

5.1 Magnetite Carbon Nanomaterials for Monitoring Arsenic in Environment

Inorganic compounds of arsenic have more toxicity than the organic ones. Arsenic concentration is increasing in our environment by using it in the form of pesticide, and also it was used in treatment for various diseases like asthma, syphilis, etc. Long-time exposure of arsenic is from intake of food and water, where its concentration is increasing due to above activities. Its exposure in the form of dust or fume in mining

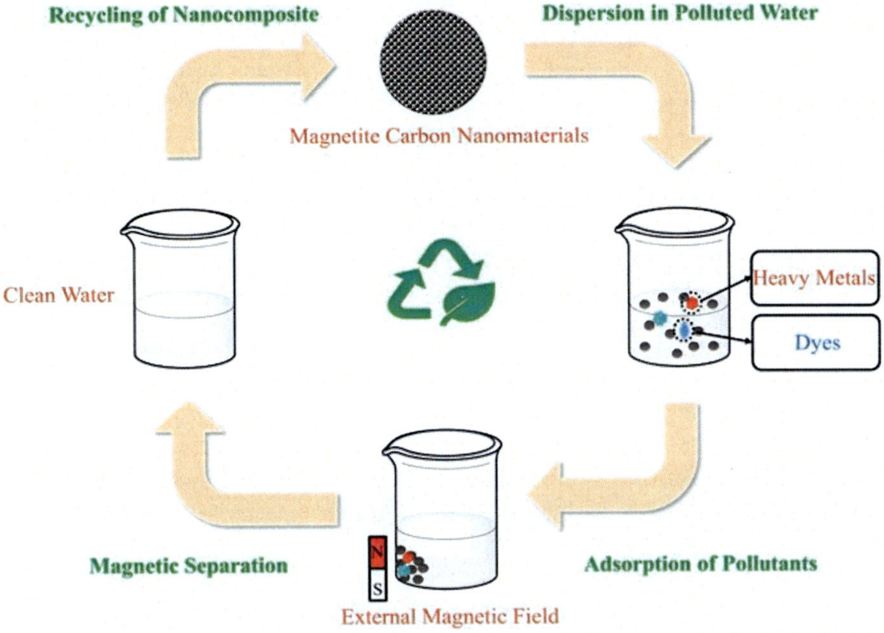

Fig. 7 Removal of pollutants using magnetite carbon nanomaterials

and industrial areas can lead to asthma or other respiratory diseases. Arsenic interacts with thiol group of intercellular sites and affects various cellular processes like mitosis and cell respiration. It can cause skin cancer and can affect various organs like liver, where it can concentrate and affect various oxidation processes. It can also block enzyme activity and transportation of glucose [95]. As per World Health Organization (WHO) guidelines, maximum permissible limit for arsenic in drinking water is 10 $\mu g\ L^{-1}$ [128]. There are many regions where arsenic exceeds this level due to geographical and industrial reasons. So, proper remediation methods are required to prevent health hazards caused by arsenic. To detect and remove arsenic from environment, huge efforts have been put by researchers such as Sahu et al. who developed a reusable nanocomposite, ORMOSIL/Fe_3O_4/reduced GO for on-site arsenic remediation in water. It showed a maximum adsorption capacity of 38 mg g^{-1} using AAS, which was much higher than ORMOSIL and Fe_3O_4/reduced GO separately. The adsorption process was following Langmuir model and kinetic study revealed pseudo-second order [97]. Ye et al. synthesized three-dimensional Fe_3O_4/graphene aerogels for the adsorption of arsenic ions. The different morphological techniques showed 3-D structure where graphene is decorated with iron oxide with interconnected structure. The good adsorption capacity of 40.04 mg g^{-1} using inductively coupled plasma atomic emission spectroscopy (ICP-AES) was attributed to 3-D interconnected structure of the adsorbent. The kinetic study showed that rate of adsorption is of pseudo-second-order and follows Langmuir model. Apart from good adsorption capacity and easy separation, it is believed by authors that it can be also used in other fields like sensors, catalysis, etc. [133]. Chen et al. have developed magnetic MWCNTs by one-pot solid method instead of conventional tedious solvent methods. The nanomaterial was able to adsorb both As(III) and As(V) with maximum capacity of 24.05 and 47.41 mg g^{-1} by using ICP-OES. They concluded that their nanomaterial have highest adsorption capacities among other carbon-based nanomaterial for arsenic due to oxygen containing groups [14]. Chen et al. synthesized magnetite CNTs using a simple solid-phase method. It was further modified with glutathione to enhance arsenic adsorption. They found that CNTs were not only supporting iron oxide but also contributing in adsorption process. The maximum adsorption capacity that was obtained using ICP-OES was 19.12 mg g^{-1} for As(III). They suggested that the active adsorption sites are heterogeneous and can have practical applications [15]. The trend of using modified magnetic MWCNTs continued by Roy et al., and they reported a nanocomposite of europium-doped magnetic graphene oxide and Au NPs functionalized MWCNTs. It showed good adsorption capacity towards As(III) and As(V) of 320 mg g^{-1} and 298 mg g^{-1}, respectively. The system showed a detection limit of 0.27 and 0.99 $\mu g\ L^{-1}$ with linearity of 0.99–100 $\mu g\ L^{-1}$ and 2–85 $\mu g\ L^{-1}$, respectively, using square wave anodic stripping voltammetry (SWASV). The sensor was applied for analysis of real samples from industrial areas with good recovery (95–99%) [91]. An overview of analytical applications of magnetite carbon nanomaterials for monitoring arsenic in environment is summarized in Table 2.

5.2 *Magnetite Carbon Nanomaterials for Monitoring Cadmium in Environment*

Cadmium exposure is increasing due to anthropogenic activities like burning of fossils, metallurgical industries and discharging the industrial waste directly into water bodies from where it starts to enter our food chain. Also to meet the demands of growing population, farmers are using fertilizers in huge amount which contains high level of cadmium. Excess of cadmium in human can cause a fatal disease known as Itai-Itai. Cadmium is found to affect DNA healing mechanism, cellular respiration, activity of enzymes and antioxidants. It also affects the fertility of humans and is considered highly carcinogenic. As per WHO guidelines for safe drinking water, maximum permissible limit is 2 $\mu g\ L^{-1}$ [128]. So monitoring of cadmium in both environment and living organism is essential. There are a lot of techniques that have been developed to detect cadmium at sub-ppb level in real environment conditions like FAAS, ICP and electrochemical [83]. The treatment for the remediation of excess cadmium in the environment has been done by researchers using various magnetite carbon nanomaterials. Madannejad et al. prepared magnetic MWCNTs modified with 8-hydroxyquinoline for the removal of cadmium. The magnetic separation using external magnetic field avoids inconvenience related to centrifugation

Table 2 Analytical applications of magnetite carbon nanomaterials for monitoring arsenic in environment

Magnetite carbon nanomaterial	Techniques	pH	Temp. (°C)	Contact time (min)	Adsorption capacity (mg g^{-1})	References
Magnetite carbon nanotube-based nanocomposites						
Fe_3O_4/non-oxidative Graphene/CNTs	ICP-AES	7	25	720	As(III)—9.11 As(V)—5.21	[73]
Magnetic iron oxide nanoparticles/MWCNTs	ICP-AES	6.8 1.7	As(III)—45 As(V)—15	240	As(III)—6.95 As(V)—9.09	[50]
Fe_3O_4/Sulphydryl-functionalized-MWCNTs	ICP-OES	7	25	24 h	19.12	[15]
Magnetic iron oxide/CNTs	ICP-OES	7	25	As(III)—180 As(V)—240	As(III)—24.05 As(V)—47.41	[14]
Europium/Au NPs/MWCNTs/Fe_3O_4/GO nanohybrid	SWASV	7	30	60	As(III)—320.0 As(V)—298.0	[91]
Magnetite graphene-based nanocomposite						
ORMOSIL/Fe_3O_4/reduced GO	AAS	7	25	120	38	[97]
TiO_2/Fe_3O_4/rGO	AAS	7	25	120	147.05	[11]
Fe_3O_4/GO	ICP-AES	7	25	24 h	As(III)—57 As(V)—12	[134]
3D-Fe_3O_4/Graphene aerogels	ICP-AES	Neutral	25	720	40.04	[133]

(continued)

Table 2 (continued)

Magnetite carbon nanomaterial	Techniques	pH	Temp. (°C)	Contact time (min)	Adsorption capacity (mg g^{-1})	References
Fe_3O_4/non-oxidative graphene	ICP-AES	7	25	1440	As(III)—38 As(V)—14	[135]
Fe_3O_4/GO	ICP-MS	7 3	23	1440	As(III)—147 As(V)—113	[109]
3D-Fe_3O_4/Graphene composites	ICP-MS	4 & 7	30	24 h	–	[32]
3D Fe_3O_4/GO hydrogel	ICP-MS	7.7 6.2	25	8 h	As(III)—25.1 As(V)—74.2	[52]
SiO_2/Fe_3O_4/GO	ICP-MS	9	55	30	As(III)—7.51 As(V)—11.46	[70]
Fe_3O_4/GO	ICP-OES	8 5	25	24 h	As(III)—54.18 As(V)—26.76	[137]
Humic acid/Fe_3O_4/Graphene	ICP-OES	7	23	24 h	As(III)—7.5 As(V)—16	[77]

and filtration. The maximum adsorption capacity obtained was 60.2 mg g^{-1} at optimized conditions using AAS. Furthermore, the applicability of the developed sensor was tested in food and vegetable samples with good recovery [63]. Manoochehri and his co-workers have reported dipyridylamine-functionalized magnetic MWCNTs as nano-adsorbent for removal of Cd and other toxic metals. The Cd was determined using FAAS with a very low detection limit of 0.1 μg L^{-1}, linear dynamic range of 0.3–120 μg L^{-1} and good repeatability (RSD of 7%). It was reported that the magnetic MWCNTs modified with dipyridylamine gave good selectivity towards the target ions. This method showed good applicability in tea leaf and drinking water real samples by giving good recovery [65]. The trend of using magnetite CNTs has been followed by Taghizadeh et al., they have functionalized magnetite MWCNTs with 8-aminoquinoline and was applied for removal of Cd and other toxic metals. After optimizing variables like adsorbent amount, extraction time and pH the system showed a detection limit of 0.09 μg L^{-1}, linear range of 0.3-100 μg L^{-1} and good precision (RSD) of 4.1% for Cd(II) using FAAS. The system was tested using standard reference materials like seafood mix 02-2932 and LKSD-4 which showed little variation [112]. Madannejad et al. have synthesized 8-hydroxyquinoline functionalized magnetic MWCNTs for the determination of cadmium in food samples. The modified adsorbent was found to be cost-effective and selective material for adsorption of cadmium because of its simple modification procedure. It was reported that under optimized condition the system showed limit of detection of 0.12 μg L^{-1}, linearity over a range of 0.42–127 μg L^{-1} and RSD of 2.25% using ICP-AES. The system was applied for determination of cadmium in food samples like starch, cereals, tobacco and real water samples with recoveries ranging from 98 to 108.2% [63]. Similarly, Xu et al. synthesized sensitive and cheap electrochemical sensor for determination of both cadmium and lead simultaneously using nanocomposite consisting MWCNTs, Fe_3O_4, chitosan and graphene composite. The sensor showed a good detection limit of 0.1 μg·L^{-1} for cadmium using SWASV. The good sensitivity is attributed to high surface area of graphene and excellent conductivity of carbon nanotubes [131]. An overview of analytical applications of magnetite carbon nanomaterials for monitoring cadmium in environment is summarized in Table 3.

5.3 Magnetite Carbon Nanomaterials for Monitoring Chromium in Environment

Chromium has properties of being hard, lustrous and resistant to corrosion in air. It has a great application in the field of metallurgy as by mixing of chromium in steel gives properties like hardness and corrosion resistance. It is also used as catalyst in making hydrocarbons as well as in industries processes like dyeing, tanning of leather and for treatment of wood. Its excessive use can increase the chromium concentration in the environment to alarming levels [122]. WHO limit for total chromium in drinking water is 50 μg L^{-1} [128]. Chromium can enter our body, while having

Table 3 Analytical applications of magnetite carbon nanomaterials for monitoring cadmium in environment

Magnetite carbon nanomaterial	Techniques	pH	Temp. (°C)	Contact time (min)	Adsorption capacity (mg g^{-1})	References
Magnetite carbon nanotube-based nanocomposites						
8-hydroxyquinoline/Fe_3O_4/MWCNTs	FAAS	7	25	10	60.2	[63]
Fe_3O_4/MWCNTs	FAAS	7	25	20	39.15	[47]
8-aminoquinoline/Fe_3O_4/MWCNTs	FAAS	6.4	–	5	201	[112]
Dipyridylamine/Fe_3O_4/MWCNTs	FAAS	5	–	9	203	[65]
Fe_3O_4/MWCNTs	ICP-AES	7	20	360	28.24	[7]
Chitosan/Fe_3O_4/MWCNTs/laser-scribed graphene/GCE	SWASV	5	25	2.5	–	[131]
Magnetite graphene-based nanocomposite						
Fe_3O_4/reduced GO	FAAS	7	28	5	–	[116]
(3mercaptopropyl)trimethoxy silane/Fe_3O_4/Graphene composite	FAAS	7	–	30	125	[59]
Fe_3O_4/Graphene nanoparticles	FAAS	6		4	–	[3]
Fe_3O_4/GO nanospheres	FAAS	6	–	–	33.7	[4]
Diethylenetriamine/Fe_3O_4/GO nanocomposite	FAAS	5.5	–	10	59.88	[2]

(continued)

Table 3 (continued)

Magnetite carbon nanomaterial	Techniques	pH	Temp. (°C)	Contact time (min)	Adsorption capacity (mg g^{-1})	References
2 mercaptobenzothiazole/Fe_3O_4/GO nanocomposite	FAAS	6	–	4	164	[21]
MgAl-layered double hydroxide/Fe_3O_4/GO nanocomposite	FAAS	4	25	600	45.05	[38]
Pyrrole-thiophene/SiO_2/Fe_3O_4/GO	FAAS	5.8	25	6.5	80	[67]
Quinolinethioacetamide/Fe_3O_4/GO composite	FAAS	5.8	–	6	233	[68]
SiO_2/Fe_3O_4/Graphene nanocomposite	FAAS	7	–	5	–	[55]
Sulphanilic acid/Fe_3O_4/GO	FAAS	6	30	1440	58.22	[37]
Silver/Fe_3O_4/reduced GO nanohybrids	ICP-MS	4	25	5	386.8	[72]
Azo-phenol ligand/Fe_3O_4/GO nanosheets	ICP-OES	7	55	2	73.5	[92]
Fe_3O_4/reduced GO	SWASV	5	–	2	–	[130]
Poly(amidoamine) dendrimer/magnetic GO/GCE	SWASV	4.5	23 ± 2	2.6	–	[8]

food, liquid or even through physical contact. The concentration of chromium in air and the environment is usually very low, but due to above anthropogenic activities, the concentration level can raise which can cause health hazard like various skin diseases for both humans and other animals. High levels of chromium in water bodies can lead to serious problems for fishes and other marine animals by affecting their gills and fertility [118]. Chromium determination using magnetite carbon nanomaterials has been performed by various researchers as Fathi et al. developed magnetite nanoporous graphene for efficient removal of chromium from water. The maximum adsorption capacity of 43.5 mg g^{-1} was obtained using AAS. Adsorption process followed Freundlich model with pseudo-second-order kinetic. Thermodynamic study revealed that the process was spontaneous and endothermic in nature. The author finally suggested that the adsorbent has good potential and can be applied for treating industrial wastewater [27]. Similarly, Islam et al. functionalized magnetite graphene oxide with triethylenetetramine that was used for the removal of chromium using FAAS. The maximum adsorption capacity of 16.4 and 9.6 mg g^{-1} was obtained for Cr(VI) and Cr(III), respectively. The system was linear from 5 to 100 $\mu g\ L^{-1}$ concentration having a low detection limit of 1.4 and 1.6 $\mu g\ L^{-1}$ for Cr(VI) and Cr(III), respectively. The method was able to speciate the chromium species and was efficiently applied in real water samples [40]. Zhao et al. developed a nanocomposite consisting diethylenetriamine, Fe_3O_4 nanoparticles and graphene oxide by one step which was used to remove Cr(VI). The maximum adsorption capacity obtained was 123.4 mg g^{-1} using ICP-MS. The adsorption process followed Langmuir model with pseudo-second-order kinetics [141]. Vu et al. synthesized magnetite graphene oxide using co-precipitation method that was further encapsulated inside alginate beads. The adsorbent was used for the removal of Cr(VI) and followed Freundlich model. This nanocomposite material was able to remove Cr(VI) almost completely from the solution, which was measured using ICP-OES. The author suggested that the reduction in aggregation resulted in improved adsorption [120].

An overview of analytical applications of magnetite carbon nanomaterials for monitoring chromium in environment is summarized in Table 4.

5.4 Magnetite Carbon Nanomaterials for Monitoring Lead in Environment

Lead comes in the category of toxic heavy metals and has many industrial applications, so a complete ban on use of lead is not possible in near future. Lead is being used in industrial processes like smelting, paints, batteries, book printing and as anti-knocking agents in fuel. Lead enters our body either through water or by ingestion of food. It can be easily absorbed through blood and reaches to various parts of the body. It generally affects the immunity, nervous system, kidneys and has potency to cause anaemia. Lead toxicity is more prominent in children as their nervous system is in developing stage, and even a small dose can be fatal. The

Table 4 Analytical applications of magnetite carbon nanomaterials for monitoring chromium in environment

Magnetite Carbon nanomaterial	Techniques	pH	Temp. (°C)	Contact time (min)	Adsorption capacity (mg g^{-1})	References
Magnetite graphene-based nanocomposite						
Fe_3O_4/Nanoporous graphene	FAAS	3	25	60	43.5	[27]
Fe_3O_4/3D-GO foam nanocomposite	FAAS	2	25	20	258.6	[51]
Diatomite/Fe_3O_4/GO	FAAS	–	–	1440	90.29	[22]
Pyrrole-thiophene/SiO_2/Fe_3O_4/GO	FAAS	5.8	25	6.5	98	[67]
Triethylenetetramine/Fe_3O_4/GO	FAAS	8 2	30 ± 2	Cr(III)—30 Cr(VI)—10	Cr(III)—9.6 Cr(VI)—16.4	[40]
2-mercaptobenzothiazole/Fe_3O_4/GO	GFAAS	6.7	25	118.6	100	[104]
SiO_2/Fe_3O_4/GO	ICP	5.8	25	50	4.7	[53]
Diethylenetriamine/Fe_3O_4/GO	ICP-MS	2	25	720	123.4	[141]
Alginate bead/Fe_3O_4/GO	ICP-OES	5	20	1440	–	[120]

toxicity mechanism follows the displacing of essential metal ions like Ca, Fe and Mg which further affect various biological processes essential for the body [126]. According to WHO, maximum permissible limit for Pb(II) in drinking water is 2 μg L^{-1} [128]. But continuous activities like burning of fossils and industrial production are causing an increase in the level which is hazardous. So developing efficient and cheap methods for monitoring Pb(II) in water is highly appreciated. A lot of research work has been published using nanomaterials for effective removal of lead [101]. Among them magnetite carbon nanomaterials are having additional benefits of easy separation and high adsorption capacity over other as discussed below.

Lead determination in the environment using magnetite carbon nanomaterials has been performed by various researchers as Jiang et al. synthesized magnetic MWCNTs nanocomposite having thiol and amino-functional groups introduced by reacting ammonium ferrous sulphate, ammonium ferric sulphate, trimethoxysilyl propanethiol and hydrazine. The nanocomposite was characterized by TEM, XPS, XRD and SEM revealing coating of hydrazine and trimethoxysilyl propanethiol (MPTs) on the surface of MWCNTs. The maximum capacity obtained was 169.89 mg g^{-1} at optimized conditions using AAS. The thermodynamic study shows that the system follows Freundlich model with pseudo-second-order kinetics and is exothermic in nature [42]. Ranjan et al. synthesized MWCNTs filled with iron oxide to impart magnetic properties. Further, MWCNTs were introduced with amine groups to covalently immobilize cyanate hydratase and characterized with FT-IR. The modified magnetic nanomaterials have long-term stability and were found to efficiently remove many toxic ions including Pb(II) whose concentration was determined using AAS. It was able to reduce 34.48% Pb(II) ions simultaneously with other ions [84]. Ren et al. prepared magnetic nanocomposite consisting of triethanolamine, Fe_3O_4 and graphene oxide. The characterization confirmed the loading of Fe_3O_4 on graphene oxide, which are further encapsulated with triethanolamine. The maximum adsorption capacity obtained was 121.5 mg g^{-1} under optimized conditions using ICP. The thermodynamic studies revealed that the adsorption process was spontaneous and exothermic in nature. The adsorption kinetics was found to be of pseudo-second order [88]. Similarly, Mahmoudian et al. synthesized magnetite-reduced graphene oxide nanosheet composites having spherical morphology using simple hydrothermal method. Electrochemical sensor was developed by fabricating the nanocomposite over GCE. The system showed good detection limit of 0.082 nM for lead at optimized conditions using differential pulse voltammetry (DPV). The high conductivity of the electrochemical sensor is due to presence of small band gaps in reduced graphene oxide. In addition to this, the sensor showed a linear range of 0.05–1.5 nM [64]. Baghayeri et al. fabricated glassy carbon electrode with magnetic graphene oxide functionalized with poly(amidoamine) dendrimer and used it for the determination of lead. A good detection limit of 130 ng L^{-1} was obtained using SWASV. Interference due to other ions was negligible, and the sensor was able to detect lead simultaneously with other ions in real water samples [8]. An overview of analytical applications of magnetite carbon nanomaterials for monitoring lead in environment is summarized in Table 5.

Table 5 Analytical applications of magnetite carbon nanomaterials for monitoring lead in environment

Magnetite carbon nanomaterial	Techniques	pH	Temp. (°C)	Contact time (min)	Adsorption capacity (mg g^{-1})	References
Magnetite carbon nanotube-based nanocomposites						
Fe_3O_4/oxidized MWCNTs	AAS	5	25	360	67.25	[43]
Hydrazine/trimethoxysilylpropanethiol/Fe_3O_4/O-MWCNTs	AAS	6	25	720	169.89	[42]
Recombinant cyanate hydratase/Fe_3O_4/MWCNTs	AAS	8	25	30	–	[84]
Magnetic hydroxyapatite-immobilized oxidized MWCNTs	EDX spectroscopy	4.1	25	40	698.4	[125]
Chitosan/Fe_3O_4/MWCNTs/laser scribed graphene–GCE	SWASV	5	25	150	–	[131]
Magnetite graphene-based nanocomposite						
Fe_3O_4/GO	AAS	6	25	1440	58.43	[140]
Fe_3O_4/reduced GO	AAS	7	28	5	–	[116]
Fe_3O_4/reduced GO	AAS	5.2	25	120	51.02	[82]
Fe_3O_4/reduced GO	AAS	5	28	120	48	[80]
Fe_3O_4/porous graphene nanocomposites	AAS	7	28	210	460	[13]
Diatomite/Fe_3O_4/graphite oxide	AAS	–	–	1440	113.5	[22]
Diethylenetriaminepentaacetic acid/Fe_3O_4/GO	AAS	3	20	1440	387.6	[56]
Ethylenediamine tetra acetic acid/Fe_3O_4/GO	AAS	4.2	25	40	508.4	[20]
Hydroxypropyl/β–Cyclodextrin/Fe_3O_4/Graphene	AAS	5	30	15	50.39	[113]
Polyamidoamine nanosheets/Fe_3O_4/GO	AAS	6	25	90	326.7	[78]
Polystyrene/Fe_3O_4/GO	AAS	6	35	480	73.52	[86]
Fe_3O_4/partially reduced GO composite	FAAS	6	20	10	373.14	[34]

(continued)

Table 5 (continued)

Magnetite carbon nanomaterial	Techniques	pH	Temp. (°C)	Contact time (min)	Adsorption capacity (mg g^{-1})	References
Alginate/Fe_3O_4/GO	FAAS	5	25	180	322.58	[103]
Diethylenetriamine/Fe_3O_4/GO	FAAS	5.5	–	10	172.41	[2]
2 mercaptobenzothiazole/Fe_3O_4/GO	FAAS	6	–	4	179	[21]
Fe_3O_4-encapsulated $C_3N_3S_3$ polymer/reduced GO composite	ICP	6	25	360	270.3	[29]
Triethanolamine/Fe_3O_4/GO composite	ICP	5	20.15	120	121.5	[88]
Fe_3O_4/GO nanocomposites	ICP-AES	7	30	600	20.8	[10]
Mg3Al–OH-layered double hydroxide/Magnetite/GO composites	ICP-AES	5	25	1440	173	[138]
Sodium/Fe_3O_4/reduced GO	ICP-AES	7	30	300	1666.6	[49]
poly(*N*-isopropylacrylamide-co-benzo-18-crown-6acrylamide) microgels/magnetic GO	ICP-MS	5	25	20	11.76	[71]
Lauric acid/ethylenediamine tetra-acetic acid/Fe_3O_4/GO	ICP-OES	3	20	105	161.8	[23]
Fe_3O_4/reduced GO nanosheet composites	DPV	6	–	–	–	[64]
Poly(amidoamine) dendrimer/magnetic GO-GCE	SWASV	4.5	23 ± 2	2.6	–	[8]
Fe_3O_4/Graphene composite	SWV	5.5	–	3	–	[36]

5.5 *Magnetite Carbon Nanomaterials for Monitoring Mercury in Environment*

Mercury is one of the most toxic elements present on earth, and it exists in various forms. Inorganic salts of mercury are not that much toxic as they only affect the kidney and gastrointestinal tract while organic mercury, where mercury is bonded to methyl, ethyl and phenyl groups, is highly toxic. They easily react with thiol groups of biomolecules in the body which interferes with the cellular and subcellular processes. It was also found that they interrupt DNA functioning, protein synthesis, heme synthesis and central nervous system. Exposure to mercury can be due to its application in dental treatment, ingestion of fish from polluted area, volcanic activities, coal burning and mining activities. Generally, mercury is found in its elemental form which after entering water bodies is converted to organic mercury by microorganisms which is later taken as food by small fishes and enters our food chain [12]. According to WHO, maximum permissible limit for mercury in drinking water is 2 $\mu g\ L^{-1}$ [128]. There is an urgent need of developing new methods for removing the mercury, and various magnetite carbon nanomaterials have been used for this purpose. Recently, a lot of development in using magnetite carbon nanomaterials has been evolved for treatment of mercury in the environment as Sadegh et al. have developed magnetic CNTs showing entangled network of Fe_3O_4 clusters and oxidized MWCNTs with high specific area of 92 $m^2\ g^{-1}$. The maximum adsorption capacity obtained was 238.78 mg g^{-1} under optimized conditions using AAS. The adsorption follows Langmuir model, and the kinetics of adsorption was found to be pseudo-second order. They suggested that such a high adsorption capacity can have good application in the environment remediation field [94]. Seidi and Fotouhi synthesized magnetic graphene oxide functionalized with polythiophene for the determination of mercury using cold vapour—atomic absorption spectroscopy (CV-AAS). At optimized conditions, the adsorbent showed a detection limit of 0.025 $\mu g\ L^{-1}$ with a linear concentration range from 1 to 85 $\mu g\ L^{-1}$. Finally, its applicability was checked in different seafood samples for the determination mercury [102]. Liu et al. synthesized magnetite graphene oxide functionalized with ethylenediamine using simple one-pot solvothermal method. The adsorbent showed a maximum capacity of 127.23 mg g^{-1} for Hg and easy separation using FAAS. It was concluded that the graphene oxide sheets helped in preventing in agglomeration of Fe_3O_4 and good dispersion was obtained for efficient removal of mercury. The adsorption kinetics was found to follow pseudo-second order and Langmuir model. Thermodynamic studies revealed that the adsorption process was spontaneous and endothermic in nature [60]. In another work, Alvand and Shemirani synthesized a nanocomposite containing SiO_2 and magnetite graphene quantum dots and concluded that the magnetite silica was covalently bonded to graphene quantum dots. The prepared nanocomposite showed strong fluorescence which is selectively quenched by Hg(II) ions due to electron–hole recombination. The sensor was linear up to concentration range from 0.1 to 70 μM with a detection limit of 30 nM. It was also suggested that high adsorption capacity (68 mg g^{-1}) using ICP-AES is due to high adsorption area and binding sites

Table 6 Analytical applications of magnetite carbon nanomaterials for monitoring mercury in environment

Magnetite carbon nanomaterial	Techniques	pH	Temp. (°C)	Contact time (min)	Adsorption capacity (mg g^{-1})	References
Magnetite carbon nanotube-based nanocomposites						
Fe_3O_4/MWCNTs	AAS	7	25	60	238.78	[94]
Fe_3O_4/MWCNTs	FAAS	3	25	60	172.83	[35]
Fe_3O_4/MWCNTs	GC-AFS	3.9	25	30	–	[89]
Magnetite graphene-based nanocomposite						
Ternary hydrosulphonyl-deep eutectic solvent/magnetic GO	Atomic fluorescence spectrometer	6	25	60	215.1	[17]
Ethylene diamine tetra-acetic acid/magnetic GO	AAS	7	25	50	268.4	[20]
Ag NPs/Fe_3O_4/reduced GO	CV-AAS	–	100	180	–	[62]
Chitosan/Mercapto/magnetic GO	CV-AAS	6.5	28	10	–	[144]
Polymerized thiophene/Fe_3O_4/GO	CV-AAS	6.5	–	21	1	[102]
Ethylenediamine/Fe_3O_4/GO	FAAS	5.3	25	10	127.23	[60]
SiO_2/Fe_3O_4/graphene quantum dots	ICP-AES	6	25	1	68	[5]
Fe_3O_4/Reduced GO	SWASV	5	29	2	–	[130]

of graphene quantum dots. The nanocomposite was easily separable using external magnetic field and can be reused by treating with EDTA solution. Finally, it was successfully tested in real water samples [5]. Xiong et al. synthesized a nanocomposite consisting of Fe_3O_4 and reduced graphene oxide using one-pot synthetic method for the simultaneous detection of toxic metals including mercury. The analysis of mercury concentration was done using SWASV under optimized conditions. Finally, the applicability of the fabricated sensor was tested in soil samples with good recovery [130]. An overview of analytical applications of magnetite carbon nanomaterials for monitoring mercury in environment is summarized in Table 6.

5.6 *Magnetite Carbon Nanomaterials for Monitoring Dyes in Environment*

Dyes are organic compounds used to colour fabric, leather, paper, cosmetics and pharmaceuticals and can be classified as natural and synthetic dyes. Synthetic azo dyes are used mostly in industries, and they are discharged in nearby water bodies after use. The effluent contains 8-10% of dye which is highly toxic and affects the entire ecosystem for aquatic organisms once enter into food chain. The magnetite carbon nanomaterials have become popular due to the large surface area and easy separation of adsorbent by magnetic nanoparticles. Various dyes have been removed using different methods like adsorption and photocatalytic degradation of dyes. ***Adsorption*** is quite simple and provides wide range of adsorbent and easy modification of nanomaterials. The use of magnetic carbon nanotube-based nanomaterial is increased rapidly, and many researchers like Kerkez and Bayazit synthesized magnetite decorated MWCNTs for removal of Malachite Green and rhodamine B dyes. The various factors which influence adsorption like pH, contact time and temperature were optimized, and the maximum adsorption capacity was 55.25 and 37.04 mg g^{-1} for MG and RhB dyes, respectively, at pH 6 and 25 °C [46]. The adsorption of two dyes was best described by kinetic pseudo-second-order model. Cheng et al. modified magnetic carbon nanotubes with cyclodextrin and used for the removal of Methylene Blue dye. β-cyclodextrin (CD) was grafted on the surface of carbon nanotube by the reduction of oxidized CNT with the help of hydrazine hydrate. The modification improved the adsorption capacity to 196.5 mg g^{-1} for Methylene Blue. The effect of temperature was also studied, and it was found from thermodynamics parameters like $\Delta G°$ and $\Delta H°$ that the reaction was spontaneous and endothermic in nature [18]. Magnetic graphene oxide-based nanocomposites have been evolved as an effective adsorbent used for dye removal by many researchers like Zhou et al. synthesized Fe_3O_4-embedded graphene oxide nanocomposite by co-precipitation method. The prepared magnetic nanomaterial was used for adsorptive removal of Methylene Blue dye. The factors affecting adsorptions like pH, contact time and temperature were optimized, and it was found that the maximum adsorption (246 mg g^{-1}) occurred at pH 10 and temperature 20 °C in 5 min of contact time [143]. Modified grapheme-based nanomaterials were also used for adsorption of dyes as Lin and Chen synthesized arginine-capped iron oxide/reduced graphene oxide nanocomposite by a simple method in which arginine used as reducing agent and capping agent. The obtained nanocomposite was successfully applied for the removal of Crocein Orange G and acid green 25 dyes. The optimum pH, contact time and temperature were, 2, 2–3 h and 30 °C, respectively. Under optimized conditions, the adsorption capacity obtained was 131.6 and 185.2 mg g^{-1} for Crocein Orange G and acid green 25 dyes, respectively. The force of attraction took place during adsorption was reported as electrostatic interaction, van der Waals forces or π-π interaction between the dyes and the magnetic nanocomposite [58].

Photocatalytic degradation has its own advantages like no production of sludge, reuse of catalytic material and less use of reagents; so many researchers studied photocatalytic degradation of dyes as Tarigh et al. prepared magnetic multiwalled carbon nanotube–TiO_2 nanocomposite and used it for photocatalytic degradation of Malachite Green dye. The results reveal that the dye degradation using magnetic MWCNTs-TiO_2 nanocomposite took place at pH 5 under UV irradiation for 240 min and the catalyst showed high reusability [115]. Photocatalytic degradation of Methylene Blue dye using graphene oxide–metal oxide (TiO_2/Fe_3O_4)-based nanocomposites was studied by Benjwal. The complete dye degraded in just 5 min under UV-light [11]. An overview of analytical applications of magnetite carbon nanomaterials for monitoring dyes in environment is summarized in Table 7.

6 Conclusion and Future Perspectives

It is clear that the complete ban on the use of these toxic pollutants is not possible due to lack of other alternatives. So, the best way to preserve our eco-system is either by reducing their use or by remediation. The magnetite carbon nanomaterials prepared using different methods as discussed in the chapter have shown great possibilities in remediation of the environmental problems. The cost-effective synthesis not only limits their application in treatment of wastewater as it is extended to food samples also. It is evident from the above discussion that one of the major factors in reducing the cost is its easy separation with the help of external magnetic field. Functionalization of these magnetite carbon nanomaterials has resulted in improved selectivity and dispersibility for effective removal of pollutants. They have been widely used for the treatment of both organic and inorganic pollutants simultaneously. There is already a huge development in techniques for the quantification of these toxicants which have their own drawbacks like high cost, bulky, require expertise for conducting analysis, etc. In future, more emphasis can be given on further development of such techniques. The nanocomposites with high adsorption capacity, recyclability, chemical and thermal stability should be prioritized accordingly for their wide-scale application. Development of these materials for the removal of other pollutants like radionuclides is still under progress and requires more effort. Further, more development is required in treatment of other environmental issues like air pollution and oil spillage in oceans. Finally, magnetite carbon nanomaterials functionalized with different functional groups possess huge potential to deal with current and future environmental issues.

Table 7 Analytical applications of magnetite carbon nanomaterials for monitoring dyes in environment

Magnetite carbon nanomaterial	Dyes	pH	Temp. (°C)	Contact time (min)	Adsorption capacity (mg g^{-1})/Degradation %	References
Adsorption						
Magnetite carbon nanotube-based nanocomposites						
Fe_3O_4/CNT	Sudan I, Sudan II, Sudan III Sudan IV	4.0	25	60	26.52 24.03 23.04 22.77	[111]
Fe_3O_4-decorated MWCNTs	Malachite Green Rhodamine B	6.0	25	80	55.25 37.04	[46]
κ-carrageenan/Fe_3O_4/CNT nanocomposite	Crystal Violet reactive Black 5	2.0 9.0	45	300	1.34×10^{-4a} 0.298×10^{-4a}	[26]
Chitin/Fe_3O_4/MWCNTs nanocomposite	Rose Bengal	8.0	25	120	6.19	[98]
Cyclodextrin/Fe_3O_4/CNT composite	Methylene Blue	–	25	1440	196.5	[18]
Gelatin/CNT/Fe_3O_4 nanocomposite beads	Direct Red 80 Methylene Blue	–	25	360	380.7 465.5	[93]
Fe_3O_4/graphene-CNT composite	Methylene Blue	7.0	10	30	65.79	[123]
MWCNTs/Fe_3O_4/PANI magnetic composite	Methyl Orange Congo Red	4.5	25	1440	446.25 417.38	[142]
O-MWCNT-Fe_3O_4 & O-MWCNTs-κ-carrageenan-Fe_3O_4 nanocomposite	Methylene Blue	6.5	25	300	1.11×10^{-4a} 1.24×10^{-4a}	[25]
Hydroxyapatite/Fe_3O_4/O-MWCNTs	Methylene Blue	8.1	25	50	328.4	[125]

(continued)

Table 7 (continued)

Magnetite carbon nanomaterial	Dyes	pH	Temp. (°C)	Contact time (min)	Adsorption capacity (mg g^{-1})/Degradation %	References
Fe_3O_4/chitosan/SiO_2/CNT nanocomposite	Direct Blue 71 Reactive Blue 19	6.8 2.0	25	120	61.35 97.08	[1]
Magnetite graphene-based nanocomposite						
Fe_3O_4/porous graphene nanocomposites	Methyl Violet	–	28 ± 0.5	5	460.0	[13]
Fe_3O_4/Graphene sponge	Methylene Blue	6.0	–	300	526.0	[136]
Fe_3O_4/Graphene nanocomposite	Rhodamine B	–	–	90	186.4	[61]
Fe_3O_4/GO magnetic Nanocomposite	Methylene Blue Methyl Violet	7.0	25	5	188.3	[19]
Fe_3O_4/GO magnetic nanohybrids	Methylene Blue, Rhodamine B Methyl Blue	6.0	25	30	33.3 30.2 20.2	[45]
Fe_3O_4-embedded GO	Methylene Blue	10	20	5	246.0	[143]
Fe_3O_4/GO-containing nanocomposite hydrogels	Crystal Violet	7.0	25	120	769.2	[79]
Fe_3O_4/Graphene-calcium alginate	Methylene Blue	7.0	40	60	51.6	[107]
Fe_3O_4/Graphene/chitosan	Acid Orange 7	3.0	25	120	30.4	[105]
Fe_3O_4/Graphene-modified polypyrrole nanocomposite	Methylene Blue	6.0	30	60	270.3	[9]
Acetone Fe_3O_4 reduced GO	Rhodamine 6G	5.0	20	120	93.4	[74]

(continued)

Table 7 (continued)

Magnetite carbon nanomaterial	Dyes	pH	Temp. (°C)	Contact time (min)	Adsorption capacity (mg g^{-1})/Degradation %	References
Arginine-capped Fe_3O_4/reduced GO nanocomposite	Crocein Orange G Acid green 25	2.0	30	120-180	131.6 185.2	[58]
Carboxylate/Fe_3O_4/GO nanostructures	Rhodamine B Methylene Blue	–	25	150 20	22.1 35.9	[33]
Cellulose/Fe_3O_4/GO	Methylene Blue	6.0	25	840	70.0	[106]
Chitosan/Fe_3O_4/GO	Methyl Orange	4.0	25	1440	398.1	[44]
Citric acid-functionalized Fe_3O_4/GO nanocomposite	Methylene Blue	6.0	25	30	112.0	[69]
Citric acid-functionalized Fe_3O_4/GO coated corn straw	Methylene Blue	12.0	25	180	315.5	[30]
Cyclodextrin/Fe_3O_4/GO	Methylene Blue	11.0	30	50	273.4	[54]
β-cyclodextrin/Fe_3O_4/GO nanocomposites	Malachite Green	7.0	25 35 45	120	740.7 900.9 990.1	[121]
Magnetic hydrogel beads based on modified gum tragacanth/GO	Crystal Violet Congo Red	8.0 5.0	25	720	94.0 101.7	[96]
Poly(acrylic acid) functionalized Fe_3O_4/GO nanocomposite	Methylene Blue	7.0	–	1440	291.0	[139]
Fe_3O_4-reduced GO/zeolitic imidazolate framework	Malachite Green	7.0	20 30 40	150	2328 2682 3165	[57]
Fe_3O_4/GO@SiO_2 nanocomposites	Methylene Blue	8.0	–	60	44.1	[132]

(continued)

Table 7 (continued)

Magnetite carbon nanomaterial	Dyes	pH	Temp. (°C)	Contact time (min)	Adsorption capacity (mg g^{-1})/Degradation %	References
Fe_3O_4-Graphene@mesoporous SiO_2 nanocomposites	Methylene Blue	11.0	20 30 40	1440	139.6 166.3 178.5	[129]
Photocatalytic degradation						
Magnetite carbon nanotube-based nanocomposites						
Fe_3O_4/MWCNTs/TiO_2 nanocomposite	Malachite Green	5.0	–	240	–	[115]
Magnetite graphene-based nanocomposite						
Fe_3O_4/GO/Immobilized laccase	Crystal Violet Malachite Green Brilliant Green	3.0	35	–	94.7 95.6 91.4	[16]
Fe_3O_4/reduced GO/TiO_2/based nanocomposites	Methylene Blue	–	–	5	~100	[11]
Fe_3O_4/Graphene/ZnO@SiO_2 nanocomposites	Methylene Blue, Methyl Orange Rhodamine B	11.0	30	60	–	[6]

[a]$molg^{-1}$

References

1. Abbasi M (2017) Synthesis and characterization of magnetic nanocomposite of chitosan/SiO_2/carbon nanotubes and its application for dyes removal. J Clean Prod 145:105–113
2. Aliyari E, Alvand M, Shemirani F (2015) Simultaneous separation and preconcentration of lead and cadmium from water and vegetable samples using a diethylenetriamine-modified magnetic graphene oxide nanocomposite. Anal Methods 7(18):7582–7589
3. Alvand M, Shemirani F (2014) Preconcentration of trace cadmium ion using magnetic graphene nanoparticles as an efficient adsorbent. Microchim Acta 181(1–2):181–188
4. Alvand M, Shemirani F (2016) Fabrication of Fe_3O_4@graphene oxide core-shell nanospheres for ferrofluid-based dispersive solid phase extraction as exemplified for Cd(II) as a model analyte. Microchim Acta 183(5):1749–1757
5. Alvand M, Shemirani F (2017) A Fe_3O_4@SiO_2@graphene quantum dot core-shell structured nanomaterial as a fluorescent probe and for magnetic removal of mercury (II) ion. Microchim Acta 184(6):1621–1629
6. Areerob Y, Cho JY, Jang WK, Oh WC (2018) Enhanced sonocatalytic degradation of organic dyes from aqueous solutions by novel synthesis of mesoporous Fe_3O_4-graphene/ZnO@SiO_2 nanocomposites. Ultrason Sonochem 41:267–278
7. Azimi S, Es'haghi Z (2017) A magnetized nanoparticle based solid-phase extraction procedure followed by inductively coupled plasma atomic emission spectrometry to determine arsenic, lead and cadmium in water, milk, Indian rice and red tea. Bull Environ Contam Toxicol 98(6):830–836
8. Baghayeri M, Alinezhad H, Fayazi M, Tarahomi M, Ghanei-Motlagh R, Maleki B (2019) A novel electrochemical sensor based on a glassy carbon electrode modified with dendrimer functionalized magnetic graphene oxide for simultaneous determination of trace Pb(II) and Cd(II). Electrochim Acta 312:80–88
9. Bai L, Li Z, Zhang Y, Wang T, Lu R, Zhou W, Zhang S (2015) Synthesis of water-dispersible graphene-modified magnetic polypyrrole nanocomposite and its ability to efficiently adsorb methylene blue from aqueous solution. Chem Eng J 279:757–766
10. Bai X, Feng R, Hua Z, Zhou L, Shi H (2015) Adsorption of 17β-estradiol (E2) and Pb(II) on Fe_3O_4/graphene oxide (Fe_3O_4/GO) nanocomposites. Environ Eng Sci 32(5):370–378
11. Benjwal P, Kumar M, Chamoli P, Kar KK (2015) Enhanced photocatalytic degradation of methylene blue and adsorption of arsenic (iii) by reduced graphene oxide (rGO)–metal oxide (TiO_2/Fe_3O_4) based nanocomposites. RSC Adv 5(89):73249–73260
12. Bernhoft RA (2012) Mercury toxicity and treatment: a review of the literature. J Environ Pub Health 2012
13. Bharath G, Alhseinat E, Ponpandian N, Khan MA, Siddiqui MR, Ahmed F, Alsharaeh EH (2017) Development of adsorption and electrosorption techniques for removal of organic and inorganic pollutants from wastewater using novel magnetite/porous graphene-based nanocomposites. Sep Purif Technol 188:206–218
14. Chen B, Zhu Z, Ma J, Yang M, Hong J, Hu X, Chen J (2014) One-pot, solid-phase synthesis of magnetic multiwalled carbon nanotube/iron oxide composites and their application in arsenic removal. J Colloid Interface Sci 434:9–17
15. Chen B, Zhu Z, Ma J, Qiu Y, Chen J (2015) Iron oxide supported sulfhydryl-functionalized multiwalled carbon nanotubes for removal of arsenite from aqueous solution. ChemPlusChem 80(4):740–748
16. Chen J, Leng J, Yang X, Liao L, Liu L, Xiao A (2017) Enhanced performance of magnetic graphene oxide-immobilized laccase and its application for the decolorization of dyes. Molecules 22(2):221
17. Chen J, Wang Y, Wei X, Xu P, Xu W, Ni R, Meng J (2018) Magnetic solid-phase extraction for the removal of mercury from water with ternary hydrosulphonyl-based deep eutectic solvent modified magnetic graphene oxide. Talanta 188:454–462

18. Cheng J, Chang PR, Zheng P, Ma X (2014) Characterization of magnetic carbon nanotube–cyclodextrin composite and its adsorption of dye. Ind Eng Chem Res 53(4):1415–1421
19. Cheng Z, Liao J, He B, Zhang F, Zhang F, Huang X, Zhou L (2015) One-step fabrication of graphene oxide enhanced magnetic composite gel for highly efficient dye adsorption and catalysis. ACS Sustain Chem Eng 3(7):1677–1685
20. Cui L, Wang Y, Gao L, Hu L, Yan L, Wei Q, Du B (2015) EDTA functionalized magnetic graphene oxide for removal of Pb(II), Hg(II) and Cu(II) in water treatment: adsorption mechanism and separation property. Chem Eng J 281:1–10
21. Dahaghin Z, Mousavi HZ, Sajjadi SM (2017) Trace amounts of Cd(II), Cu(II) and Pb(II) ions monitoring using Fe_3O_4@graphene oxide nanocomposite modified via 2-mercaptobenzothiazole as a novel and efficient nanosorbent. J Mol Liq 231:386–395
22. Dalagan JQ, Ibale RA (2016) Adsorption behaviour of heavy metal ions (Cr^{3+}, Pb^{2+} and Cu^{2+}) into magnetite-graphite oxide-diatomite. Asia Pac High Educ Res J 3(1)
23. Danesh N, Hosseini M, Ghorbani M, Marjani A (2016) Fabrication, characterization and physical properties of a novel magnetite graphene oxide/Lauric acid nanoparticles modified by ethylenediaminetetraacetic acid and its applications as an adsorbent for the removal of Pb(II) ions. Synth Met 220:508–523
24. Dhir B (2013) Aquatic plant species and removal of contaminants. Phytoremediation: role of aquatic plants in environmental clean-up. Springer, India, pp 21–50
25. Duman O, Tunç S, Polat TG, Bozoğlan BK (2016) Synthesis of magnetic oxidized multi-walled carbon nanotube-κ-carrageenan-Fe_3O_4 nanocomposite adsorbent and its application in cationic Methylene Blue dye adsorption. Carbohydr Polym 147:79–88
26. Duman O, Tunç S, Bozoğlan BK, Polat TG (2016) Removal of triphenylmethane and reactive azo dyes from aqueous solution by magnetic carbon nanotube-κ-carrageenan-Fe_3O_4 nanocomposite. J Alloys Compd 687:370–383
27. Fathi S, Rezaei KR, Rashidi A, Karbassi A (2016) Hexavalent chromium adsorption from aqueous solutions using nanoporous graphene/Fe_3O_4(NPG/Fe_3O_4: modeling and optimization). Desalin Water Treat 57(58):28284–28293
28. Fernandes LF, Bruch GE, Massensini AR, Frézard F (2018) Recent advances in the therapeutic and diagnostic use of liposomes and carbon nanomaterials in ischemic stroke. Front Neurosci 12:453
29. Fu W, Wang X, Huang Z (2019) Remarkable reusability of magnetic Fe_3O_4-encapsulated $C_3N_3S_3$ polymer/reduced graphene oxide composite: a highly effective adsorbent for Pb and Hg ions. Sci Total Environ 659:895–904
30. Ge H, Wang C, Liu S, Huang Z (2016) Synthesis of citric acid functionalized magnetic graphene oxide coated corn straw for methylene blue adsorption. Bioresour Technol 221:419–429
31. Giubileo F, Di BA, Iemmo L, Luongo G, Urban F (2018) Field emission from carbon nanostructures. Appl Sci 8(4):526
32. Guo L, Ye P, Wang J, Fu F, Wu Z (2015) Three-dimensional Fe_3O_4-graphene macroscopic composites for arsenic and arsenate removal. J Hazard Mater 298:28–35
33. Guo R, Jiao T, Li R, Chen Y, Guo W, Zhang L, Peng Q (2017) Sandwiched Fe_3O_4/carboxylate graphene oxide nanostructures constructed by layer-by-layer assembly for highly efficient and magnetically recyclable dye removal. ACS Sustain Chem Eng 6(1):1279–1288
34. Guo T, Bulin C, Li B, Zhao Z, Yu H, Sun H, Zhang B (2018) Efficient removal of aqueous Pb(II) using partially reduced graphene oxide-Fe_3O_4. Adsorpt Sci Technol 36(3–4):1031–1048
35. Homayoon F, Faghihian H, Torki F (2017) Application of a novel magnetic carbon nanotube adsorbent for removal of mercury from aqueous solutions. Environ Sci Pollut Res 24(12):11764–11778
36. He B, Shen XF, Nie J, Wang XL, Liu FM, Yin W, Fa HB (2018) Electrochemical sensor using graphene/Fe_3O_4 nanosheets functionalized with garlic extract for the detection of lead ion. J Solid State Electrochem 22(11):3515–3525
37. Hu XJ, Liu YG, Zeng GM, You SH, Wang H, Hu X, Guo FY (2014) Effects of background electrolytes and ionic strength on enrichment of Cd(II) ions with magnetic graphene oxide–supported sulfanilic acid. J Colloid Interface Sci 435:138–144

38. Huang Q, Chen Y, Yu H, Yan L, Zhang J, Wang B, Xing L (2018) Magnetic graphene oxide/MgAl-layered double hydroxide nanocomposite: one-pot solvothermal synthesis, adsorption performance and mechanisms for Pb^{2+}, Cd^{2+}, and Cu^{2+}. Chem Eng J 341:1–9
39. Huang Y, Fulton AN, Keller AA (2016) Simultaneous removal of PAHs and metal contaminants from water using magnetic nanoparticle adsorbents. Sci Total Environ 571:1029–1036
40. Islam A, Ahmad H, Zaidi N, Kumar S (2016) A graphene oxide decorated with triethylenetetramine-modified magnetite for separation of chromium species prior to their sequential speciation and determination via FAAS. Microchim Acta 183(1):289–296
41. Jan A, Azam M, Siddiqui K, Ali A, Choi I, Haq Q (2015) Heavy metals and human health: mechanistic insight into toxicity and counter defense system of antioxidants. Int J Mol Sci 16(12):29592–29630
42. Jiang L, Li S, Yu H, Zou Z, Hou X, Shen F, Yao X (2016) Amino and thiol modified magnetic multi-walled carbon nanotubes for the simultaneous removal of lead, zinc, and phenol from aqueous solutions. Appl Surf Sci 369:398–413
43. Jiang L, Yu H, Zhou X, Hou X, Zou Z, Li S, Yao X (2016) Preparation, characterization, and adsorption properties of magnetic multi-walled carbon nanotubes for simultaneous removal of lead(II) and zinc(II) from aqueous solutions. Desalin Water Treat 57(39):18446–18462
44. Jiang Y, Gong JL, Zeng GM, Ou XM, Chang YN, Deng CH, Huang SY (2016) Magnetic chitosan–graphene oxide composite for anti-microbial and dye removal applications. Int J Biol Macromol 82:702–710
45. Jiao T, Liu Y, Wu Y, Zhang Q, Yan X, Gao F, Li B (2015) Facile and scalable preparation of graphene oxide-based magnetic hybrids for fast and highly efficient removal of organic dyes. Sci Rep 5:12451
46. Kerkez Ö, Bayazit ŞS (2014) Magnetite decorated multi-walled carbon nanotubes for removal of toxic dyes from aqueous solutions. J Nanopart Res 16(6):2431
47. Khan TA, Nazir M, Khan EA (2016) Magnetically modified multiwalled carbon nanotubes for the adsorption of bismarck brown R and Cd(II) from aqueous solution: batch and column studies. Desalin Water Treat 57(41):19374–19390
48. Kim C, Eom J, Jung S, Ji T (2016) Detection of organic compounds in water by an optical absorbance method. Sensors 16(1):61
49. Kireeti KV, Chandrakanth G, Kadam MM, Jha N (2016) A sodium modified reduced graphene oxide–Fe_3O_4 nanocomposite for efficient lead (II) adsorption. RSC Adv 6(88):84825–84836
50. Lee CG, Kim SB (2016) Removal of arsenic and selenium from aqueous solutions using magnetic iron oxide nanoparticle/multi-walled carbon nanotube adsorbents. Desalin Water Treat 57(58):28323–28339
51. Lei Y, Chen F, Luo Y, Zhang L (2014) Three-dimensional magnetic graphene oxide foam/Fe_3O_4 nanocomposite as an efficient absorbent for Cr(VI) removal. J Mater Sci 49(12):4236–4245
52. Liang J, He B, Li P, Yu J, Zhao X, Wu H, Fan Q (2019) Facile construction of 3D magnetic graphene oxide hydrogel via incorporating assembly and chemical bubble and its application in arsenic remediation. Chem Eng J 358:552–563
53. Li H, Chi Z, Li J (2014) Covalent bonding synthesis of magnetic graphene oxide nanocomposites for Cr(III) removal. Desalin Water Treat 52(10–12):1937–1946
54. Li L, Fan L, Duan H, Wang X, Luo C (2014) Magnetically separable functionalized graphene oxide decorated with magnetic cyclodextrin as an excellent adsorbent for dye removal. RSC Adv 4(70):37114–37121
55. Li WN, Zhang HT, Dong YL, Li JC, Ma JJ (2014) Rapid magnetic solid phase extraction based on Fe_3O_4@SiO_2/graphene nanocomposite for the preconcentration of cadmium and its determination by flame atomic absorption spectrometry. J Chem Soc Pak 36(5):952–955
56. Li X, Wang S, Liu Y, Jiang L, Song B, Li M, Ding Y (2016) Adsorption of Cu(II), Pb(II), and Cd(II) ions from acidic aqueous solutions by diethylenetriaminepentaacetic acid-modified magnetic graphene oxide. J Chem Eng Data 62(1):407–416
57. Lin KYA, Lee WD (2016) Highly efficient removal of Malachite green from water by a magnetic reduced graphene oxide/zeolitic imidazolate framework self-assembled nanocomposite. Appl Surf Sci 361:114–121

58. Lin TY, Chen DH (2014) One-step green synthesis of arginine-capped iron oxide/reduced graphene oxide nanocomposite and its use for acid dye removal. RSC Adv 4(56):29357–29364
59. Liu J, Du H, Yuan S, He W, Liu Z (2015) Synthesis of thiol-functionalized magnetic graphene as adsorbent for Cd(II) removal from aqueous systems. J Environ Chem Eng 3(2):617–621
60. Liu M, Tao Z, Wang H, Zhao F, Sun Q (2016) Study on the adsorption of Hg(II) by one-pot synthesis of amino-functionalized graphene oxide decorated with a Fe_3O_4 microsphere nanocomposite. RSC Adv 6(88):84573–84586
61. Lü W, Wu Y, Chen J, Yang Y (2014) Facile preparation of graphene–Fe_3O_4 nanocomposites for extraction of dye from aqueous solution. CrystEngComm 16(4):609–615
62. Ma Y, Mu B, Zhang X, Xu H, Qu Z, Gao L, Tian J (2019) Ag-Fe_3O_4@ rGO ternary magnetic adsorbent for gaseous elemental mercury removal from coal-fired flue gas. Fuel 239:579–586
63. Madannejad S, Shemirani F, Fasih RN (2016) Facile synthesis of magnetic MWCNT functionalised 8-hydroxyquinoline: characterisation and application for selective enrichment of cadmium ions in food samples. Int J Environ Anal Chem 96(6):595–607
64. Mahmoudian MR, Alias Y, Basirun WJ, Woi PM, Sookhakian M, Jamali-Sheini F (2015) Synthesis and characterization of Fe_3O_4 rose like and spherical/reduced graphene oxide nanosheet composites for lead (II) sensor. Electrochim Acta 169:126–133
65. Manoochehri M, Naghibzadeh L (2017) A nanocomposite based on Dipyridylamine functionalized magnetic Multiwalled carbon nanotubes for separation and Preconcentration of toxic elements in black tea leaves and drinking water. Food Anal Methods 10(6):1777–1786
66. Martinez-Huitle CA, Rodrigo MA, Sires I, Scialdone O (2015) Single and coupled electrochemical processes and reactors for the abatement of organic water pollutants: a critical review. Chem Rev 115(24):13362–13407
67. Molaei K, Bagheri H, Asgharinezhad AA, Ebrahimzadeh H, Shamsipur M (2017) SiO_2-coated magnetic graphene oxide modified with polypyrrole–polythiophene: a novel and efficient nanocomposite for solid phase extraction of trace amounts of heavy metals. Talanta 167:607–616
68. Naghibzadeh L, Manoochehri M (2018) Determination of nickel and cadmium in fish, canned tuna, black tea, and human urine samples after extraction by a novel quinoline thioacetamide functionalized magnetite/graphene oxide nanocomposite. Carbon Lett 26:43–50
69. Namvari M, Namazi H (2014) Synthesis of magnetic citric-acid-functionalized graphene oxide and its application in the removal of methylene blue from contaminated water. Polym Int 63(10):1881–1888
70. Nodeh HR, Ibrahim WAW, Ali I, Sanagi MM (2016) Development of magnetic graphene oxide adsorbent for the removal and preconcentration of As(III) and As(V) species from environmental water samples. Environ Sci Pollut Res 23(10):9759–9773
71. Pan L, Zhai G, Yang X, Yu H, Cheng C (2019) Thermosensitive microgels-decorated magnetic graphene oxides for specific recognition and adsorption of Pb(II) from aqueous solution. ACS Omega 4(2):3933–3945
72. Park CM, Wang D, Han J, Heo J, Su C (2019) Evaluation of the colloidal stability and adsorption performance of reduced graphene oxide–elemental silver/magnetite nanohybrids for selected toxic heavy metals in aqueous solutions. Appl Surf Sci 471:8–17
73. Park WK, Yoon Y, Kim S, Yoo S, Do Y, Kang JW, Yang WS (2016) Feasible water flow filter with facilely functionalized Fe_3O_4-non-oxidative graphene/CNT composites for arsenic removal. J Environ Chem Eng 4(3):3246–3252
74. Parmar KR, Patel I, Basha S, Murthy ZVP (2014) Synthesis of acetone reduced graphene oxide/Fe_3O_4 composite through simple and efficient chemical reduction of exfoliated graphene oxide for removal of dye from aqueous solution. J Mater Sci 49(19):6772–6783
75. Patil SS, Shedbalkar UU, Truskewycz A, Chopade BA, Ball AS (2016) Nanoparticles for environmental clean-up: a review of potential risks and emerging solutions. Environ Technol Innov 5:10–21
76. Patra JK, Baek KH (2014) Green nanobiotechnology: factors affecting synthesis and characterization techniques. J Nanomater 2014:219

77. Paul B, Parashar V, Mishra A (2015) Graphene in the Fe_3O_4 nano-composite switching the negative influence of humic acid coating into an enhancing effect in the removal of arsenic from water. Environ Sci Water Res Technol 1(1):77–83
78. Peer FE, Bahramifar N, Younesi H (2018) Removal of Cd(II), Pb(II) and Cu(II) ions from aqueous solution by polyamidoamine dendrimer grafted magnetic graphene oxide nanosheets. J Taiwan Inst Chem Eng 87:225–240
79. Pourjavadi A, Nazari M, Hosseini SH (2015) Synthesis of magnetic graphene oxide-containing nanocomposite hydrogels for adsorption of crystal violet from aqueous solution. RSC Adv 5(41):32263–32271
80. Prasad C, Murthy PK, Krishna RH, Rao RS, Suneetha V, Venkateswarlu P (2017) Bio-inspired green synthesis of RGO/Fe_3O_4 magnetic nanoparticles using Murrayakoenigii leaves extract and its application for removal of Pb(II) from aqueous solution. J Environ Chem Eng 5(5):4374–4380
81. Pratt A (2014) Environmental applications of magnetic nanoparticles. Frontiers of Nanoscience, vol 6. Elsevier, Amsterdam, pp 259–307
82. Qi T, Huang C, Yan S, Li XJ, Pan SY (2015) Synthesis, characterization and adsorption properties of magnetite/reduced graphene oxide nanocomposites. Talanta 144:1116–1124
83. Rahimzadeh MR, Rahimzadeh MR, Kazemi S, Moghadamnia AA (2017) Cadmium toxicity and treatment: an update. Caspian J Intern Med 8(3):135
84. Ranjan B, Pillai S, Permaul K, Singh S (2019) Simultaneous removal of heavy metals and cyanate in a wastewater sample using immobilized cyanate hydratase on magnetic-multiwall carbon nanotubes. J Hazard Mater 363:73–80
85. Rashed MN (2013) Adsorption technique for the removal of organic pollutants from water and wastewater. In: Organic pollutants-monitoring, risk and treatment. IntechOpen. https://doi.org/10.5772/54048
86. Ravishankar H, Wang J, Shu L, Jegatheesan V (2016) Removal of Pb(II) ions using polymer based graphene oxide magnetic nano-sorbent. Process Saf Environ Prot 104:472–480
87. Rekhi H, Rani S, Sharma N, Malik AK (2017) A review on recent applications of high-performance liquid chromatography in metal determination and speciation analysis. Crit Rev Anal Chem 47(6):524–537
88. Ren HS, Cao ZF, Wen X, Wang S, Zhong H, Wu ZK (2019) Preparation of a novel nano-Fe_3O_4/triethanolamine/GO composites to enhance Pb^{2+}/Cu^{2+} ions removal. Environ Sci Pollut Res 26(10):10174–10187
89. Ricardo AIC, Sánchez-Cachero A, Jiménez-Moreno M, Bernardo FJG, Martín-Doimeadios RCR, Ríos Á (2018) Carbon nanotubes magnetic hybrid nanocomposites for a rapid and selective preconcentration and clean-up of mercury species in water samples. Talanta 179:442–447
90. Robinson T, McMullan G, Marchant R, Nigam P (2001) Remediation of dyes in textile effluent: a critical review on current treatment technologies with a proposed alternative. Bioresour Technol 77(3):247–255
91. Roy E, Patra S, Madhuri R, Sharma PK (2016) Europium doped magnetic graphene oxide-MWCNT nanohybrid for estimation and removal of arsenate and arsenite from real water samples. Chem Eng J 299:244–254
92. Sa'adi A, Es'haghi Z (2019) Azo-phenol ligand surface-active magnetic graphene oxide nanosheets as solid-phase adsorbents for extraction of cadmium in food samples. J Food Meas Charact 13(1):579–591
93. Saber-Samandari S, Saber-Samandari S, Joneidi-Yekta H, Mohseni M (2017) Adsorption of anionic and cationic dyes from aqueous solution using gelatin-based magnetic nanocomposite beads comprising carboxylic acid functionalized carbon nanotube. Chem Eng J 308:1133–1144
94. Sadegh H, Ali GA, Makhlouf ASH, Chong KF, Alharbi NS, Agarwal S, Gupta VK (2018) MWCNTs-Fe_3O_4 nanocomposite for Hg(II) high adsorption efficiency. J Mol Liq 258:345–353

95. Saha JC, Dikshit AK, Bandyopadhyay M, Saha KC (1999) A review of arsenic poisoning and its effects on human health. Crit Rev Environ Sci Technol 29(3):281–313
96. Sahraei R, Pour ZS, Ghaemy M (2017) Novel magnetic bio-sorbent hydrogel beads based on modified gum tragacanth/graphene oxide: removal of heavy metals and dyes from water. J Clean Prod 142:2973–2984
97. Sahu TK, Arora S, Banik A, Iyer PK, Qureshi M (2017) Efficient and rapid removal of environmental malignant arsenic(III) and industrial dyes using reusable, recoverable ternary iron oxide-ORMOSIL-reduced graphene oxide composite. ACS Sustain Chem Eng 5(7):5912–5921
98. Salam MA, El-Shishtawy RM, Obaid AY (2014) Synthesis of magnetic multi-walled carbon nanotubes/magnetite/chitin magnetic nanocomposite for the removal of Rose Bengal from real and model solution. J Ind Eng Chem 20(5):3559–3567
99. Santhosh C, Velmurugan V, Jacob G, Jeong SK, Grace AN, Bhatnagar A (2016) Role of nanomaterials in water treatment applications: a review. Chem Eng J 306:1116–1137
100. Saratale RG, Saratale GD, Chang JS, Govindwar SP (2011) Bacterial decolorization and degradation of azo dyes: a review. J Taiwan Inst Chem Eng 42(1):138–157
101. Saxena R, Madaan K, Bansal S, Saxena M, Sharma N (2018) A review on nanomaterials as solid phase extractants for determination of lead in environmental samples. IOSR J Appl Chem 11(9):27–38
102. Seidi S, Fotouhi M (2017) Magnetic dispersive solid phase extraction based on polythiophene modified magnetic graphene oxide for mercury determination in seafood followed by flow-injection cold vapor atomic absorption spectrometry. Anal Methods 9(5):803–813
103. Sharif A, Khorasani M, Shemirani F (2018) Nanocomposite bead (NCB) based on bio-polymer alginate caged magnetic graphene oxide synthesized for adsorption and preconcentration of Lead(II) and Copper(II) Ions from urine, saliva and water samples. J Inorg Organomet Polym Mater 28(6):2375–2387
104. Sheikhmohammadi A, Mohseni SM, Hashemzadeh B, Asgari E, Sharafkhani R, Sardar M, Almasiane M (2019) Fabrication of magnetic graphene oxide nanocomposites functionalized with a novel chelating ligand for the removal of Cr(VI): modeling, optimization, and adsorption studies. Desalin Water Treat 160:297–307
105. Sheshmani S, Ashori A, Hasanzadeh S (2014) Removal of Acid Orange 7 from aqueous solution using magnetic graphene/chitosan: a promising nano-adsorbent. Int J Biol Macromol 68:218–224
106. Shi H, Li W, Zhong L, Xu C (2014) Methylene blue adsorption from aqueous solution by magnetic cellulose/graphene oxide composite: equilibrium, kinetics, and thermodynamics. Ind Eng Chem Res 53(3):1108–1118
107. Song N, Wu XL, Zhong S, Lin H, Chen JR (2015) Biocompatible G-Fe_3O_4/CA nanocomposites for the removal of Methylene Blue. J Mol Liq 212:63–69
108. Silva S, Almeida AJ, Vale N (2019) Combination of cell-penetrating peptides with nanoparticles for therapeutic application: a review. Biomolecules 9(1):22
109. Su H, Ye Z, Hmidi N (2017) High-performance iron oxide–graphene oxide nanocomposite adsorbents for arsenic removal. Colloids Surf A Physicochem Eng Asp 522:161–172
110. Su S, Wu W, Gao J, Lu J, Fan C (2012) Nanomaterials-based sensors for applications in environmental monitoring. J Mater Chem 22(35):18101–18110
111. Sun X, Ou H, Miao C, Chen L (2015) Removal of Sudan dyes from aqueous solution by magnetic carbon nanotubes: equilibrium, kinetic and thermodynamic studies. J Ind Eng Chem 22:373–377
112. Taghizadeh M, Asgharinezhad AA, Samkhaniany N, Tadjarodi A, Abbaszadeh A, Pooladi M (2014) Solid phase extraction of heavy metal ions based on a novel functionalized magnetic multi-walled carbon nanotube composite with the aid of experimental design methodology. Microchim Acta 181(5–6):597–605
113. Tahir MU, Su X, Zhao M, Liao Y, Wu R, Chen D (2019) Preparation of hydroxypropyl-cyclodextrin-graphene/Fe_3O_4 and its adsorption properties for heavy metals. Surf Interfaces 16:43–49

114. Tan KB, Vakili M, Horri BA, Poh PE, Abdullah AZ, Salamatinia B (2015) Adsorption of dyes by nanomaterials: recent developments and adsorption mechanisms. Sep Purif Technol 150:229–242
115. Tarigh GD, Shemirani F, Maz'hari NS (2015) Fabrication of a reusable magnetic multi-walled carbon nanotube–TiO_2 nanocomposite by electrostatic adsorption: enhanced photodegradation of malachite green. RSC Adv 5(44):35070–35079
116. Thakur S, Karak N (2014) One-step approach to prepare magnetic iron oxide/reduced graphene oxide nanohybrid for efficient organic and inorganic pollutants removal. Mater Chem Phys 144(3):425–432
117. Tiwari S, Sharma N, Saxena R (2015) A review on application of hyphenated flow injection system for determination of heavy metal ions in water samples. Int J Adv Technol Eng Sci 3(1):692–702
118. Tiwari S, Sharma N, Saxena R (2017) Modified carbon nanotubes in online speciation of chromium in real water samples using hyphenated FI-FAAS. New J Chem 41(12):5034–5039
119. Verma AK, Dash RR, Bhunia P (2012) A review on chemical coagulation/flocculation technologies for removal of colour from textile wastewaters. J Environ Manage 93(1):154–168
120. Vu HC, Dwivedi AD, Le TT, Seo SH, Kim EJ, Chang YS (2017) Magnetite graphene oxide encapsulated in alginate beads for enhanced adsorption of Cr(VI) and As(V) from aqueous solutions: role of crosslinking metal cations in pH control. Chem Eng J 307:220–229
121. Wang D, Liu L, Jiang X, Yu J, Chen X (2015) Adsorption and removal of malachite green from aqueous solution using magnetic β-cyclodextrin-graphene oxide nanocomposites as adsorbents. Colloids Surf A Physicochem Eng Asp 466:166–173
122. Wang H, Na C (2014) Binder-free carbon nanotube electrode for electrochemical removal of chromium. ACS Appl Mater Interfaces 6(22):20309–20316
123. Wang P, Cao M, Wang C, Ao Y, Hou J, Qian J (2014) Kinetics and thermodynamics of adsorption of methylene blue by a magnetic graphene-carbon nanotube composite. Appl Surf Sci 290:116–124
124. Wang Y, Hu S (2016) Applications of carbon nanotubes and graphene for electrochemical sensing of environmental pollutants. J Nanosci Nanotechnol 16(8):7852–7872
125. Wang Y, Hu L, Zhang G, Yan T, Yan L, Wei Q, Du B (2017) Removal of Pb(II) and methylene blue from aqueous solution by magnetic hydroxyapatite-immobilized oxidized multi-walled carbon nanotubes. J Colloid Interface Sci 494:380–388
126. Wani AL, Ara A, Usmani JA (2015) Lead toxicity: a review. Interdiscip Toxicol 8(2):55–64
127. Warner MG, Warner CL, Addleman RS, Yantasee W (2009) Magnetic nanomaterials for environmental applications. In: Magnetic nanomaterials
128. World Health Organization (4th Edition) (2017) Guidelines for drinking-water quality, 4th edn. Incorporating the First Addendum, Geneva
129. Wu XL, Shi Y, Zhong S, Lin H, Chen JR (2016) Facile synthesis of Fe_3O_4-graphene@ mesoporous SiO_2 nanocomposites for efficient removal of Methylene Blue. Appl Surf Sci 378:80–86
130. Xiong S, Yang B, Cai D, Qiu G, Wu Z (2015) Individual and simultaneous stripping voltammetric and mutual interference analysis of Cd^{2+}, Pb^{2+} and Hg^{2+} with reduced graphene oxide-Fe_3O_4 nanocomposites. Electrochim Acta 185:52–61
131. Xu Z, Fan X, Ma Q, Tang B, Lu Z, Zhang J, Ye J (2019) A sensitive electrochemical sensor for simultaneous voltammetric sensing of cadmium and lead based on Fe_3O_4/multiwalled carbon nanotube/laser scribed graphene composites functionalized with chitosan modified electrode. Mater Chem Phys 238:121877
132. Yang S, Zeng T, Li Y, Liu J, Chen Q, Zhou J, Tang B (2015) Preparation of graphene oxide decorated Fe_3O_4@ SiO_2 nanocomposites with superior adsorption capacity and SERS detection for organic dyes. J Nanomater 16(1):337
133. Ye Y, Yin D, Wang B, Zhang Q (2015) Synthesis of three-dimensional Fe_3O_4/graphene aerogels for the removal of arsenic ions from water. J Nanomater 16(1):250
134. Yoon Y, Park WK, Hwang TM, Yoon DH, Yang WS, Kang JW (2016) Comparative evaluation of magnetite–graphene oxide and magnetite-reduced graphene oxide composite for As(III) and As(V) removal. J Hazard Mater 304:196–204

135. Yoon Y, Zheng M, Ahn YT, Park WK, Yang WS, Kang JW (2017) Synthesis of magnetite/non-oxidative graphene composites and their application for arsenic removal. Sep Purif Technol 178:40–48
136. Yu B, Zhang X, Xie J, Wu R, Liu X, Li H, Yang ST (2015) Magnetic graphene sponge for the removal of methylene blue. Appl Surf Sci 351:765–771
137. Yu F, Sun S, Ma J, Han S (2015) Enhanced removal performance of arsenate and arsenite by magnetic graphene oxide with high iron oxide loading. Phys Chem Chem Phys 17(6):4388–4397
138. Zhang F, Song Y, Song S, Zhang R, Hou W (2015) Synthesis of magnetite–graphene oxide-layered double hydroxide composites and applications for the removal of Pb(II) and 2, 4-dichlorophenoxyacetic acid from aqueous solutions. ACS Appl Mater Interfaces 7(13):7251–7263
139. Zhang J, Azam MS, Shi C, Huang J, Yan B, Liu Q, Zeng H (2015) Poly (acrylic acid) functionalized magnetic graphene oxide nanocomposite for removal of methylene blue. RSC Adv 5(41):32272–32282
140. Zhang J, Gong JL, Zenga GM, Ou XM, Jiang Y, Chang YN, Liu HY (2016) Simultaneous removal of humic acid/fulvic acid and lead from landfill leachate using magnetic graphene oxide. Appl Surf Sci 370:335–350
141. Zhao D, Gao X, Wu C, Xie R, Feng S, Chen C (2016) Facile preparation of amino functionalized graphene oxide decorated with Fe_3O_4 nanoparticles for the adsorption of Cr(VI). Appl Surf Sci 384:1–9
142. Zhao Y, Chen H, Li J, Chen C (2015) Hierarchical MWCNTs/Fe_3O_4/PANI magnetic composite as adsorbent for methyl orange removal. J Colloid Interface Sci 450:189–195
143. Zhou C, Zhang W, Wang H, Li H, Zhou J, Wang S, Zhou J (2014) Preparation of Fe_3O_4-embedded graphene oxide for removal of methylene blue. Arab J Sci Eng 39(9):6679–6685
144. Ziaei E, Mehdinia A, Jabbari A (2014) A novel hierarchical nanobiocomposite of graphene oxide–magnetic chitosan grafted with mercapto as a solid phase extraction sorbent for the determination of mercury ions in environmental water samples. Anal Chim Acta 850:49–56
145. Zulkifli SN, Rahim HA, Lau WJ (2018) Detection of contaminants in water supply: a review on state-of-the-art monitoring technologies and their applications. Sens Actuators B Chem 255:2657–2689

Volatile Organic Compounds Detection Using Carbon Nano Composites

Bhupinder Kumar, Vaneet Kumar, Saruchi, and Ashvinder Kumar Rana

Abstract Volatile organic compounds (VOCs) are the compounds carrying many adverse and ill effects to the environment as well as human beings. There are a variety of compounds that come under the name of these compounds (VOCs). Depending upon the nature of these compounds, VOCs can have many harmful effects on human health which may be short term or long term. Some health effects of these VOCs are throat infection, deep pain in head, coordination lose, a feeling of sickness and vomit, liver and kidney spoil, and damage of central nervous system. Some VOCs can also cause carcinogenic as well as toxic effect in animals and human being. Hence, it is the need of hour to remove such potentially harmful chemicals from our environment. In order to achieve this goal, various removal/detection techniques and applications have been discussed in this chapter. These techniques may be very helpful for reduction in toxic materials from the environment. An extensive literature survey has been done for completing this task.

Keywords VOCs · Carbon Nano Tuble · Composites · Environment

1 Introduction

For a healthy life, we human beings require a healthy and clean environment but as we develop in the field of science and technology, the problems associated with it are also growing day by day. Hence, in today's era, we are surrounded by very dangerous and harmful chemicals which in turn deteriorate our health day by day. There is a variety of the most affecting compounds present which can be collectively named as volatile organic compounds (VOCs). Various severities have been directly linked with these compounds like toxicity, mutagenicity, and carcinogenicity which

B. Kumar · V. Kumar (✉) · Saruchi
CT Group of Institutions, Shahpur Campus, Jalandhar, Punjab 144020, India
e-mail: vaneet2106@gmail.com

A. K. Rana
Sri Sai University, Palampur, India

M. Jawaid et al. (eds.), *Environmental Remediation Through Carbon Based Nano Composites*, Green Energy and Technology,
https://doi.org/10.1007/978-981-15-6699-8_6

can play a dominating role for being an unadorned hazard to human health and our ecosystem. VOCs are composed different chemicals which are simple as well complex and present in our environment as gases and vapors; these compounds can be classified as methyl alcohol, ethyl alcohol, isopropyl alcohol, butyl alcohol, acetone, toluene, xylene, and compounds containing halogens like 1,2-trichloroethylene, 1,2-dichlorobenzene, and chlorobenzene [1–10]. It is the need of the hour to reduce such harmful chemical from the environment to make our environment suitable for us. We can see the increasing concentrations of VOC in the environment because of modernization and industrialization. Sources of VOCs in our environment are human activities and natural emissions. There are many processes such as building manufacturing, industrialized processes, interior generation, and transportation from which a large concentration of VOCs has been generated [11]. One of the largest emitters of VOCs is petrochemical industries such as butane, propene, isobutene, alcohols, ketones, benzene, toluene, xylene, chlorinated compounds, and carcinogenic polycyclic aromatic hydrocarbons (PAHs) can also be generated in large amounts by industry as described by Yen and Horng [12]. The localities near such petrochemical plants are more prone to the adverse health effects caused by VOC emission. Wu et al. in [13] found adverse health effects caused by VOCs such as irritation in skin, bluish skin, violent behavior, environmental toxicity, cancer, failure of kidney, damage of liver, brain injure, and breathing problem. The presence of VOCs in atmosphere even at low concentration 1 ppm may cause ozone layer depletion, climate change, global warming, acid rain, and photochemical smog formation [14]. These effects do not only harm human beings and their health, but also harm biodiversity of environment, ecosystems such as marine life, crops, and vegetation [15].

In this chapter, we have used different carbon-based nano materials for removal of VOCs from the environment. It has become center of attention in growing new techniques to extract VOCs from interior and exteriors due to supple properties of nanocarbon-based materials. To protect our environment from the adverse effects, the removal of VOCs is therefore very important. Therefore, this chapter discussed various removal techniques and applications.

2 Nanocarbon-based composite materials

2.1 Volatile Organic Compounds Removal Using Carbon Nanofibers

Carbon nanofibers (CNFs) have attracted the interest of researchers of various fields due to their unique physicochemical and electrical properties [16–18]. Traditionally, CNFS have been used for the fabrication of polymer composites through dispersion technique [19]. But nowadays, researchers are also focusing their attention on the utility of CNFs in the adsorption of volatile organic compounds (VOC) such as alkanes, aromatics derivatives, and chlorinated hydrocarbons present in air. Since different VOC molecules have different molecular structure and polarity, so the adsorption capacity of CNFs might be affected by VOCs and its different properties. In general, the activated CNFS shows better properties than un-activated one.

CNFs are hydrophobic fibers with diameter ranging from 50 to 200 nm and have confined number of functional groups onto its surface. So, the surface of carbon nanofibers has been tailored by various researchers as per the applications demands. Bikshapathi et al. [20] have doped carbon nanofibers with Fe particles using different surfactants such as sodium dodecyl sulfate (SDS), tri-noctylphosphine (TOPO), and triton X-100, with an aim to adsorb carbon tetrachloride from atmospheric air. In this study, they reported that the prepared Fe-CNFSDS samples were very effective for adsorption of VOC.

Ahmed et al. [21] synthesized carbon nanofibers (CNFs) through nickel ion (Ni^{2+}) impregnation of oil palm kernel shell-based powdered activated carbon (PAC) by using acetylene as carbon source. For the successful dispersion of Ni^{2+} catalyst, they sonicated nickel (II) nitrate hexahydrate in acetone for the achievement of fruitful development of CNF. Various compositions of Ni^{2+} catalysts were examined and at 3% Ni^{2+}(w/w), and in this case they get the best growth of CNF. They have characterized samples by using spectroscopy techniques like FESEM and TEM and reported PAC-CNF graphitic structure, surface area of BET as 836.7 m^2/g, and zeta potential was −24.9 mV [21].

In order to increase the performance of activated carbon (AC) for detection of VOCs, their composites with carbon nanofiber (CNF) were prepared by Jahangiri et al. [22]. They firstly activated the carbon by impregnating with a nickel nitrate catalyst followed by deposition of carbon nanofibers (CNF) directly on the AC surface using catalytic chemical vapor deposition technique. Deposited CNFs were then activated by CO_2 to reclaim the surface area and micropores. Adsorption capacities of AC, AC/CNF, and CO_2 activated AC/CNF composites for the VOCs removal have been also evaluated and were found better in case of AC/CNF activated by CO_2 followed by AC and AC/CNF. A high adsorption capacity for activated carbon nanofibers (ACNFs) as compared to activated carbon fibers has been reported by Bai et al. [23, 24]. They synthesized the ACNFs by electro spinning of polyacrylonitrile solutions followed by steam activation process. SEM, X-ray spectroscopy, and N_2 adsorption at 77 K are the different technique to characterize ACNFs. The elevated adsorption capacity of ACNFs for VOCs is compared to activated carbon fibers (ACFs), because of slighter width and further specific available adsorption sites on the surface. Effect of CNFs surface modification onto adsorption of different VOCs such as n-decane, cyclohexane, benzene, toluene, dichloromethane (DCM), trichloroethylene (TCE), and tetrachloroethylene (TTCE) has also been evaluated by Cuervo et al. [25]. Capacity and enthalpy of adsorption were found out to be decreased after the oxidation of CNFs using HNO_3 as oxidizing agent.

Young-Wan and Gil-Young [26] have also evaluated the adsorption capacities of CNFs synthesized by electro spinning of poly acrylic nitrile (PAN) and cellulose acetate (CA) solution. The relative ratios of PAN and CA solution were varied as 9:1, 8:2, 7:3 (PAN:CA) by weight. They reported increase in adsorption performance of resulted CNFs with increase in CA contents up to 20%, which may be due to better pore size and after that decrease in adsorption performance. Furthermore, the CNFs prepared by using 20% (8:2 by wt.) of CA contents have been also found to have better adsorption capacities as compared to neat PAN and CNFs synthesized in the ratio of 9:1 and 7:3 (PAN:CA) by weight.

Lee et al. [27] have also synthesized ACNFs by first stabilization of PAN nanofiber in air from temperature ranging from 37 to 270 °C at heating rate of 0.5 °C/min, in order to construct infusible ladder form, followed by carbonization at 600 °C for 1 h in the presence of steam and He gas. The steam was added in the humidity range of 0–90% into the helium gas flow with an aim to produce ACNFs of controlled microporosity and sufficient nitrogen atom containing groups, which act as highly efficient adsorption sites. The ACNF thus synthesized has been found to have enhanced capacity in formaldehyde removal than the conventional ACF in humid atmosphere.

The amount of nitrogen contents as well as porosity on the adsorbent is the most decisive factor which affects the capacity of adsorption of formaldehyde. BET surface area onto ACNFs has been found to increase with the increase in relative humidity percentage during activation and more homogeneous micropores 94.7% were also found when relative humidity reached 90%.

The sensitive coating formed by the mixture of poly vinylidene fluoride (PVDF) and active carbon electro spun nanofiber was deposited by Zamarreno et al. onto gold-coated screen-printed PZT cantilevers for the detection of VOCs [28] The larger SSA of carbon nanofibers in the sensitive coating has resulted in maximized surface interaction and hence better diffusion of the VOCs molecules in sensitive layer. On exposure of sensitive coating to VOCs, the absorption and desorption processes induce shifts in the cantilever resonance frequency value and resonance shift was observed on exposure to acetone.

2.2 *Adsorption of Volatile Organic Compounds Using Graphene*

Graphene oxide is itself a hydrophilic material [29]; however, graphene on the other hand shows hydrophobic character and in aqueous solutions it undergoes agglomeration to form graphite via Van der Waals interactions [30]. The ability of graphene derivatives to undergo interactions with VOCs depends upon number of factors such as density of π electrons in each of the interacting species, their structure and geometry, and degree of covalent functionalization [31, 32]. Greater is the number of aromatic rings in the molecules of adsorbate, more will be $\pi - \pi$ interactions with graphene derivatives (adsorbents), and hence, adsorption will be better.

Also, graphene has an impressive theoretical specific surface area (SSA) of 2630 m^2g^{-1}, which makes it a potential candidate for adsorption of VOCs [33]. In general, larger the SSA, more will be number of adsorption sites, and hence, better will be adsorption of VOCs. The SSA can be easily retailored by different chemical reactions that will control the surface characteristics, such as the number of oxygen groups, and can be fabricated by using tape [34], chemical vapor deposition technique Reina et al. [35], hydrothermal processes Xu et al. [36], etc. Therefore, graphene has been considered to be a gifted adsorbent material due to its large SSA, availability of a large number of production techniques, and ease with which it can be easily modified. Various chemical modifications techniques have been adopted by different researchers for surface modification of graphene.

Lim et al. [37] have used both graphene oxide (GO) powder and thermally expanded graphene oxide powder (TEGP800, TEGP500, and TEGP200 samples), with a mesoporous structure, to adsorb VOCs such as toluene and xylene. Polypropylene filter was used for adsorption test, filled with adsorbents (0.25 g). The SSA of graphite oxide (GO) has been found to be increased significantly up to 542 m^2 g^{-1} after thermal expansion of GO, which is also accompanied by change in its chemical behavior from polar to nonpolar. Thermal expansion (Fig. 1) was carried out by placing the GO powder in furnace, composed of argon atmosphere using vacuum pump and argon bombe, for different temperatures 200, 500, and 800 °C at heating rate of 5 °C/min. Further, thermally expanded graphene powders at 800 °C (TEGP800) have been found to have a better adsorption capacity (Fig. 2) for toluene (92.7–98.3%) and xylene (96.7–98%), and its reusability is also remarkable, being at least 91%.

Liu et al. have used eight different amines (ethylamine hydrochloride, hexylamine, octylamine, benzylamine, 2-(4-chlorophenyl) ethylamine, 1-(2-aminoethyl) piperidine, tyramine, and 1,3-diaminopropane) for the functionalization of GO, and these reduced GO samples were subsequently utilized for the synthesis of electronic nose for the identification of VOCs. A linear response was observed by electronic nose against four cancer-related VOCs (ethanol, 2-ethylhexanol, nonanal, and ethylbenzene) with high sensitivity of 25 ppm.

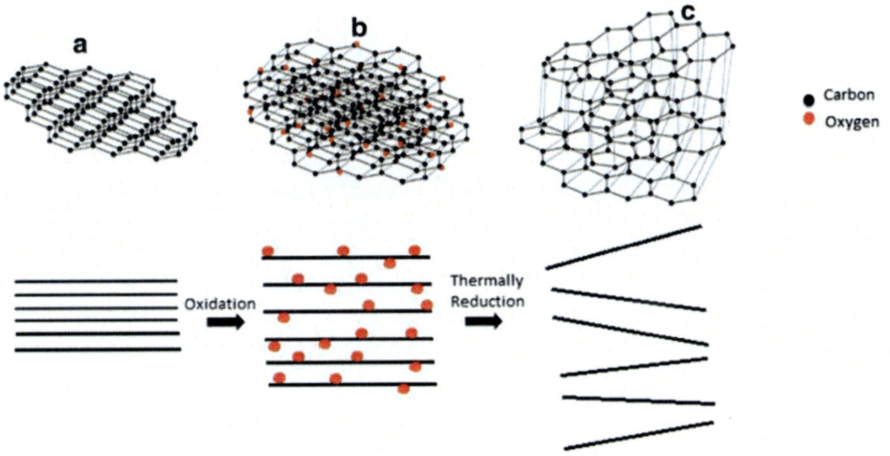

Fig. 1 Schematic representation of thermal expansion of **a** graphite powder, **b** GO powder, and **c** TEGP. Image was copied from research article published by Lim et al. [37] in Scientific Reports

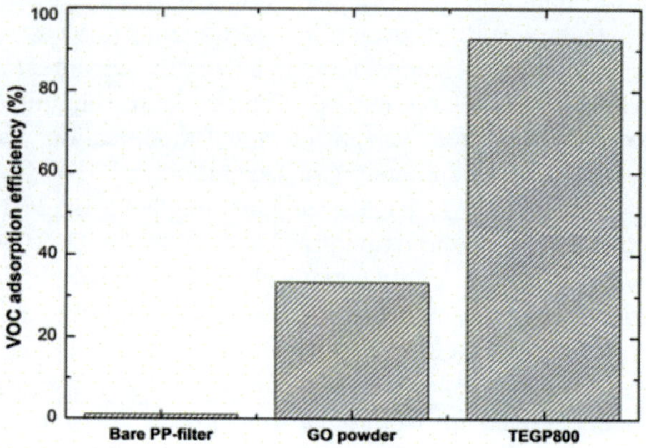

Fig. 2 Adsorption efficiency of bare PP filter, GO powder, and TEGP800 for toluene. Image was copied from research article published by Lim et al. [37] in Scientific Reports

Yu et al. [38] have also functionalized the GO with an aim to increase its absorption capacity toward VOCs. GO particles were synthesized in laboratory by modified Hummer's method and were subsequently reduced by adding a reducing agent, hydrazinium hydroxide (H_6N_2O) in colloidal solution of GO in water. Reduced graphene oxide (rGO) has been found to have higher capacity of adsorption and burst through times as compared to GO. The capacity of adsorption in the case of GO and rGO for both C_6H_6 and $C_6H_5CH_3$ was found to be 216.2 and 240.6 mg/g, and 276.4 and 304.4 mg/g, respectively. Further increase in SSA (236.4 m^2/g) in case of rGO after the reduction of GO has been also found. Nag et al. [39] have studied various types of rGO-based sensors and found that cyclodextrin functionalized rGO (CD-rGO) had a major nonpolar interaction with squat capability of disconnection. The attractions between functional groups on the CD-rGO nanocomposites and VOC can be associated with chemo-resistive response of amplitude. Further, CD-rGO has been found to have high sensitivity to detect the presence of VOC [40].

Some et al. [41] deposited GO and rGO flakes (which were obtained by exposing GO flakes to sunlight) on an optical fiber and monitored their reflectance after their exposure to different VOCs. Due to the hydrophilic and hydrophobic character of GO and rGO flakes, respectively, some VOCs have been found to adsorb better on GO and some on the rGO. Combining the GO and rGO responses, the so-fabricated sensor array has been found to a good detector to distinguish between tetrahydrofuran (THF) and dichloromethane (MC).

Four different porphyrins, i.e., TPP(NH2)4, CuTPP, ZnTPP, and CoTPP were used for functionalization of rGO film by spin coating method [42] and for the subsequent formation of multifunctional wearable sensing device for detection of eight different VOC biomarkers. The functionalized rGO array of sensor has shown a better result to vapors of VOC, and 08 various VOC biomarkers have been easily found and verified by using organized array sensor.

Bhai et al. have synthesized rGO/carbon composite ultrafine fibers (PCGF) by electro spinning of a mixture of phenolic resin and graphene oxide (GO) followed by carbonization under nitrogen or in argon/hydrogen atmosphere, and the samples were subsequently evaluated for their performance for VOC adsorption. The VOCs absorption capacity of rGO/carbon composite ultrafine fibers prepared under hydrogen atmosphere (PCGF-H) has been found to be higher than rGO/carbon composite ultrafine fibers prepared under nitrogen atmosphere (PCGF-N) and pristine fiber (PCF). The better the adsorption capacities in case of PCGF-H are because of the development of high SSA and consecutive network structure with ultrafine fiber diameter after the carbonization.

Pristine graphene after functionalization with TFQ (2,3,5,6,-Tetrafluorohydroquinone) organic molecules or amine groups and/or a thin layer of HfO_2 has been also found to be a better sensor for sensing formaldehyde as these functional groups work as chemical recognition links between graphene and formaldehyde [43]. In addition, there are a large number of graphene-based sensors which have been developed by researchers for detecting ethanol Meng et al. [44], dimethyl methylphosphonate (DMMP) [45], NO_2, SO_2 [46], and tumor markers [47].

Park et al. [48] found DMMP-based gas sensor with high flexibility and sensitivity. They fabricated it by coating polypyrrole onto graphene surface and reported a high degree of selectivity for this sensor regardless of the types of VOCs molecules (acetone, methanol, water, and tetradecane).

2.3 VOCs Removal Using Carbon Nanotube-Based Nanocarbon Materials

Liu et al. [49] have used biosensors based on single-walled carbon nanotube for detection of volatile organic compounds causing lung cancer. They have developed a highly selective SWNTs biosensor, coated with non-polymeric organic materials such as tricosane ($C_{23}H_{48}$) and pentadecane ($C_{15}H_{32}$). Their results showed a significant change in the resistance in SWNTs biosensors when tested with 1,2,4-trimethylbenzene. The results of their study present that tricosane functionalized SWNTs have shown marked sensitivity toward VOCs molecules.

Badhulika et al. [50] developed a single-walled nanotube-based conductive gas sensor polymer to detect VOCs. Their research includes production and thorough evaluation of poly (3,4-ethylene dioxythiophene) (PED) single-walled carbon nanotube (SWNTs) sensors doped with poly styrene sulfonic acid (PSS) for industrial interest. To check the existence of PED, PSS covered on SWNTs, their electrical characteristics were evaluated by cyclic voltammetry, resistance adjustment, and field-effect transistor measurement. The engineered sensor material exhibited good sensing properties for saturated vapors of volatile organic compounds (VOCs) such as methyl alcohol, ethyl alcohol, and methyl ethyl ketone (MEK) at room temperature over a wide range of VOCs concentrations, and detection limit of this sensor was found to be 1.3% in case of methyl alcohol, 5.95% for ethyl alcohol, and 3% in case of MEK. They observed that these hybrid sensors demonstrated greater flexibility in terms of sensing ability when compared with SWNTs.

Chatterjee et al. [51] used carbon nanotube sensors sprayed with surfactants for detecting eleven lung cancer biomarkers. The group observed that they can modify the selectivity and efficiency of CNTs by changing the nature of the surfactants. The efficiency of surfactant–CNT sensors, hierarchically arranged by spray layer by layer with $C_{24}H_{39}NaO_4$ (DOC), $C_{18}H_{29}NaO_3S$(SDBS), $CH_3(CH_2)_{15}N(Br)(CH_3)_3$(CTAB), $C_{27}H_{42}NO_2Cl$ (BnzlkCl), and triton x-405 (TX405) was found to rely on the contacts between the surfactants and the analytes, on their supramolecular alignment with CNT, but also on the initial resistance R_0 that can be adjusted by CNT content, surfactant: CNT ratio or the surfactant concentration over its CMC. Sensors CNT-DOC has been shown to be responsive to methanol and other alcohols but also water. Chemicals like benzene, n-pentane, and chloroform were found to be prone to TX405-CNT sensors. With the exception of isopropanol, SDBS-CNT sensors may identify ammonia, acetone, chloroform, and water but not many biomarkers. N-pentane, isoprene, acetone, and ethanol were sensitive to the BnzlkCl-CNT sensors. CTAB-CNT sensors were mildly sensitive to most VOCs but displayed no extreme selectivity, whereas pristine CNT sensors were found to be good at detecting most of the set's aromatic VOCs.

The impact of the analyte's physicochemical properties on polymethylmethacrylate selectivity: functionalized composite sensor multiwalled carbon nanotubes (PMMA: f-MWCNTs) was shown by Kaur et al. [52]. They developed composite sensor and optimized their selectivity toward different organic VOCs followed by studying different parameters such as response time, possible reaction, and recovery time. In this study, they observed that the selectivity is due to the electronic and structural properties of both the reacting species, i.e., the organic vapors and sensing material. Discrimination against different organic vapors has been found to be dependent upon the analyte's adsorption capacity and their chemical properties. Further, the sensor was found to be extremely selective for CH_3OH vapor because the molecular size and electro negativity are very good.

Li et al. [53] used multiwall carbon nanotubes (MWCNTs) as a gas/solid partitioning adsorbent of chosen volatile organic compounds (VOCs). In this analysis, they tested 15 VOCs utilizing inverse gas chromatography to evaluate their fluid/solid partitioning coefficient (Log Kd) at various relative humidity (RH) levels. They analyzed the interactions between MWCNTs and VOCs by reversing the observed Log Kd with the linear energy solvation relation (LSER). The results show that because of the electron pair interactions and hydrogen bond acidity, the MWCNT carbonyl and carboxyl groups provide the VOCs with strong adsorption power. The resulted LSER equations give good value of Log Kd. This technique for VOC removal and prediction of VOC pollutant by MWCNT adsorption is very essential from environmental point of view.

Ghanbarian et al. [54] developed a new nanocomposite-based resistive sensor MIL-53(Cr–Fe)/Ag/CNT for detecting volatile organic compounds. They synthesized, by a sono-chemical method, MIL-53(Cr–Fe) nanoparticles as a bimetallic type of organic metal frameworks (MOFs). Such nanoparticles were used to create a ternary nanocomposite MIL-53(Cr–Fe)/Ag/CNT to develop a resistive gas sensor system to track volatile organic compounds such as methyl alcohol, ethyl alcohol, and isopropyl alcohol under environmental conditions (10% relative humidity and 25 °C). This ternary nanocomposite shows high reaction, precisely for methyl alcohol, to polar VOCs. This scientist used different spectroscopy technique like XRD, SEM, TEM, FTIR, and specific surface area measurement to identify the nanocomposites.

3 Conclusion and Forthcoming Prospects

The interior and exterior air quality is highly affected by harmful pollutants for instance VOCs, and the chief cause is burning of coal, oil, bio fuels as well as the compounds used in our day to day life such as petroleum products, paints, adhesives, preservatives, cosmetics, antiseptics, pharmaceuticals, decompositions of bio-waste, perfumes, and synthetic resins. Inhalation and exposure of these chemicals in minor or major quantities play a significant role in human health. Because of their poisonous, carcinogenic, or mutagenic characteristics, they are regarded toxic pollutants. Over the past few centuries, various techniques for the elimination of VOCs have been created. Due to some disadvantages such as low effectiveness, high energy consumption, or the manufacturing of serious toxic by-products, many of these techniques are not appropriate at the business stage. Adsorption is one of the most effective and easy techniques of removing VOCs in practical apps among all these techniques. The selection of adsorbents is the most important aspect for the adsorption of these volatile organic compounds from different sources. The innovative VOC adsorbents as nano carbon-based products (e.g., carbon nanotubes, graphene, nanofibers, etc.) have an outstanding ability to extract contaminants from the atmosphere owing to their big surface area, elevated strength, strong permeable composition, and comparatively small price. The primary focus in this section was the use of nanocarbon-based materials as effective adsorbents to remove different VOCs from the atmosphere. Nowadays, more attention is being paid to nanocarbon materials as adsorbent or sorbent materials throughout the globe. Nanocarbon products are commonly used to remove and purify natural samples as the most successful, cheapest, and most efficient components. However, to enhance the effectiveness of extraction, more study should be held out. In addition, nanocarbon-based nanomaterials not only extract contaminants from the atmosphere from VOCs, nevertheless also demonstrate distinct applications in multiple areas (e.g., energy storage, biomedical studies, electrochemical capacitors, detectors, catalysts, fuel cells, solar cells, electronics, filters, etc.). Consequently, ongoing commitment to this sector continues imperatively based on the multifunctional implementation of nanocarbon-based nanomaterials.

Acknowledgements Author Bhupinder Kumar and Vaneet Kumar are thankful to CT group of Institutions, Shahpur, Jalandhar, providing facilities to completing this task.

References

1. Hewitt CN (1999) Reactive hydrocarbons in the atmosphere. Academic Press, San Diego, CA
2. Li Y, Lee C, Gullett B (2002) The effect of activated carbon surface moisture on low temperature mercury adsorption. Carbon 40:65–72
3. Lovelock JE (1975) Natural halocarbons in the air and in the sea. Nat 256:193–194
4. Muller JF (1992) Geographical distribution and seasonal variation of surface emissions and deposition velocities of atmospheric trace gases. J Geophys Res Atmos 97:3787–3804
5. McInnes G (1996) Atmospheric emission inventory guidebook: a joint emep/corinair production. EC
6. Nemecek-Marshall M, Wojciechowski C, Kuzma J, Silver GM, Fall R (1995) Marine vibrio species produce the volatile organic compound acetone. Appl Environ Microbiol 61:44–47
7. Qiu K, Yang L, Lin J, Wang P, Yang Y, Ye D (2014) Historical industrial emissions of non-methane volatile organic compounds in China for the period of 1980–2010. Atmos Environ 86:102–112
8. Singh H, Chen Y, Staudt A, Jacob D, Blake D, Heikes B (2001) Evidence from the pacific troposphere for large global sources of oxygenated organic compounds. Nat 410:1078–1081
9. Singh HB, Tabazadeh A, Evans MJ, Field BD, Jacob DJ, Sachse G (2003) Oxygenated volatile organic chemicals in the oceans: inferences and implications based on atmospheric observations and air-sea exchange models. Geophys Res Lett 30:1862
10. Wang H, Nie L, Li J, Wang Y, Wang G, Wang J (2013) Characterization and assessment of volatile organic compounds (VOCs) emissions from typical industries. Chin Sci Bull 58:724–730
11. Massolo L, Rehwagen M, Porta A, Ronco A, Herbarth O, Mueller A (2010) Indoor-outdoor distribution and risk assessment of volatile organic compounds in the atmosphere of industrial and urban areas. Environ Toxicol 25:339–349
12. Yen CH, Horng JJ (2009) Volatile organic compounds (VOCs) emission characteristics and control strategies for a petrochemical industrial area in middle Taiwan. J Environ Sci Health A 44:1424–1429
13. Wu XM, Fan ZT, Zhu X, Jung KH, Ohman-Strickland P, Weisel CP, Lioy PJ (2012) Exposures to volatile organic compounds (VOCs) and associated health risks of socio-economically disadvantaged population in a "hot spot" in Camden. New Jersey. Atmos Environ 57:72–79
14. Mohan N, Kannan GK, Upendra S, Subha R, Kumar NS (2009) Breakthrough of toluene vapours in granular activated carbon filled packed bed reactor. J Hazard Mater 168:777–781
15. Mohamed EF, El-Hashemy MA, Abdel-Latif NM, Shetaya WH (2015) Production of sugarcane bagasse-based activated carbon for formaldehyde gas removal from potted plants exposure chamber. J Air Waste Manage Assoc 65:1413–1420
16. Chatterjee A, Deopura BL (2002) Carbon nanotubes and nanofibre: an overview. Fibers Polym 3:134–139
17. Inagaki M, Radovic LR (2002) Nanocarbons. Carbon 40:2279–2282
18. Song H, Shen W (2014) Carbon nanofibers: synthesis and applications. J Nanosci Nanotechnol 14:1799–1810
19. Hammel E, Tang X, Trampert M (2004) Carbon nanofibers for composite applications. Carbon 42:1153–1158
20. Bikshapathia M, Singha S, Bhaduria B, Mathura GN, Sharma A, Verma N (2012) Fe-nanoparticles dispersed carbon micro and nanofibers: Surfactant-mediated preparation and application to the removal of gaseous VOCs. Colloids Surf A Physicochem Eng Aspects 399:46–55
21. Ahmed YM, Al-Mamun A, Jameel AT, AlKhatib MFR, Amosa MK, AlSaadi MA (2016) Synthesis and characterization of carbon nanofibers grown on powdered activated carbon. J Nanotechnol 2016:1–10
22. Jahangiri M, Adl J, Shahtaheri SJ, Rashidi A, Ghorbanali A, Kakooe H, Forushani AR, Ganjal MR (2013) Preparation of a new adsorbent from activated resin-based carbon composite ultra-fine fibers and their adsorption performance for volatile organic compounds and water. J Mater Chem A 1:9536–9543

23. Bai Y, Huang ZH, Kang F (2013) Synthesis of reduced graphene oxide/phenolic carbon and carbon nanofiber (AC/CNF) for manufacturing organic-vacbpour respirator cartridge. Iran J Environ Health Sci Eng 10:10–15
24. Bai Y, Huang ZH, Wang MX, Kang F (2013) Adsorption of benzene and ethanol on activated carbon nanofibers prepared by electrospinning. Adsorpt 19:1035–1043
25. Cuervo MR, Asedegbega-Nieto E, Dıaz E, Vega A, Ordonez S, Lopez EC, Rodrıguez-Ramos I (2008) Effect of carbon nanofiber functionalization on the adsorption properties of volatile organic compounds. J Chromatogr A 1188:264–273
26. Young-Wan J, Gil-Young O (2017) Behavior of toluene adsorption on activated carbon nanofibers prepared by electrospinning of a polyacrylonitrile-cellulose acetate blending solution. Korean J Chem Eng 34:2731–2737
27. Lee KJ, Shiratori N, Lee GH, Miyawaki J, Mochida I, Yoon S, Jang J (2010) Activated carbon nanofiber produced from electrospun polyacrylonitrile nanofiber as a highly efficient formaldehyde adsorbent. Carbon 48:4248–4255
28. Zamarreno CR, Arregui FJ, Garcia1 YR, Vazquez R, Llobet E, Lakhmi R, Clement P, Debéda H (2016) Electrospun nanofibers fabricated on screen-printed PZT energy relationship. J Hazard Mater 315:35–41
29. Saleem H, Haneef M, Abbasi HY (2018) Synthesis route of reduced graphene oxide via thermal reduction of chemically exfoliated graphene oxide. Mater Chem Phys 204:1–7
30. Ren X, Li J, Chen C, Gao Y, Chen D, Sue M, Alsaedi A, Hayat T (2018) Graphene analogues in the aquatic environment and porous media: Dispersion, aggregation, deposition and transformation. Environ Sci: Nano 5:1298–1340
31. Bottari G, Herranz MA, Wibmer L, Volland M, Rodríguez-Pérez L, Guldi DM, Hirsch A, Martín N, D'Souza F, Torres T (2017) Chemical functionalization and characterization of graphenebased materials. Chem Soc Rev 46:4464–4500
32. Georgakilas V, Otyepka M, Bourlinos AB, Chandra V, Kim N, Kemp KC, Hobza P, Zboril R, Kim KS (2012) Functionalization of graphene: covalent and non-covalent approaches, derivatives and applications. Chem Rev 112:6156–6214
33. Stoller MD, Park S, Zhu Y, An J, Ruoff RS (2008) Graphene-based ultracapacitors. Nano Lett 8:3498–3502
34. Compton OC, Nguyen ST (2010) Graphene oxide, highly reduced graphene oxide, and graphene: versatile building blocks for carbon-based materials. Small 6:711–723
35. Reina A, Jia X, Ho J, Nezich D, Son H, Bulovic V, Dresselhaus MS, Kong J (2008) Large area, few-layer graphene films on arbitrary substrates by chemical vapor deposition. Nano Lett 9:30–35
36. Xu Y, Sheng K, Li C, Shi G (2010) Self-assembled graphene hydrogel via a one-step hydrothermal process. ACS Nano 4:4324–4330
37. Lim ST, Kim JH, Lee CY, Koo S, Jerng DW, Wongwises S, Ahn HS (2019) Mesoporous graphene adsorbents for the removal of toluene and xylene at various concentrations and its reusability. Sci Rep 9:10922
38. Yu L, Wang L, Xu W, Chen L, Fu M, Wu J, Ye D (2018) Adsorption of VOCs on reduced graphene oxide. J Environ Sci 67:171–178
39. Nag S, Duarte L, Bertrand E, Celton V, Castro M, Choudhary V, Guegan P, Feller JF (2014) Ultrasensitive QRS made by supramolecular assembly of functionalized cyclodextrins and graphene for the detection of lung cancer VOC biomarkers. J Mater Chem B 2:6571–6579
40. Kumar B, Castro M, Feller JF (2012) Materials self assembly and fabrication in confined spaces. J Mater Chem 22:10382–10405
41. Some S, Xu Y, Kim Y, Yoon Y, Qin H, Kulkarni A, Kim T, Lee H (2013) Highly sensitive and selective gas sensor using hydrophilic and hydrophobic graphenes. Sci Rep 3:1868
42. Xu H, Xiang JX, Lu YF, Zhang MK, Li JJ, Gao B, Zhao Y, Gu Z (2018) Multifunctional wearable sensing devices based on functionalized graphene films for simultaneous monitoring of physiological signals and VOC biomarkers. ACS Appl Mater Interfaces 10:11785–11793
43. Tang X, Mager N, Vanhorenbeke B, Hermans S, Raskin JP (2017) Defect-free functionalized graphene sensor for formaldehyde detection. Nanotechnol 28:1–11

44. Meng F, Zheng H, Chang Y, Zhao Y, Li M, Wang C et al (2018) One-step synthesis of Au/SnO2/RGO nanocomposites and their VOC sensing properties. IEEE Trans Nanotechnol 17:212–219
45. Park J, Choi S, Janardhan AH, Lee S-Y, Raut S, Soares J et al (2016) Electromechanical cardioplasty using a wrapped elasto-conductive epicardial mesh. Sci Transl Med 8:344ra86–344ra86
46. Cui H, Zheng K, Zhang Y, Ye H, Chen X (2018) Superior selectivity and sensitivity of C_3N sensor in probing toxic gases NO_2 and SO_2. IEEE Electron Device Lett 39:284–287
47. Barash O, Zhang W, Halpern JM, Hua Q-L, Pan Y-Y, Kayal H et al (2015) Differentiation between genetic mutations of breast cancer by breath volatolomics. Oncotarget 6:1–13
48. Park J, Kim J, Kim K, Kim S-Y, Cheong WH, Park K et al (2016) Wearable, wireless gas sensors using highly stretchable and transparent structures of nanowires and graphene. Nanoscale 8:10591–10597
49. Liu FL, Xiao P, Fang HL, Dai HF, Qiao L, Zhang YH (2011) Single-walled carbon nanotube-based biosensors for the detection of volatile organic compounds of lung cancer. Phys E 44:367–372
50. Badhulika S, Myung NV, Mulchandani A (2014) Conducting polymer coated single-walled carbon nanotube gas sensors for the detection of volatile organic compounds. Talanta 123:109–114
51. Chatterjee S, Castro M, Feller JF (2015) Tailoring selectivity of sprayed carbon nanotube sensors (CNT) towards volatile organic compounds (VOC) with surfactants. Sens Actuators B 220:840–849
52. Kaur A, Singh I, Kumar A, Rao PK, Bhatnagar PK (2016) Effect of physicochemical properties of analyte on the selectivity of polymethylmethacrylate: carbon nanotube based composite sensor for detection of volatile organic compounds. Mater Sci Semicond Process 41:26–31
53. Li M-S, Wu SC, Shih Y-H (2016) Characterization of volatile organic compound adsorption on multi wall carbon nanotubes under different levels of relative humidity using linear solvation energy relationship. J Hazard Mater 315:35–41
54. Ghanbarian M, Zeinali S, Mostafavi A (2018) A novel MIL-53(Cr-Fe)/Ag/CNT nanocomposite based resistive sensorfor sensing of volatile organic compounds. Sens Actuators B 267:381–391

Nanocomposites Materials as Environmental Cleaning

Kirtanjot Kaur, Vaneet Kumar, and Saruchi

Abstract Remediation of the environment with the help of nanocomposites is a thrust area. Nanocomposites are high-performance materials with an extensive range of applications not only in engineering, plastics, elastomers, pest detection control, and agricultural productivity but also in the remediation of the environment from various hazardous pollutants. This chapter will provide an insight into the various types of nanocomposites, its composition, and its application in the removal of specific contaminants. A good attempt has been made to provide brief of various types of nanocomposites used for remediation of soil, groundwater, and air. Details of different technologies that are commonly employed for this are adsorption, absorption, chemical reactions, filtration, and photocatalysis will be elaborated. The chapter entails an overview of the treatment of pollutants like heavy metals, dyes, chlorinated organic compounds, halogenated herbicides, organophosphorus compounds with a smarter and greener approach using nanocomposites as compared to conventional methods, their applications in building, slow release of fertilizers and medicinal applications highlighting their role in the remediation of the environment.

Keywords Nanocomposites · Types · Environment remediation · Graphene-based nanomaterials · Silica nanomaterials · Polymer-based nanomaterials · Building and environment · Slow-release of fertilizers

K. Kaur · Saruchi
Department of Chemistry, Sri Guru Gobind Singh College, Sector 26, Chandigarh 160019, India

V. Kumar (✉)
Department of Applied Sciences, CT Group of Institutions, Shahpur Campus, Jalandhar, Punjab 144020, India
e-mail: vaneet2106@gmail.com

Department of Biotechnology, CT Group of Institutions, Shahpur Campus, Jalandhar, Punjab 144020, India

M. Jawaid et al. (eds.), *Environmental Remediation Through Carbon Based Nano Composites*, Green Energy and Technology,
https://doi.org/10.1007/978-981-15-6699-8_7

1 Introduction

Nanocomposites can be defined as solid material having multiple phases in which one of the phases should be in the nanometer range or structure that will have a repeat distance of nanoscale [9, 48]. The ideal size of less than 5 nm makes them a suitable material for catalytic activity and of 100 nm for achieving superparamagnetism and mechanical strengthening. They are the molecules with high potential of applications in distinct fields because of their small size and surface area is very high [21]. Moreover, their properties can be tailored according to the requirement but it is not as easy as it seems there are many challenges too like they are inherently unstable under normal conditions, agglomeration, cost-effectiveness, nontoxicity, and biodegradability. So, different approaches are designed to meet these difficulties and make them appropriate material for target pollutants [14]. Recent advances in the field of nanocomposites have opened new opportunities in all zones of the industry but remarkably significant in the zone of the environment which is the concern of today [2]. Before going into details of how nanocomposites can be used for environmental remediation there is utmost need to understand the basic composition of different types of nanocomposites, its properties, highlighting their need followed by its potential applications as environment cleaner.

2 Types of Nanocomposites

2.1 Ceramic Matrix Composites

These are composites formed by inserting ceramic matrix in ceramic fibre. The fibre used here is carbon fibre and the ceramic is from the group of borides, oxides, silicides, and nitrides. However, metal is used as the second component [25]. Preparation of this involves dispersal of both components by vapour techniques and chemical methods. In chemical method, sol-gel process is most commonly employed. Colloidal, template synthesis, polymer precursor route and Spray pyrolysis are other methods that are used to prepare ceramic matrix composites. These matrix composites in turn have better crack resistance, corrosion-resistance, fracture toughness, electrical and magnetic properties than conventional technical ceramics like zirconia, aluminium nitrides, etc. [62].

2.1.1 Ceramic Matrix-Discontinuous Reinforcement Nanocomposite Systems

This type of nanocomposite system comprises the introduction of different reinforcements and also the phases at the phase boundaries of ceramics which leads to enhancement in hardness and fracture toughness properties of ceramics [3, 31]. The mechanical properties and strength of ceramics is an important aspect. It has been observed that when a metal phase is introduced into matrix its mechanical properties can be changed this is the reason that in ceramic matrix nanocomposites of Al_2O_3 and Fe_2O_3 distribution of Co and Ni can be seen. Even by incorporating another ceramic can change the properties like by adding 10% SiC in Si_3N_4 increases its strength to such an extent that it does not fail even after 1000 h at strain of 1.5%. This result is much more improved because Si_3N_4 alone fails after 0.4 h at strain of 0.3% [37]. Further modification has been done where the preparation of advanced nanocomposites with high toughness and superior characteristics has been prepared as compared to the conventional ceramic materials which lead to the sudden failure properties [35].

2.1.2 Ceramic Matrix-Carbon Nanotube Systems

By incorporating carbon nanotubes in the ceramic matrix, the mechanical properties of matrix can be enhanced like there is an increase in fracture toughness of 194% over pure alumina when single-walled carbon nanotubes are applied as a reinforcement of ceramic composites with the help of technique called spark-plasma sintering. On the other hand, 24% increase in fracture toughness over the matrix was observed in nanograined Al_2O_3 composite when 10 vol.% multi-walled carbon nanotubes were introduced. This theory of short fibre-reinforced composites comprises of this observation that there is increase in mechanical properties like bending strength and fracture toughness when the volume content of carbon nanotubes is less than 5 vol.% and above this much concentration it tends to decrease. This decrease is probably due to agglomeration [24, 51].

2.2 *Metal Matrix Nanocomposites (MMNC)*

MMNC is also called a reinforced metal matrix composites. These composites are classified as Continuous and Non-continuous reinforced materials. Metal-metal nanocomposites acquire strong resistance to growth of grains and their thermal stability. Actually, new properties can be obtained by simple mixing of two different metal nanocomposites. Various parameters like microstructural, compositional as well as porosity, impurity, distribution of grain size and texture are also taken into account during the formation of this type of nanocomposites. The techniques which are being used for the preparation of these type of nanocomposites are Spray pyrolysis, rapid solidification, Vapour techniques, electrodeposition, colloidal and sol-gel processes. These nanocomposites shows manyfold increase in hardness and Young's modulus. For instance, Al/Pb nanocomposites exhibit much improved frictional features. Another application can be seen in rocket propellants which are prepared from a polymer of Al/Al_2O_3 nanocomposite and they showed improvement in the ballistic performance [22].

2.3 Carbon Nanotube Metal Matrix Composites (CNT-MMC)

CNT-MMC which is a coming up new material has the superior characteristics of high tensile strength and electrical conductivity of carbon nanotube materials. To achieve this, the techniques for their synthesis should be such so as to provide homogeneous dispersion of nanotubes in the matrix of metal. This leads to strong interfacial adhesion between both and also it is economically producible. Two different ways are in situ preparation which helps in improving dispersion in the case of carbon nanotube-reinforced polymer composites and other is ex-situ techniques where alignment of carbon nanotubes can be achieved easily [39]. Other areas of research apart from carbon nanotube metal matrix composites are boron nitride reinforced metal matrix composites (BN-MMC) and carbon nitride metal matrix composites [12].

2.4 Polymer Matrix Nanocomposites

These types of nanocomposites can be obtained by adding nanoparticles to a polymer matrix that can lead to enhancement in the properties depending on the nature of the nanoscale filler. This is the reason that these materials are termed as nano filled polymer matrix composites [44]. Nanoparticles that are used for this purpose are graphene carbon nanotubes or molybdenum and tungsten disulphide. Even very low concentrations of nanocomposites addition can lead to significant improvement in the bending strength as well as compressive properties of polymeric nanocomposites as observed by Lalwani et al. [27]. They are basically strengthening agents for the formation of strong biodegradable polymeric nanocomposites which has got an application in bone tissue engineering.

2.5 Magnetic Nanocomposites

It comprises two components one inorganic magnetic component in the form of fibres or particles which is embedded in an organic polymer and at least one dimension should be in nanometer range. The most commonly used methods for synthesizing magnetic NPs are in situ and ex-situ preparation. In situ method involves the coprecipitation of Fe^{2+} and Fe^{3+} ions by a base or thermal decomposition of metal precursors such as metal carbonyls ($Co_2(CO)_8$, $Fe(CO)_5$, $Ni(CO)_4$) and metal oleates. Ex-situ method involves the blending of polymer with pre-synthesized nanostructures by special techniques like ball milling, thermal curing and also melt blending. One such example is the preparation of superparamagnetic nickel ferrite/polypropylene nanocomposites with ball milling process by the incorporation of the previously synthesized nickel ferrite NPs. But the in situ method has the advantage of particle distribution which is easier to manipulate by this method. There are a number of applications of magnetic nanocomposites in a catalytic, medical, and technical field. Instead of using palladium alone magnetic nanoparticle supported palladium complexes can be used in catalysis so that the efficiency of the palladium can be increased in the reaction. Though another applications of magnetic nanocomposites can also be seen in electronics especially in sensors and also in wastewater treatment. The carbon-coated magnetic NPs and graphene-coated magnetic NP have been widely used for Cr (VI) removal from wastewater [63].

2.6 *Polymer Nanocomposites with Layered Reinforcements*

A wide range of nanoparticles including ceramic, polymeric, metal oxide and carbon-based nanomaterials are introduced within the polymeric network to obtain desired property combinations. There is a special and distinctive interaction between polymer and nanocomposites. This is the reason why the range of property combinations can be tailored so as to imitate vernacular tissue structure and their properties. Therefore, a wide range of natural as well as synthetic polymers such as starch, cellulose, alginate, chitosan, collagen, gelatin, fibrin and PVA, PEG, poly(caprolactone) (PCL), poly(lactic-co-glycolic acid) (PLGA), and poly(glycerol sebacate) (PGS) are used to make polymeric nanocomposites for biomedical applications such as tissue engineering, target drug delivery, cellular therapies [13].

2.7 *Methods of Characterizations*

The characterization of nanocomposites can be done by various spectroscopic techniques like AFM, STM, FTIR, XPS, NMR, DSC, SEM/TEM, and XRD.

3 Environment Remediation

There are some of the harmful pollutants which are very complex and cannot be degraded easily due to their high volatility and low reactivity. Other than the conventional methods used earlier, now the use of the nanocomposites, its different forms, and its uses are the current areas of research for the remediation of environment [53]. There are many aspects of environmental pollution which are proving very fatal to the health of living beings and many of them are not easily detected and degraded. Recent studies focus on the use of nanocomposites as the remediation for soil, air, and water.

3.1 Conventional Methods of Environment Remediation

Remediation of heavy metal-contaminated soil, water, and air is the need of the hour. Conventional methods of remediation with the newer methods are listed in Fig. 1. Conventional methods have been in use for decades which were very good but have their own limitations. One of the limitations is they release the toxic material as by-product and secondly the biological process is very slow and time-consuming.

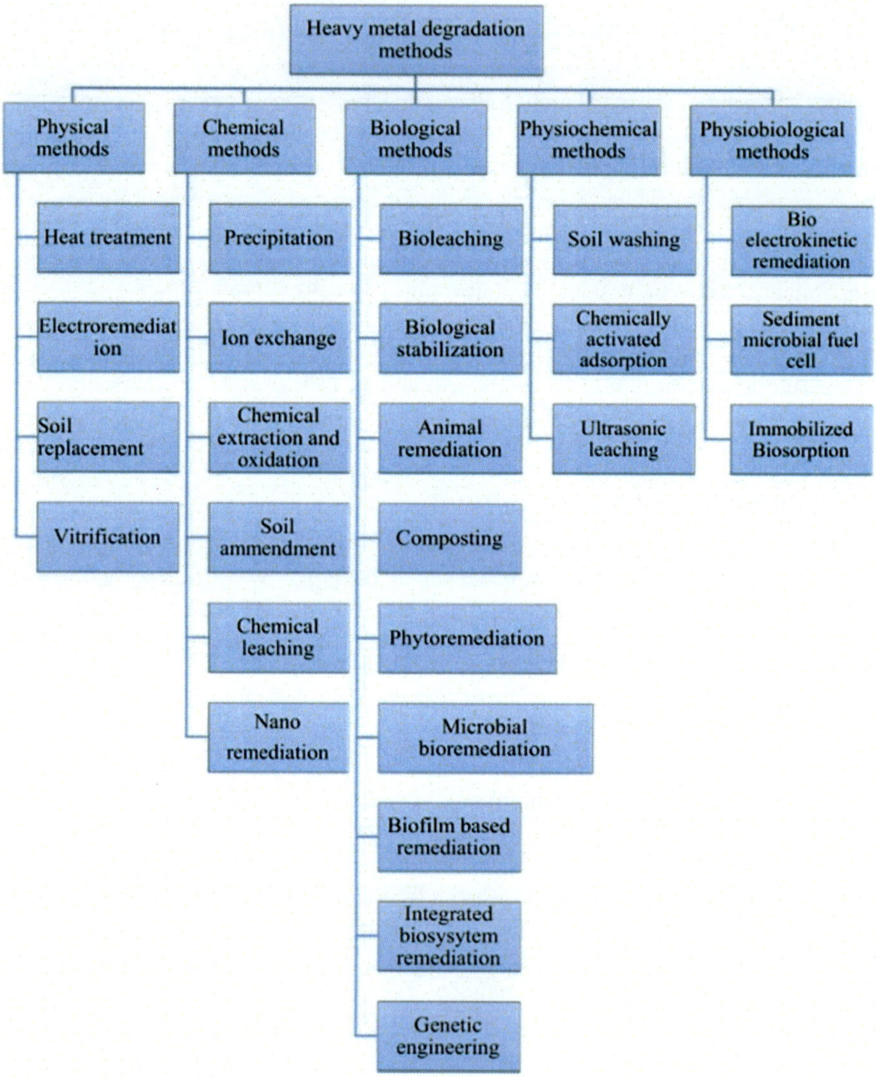

Fig. 1 Conventional methods of environment remediation

Therefore new methods of biosorption, microbial techniques, and most popular are nano remediation. The benefits of using nanocomposites for remediation is that their efficiency is high as well as they are cost-effective [50].

4 Nanocomposites as Newer Approach

Furthermore, there are many more advantages of nanocomposites. Prime advantage is in their physical properties like small size, varied morphology, high porosity, and the different chemical composition which offers wide scope where properties can be made suitable according to the type of pollutant. These tailoring properties give interesting aspects to the study of nanocomposites which offer significant advantages over conventional methods in the remediation of the environment. So, the methods that are developed as a combination of several different materials (hybrids/composites) show very high efficiency, it is very selective for the different contaminants and moreover, it is a very stable method than earlier conventional methods based upon a single strategy. For example, instead of using nanoparticles alone the alternate way of attaching nanoparticles to a moiety as explained in the earlier section leading to the formation of nanocomposites can increase the stability of the material. This approach of functionalizing nanomaterials with specific chemicals can help to enhance the efficiency and its selectivity [17] which are responsible for targeting contaminant molecules of interest. But there are many challenges also to meet this requirement, which are given as follows:

- target-specific capture
- cost-effectiveness
- nontoxicity and biodegradability
- recyclability and regeneration
- unstable in normal conditions
- prevent agglomeration, enhance monodispersity.

5 Different Categories of Nanomaterials Used

Three broad categories of nanomaterials are used for the remediation of different environmental contaminant.

5.1 Inorganic Nanomaterials (Metal and Metal Oxide-Based Nanomaterials)

The nanomaterials of this type have numerous applications as environment cleaner especially in aqueous systems as they are highly adaptable toward applications whether ex-situ or in situ applications. They follow fast kinetics and have high capacity of adsorption [49]. Systematic synthetic methods were designed to obtain nanomaterials whose shape can be controlled, monodisperse metal/metal oxide, and also very stable. The synthetic conventional methods like thermal decomposition, hydrothermal method, reduction, and coprecipitation [8] are extensively used and can be easily scaled for better results in terms of yields. Some of the examples of this type are explained below with their applications.

5.1.1 Silver Nanoparticles (AgNPs)

These nanoparticles are used as water disinfectants due to their remarkable activities whether they are antiviral or antifungal and antibacterial activity [7]. It has been observed that the size of AgNPs plays an important role such as if the size is less than 10 nm in diameter they were found to be more toxic to gram-negative bacteria like *Escherichia coli* and *Pseudomonas aeruginosa*. They bind to virus glycoproteins hence prevent binding of the virus to the host cells. The particle size greater than 10 nm has lesser antibacterial activity [15]. Among different forms, triangular AgNPs manifests better properties than Ag nanorods and Ag nanospheres. This observation demonstrates the importance of shape of the particles in drawing out their appropriate role [38]. Silver nanoparticles have been coupled with many materials like polymer and mostly metal oxides to improve the efficiency of the resulting nanocomposite.

5.1.2 Titanium Oxides (TiO_2)

Another frequent metal-based material that is investigated is the titanium oxides which have the advantages of low cost, photocatalytic, nontoxicity, gas sensing, and semiconducting properties. Because of these advantages, TiO_2NPs have been studied for treatment of waste material and as well as in purification of air [1, 28]. They are also used as a photocatalyst in purification of wastewater. Actually, TiO_2NPs are activated by light which in turn produces oxidants such as hydroxyl radicals. These hydroxyl radicals are very reactive and act as a disinfectant for microorganisms [60]. There is a significant increase in the performance of photocatalytic ability if TiO_2 is doped with another transition metal. Ag-doped TiO_2 nanofibers formed with the help of sol-gel electrospinning technique. Ag-doped TiO_2 nanofibers act as photo catalysts for the photocatalytic degradation of substituted phenols specifically 2-chlorophenol when irradiated with UV radiation. They showed an increase in photodegradation as compared to the TiO_2 nanofibers alone because of the availability of more silver on the surface that produces photo-induced electrons and photo-induced holes. There is a quick transference of photo-induced electrons to the oxygen which is adsorbed results in an increased amount of surface hydroxyl groups present on the surface of the nanofibers [41].

5.1.3 Titanates

This is another very important class of inorganic compounds of titanium oxide. The best method of formation is hydrothermal method by which we can synthesize basic titanate nanotubes (TNTs), acidic titanate nanotubes (TNTs), and neutral titanate nanotubes [6]. These TNTs showed remarkable catalytic reduction properties of NO with ammonia. These TNT formulations can be loaded with manganese oxide resulting in Mn-doped titanate nanosheets in the case of basic, titanate nanorods in acidic and titanate nanotubes in neutral pH medium. It has been observed that out of these three TNTs, the neutral Mn/TNTs exhibits the best results because they have the greatest surface area.

5.1.4 Mixed Oxide Materials

These materials which are mixed oxides like TiO_2 and SiO_2 can be synthesized using titanium isopropoxide or titanium butoxide and bamboo as a silica source [45]. These materials were further tested for the degradation of various dyes such as methylene blue and they manifest outstanding photoactivity and varied treatment times. It is seen that these oxides can be better materials for removal of varied variety of pollutants. The only disadvantage of these materials is it can be used for selected contaminants and these have applications in industrial wastewater treatment system which are on smaller scale.

5.1.5 Magnetic Metallic Nanoadsorbents

This type of nanoadsorbents basically includes iron and iron oxides NPs. Extensive literature data is there for magnetic metallic nanoadsorbents and its applications. Most significant application is that they are used in the remediation of environment by removal of different heavy metals like Cd^{2+}, Ni^{2+}, Cu^{2+}, and Co^{2+} [11]. Also, chlorinated organic solvents can be removed from the environment with the help of magnetic metallic nanoadsorbents [18]. The structure of Iron NPs is a core-shell structure with the core consists of elemental iron (Fe^0) and the shell consist of Fe(II) and Fe(III) oxides. Therefore, these can be reduced easily from donation of electron from the core containing Fe^0 to mixed valent iron oxides in the shell as depicted in Fig. 2.

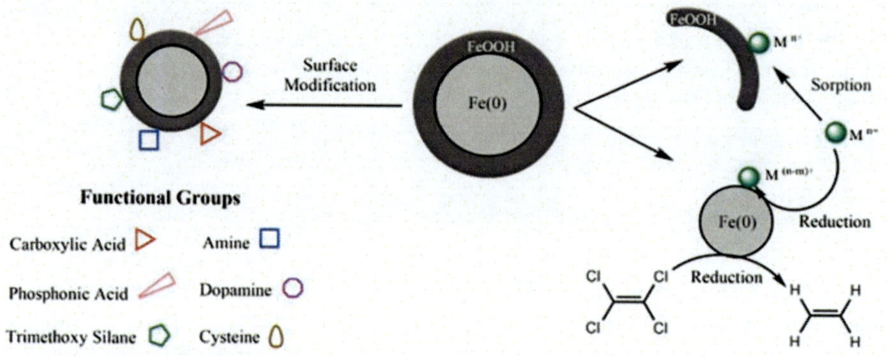

Fig. 2 Demonstration of mechanism of magnetic metallic nano adsorbents [16, 30]

There are various methods to improve the efficiency of nanoparticles. First method is sonication which is generally employed to the iron NP solution to avoid the aggregation and enhance the removal of Ni^{2+} and Co^{2+}. It has been observed that the time required for removal of nickel is 20 min and for removal of cobalt is 30 min. The results of removal are for nickel it is 35–40% and for cobalt it is 55–65% [19]. Second method is to enhance the stability of zerovalent iron nanoparticles by blending of a second metal such as Pd, Ni, or Cu [58]. Third is to introduce noble metals that are resistant to corrosion and oxidation in moist air with zerovalent iron nanoparticles (nZVI) to catalyze dechlorination and hydrogenation reactions with contaminants. Fourth is the green synthesis of zerovalent iron nanoparticles (nZVI) as demonstrated by some workers from natural resources like oak and mulberry leaf extracts which were obtained from waste and they provide adsorbent which is cheapest for the cleaning of water from contaminants [10]. Synthesis of these zerovalent iron nanoparticles (nZVI) can be done by microemulsion method. Emulsified zero-valent iron (EZVI) nanoparticles can be formed by loading of Ni/Fe nanoparticles with lecithin. These EZVI are effective in removal of major pollutants of environment like 3,3′,4,4′-tetrachlorobiphenyl (PCB77) and chlorinated volatile organic compounds (CVOCs) and hence act as environment cleaner and total CVOC mass decrease of 86% has been estimated by the end of 2.5 years of monitoring period [52].

5.2 *Silica Nanomaterials*

Mesoporous silica materials have gained so much attention for their application in adsorption and catalysis. The versatility of these materials is due to the presence of –OH groups on the surface of these materials. As illustrated in Fig. 3 covalent tethering, surface modification, incorporation of more functionalities, gas adsorption, physical impregnation, and in situ polymerization can be done [20].

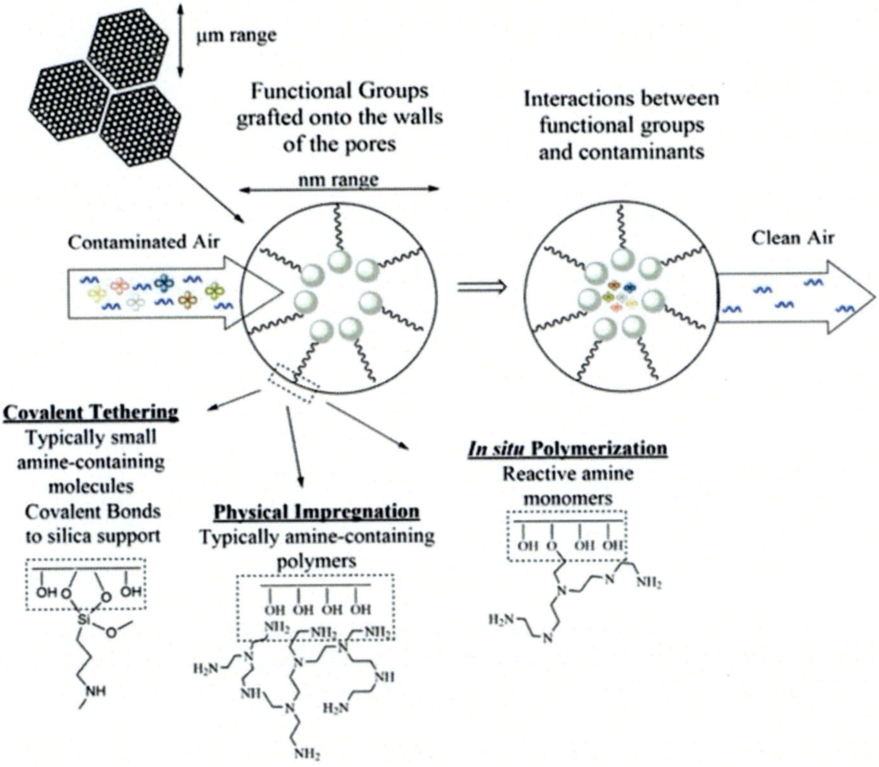

Fig. 3 Surface characteristics of mesoporous silica materials [16]

5.2.1 Amine-Surface-Modified Silica Xerogels

These silica xerogels are used for the selective removal of CO_2 and H_2S from natural gas. The MCM-48 ordered mesoporous silica showed a high adsorption rate. The presence of $-NH_2$ groups on the surface of the silica materials are large which increases the rate of adsorption to almost 80% as observed by group of researchers [20].

5.2.2 Amine-Modified Aluminosilicates

These are used for the absorption of CO_2 specifically in compounds containing carbonyl moiety such as aldehydes and ketones. They do so by forming imine or hemi-aminal formation. Sometimes with reversible adsorption of the gaseous molecule [4, 26]. Consequently, these materials have advantages over traditional CO_2 capture as explained earlier with aqueous amines and other silica-supported amines in the way that they have low cost, synthesis is not complex, show greater stability and better performance [43]. There is one limitation that the amine functionality is introduced during the process of synthesis by the ring-opening polymerization of aziridine so makes these materials not suitable for pollutants that react with amines. The aziridine monomer also has difficulty in handling without proper equipment.

5.2.3 Amine-Functionalized Porous Silica

These are particularly used for removal of low-molecular-weight aldehydes like formaldehyde. 1° and 2° amines are more appropriate for capturing aldehydes as compare to 3° amines by the formation of imine and hemiaminal intermediates. Nomura and Jones [36] studied amine-functionalized porous silica in detail and also on higher molecular weight less volatile aldehydes but observed that it is not suitable for them as reaction time required in that case exceeded more than 10 h. Much longer reaction time makes them unsuitable for industrial applications.

5.2.4 Mesoporous Silica with –COOH Groups

These are used for compounds that have the tendency to form hydrogen bonds. This is known as hydrogen bonding capture model. Various compounds like heavy metal ions, dyes, and contaminants which can form hydrogen bonds with a carboxylic acid group of mesoporous silica can be removed by this method. But the only limitation is that interactions occur at specific pH values as it has been observed that the maximum uptake for methylene blue was obtained at pH = 9 [54].

Similarly, there are some other groups which work well as remediation of environment like [47] studied amino-functionalized polycarboxylic acid, Nakanishi et al. [34] studied amino-functionalized silica materials for the removal of different metal ions such as particularly transition metal ions like Cd^{2+}, Co^{2+}, Cu^{2+}, Zn^{2+}, Ni^{2+} and other like Al^{2+}, Cr^{3+}, Pb^{2+}, Hg^{2+}, and U^{6+}. Apart from this chitosan-based silica materials [55] and thiol-functionalized silica materials [56] are also used in the removal of heavy metal ions as well as volatile organic compounds from the environment.

5.3 *Carbon-Based Nanomaterials*

The elemental carbon is well-known for its distinctive physical properties, peculiar chemical properties, and multiple electronic properties. So, introduction of elemental carbon to these nanomaterials [46] gives better results than metal-based nanomaterials. Moreover, the mutable hybridization states of carbon can result in different structural configurations such as fullerene (C_{60}), multi-walled, and single-walled carbon nanotubes (MWCNTs and SWCNTs) and also graphene which is proved to be the best materials. The primary requirement is the treatment of its surface which means its activation as well as the functionalization of the material. Multi-walled and single-walled carbon nanotubes (MWCNTs and SWCNTs) have gained much attention presently due to their remarkable photocatalytic approaches and adsorption properties [40]. The phenomenon of their action can be well visualized from Fig. 4. We can observe that with UV irradiation on any of these forms may be CNTs, fullerene, and nanocomposites there is the absorption of photons of energy greater than or equal to the bandgap of these forms. With the absorption of photons, they generate valence band holes (h^+) and conduction band electrons (e^-). The electrons which are produced form superoxide radicals which result in the reduction of heavy metal contaminants. Holes form hydroxyl radicals which result in the oxidation of chlorinated organic compounds. In SWCNT there are four different sites. One for adsorption, second is internal site having less adsorption energy, third is external site present on the surfaces of the external CNTs having adsorption energy which is very high, fourth is present in between two neighboring tubes. The external sites show greater rate of adsorption as the equilibrium can be attained much faster as compared to internal sites because of the direct exposure to the adsorbing material. When specific methods of

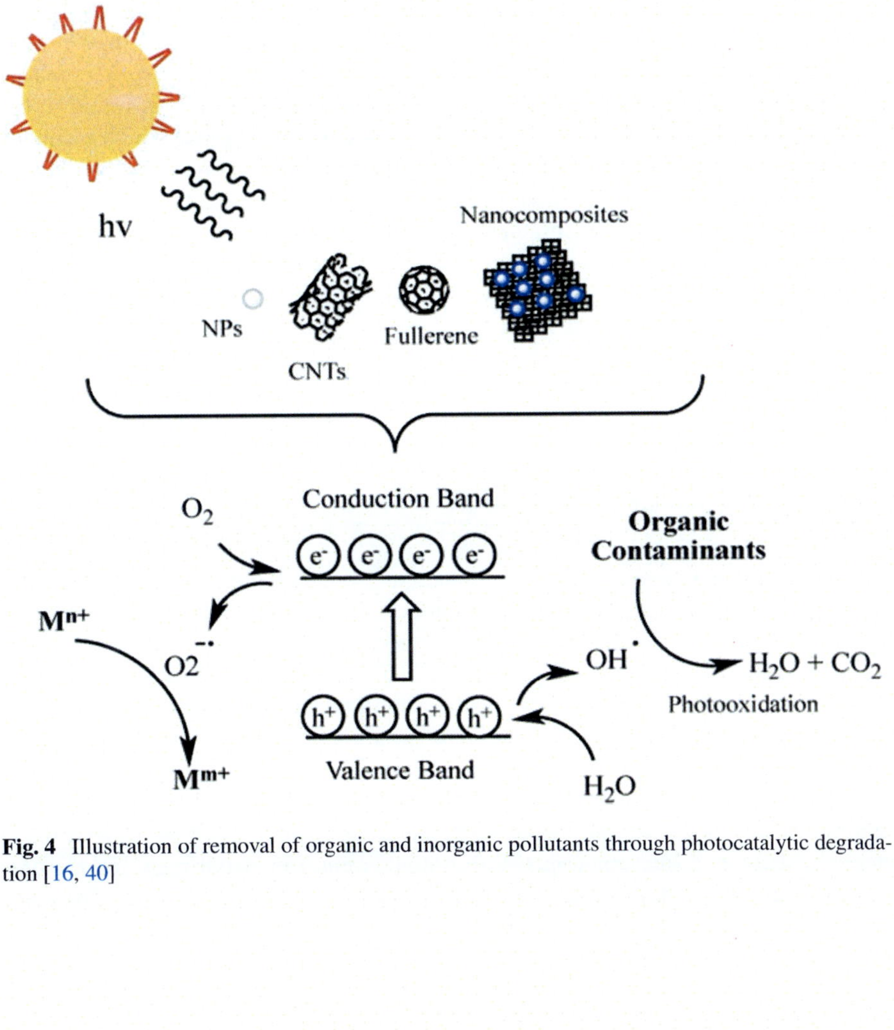

Fig. 4 Illustration of removal of organic and inorganic pollutants through photocatalytic degradation [16, 40]

preparation are used multi-walled carbon nanotubes exist as bundles [46]. The carbon nanotubes may contain hydroxyl, carbonyl, and carboxylic acid groups which can increase adsorption capacities because of an increase in oxygen content. When these nanotubes are oxidized with nitric acid as well as with other oxidizing agents there is an increase in adsorption capabilities of heavy metal ions [23]. Adsorption of cationic dyes also increases with the increase in pH due to the increase in electrostatic attraction between dyes and active carbon forms. Some more properties like molecular weight, dipole moment, critical temperature of the adsorbate gas, these physicochemical properties can remarkably alter the rate of adsorption.

5.4 *Graphene Materials*

Another class of materials that are promising nanocomposites is the use of graphene to fabricate photocatalytic nanocomposites that showed an increase in photocatalytic activity in graphene composites containing TiO_2NPs as compared to bare TiO_2NPs attributed to an increase in conductivity [61]. The pristine graphene was used earlier for the removal of fluoride from an aqueous solution as an effective adsorbent material. The monolayer adsorption capacity of fluoride by graphene was found to be 35.59 mg/g at 298 K and pH = 7.0 [29]. The Graphene oxide (GO) which is modified graphene is used for the remediation of environment by adsorption of a variety of gases like SO_x, H_2S, NH_3. Actually, carbon surface of graphene oxides (GOs) bears several oxygen-containing functional groups such as carboxylic acids, epoxides, and hydroxyls. There is strong acid-base interactions layered GO structure offers acidity and ammonia offers basicity. However, GO can be used for the removal of anionic metals but it requires the modification of GO with organic or metal oxides. Moreover, it decreases the aggregation of the graphene layers and also increases the effective surface area, making it more suitable material than pristine graphene [57].

5.5 *Polymer-Based Nanomaterials*

Polymers are mostly used for the detection and removal of contaminant chemicals, gases, and organic pollutants such as manganese, nitrate, iron, arsenic, heavy metals, CO, SO_2, NO_x like aliphatic and aromatic hydrocarbons, pharmaceuticals or VOCs. Polymeric hosts like surfactants, emulsifiers, stabilizing agents, and surface functionalized ligands can be incorporated to increase stability, mechanical strength, recyclability and overcome some of the limitations of pristine NPs. Here polymers are used as host materials and NPs are responsible for the contaminant remediation. Amphiphilic polyurethane (APU)NPs removed phenanthrene and polynuclear aromatic hydrocarbons (PAHs) from contaminated aquifer sand but there is an issue regarding biodegradability of such materials. Poly (amidoamine) or dendrimers (PAMAM) have been utilized in wastewater remediation for water samples contaminated with

Table 1 List of nanocomposites with removal of target pollutants

S. No.	Type of nanoparticles	Removal target
1	PLA/PEG incorporated with PEI	VOCs
2	SiO_2 NPs with poly (acrylic acid-co-acrylamide) nanocomposite	methylene blue (MB)
3	Wheat xylan/poly(acrylic acid) NP hydrogel with Fe_3O_4 nanoparticles	methylene blue (MB)
4	Gold-coated with chitosan polymer	Zn^{2+} aq, Cu^{2+} aq
5	Poly (methacrylic acid)-grafted chitosan/bentonite	Th^{4+}
6	Fe_3O_4 (sodium alginate with tetrasodium thiacalixarene tetrasulfonate)	Ni^{2+}, Co^{2+}, Pb^{2+}, Cd^{2+}, Cu^{2+}, and Cr^{2+} ions
7	Ag-doped TiO_2 nanofibers	Methylene blue dye
8	AgNPs and Ag^+ (mixture of polymers)	*Escherichia coli*, *Staphylococcus aureus*, *Aspergillus niger*, and *Salmonella enterica*
9	Carbon nanotubes/Al_2O_3 nanocomposite	Fluoride
10	Cu/Fe/Ag-doped TiO_2	Nitrate (NO_3^-)
11	Multiwall carbon nanotube (MWCTs)	Zn^{2+}

metal ions such as Cu^{2+}, Ag^+, Au^+, Fe^{2+}, Fe^{3+}, Ni^{2+}, Zn^{2+}, and U^{6+}. The reason behind this dendritic nanopolymers is that they contain functional groups which are able to encapsulate a broad range of solutes in water. Chitosan-based carbon nanofibres (CNFs) incorporated in iron oxide nanoparticles along with polyvinyl alcohol nanocomposite films have efficient adsorption capacity of Cr^{6+} from water.

Another modification in which Fe_3O_4 magnetic NP with 3-aminopropyltriethoxysilane and (acrylic/crotonic acid) copolymers was prepared. This modification is used for removal of Cu^{2+}, Cd^{2+}, Pb^{2+}, and Zn^{2+} from metal-contaminated water.

Potential antibacterial property has been found when AgNPs are embedded in cellulose acetate fibers (Table 1).

6 Applications

6.1 *Building and Environment*

TiO_2 which is excellent photocatalytic material has the ability to produce self-cleaning as well as de-polluting building materials. Using sol-gel synthesis, Au-TiO_2 photocatalysts were blended into silica thereby forming TiO_2–SiO_2 nanocomposite material which is sprayed inside the pore structure of a very friable carbonate stone and a non-ionic surfactant generally *n*-octylamine [32]. The nanomaterial which is produced as a result has superior characteristics like good adhesiveness, crack-free surface layer to the stone, and self-cleaning properties. Apart from this, there is an increase in mechanical resistance as it has a greater penetrating property which gets into the pores of the stone. Other important benefit of the nanocomposite is that it improves protection against salt crystallization degradation mechanisms. In a trial, it has been observed that after three cycles of $NaSO_4$ crystallization degradation, the stone which is not treated is reduced to a completely powdered material, whereas the stone which is treated with this novel product remains practically unchanged even after thirty cycles. For the sake of comparison, two industrial products were also checked and they resulted in a crack of coatings and less mechanical resistance to the stone as compared to nanocomposite products. So, it has got excellent applications in building and the environment. While in its designing two factors one is TiO_2 light absorption restricted to UV and second is its poor adhesion to the substrates plays a crucial role.

6.2 Removal of Dye in Water

Water is polluted with dyes released from industries thereby threatening issues for water resources. So, their removal is the current area of interest. Cellulose–clay hydrogel with nanocomposites have superabsorbent properties and superior mechanical performance. These properties can be employed for the removal of dyes from water. The synthesis of these superabsorbent hydrogels nanocomposites can be done in NaOH/urea aqueous solution by chemical cross-linking of nanocomposites with carboxymethyl cellulose (CMC) and the intercalated clay. These hydrogels exhibited high absorption capacity for methylene blue (MB) solution. The removal efficiencies were observed around 96.6–98%, in concentration range of $10-100$ mg L^{-1} of these hydrogel samples. These results gave a new platform for dye decontamination where cellulose–clay nanocomposite hydrogels are used as water cleaners [42].

6.3 Slow Release of Fertilizers

As we know that two important macronutrients responsible for the growth and yield of agricultural crops are Nitrogen (N) and Phosphorus (P). Generally, Nitrogen and Phosphorus applied as normal fertilizers are lost to the environment and these losses come at a large environmental cost. The reason for this low efficiency of P fertilizers is especially in tropical soils due to the formation of Fe- and Al-based oxides. Most of the phosphates released from organic matter and that added as fertilizer is rapidly scavenged by soil minerals which in turn changes into fixed or insoluble inorganic compounds that are not susceptible to leaching. As a result, concentrations of soil phosphate are very low. Secondly, due to high NH_3 volatilization and fast hydrolysis of urea ultimately leads to an accumulation of NH_4^+ resulting in increase in pH of soil. The slow release of fertilizers (SRF) has been considered to be a great strategy to improve the utilization of macronutrients. Slow-release fertilizers have many advantages over conventional fertilizers like better fertilizer use in potato, better matching of nutrient demand in crops, and increased phosphorus recovery in barley. In this application, a novel series of hydrogels composed of polyacrylamide (PAAm), methylcellulose (MC) and calcic montmorillonite (MMt) were synthesized which are appropriate for the controlled release of fertilizers where the components presented a synergistic effect, giving very high fertilizer loading in their structure [5].

Other method is to produce nanocomposites from urea (Ur) which act as a matrix in which hydroxyapatite particles (Hap) were blended. The nutrients are released by slow-release fertilizers gradually in order to coincide with the nutrient requirement of plants. Urea is considered as the important nitrogen-containing fertilizer and its low cost makes it more convenient. Due to the cheap rate of starch, it is used as an encapsulating matrix of agrochemicals. When plasticized by alcohol or even by urea, starch is known as thermoplastic starch (TPS). The slow-release nanofertilizers will tend to decrease the use of the chemical fertilizers which needs regular spraying in fields. The bi-product released will be nontoxic which will not affect the soil parameters.

6.4 Biological Applications

Large quantities of nanocomposites can be produced from transition metals such as Cu, Ag, In and Fe in aqueous media using a polymer which is biodegradably named carboxymethyl cellulose (CMC). Generally, sodium salt of CMC is used for this purpose [33]. These nanocomposites exhibit broader decomposition temperatures. Ag-based CMC nanocomposites exhibit greater luminescent property at longer wavelengths. The noble metals like Au and Pt, react under microwave irradiation (MW) conditions at 100 °C and do not react at room temperature with aqueous solutions of carboxymethyl cellulose. This environmentally friendly method gives many technological and medicinal applications rather than using any toxic reducing agent such as sodium borohydride ($NaBH_4$). Magnetic chitosan–iron (III) hydrogel (MCh-Fe) was synthesized and used to remove toxic Cr^{VI} from aqueous solution and characterized using spectroscopic techniques using SEM, TG, and FT-IR [59].

7 Conclusion

This chapter provides a brief outline of the types of nanocomposites followed by different methods for environmental remediation. The extensive details of metal matrix, silica, graphene, and polymeric nanocomposites have been provided with their target pollutant and their mode of action. These materials are better materials and they earned huge success in environmental remediation. For reference table of nanocomposite with target pollutant is also given. The approach is to select the best nanocomposite for a particular pollutant in a given environment that requires a complete understanding of its mechanism. Every nanocomposite has its own advantages as well as challenging. The potential applications of nanocomposites like building and environment, removal of dye in water, slow release of fertilizers, and also biological applications were also discussed. Even though the recyclability of some materials is an important issue which makes them no longer useful. Nanocomposites provide newer route and plan for making our environment free from contaminants of air, water, and soil and make it better for living.

References

1. Adesina AA (2004) Industrial exploitation of photocatalysis: progress, perspectives and prospects. Catal Surv Asia 8:265–273
2. Aruna ST, Rajam KS (2003) Synthesis, characterization and properties of Ni/PSZ and Ni/YSZ nanocomposites. Scripta Mater 48(5):507–512
3. Awaji H, Choi SM, Yagi E (2002) Mechanisms of toughening and strengthening in ceramic-based nanocomposites. Mech Mater 34(7):411–422
4. Bollini P, Didas SA, Jones CW (2011) Amine-oxide hybrid materials for acid gas separations. J Mater Chem 21:15100–15120
5. Bortolin A, Aouada A, Mattoso L, Ribeiro C (2013) Nanocomposite PAAm/methyl cellulose/montmorillonite hydrogel: evidence of synergistic effects for the slow release of fertilizers. J Agric Food Chem 613:7431–7439
6. Chen X, Cen C, Tang Z, Zeng W, Chen D, Fang P, Chen Z (2013) The key role of pH value in the synthesis of titanate nanotubes-loaded manganese oxides as a superior catalyst for the selective catalytic reduction of NO with NH_3. J Nanomater 2013:871528
7. Chou KS, Lu YC, Lee HH (2005) Effect of alkaline ion on the mechanism and kinetics of chemical reduction of silver. Mater Chem Phys 94:429–433
8. Cushing BL, Kolesnichenko VL, O' Connor CJ (2004) Recent advances in the liquid-phase syntheses of inorganic nanoparticles. Chem Rev 104:3893–3946
9. Dalton AB, Coolins S, Munoz E (2003) Super-tough carbon-nanotube fibres: these extraordinary composite fibres can be woven into electronic textiles. Nature 423(6941):703–706
10. Ding S, Zhao L, Qi Y, Lv Q (2014) Preparation and characterization of lecithin-nano Ni/Fe for effective removal of PCB77. J Nanomater 2014:478–489
11. Ebrahim SE, Sulaymon AH, Saad Alhares H (2016) Competitive removal of Cu^{2+}, Cd^{2+}, Zn^{2+}, and Ni^{2+} ions onto iron oxide nanoparticles from wastewater. Desalin Water Treat 57:20915–20929
12. Evangelos M (2007) Nanocomposites: stiffer by design. Nat Mater 6(1):9–11
13. Gaharwar AK, Peppas NA, Khademhosseini A (2014) Nanocomposite hydrogels for biomedical applications. Biotechnol Bioeng 111(3):441–453

14. Gangopadhyay R, Amitabha D (2000) Conducting polymer nanocomposites: a brief overview. Chem Mater 12(7):608–622
15. Gogoi SK, Gopinath P, Paul A, Ramesh A, Ghosh SS, Chattopadhyay A (2006) Green fluorescent protein-expressing Escherichia coli as a model system for investigating the antimicrobial activities of silver nanoparticles. Langmuir 22:9322–9328
16. Guerra FD, Attia MF, Whitehead DC, Alexis F (2018) Nanotechnology for environmental remediation: materials and applications. Molecules 23(7):1760
17. Guerra FD, Campbell ML, Whitehead DC, Alexis F (2017) Tunable properties of functional nanoparticles for efficient capture of VOCs. ChemistrySelect 2:9889–9894
18. Han Y, Yan W (2017) Reductive dechlorination of trichloroethene by zero-valent iron nanoparticles: reactivity enhancement through sulfidation treatment. Environ Sci Technol 50:12992–13001
19. Hooshyar Z, Rezanejade BG, Ghayeb Y (2013) Sonication enhanced removal of nickel and cobalt ions from polluted water using an iron based sorbent. J Chem 2013:786–954
20. Huang HY, Yang RT, Chinn D, Munson CL (2003) Amine-grafted MCM-48 and silica xerogel as superior sorbents for acidic gas removal from natural gas. Ind Eng Chem Res 42:2427–2433
21. Iijima S (1991) Helical microtubes of graphitic carbon. Nature 354(6348):56–58
22. Jordan J, Jacob KI, Tannenbaum R (2005) Experimental trends in polymer nanocomposites, a review. Mater Sci Eng A 393(1–2):1–11
23. Khin MM, Nair AS, Babu VJ, Murugan R, Ramakrishna S (2012) A review on nanomaterials for environmental remediation. Energy Environ Sci 5:8075–8109
24. Kojima Y, Usuki A, Kawasumi M (1993) Mechanical properties of nylon-6-clay hybrid. J Mater Res 8(5):1185–1189
25. Kruis FE, Fissan H, Peled A (1998) Synthesis of nanoparticles in the gas phase for electronic, optical and magnetic applications, a review. J Aerosol Sci 29(5–6):511–535
26. Kuwahara Y, Kang DY, Copeland JR, Bollini P, Sievers C, Kamegawa T, Yamashita H, Jones CW (2012) Enhanced CO_2 adsorption over polymeric amines supported on heteroatom-incorporated SBA-15 silica: impact of heteroatom type and loading on sorbent structure and adsorption performance. Chem A Eur J 18:16649–16664
27. Lalwani G, Henslee AM, Farshid B, Lin L, Kasper F, Kurtis Qin Y-X, Xian M, Antonios G, Balaji S (2013) Two-dimensional nanostructure-reinforced biodegradable polymeric nanocomposites for bone tissue engineering. Biomacromolecules 14(3):900–909
28. Li Q, Mahendra S, Lyon DY, Brunet L, Liga MV, Li D, Alvarez PJJ (2008) Antimicrobial nanomaterials for water disinfection and microbial control: potential applications and implications. Water Res 42:4591–4602
29. Li Y, Zhang P, Du Q, Peng X, Liu T, Wang Z, Xia Y, Zhang W, Wang K, Zhu H (2011) Adsorption of fluoride from aqueous solution by graphene. J Colloid Interface Sci 363:348–354
30. Li XQ, Zhang WX (2006) Iron nanoparticles: the core–shell structure and unique properties for Ni (II) sequestration. Langmuir 22:4638–4642
31. Liu TX, Phang IY, Shen L (2004) Morphology and mechanical properties of multiwalled carbon nanotubes reinforced nylon-6 composites. Macromolecules 37(19):7214–7222
32. Luna M, Mosquera MJ, Vidal H, Gatica JM (2019) Au-TiO_2/SiO_2 photocatalysts for building materials: self-cleaning and de-polluting performance. Build Environ 164:106347
33. Nadagouda MN, Varma RS (2007) Synthesis of thermally stable carboxymethyl cellulose/metal biodegradable nanocomposites for potential biological applications. Biomacromolecules 89:2762–2767
34. Nakanishi K, Tomita M, Kato K (2015) Synthesis of amino-functionalized mesoporous silica sheets and their application for metal ion capture. J Asian Ceram Soc 3:70–76
35. Niihara K (1991) New design concept of structural ceramics-ceramic nanocomposite. J Ceram Soc Jpn 99(6):974–982
36. Nomura A, Jones CW (2013) Amine-functionalized porous silica's as adsorbents for aldehyde abatement. ACS Appl Mater Interfaces 5:5569–5577
37. Ogawa M, Kuroda K (1997) Preparation of inorganic composites through intercalation of organoammonium ions into layered silicates. Bull Chem Soc Jpn 70(11):2593–2618

38. Pal S, Tak YK, Song JM (2007) Does the antibacterial activity of silver nanoparticles depend on the shape of the nanoparticle? A study of the gram-negative bacterium *Escherichia coli*. Appl Environ Microbiol 73:1712–1720
39. Pandey JK, Kumar AP, Misra M (2005) Recent advances in biodegradable nanocomposites. J Nanosci Nanotechnol 5(4):497–526
40. Paola Di A, García-Lopez E, Marcì G, Palmisano L (2012) A survey of photocatalytic materials for environmental remediation. J Hazard Mater 211–212:3–29
41. Park JY, Lee IH (2014) Photocatalytic degradation of 2-chlorophenol using Ag-doped TiO_2 nanofibers and a near-UV light-emitting diode system. J Nanomater 2014:250803
42. Peng Na, Hu Danning, Zeng J, Li Yu, Liang Lei, Chang Chu (2016) Superabsorbent cellulose-clay nanocomposite hydrogels for highly efficient removal of dye in water. Sustainable Chem Eng 412:7217–7224
43. Qi G, Wang Y, Estevez L, Duan X, Anako N, Park AHA, Li W, Jones CW, Giannelis EP (2011) High efficiency nanocomposite sorbents for CO_2 capture based on amine-functionalized mesoporous capsules. Energy Environ Sci 4:444–452
44. Rafiee MA (2009) Enhanced mechanical properties of nanocomposites at low graphene content. ACS Nano 3(12):3884–3890
45. Rasalingam S, Peng R, Koodali RT (2014) Removal of hazardous pollutants from wastewaters: applications of TiO_2-SiO_2 mixed oxide materials. J Nanomater 2014:617405
46. Ren X, Chen C, Nagatsu M, Wang X (2011) Carbon nanotubes as adsorbents in environmental pollution management: a review. Chem Eng J 170:395–410
47. Repo E, Warchoł JK, Bhatnagar A, Mudhoo A, Sillanpaa M (2013) Aminopolycarboxylic acid functionalized adsorbents for heavy metals removal from water. Water Res 47:4812–4832
48. Roy R, Roy RA, Roy DM (1986) Alternative perspectives on "quasi-crystallinity": non-uniformity and nanocomposites. Mater Lett 4(8–9):323–328
49. Santhosh C, Velmurugan V, Jacob G, Jeong SK, Grace AN, Bhatnagar A (2016) Role of nanomaterials in water treatment applications: a review. Chem Eng J 306:1116–1137
50. Sharma S, Tiwari S, Hasan A, Saxena V, Pandey LM (2018) Recent advances in conventional and contemporary methods for remediation of heavy metal-contaminated soils. 3 Biotech 8(4):216
51. Stearns LC, Zhao J, Martin P (1992) Harmer Processing and microstructure development in Al_2O_3-SiC nanocomposites. J Eur Ceram Soc 10(3):473–477
52. Su C, Puls RW, Krug TA, Watling MT, O'Hara SK, Quinn JW, Ruiz NE (2012) A two and half-year-performance evaluation of a field test on treatment of source zone tetrachloroethene and its chlorinated daughter products using emulsified zero valent iron nanoparticles. Water Res 46:5071–5084
53. Tratnyek PG, Johnson RL (2006) Nanotechnologies for environmental cleanup. Nano Today 1:44–48
54. Tsai CH, Chang WC, Saikia D, Wu CE, Kao HM (2016) Functionalization of cubic mesoporous silica SBA-16 with carboxylic acid via one-pot synthesis route for effective removal of cationic dyes. J Hazard Mater 309:236–248
55. Vunain E, Mishra AK, Mamba BB (2016) Dendrimers, mesoporous silicas and chitosan-based nanosorbents for the removal of heavy-metal ions: a review. Int J Biol Macromol 86:570–586
56. Walcarius A, Delacote C (2005) Mercury(II) binding to thiol-functionalized mesoporous silicas: critical effect of pH and sorbent properties on capacity and selectivity. Anal Chim Acta 547:3–13
57. Wang S, Sun H, Ang HM, Tade MO (2013) Adsorptive remediation of environmental pollutants using novel graphene-based nanomaterials. Chem Eng J 226:336–347
58. Wu L, Ritchie SMC (2006) Removal of trichloroethylene from water by cellulose acetate supported bimetallic Ni/Fe nanoparticles. Chemosphere 63:285–292
59. Yu Z, Xiaodan Z, Huang Y (2013) Magnetic chitosan-iron (III) hydrogel as a fast and reusable adsorbent for chromium (VI) removal. Ind Eng Chem Res 5234:11956–11966
60. Zan L, Fa W, Peng T, Gong Z (2007) Photocatalysis effect of nanometer TiO_2 and TiO_2-coated ceramic plate on Hepatitis B virus. J Photochem Photobiol B Biol 86:165–169

61. Zhang Y, Tang Z-R, Fu X, Xu Y-J (2010) TiO_2 graphene nanocomposites for gas-phase photocatalytic degradation of volatile aromatic pollutant: is TiO_2 graphene truly different from other TiO_2 carbon composite materials? ACS Nano 4:7303–7314
62. Zhang S, Sun D, Fu Y, Du H (2003) Recent advances of superhard nanocomposite coatings: a review. Surf Coat Technol 167(2–3):113–119
63. Zhu J, Wei S, Chen M, Gu H, Rapole SB, Pallavkar S, Ho CT, Hopper J, Guo Z (2013) Magnetic nanocomposites for environmental remediation. Adv Powder Technol 24(2):459–467

Composition and Arrangement of Carbon-Derived Membranes for Purifying Wastewater

Ritu Painuli, Pallavi Jain, Sapna Raghav, and Dinesh Kumar

Abstract Wastewater can be treated in many ways, out of which membrane separation technology is considered the most effective and unique one. Especially, carbon nanotubes (CNTs)-based membranes are getting noteworthy attention owing to the combined merits of CNTs and membrane separation. This results in offering superior membrane properties. This chapter discusses the classification and characterization of CNTs based membranes. It also reviews the fabrication methods for mixed CNTs based membranes in detail. Furthermore, the future direction and challenges related to CNTs based membranes are also briefly outlined.

Keywords Carbon nanotubes · Classification · Preparation · Characterization · Challenges

1 Introduction

Freshwater is an important and vital part of human's life. It also acts as an important storage unit for various other industries. According to a report, 75 percent of the world population could be underwater shortage conditions by 2025 [32, 34, 35, 38, 83]. It is known that millions of people will suffer from water scarcity conditions by 2050 [27]. Extensive efforts are being made to protect the world from this blooming water crisis.

R. Painuli · S. Raghav
Department of Chemistry, Banasthali Vidyapith, Banasthali, Tonk 304022, India
e-mail: ritsjune8.h@gmail.com

S. Raghav
e-mail: sapnaraghav04@gmail.com

P. Jain
Department of Chemistry, SRM Institute of Science & Technology, Delhi-NCR Campus, Modinagar 210204, India
e-mail: palli24@gmail.com

D. Kumar (✉)
School of Chemical Sciences, Central University of Gujarat, Gandhinagar, India
e-mail: dinesh.kumar@cug.ac.in

M. Jawaid et al. (eds.), *Environmental Remediation Through Carbon Based Nano Composites*, Green Energy and Technology,
https://doi.org/10.1007/978-981-15-6699-8_8

The three Rs, reuse, recycle, and recovery, for water have proved to be beneficial in generating freshwater with no side effects on human health. The most prevalent technology is membrane filtration, which is used to purify all kinds of water, including waste, sea, and brackish [33, 36, 37, 83]. Membranes are categorized with the classifications based on the compositions and the cut-off molecular weight. Membrane techniques like ultrafiltration, microfiltration, reverse osmosis, nanofiltration, pervaporation, and distillation of membranes are the most extensively used techniques for water purification. Polymers, ceramic, and hybrid materials are the main elements from which membranes are composed [32, 34, 35]. Polymeric membranes find their usage in purification and desalination of water because of their greater selectivity and high mechanical strength Ceramic membranes are normally used for challenging water purification processes owing to their better thermal and chemical stability. Both these membranes have a lot of setbacks and can still be modified for better performance [32, 34, 35]. In contrast to ceramic membranes, the polymer membranes are lesser chemically stable and have low resistance toward fouling but are cheaper than ceramic ones [76]. Hence Ceramic membranes are considered only for small-scale industries. In modern times, a lot of modifications in nanomaterials like nanoparticles, metal/metal-oxide, and carbon nanoparticles, dendrimers, and zeolites have been employed for the water purification [43–45]. But because of the high surface area, better mechanical strength, and high thermal stability, CNTs have received much attention in this industry. They are used in removing a lot of impure particles present in the solution [4, 5, 32–37]. Carbon nanotubes have also contributed in the development of modified membranes for water decontamination [13, 25, 46, 50, 52, 53, 56, 84, 88, 95, 96, 100]. The significant properties that make CNTs as an excellent material in the water purification are their enhanced surface area along with high aspect ratio, rapid water transport, and ease of modification [52, 53]. For improvising its efficacy, the carbon nanotubes can also be utilized as filler/packing components. This chapter explores the classification, characterization (Table 1) as

Table 1 Carbon nanotubes characterization

S. No.	Characterization techniques	Major aims	References
1	SEM/TEM	Analysis of morphology (diameter, defects, length, and purity), state of arrangement (SWCNTs and MWCNTs), several layers, and distance between multi-walled nanotubes)	[30]
2	Energy-dispersive spectroscopy (EDS)	Elemental composition, functionalization	[7]
3	Fourier transform infrared spectroscopy (FT-IR)	Functionalization	[7]
4	TGA	Purity, functionalization	[55]
5	XPS	Elemental composition, functionalization	[91]

well as the composition of the CNTs based membranes. The challenges related to the future of the CNTs based membranes are also discussed at the end of the chapter.

2 Classification of Carbon Nanotube Membranes

CNTs based membranes are divided based on its implementation in fabrication processes, but broadly there are two main categories:

1. Freestanding carbon nanotube membranes
2. Mixed-carbon nanotube membranes

The freestanding membrane is further classified as vertically aligned carbon nanotubes membranes and bucky paper membranes. They are used in removing salt from the water and other wastewater treatment implementations [16, 69]. Carbon nanotubes are arranged as cylindrical pores in a vertically aligned carbon nanotube to force the liquid to cross the holes [29, 61]. Bucky paper CNTs based membranes have a 3D network with large pores that have an enhanced surface area. Mixed-carbon nanotube membrane has a design like that of the reverse osmosis structured membranes. In this arrangement, the top layer is assorted with a carbon nanotube and another polymer. The vertically aligned carbon nanotubes have a profound change in the rate of flow of water because of the small length of nanochannel and dense forest of the nanotube. Therefore, these membranes are more beneficial over bucky membranes. Moreover, tedious fabrication methods are the major challenge in the preparation of these membranes for large-scale applications. Whereas, the mixed-carbon nanotube membranes possess the benefit of the simpler fabrication process, but in contrast with the vertically aligned membranes, these membranes have a lower flux rate.

3 Aligned Carbon Nanotubes (ACNTs) Membranes

Aligned CNT membranes are composed of a single carbon nanotube arranged in high order and a vertically aligned array. Because of this, they have a porous structure composed of tiny spaces existing internally within the single tubes. These cavities are $\approx$5 nm in multi-walled nanotubes [31]. This diameter is similar to the size of many biomolecules and other macromolecules, which shows that the vertically aligned carbon nanotube membranes are very well be fitted for various filtration processes [22]. A vital property of ACNT membranes is that their pore dimension can be determined by managing the dimensions of the catalytic particles used during the growth of nanotube. This gives out a method by which the membrane selectivity can be customized according to the particular separation application. It is also necessary to make small adjustments in the selectivity of these substances by covalently functionalizing the edges of the carbon nanotubes with certain moieties or groups [66, 67]. It was also seen that in these membranes, it is probable to adjust the pores' diameters between 38 and 7 mm. This adjustment can be made by applying an upright outward force across the parallel dimensions of the carbon nanotubes [51]. This causes compression in nanotubes, and the permeability increases, which is higher than that in other carbon nanotube membranes. The membrane also reduces the adhesion of bacteria, demonstrating its benefit over other membranes by being less affected by the formation of biofilm and fouling. Aligned carbon nanotube membranes are made by implanting carbon nanotubes into a matrix. They can also be made by developing them on a substrate using a chemical vapor deposition (CVD) process. While growing them on the substrate, the aligned CNTs must be treated with packing material like polystyrene or Si_3N_4 so as to furnish the interstitial spaces among the individual carbon nanotubes [59, 68]. This opens a lot of entries of solvent, solute, and gas molecules to the openings of nanotubes. Free ACNT membranes can also be produced in the absence of any holding substance [98]. The CNTs that are manufactured by this process have large spaces across the structure that can be stretched up to tens of nanometers in diameter. These membranes can filter selective solute molecules that are available in the watery solution. In a study, macroscopic hollow cylinders were made that had multi-walled nanotubes aligned radially [93]. These were shown to retain the heavy constituents of a hydrocarbon mixture along with some microorganisms such as bacteria and viruses. Compared to UF membranes, ACNT membranes supply a better water flux, which is three times more than that of the ultrafiltration membrane [6]. The aligned carbon nanotube also shows a better and higher biofouling resistance along with low levels of bacterial adhesion [6]. In another study, a new modified ultrafiltration membrane was used with the help of multi-walled nanotube and polyethersulfone [56]. The arrangement of multi-walled is ordered within the PES matrix. It provides a path for transport of water, thus causing a change of water flux rate, which was thrice greater than that given by multi-walled/polyethersulfone membrane. The flux rate was ten times more than that of the pure PES membrane and antifouling properties [56]. The pores that are present have very small diameters in the ACNT membranes and have been receiving significant importance due to their prospective implementations in the removal of salt from water. The permeable properties of aligned carbon nanotube membranes are comparable to that of nanofiltration and ultrafiltration membranes. The drawback associated with this is that the aligned carbon nanotube's forest must be eliminated from the underlying substrate, which can comprise rigorous chemical embedding processes using harmful reagents. An additional drawback of carbon nanotube usage is that their ends must be open properly, which again needs strict conditions like plasma oxidation. Both steps are confusing and expensive. Most aligned carbon nanotube membranes produced till now posses smaller surface area, thus requiring a long step of fabrication. It has a lesser packing density, reduced mechanical stability, and has very little resistance to fouling [43, 45, 75]. Thus, numerous substitutes are being developed that are less complex and have lesser harmful steps, which can be again modified for further advancements.

4 Bucky Paper Membranes Buckypapers (BPs)

Bucky paper membranes have a simpler structure and comprise an array of individual carbon nanotubes supporting themselves [24, 47]. Bucky paper membranes are flexible and have considerable chemical and physical stability [92]. Because of their inherent thermal, mechanical and electrical properties, bucky paper is suggested for various implementations like in microscopic servomechanism, nanosensors, electronic filters, for mimicking natural muscles, and cathodes field-emission electron gun [17, 48, 80, 99]. They are made from carbon nanotube dispersions, which are developed by involving extremely high energy samples comprising nanotubes along with the prospective dispersant. When the dispersions are filtered on a holding membrane in the vacuum, then the bucky paper membranes are fabricated [26, 94].

Due to the simple and cheaper manufacturing mechanisms of bucky paper, it is possible to make bucky paper for large-scale industries in contrast to aligned membranes. A close observation of the buck paper surfaces with the help of scanning electron microscopy tells about a highly disarranged structure including carbon nanotubes held together by weak forces along with pi–pi interactions [101]. The interior assembly of bucky paper membranes consists of pores varying from small to large is in correlation with the spaces in between and the bundles of carbon nanotubes, respectively. The pores in bucky paper accord to 60–70% of their total volume, thus befitting as a medium for filtration. Apart from this, the filtration characteristics of bucky paper have also been observed but only in small numbers because of their weak mechanical properties owing to their brittle nature. A method to overcoming this is to strengthen bucky paper membranes with the help of polymer intercalation [15]. The infiltration of various polymers, for instance, polystyrene, polyvinyl acetate into bucky paper membranes gives rise in the tensile strength, Young's modulus, tough character, and straining to crack values [15]. The addition of biopolymers like proteins and polysaccharides into bucky papers comprised of single-walled nanotubes can improvise their mechanical abilities [8]. A detailed analysis has shown that some biopolymers were left in the bucky paper membranes after vacuum filtration because of their ability to non-covalently interact with the nanotube. Improvising the mechanical properties of bucky paper membranes is again crucial as it reduces the risk which occurs because of the excretions of single carbon nanotubes into the environment.

There have been observations into the biological consequences of exposure to CNTs due to the similarity of these materials to asbestos elements. These studies have also shown that carbon nanotubes provide a specific effect like oxidative stress, disruption of membrane and interference with cell signaling pathways [19, 23, 63, 70, 74, 81, 85]. Consequentially, it is crucial to consider those very small quantities of carbon nanotubes should not break from bucky paper membranes or any other carbon nanotube membrane. It can be achieved by joining the nanotubes to each other using a covalent bond in bucky paper or aligned membrane. Because of their cheap manufacturing methods, it is possible to prepare bucky papers on a larger scale than aligned.

5 Preparation of CNTs

The main techniques that are implemented to prepare considerable amounts of carbon nanotubes are laser ablation, arc discharge, gas-phase catalytic growth from carbon monoxide, and chemical vapor deposition from hydrocarbons [79]. Arc discharge and laser ablation approaches are only good to prepare small numbers of carbon nanotubes. The products prepared often have some quantity of impurity in the form of particles of catalyst and amorphous carbon [79]. Purification techniques are needed to separate the nanotubes from unwanted by-products before investigating their characteristics and prospective functions. The results observed provided prospective encouragement to explore the CNT membrane material for filtration purposes. This has been strengthened after observing the cytotoxic properties of carbon nanotube membranes. This shows that these materials are least influenced by biofouling in comparison to that of traditional polymeric membranes and also displayed enhanced membrane lifetime duration via eliminating microbes [9].

6 Production of CNTs

Purification procedures require the separation of nanotubes from unwanted byproducts before being implemented for further instigation. The gas-phase techniques that produce nanotubes at low temperatures are changeable to the non-interrupted manufacture of a vast number of CNTs as continue flowing of gas would significantly moderate the source of the preparatory material.

An additional advantage related to the fabrication of the carbon tube with the chemical vapor deposition is the enhanced purity of the getting material (Fig. 1), which reduces the requirement for accomplishing all the stages [73]. With the help

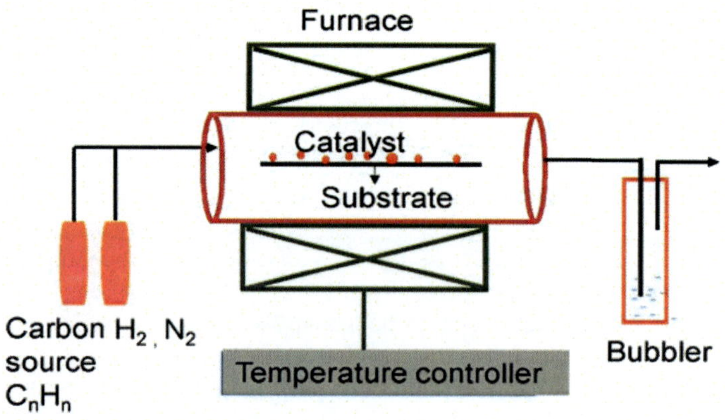

Fig. 1 Diagrammatic representation of the CVD equipment

Table 2 Methods used for preparing CNT-based composite membranes

Type of membrane	Synthesis method	References
CNT/PA	Interfacial polymerization	[43, 45]
MWCNT/PSf (C/P)	Phase inversion	[12]
MWCNT/PA	Polymer grafting	[89]
(VACNTs)/polyaniline (PANi)	In situ polymerization	[18]
MWCNTs/PAN	Phase inversion	[65]
DDA-MWNTs/PSf	Phase inversion	[40]
(TNRs)/MWCNTs/PES	Phase inversion	[90]
TFC/polysulfone (PS-20)/MWCNT	Interfacial polymerization	[2]
PSF/CNTs	Phase inversion	[41]
A-MWCNTs	Phase inversion	[102]
Zwitterionic membrane	Phase inversion	[28]
Polymer membranes	In situ polymerization	[1]
Graphene oxide-incorporated thin-film nanocomposite membrane	In situ polymerization	[49]
Thin-film nanocomposite membrane	In situ polymerization	[97]
Polyester thin-film composite membrane	In situ polymerization	[64]
Carbon nanotube/PSf	Immersion precipitation	[39]
MWCNT/PVDF/PDMS	Deposition/coating	[62]
MWCNT/PVDF	Phase inversion	[60]
Acid-modified MWCNTs/nanosilver/PSf	Interfacial polymerization and phase inversion	[42]
F-MWCNTs/PES	Phase inversion	[104]
(NCNT)/PES	Modified phase inversion	[77]
PVDF/Fe_2O_3/MWCNTs	In situ polymerization	[3]
Surface-modified polyethersulfone (PES) composite membranes	Spray-assisted layer-by-layer	[58]
VA CNTs	In situ polymerization	[47]
MWCNT/nylon6	In situ polymerization	[86]

of the chemical vapor deposition method, single-walled nanotubes with the excellent purity have been fabricated in the gaseous phase by using $Fe(CO)_5$ and carbon monoxide in the increased pressure CO disproportion method [10].

7 Techniques for the Fabrication of Mixed CNTs Membranes

The following are the methods (Table 2) used for preparing the mixed-carbon nanotube membranes:

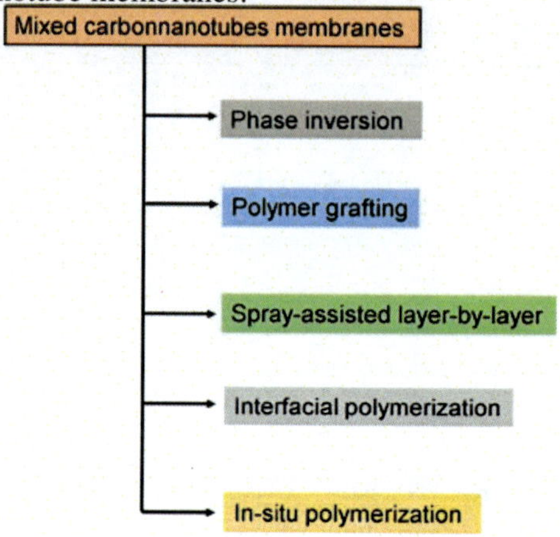

7.1 *Phase Inversion*

Multi-walled carbon nanotubes blend membranes prepared through the phase inversion process with a coagulant in the form of water [14]. A homogeneous multi-walled carbon nanotubes solution was made in N-methyl-2pyrrolidone (NMP) and blended with PSf solution. Dodecylamine functionalized multi-walled CNTs (DDA-MWNTs) were fabricated by Khalied and co-workers. The nanocomposite polysulfone/DDA-MWNts was casted by the phase inversion method. The fabricated nanocomposite membrane displayed excellent fouling resistance and flux recovery [40]. Phase inversion process with dimethylacetamide as a solvent and polyvinylpyrrolidone as a porogen was used to prepare flat sheet nanocomposite PSf/DDA-MWNTs membranes. A novel polyethersulfone (PES) membranes were prepared with the help of phase inversion method with the increased loading of the functionalized oxidized MWCNTs (OMWCNTS) together with the Arabic gum. The prepared OMWCNTs were characterized by various techniques like scanning electron microscopy and transmission electron microscopy, energy-dispersive X-ray spectroscopy [71].

7.2 *Interfacial Polymerization*

By employing interfacial polymerization, polyamide reverse osmosis membranes (RO) with the carbon nanotubes were fabricated. In this process, the functionalized CNTs were fabricated by the reaction of CNTs with the acidic mixture of sulfuric acid and nitric acid (in ratio 3:1), at different amounts of reaction conditions. The synthesized carbon nanotubes were observed to be well settled in the PA layer; this has been confirmed via various analytical techniques. The polyamide RO membranes containing well-dispersed CNTs possess an enhanced flux rate than the polyamide amide membranes devoid of CNTs [43, 45]. Polyamide thin-film membranes were prepared on polysulfone (PS-20) base by using interfacial polymerization of aqueous m-phenylenediamine (MPD) solution and 1,3,5-benzenetricarbonyl trichloride (TMC) in n-hexane organic solution. MWCNT were carboxylated by the heating of MWCNT powder in the sulfuric acid and nitric acid under continuous sonication at various intervals. Polyamide nanocomposites were then synthesized by the incorporation of MWCNT and the carboxylated MWCNT at various concentrations. The salt rejection and water flux performances of the prepared membrane revealed superior performance with that of other membranes [2]. CNT-enhanced thin-film composite membranes were fabricated by the incorporation of CNTs into the active layers of membranes for increasing its efficacy for the water treatment. MWCNT grafted via poly(methyl methacrylate) PMMA was prepared by microemulsion polymerization of methyl methacrylate(MMA) in the presence of c-MWNTS (acid-modified MWCNTS). The prepared membranes have proven significantly improved selectivity and permeability [72].

7.3 *Spray-Assisted Layer-by-Layer*

Using the spray-aided layer-by-layer method, a functionalized multi-walled CNT was fabricated by [57]. For improving the commercial polyethersulfone (PES) ultrafiltration (UF) membranes, antifouling properties negatively charged functionalized MWCNTs, mixed poly(sodium 4-styrenesulfonate) (PSS), and a positively charged poly(diallyldimethylammonium chloride) (PDDA) were deposited PES substrate through spray-assisted layer-by-layer L) method. The synthesized membrane displayed better anti-protein fouling and flux recovery [57]. Surface-modified polyethersulfone (PES) composite ultra-filtration membrane by using a spray-assisted layer-by-layer Liu and co-workers proved method. The prepared nanocomposite membrane displayed enhancement in the antifouling properties [58].

7.4 Polymer Grafting

A multi-walled carbon nanotube aromatic polyamide nanocomposite membrane fabrication was shown by Shawky and co-workers. Various instrumental techniques characterized the morphology of the surface, toughness, and roughness of the prepared nanocomposite membrane. The SEM and AFM images displayed that the MWCNTs were well dispersed in the PA (aromatic polyamide) matrix. Measurements of mechanical properties of this composite showed increasing membrane strength with increasing MWCNT content with monotonic increases in Young's modulus, toughness, and tensile strength. The prepared nanocomposite membrane displayed better salt rejection and organic matter rejection than the normal polyamide matrix membrane.

7.5 In Situ Polymerization

For the removal of natural organic matter in the water, MWCNT polyaniline (PANI)/polyethersulfone (PES) membranes were synthesized by incorporation of in situ polymerized MWCNTs/PANI complex. The prepared membrane showed enhanced permeability than that of the PES membranes. Higher rates for the rejection of the natural organic matter were also observed. This greater presentation is accredited to the synergetic effect of amplified porosity, narrow pore size distribution and hydrophilicity, and positively charged of the membranes by the inclusion of MWCNTs/PANI complex. The prepared membrane also demonstrated a cent percent water flux [52, 53]. A VACNTs/polyaniline (PANi) composite membrane was also fabricated via microwave supported in situ polymerization [18]. It was proved that with the help of a microwave, a better nanocomposite membrane could be fabricated.

8 CNTs Characterizations

Various techniques are available to analyze the characterization of carbon nanotubes. transmission electron microscopy (TEM) along with the scanning electron microscopy (SEM) are the methods that are known to observe the top of the peak along with the sidewall and with the morphology of CNTs [7, 30, 78]. The most significant tool for the characterization of the carbon nanotubes is the Raman spectroscopy technique [20, 21, 87]. It is regularly seen to check the quality as well as the pureness of the made carbon nanotubes. A Raman spectrum of carbon nanotubes shows two chiefs first-order bands, which include D band and G band. The former band is concerned with the imperfections of the carbon nanotubes and can be seen around 1350 cm^{-1}. The latter band is concerned with the amount of graphitization of carbon nanotubes that are at 1600 cm^{-1}. Therefore, the ratio of the area of both the band is found to determine the defect level in a specific carbon nanotube sample. Hence, by modifying reactants and chemical vapor deposition preparation dimensions like a catalyst, substrate, temperature, carbon precursor, pressure, time, and rate of gas flow assisted with several customizations for functional groups and characterization techniques here optimized carbon nanotubes could be gotten for various practical applications (Table 1).

9 Challenges Related to CNTs

Carbon nanotube membranes have a great prospective future in the wastewater treatment industry. However, it faces a lot of challenges to produce membranes as they are in the very first stage, and various vital issues are still to be tested. Viable readiness, reducing the cost of CNT, scaling in the industries, and assessing probable lethal effects of carbon nanotubes are some encounters that are about to be finished. Manufacturing carbon nanotubes on a large scale with a considerable pore size and the way to distribute is yet a vital challenge in implementing carbon nanotube on a great economic scale. Researchers must study more changed methods to get a more economical method to create a carbon nanotube. Another obstruction that prevents the implementation of carbon nanotubes in large-scale operation is the cost, specifically that of a single-walled carbon nanotube. Because of the high rise in the industrial manufacture of carbon nanotubes, the cost related to them will be cut down in the future. The prospective hazardous issues by carbon nanotubes on the health of humans and on the atmosphere made significant questions supposed to be answered detrimentally. It is assumed that raw carbon nanotubes are more hazardous in contrast to chemically modified carbon nanotubes. This is also because of the availability of a metal catalyst in raw carbon nanotubes. Another obstacle is the difficult growth of carbon nanotubes with good alignment in vertically aligned carbon nanotube membranes. The disarranged alignment can affect membrane properties like salt rejection and flux. The mechanisms that separated the pollutants from freshwater must be examined carefully.

10 Conclusion

Researchers were focusing on CNTs because of them showing excellent permeability. Their level of performance is the best among other membranes derived by carbon nanotubes. The latter offers good benefits like cheap cost and higher ease at production, along with the capability to be generated at a larger scale. Investigations into the applications like desalination, ultrafiltration, nanofiltration have shown that carbon nanotube membrane often showed increased resistance to biofouling in contrast to the tradition polymer. There is also a need to investigate the differences between the characteristics of filtration of bucky papers and that of composite membranes using various carbon nanotubes and agents of dispersion.

Acknowledgements The authors are thankful to the Banasthali Vidyapith and the Central University of Gujarat. One of the authors, Dr. Pallavi Jain, is grateful to the SRM Institute of Science and Technology, Delhi-NCR Campus.

References

1. Adamczak M, Kamińska G, Bohdziewicz J (2019) Preparation of polymer membranes by in situ interfacial polymerization. Int J Pol Sci 2019:1–13
2. Al-Hobaib AS, Al-Sheetan KhM, Shaik MR, Al-Suhybani MS (2017) Modification of thin-film polyamide membrane with multi-walled carbon nanotubes by interfacial polymerization. Appl Water Sci 7:4341–4350
3. Alpatova A, Meshref M, Mcphedran KN, El-din MG (2015) Composite polyvinylidene fluoride (PVDF) membrane impregnated with Fe_2O_3 nanoparticles and multiwalled carbon nanotubes for catalytic degradation of organic contaminants. J Membr Sci 490:227–235
4. Asmaly HA, Abussaud B, Ihsanullah Saleh TA, Alaadin A, Laoui T, Shemsi AM, Gupta VK, Atieh MA, Asmaly HA, Abussaud B, Saleh TA, Alaadin A (2015) Evaluation of micro- and nano-carbon-based adsorbents for the removal of phenol from aqueous solutions. Toxicol Environ Chem 97:1164–1179
5. Asmaly HA, Abussaud B, Ihsanullah Saleh TA, Gupta VK, Atieh MA (2015) Ferric oxide nanoparticles decorated carbon nanotubes and carbon nanofibers: from synthesis to enhanced removal of phenol. J Saudi Chem Soc 19:511–520
6. Baek Y, Kim C, Seo DK, Kim T, Lee JS, Kim YH, Ahn KH, Bae SS, Lee SC, Lim J (2014) High performance and antifouling vertically aligned carbon nanotube membrane for water purification. J Membr Sci 460:171–177
7. Belin T, Epron F (2005) Characterization methods of carbon nanotubes: a review. Mater Sci Eng B 119:105–118
8. Boge J, Sweetman LJ, Panhuis M, Ralph SF (2009) The effect of preparation conditions and biopolymer dispersants on the properties of SWNTs buckypapers. J Mater Chem A 19:9131–9140
9. Brady-Estévez AS, Kang S, Elimelech M (2008) A single-walled-carbon-nanotube filter for removal of viral and bacterial pathogens. Small 4:481–484
10. Bronikowski MJ, Willis PA, Colbert DT, Smith KA, Smalley RE (2001) Gas-phase production of carbon single-walled nanotubes from carbon monoxide via the Hipco process: a parametric study. Vac Sci Technol A 19:1800–1805
11. Brunet L, Lyon D, Zodrow K, Rouch J-C, Caussat B, Serp P, Remigy J-C, Wiesner M, Alvarez PJ (2008) Properties of membranes containing semi-dispersed carbon nanotubes. Environ Eng Sci 25:565–575
12. Celik E, Park H, Choi H (2011) Carbon nanotube blended polyethersulfone membranes for fouling control in water treatment. Water Res 45:274–282
13. Chen W, Chen S, Liang T, Zhang Q, Fan Z, Yin H, Huang K-W, Zhang X, Lai Z, Sheng P (2018) High-flux water desalination with interfacial salt sieving effect in nanoporous carbon composite membranes. Nat Nanotechnol 13:345–350
14. Choi J, Jegal J, Kim W (2006) Fabrication and characterization of multi-walled carbon nanotubes/polymer blend membranes. J Membr Sci 284:406–415

15. Coleman JN, Blau WJ, Dalton AB, Munoz E, Collins S, Kim BG, Razal J, Selvidge M, Vieiro G, Baughman RH (2003) Improving the mechanical properties of single-walled carbon nanotube sheets by intercalation of polymeric adhesives. Appl Phys Lett 82:1682–1684
16. Das R, Ali E, Bee S, Hamid A, Ramakrishna S, Zaman Z (2014) Carbon nanotube membranes for water purification: a bright future in water desalination. Desalination 336:97–109
17. Dharap P, Li Z, Nagarajaiah S, Barrera EV (2004) Nanotube film based on single-wall carbon nanotubes for strain sensing. Nanotechnology 15:379–382
18. Ding J, Li X, Wang X, Zhang J, Yu D, Qiu B (2015) Fabrication of vertical array CNTs/polyaniline composite membranes by microwave-assisted in situ polymerization. Nanoscale Res Lett 10:1–9
19. Dong J, Ma Q (2015) Advance sin mechanisms and signaling pathways of carbon nano tube toxicity. Nano Toxicol 9:658–676
20. Dresselhaus MS, Dresselhaus G, Jorio A (2007) Raman spectroscopy of carbon nanotubes in 1997 and 2007. J Phys Chem C 111:17887–17893
21. Dresselhaus MS, Jorio A, Saito R (2010) Characterizing grapheme graphite, and carbon nanotubes by Raman spectroscopy. Annu Rev Condens Matter Phys 1:89–108
22. Elimelech M, Phillip WA (2011) The future of seawater desalination: energy, technology, and the environment. Science 333:712–717
23. Ema M, Gamo M, Honda K (2016) A review of toxicity studies of single-walled carbon nanotubes in laboratory animals. Regul Toxicol Pharmacol 74:42–63
24. Endo M, Muramatsu H, Hayashi T, Kim YA, Terrones M, Dresselhaus MS (2005) Anotechnology: 'buckypaper' from coaxial nanotubes. Nature 433:476
25. Farahani MHDA, Vatanpour V (2018) A comprehensive study on the performance and antifouling enhancement of the PVDF mixed matrix membranes by embedding different nanoparticulates: clay, functionalized carbon nanotube, SiO_2 and TiO_2. Sep Purif Technol 197:372–381
26. Frizzell CJ, Panhuis M, Coutinho DH, Balkus KJ, Minett AI, Blau WJ, Coleman JN (2005) Reinforcement of macroscopic carbon nanotube structures by polymer intercalation: the role of polymer molecular weight and chain conformation. Phys Rev B 72:245420
27. Goh PS, Ismail AF, Ng BC (2013) Carbon nanotubes for desalination: performance evaluation and current hurdles. Desalination 308:2–14
28. Guo YS, Mi YF, Ji YL, An QF, Gao CJ (2019) One-step surface grafting method for preparing zwitterionic nanofiltation membrane via in situ introduction of initiator in interfacial polymerization. ACS Appl Polym Mater 15:1022–1033
29. Hebbar RS, Isloor AM, Inamuddin Asiri AM (2017) Carbon nanotube and graphene-based advanced membrane materials for desalination. Environ Chem Lett 15:643–671
30. Herrero-Latorre C, Alvarez-Mendez J, Barciela-Garcia J, García-Martin S, Pena-Crecente RM (2015) Characterization of carbon nanotubes and analytical methods for their determination in environmental and biological samples: a review. Anal Chim Acta 853:77–94
31. Hinds BJ, Chopra N, Rantell T, Andrews R, Gavalas V, Bachas LG (2004) A ligned multi walled carbon nanotube membranes. Science 303:62–65
32. Ihsanullah AM, Al Amer, Laoui T, Abbas A, Al-Aqeeli N, Patel F, Khraisheh M, Ali M, Hilal N (2016) Fabrication and antifouling behaviour of a carbon nanotube membrane. Mater Des 89:549–558
33. Ihsanullah T, Laoui AM, Al-Amer Khalil AB, Abbas A, Khraisheh M, Atieh MA (2015) Novel anti-microbial membrane for desalination pretreatment: a silver nanoparticle-doped carbon nanotube membrane. Desalination 376:82–93
34. Ihsanullah A Abbas, Al-Amer AM, Laoui T, Al-Marri MJ, Nasser MS, Khraisheh M, Atieh MA (2016) Heavy metal removal from aqueous solution by advanced carbon nanotubes: critical review of adsorption applications. Sep Purif Technol 157:141–161
35. Ihsanullah Al-khaldi FA, Abu-sharkh B, Mahmoud A, Qureshi MI, Laoui T, Atieh MA (2016) Effect of acid modification on adsorption of hexavalent chromium (Cr(VI)) from aqueous solution by activated carbon and carbon nanotubes. Desalin Water Treat 57:7232–7244

36. Ihsanullah Asmaly HA, Saleh TA, Laoui T, Gupta VK, Atieh MA (2015) Enhanced adsorption of phenols from liquids by aluminum oxide/carbon nanotubes: comprehensive study from synthesis to surface properties. J Mol Liq 206:176–182
37. Ihsanullah Al-Khaldi FA, Abusharkh B, Khaled M, Atieh MA, Nasser MS, Laoui T, Saleh TA, Agarwal S, Tyagi I, Gupta VK (2015) Adsorptive removal of cadmium(II) ions from liquid phase using acid modified carbon-based adsorbents. J Mol Liq 204:255–263
38. Kar S, Bindal RC, Tewari PK (2012) Carbon nanotube membranes for desalination and water purification: challenges and opportunities. Nano Today 7:385–389
39. Kar S, Subramanian M, Pal A, Ghosh AK, Bindal RC, Prabhakar S, Nuwad J, Pillai CGS, Chattopadhyay S, Tewari PK (2013) Preparation, characterisation and performance evaluation of anti-biofouling property of carbon nanotube-polysulfone nanocomposite membranes. AIP Conf Proc 1538:181–185
40. Khalid A, Al-Juhani AA, Al-Hamouz OC, Laoui T, Khan Z, AliAtieh M (2015) Preparation and properties of nanocomposite polysulfone/multi-walled carbon nanotubes membranes for desalination. Desalination 367:134–144
41. Khoshrou S, Moghbeli MR, Ghasemi E (2015) Polysulfone/carbon nanotubes asymmetric nanocomposite membranes: effect of nanotubes surface modification on morphology and water permeability. Iran J Chem Eng 12:69–83
42. Kim E, Hwang G, El-din MG, Liu Y (2012) Development of nanosilver and multi-walled carbon nanotubes thin-film nanocomposite membrane for enhanced water treatment. J Membr Sci 394–395:37–48
43. Kim HJ, Choi K, Baek Y, Kim D, Shim J, Yoon J, Lee J (2014) High-performance reverse osmosis CNT/polyamide nanocomposite membrane by controlled interfacial interactions. ACS Appl Mater Interfaces 6:2819–2829
44. Kim J, Van Der Bruggen B (2010) The use of nanoparticles in polymeric and ceramic membrane structures: review of manufacturing procedures and performance improvement for water treatment. Environ Pollut 158:2335–2349
45. Kim S, Fornasiero F, Park HG, In JB, Meshot E, Giraldo G, Stadermann M, Fireman M, Shan J, Grigoropoulos CP (2014) Fabrication of flexible, aligned carbon nanotube/polymer composite membranes by in-situ polymerization. J Membr Sci 460:91–98
46. Kim TH, Lee I, Yeon K-M, Kim J (2018) Bio catalytic membrane with acylase stabilized on intact carbon nanotubes for effective antifouling via quorum quenching. J Membr Sci 554:357–365
47. Kim YA, Muramatsu H, Hayashi T, Endo M, Terrones M, Dresselhaus MS (2006) Fabrication of high-purity, double-walled carbon nanotube buckypaper. Chem Vap Depos 12:327–330
48. Knapp W, Schleussner D (2002) Carbon buckypaper field emission investigations. Vacuum 69:333–338
49. Lai GS, Lau WJ, Goh PS, Tan YH, NgA BC, Ismail F (2019) A novel interfacial polymerization approach towards synthesis of graphene oxide-incorporated thin film nanocomposite membrane with improved surface properties. Arab J Chem 12:75–87
50. Lalia BS, Ahmed FE, Shah T, Hilal N, Hashaikeh R (2015) Electrically conductive membranes based on carbon nanostructures for self-cleaning of biofouling. Desalination 360:8–12
51. Lee B, Baek Y, Lee M, Jeong DH, Lee HH, Yoon J, Kim YH (2015) A carbon nanotube wall membrane for water treatment. Nat Commun 6:7109
52. Lee J, Jeong S, Liu Z (2016) Progress and challenges of carbon nanotube membrane in water treatment. Crit Rev Environ Sci Technol 46:999–1046
53. Lee J, Ye Y, Ward AJ, Zhou C, Chen V, Minett AI, Lee S, Liu Z, Chae S, Shi J (2016) High flux and high selectivity carbon nanotube composite membranes for natural organic matter removal. Sep Purif Technol 163:109–119
54. Lee K-J, Park H-D (2016) The most densified vertically-aligned carbon nanotube membranes and their normalized water permeability and high pressure durability. J Membr Sci 501:144–151
55. Lehman JH, Terrones M, Mansfield E, Hurst KE, Meunier V (2011) Evaluating the characteristics of multi wall carbon nanotubes. Carbon 49:2581–2602

56. Li S, Liao G, Liu Z, Pan Y, Wu Q, Weng Y, Zhang X, Yang Z, Tsui OKC (2014) Enhanced water flux in vertically aligned carbon nanotube arrays and polyethersulfone composite membranes. J. Mater. Chem. A. 2:12171–12176
57. Liu L, Son M, Chakraborty S, Bhattacharjee C (2013) Fabrication of ultra-thin polyelectrolyte/carbon nanotube membrane by spray-assisted layer-by-layer technique: characterization and its anti-protein fouling properties for water treatment. Desalin Water Treat 51:6194–6200
58. Liu L, Son M, Park H, Celik E, Bhattacharjee C, Choi H (2014) Efficacy of CNT-bound polyelectrolyte membrane by spray-assisted layer-by-layer (LbL) technique on water purification. RSC adv 4:32858–32865
59. López-Lorente AI, Simonet BM, Valcárcel M (2010) The potential of carbon nanotube membranes for analytical separations. Anal Chem 82:5399–5407
60. Ma J, Zhao Y, Xu Z, Min C, Zhou B, Li Y, Li B, Niu J (2013) Role of oxygen-containing groups on MWCNTs in enhanced separation and permeability performance for PVDF hybrid ultra-filtration membranes. Desalination 320:1–9
61. Ma L, Dong X, Chen M, Zhu L, Wang C, Yang F, Dong Y (2017) Fabrication and water treatment application of carbon nanotubes (CNTs)-based composite membranes: a review. Membranes (basel) 7:1–21
62. Madaeni SS, Zinadini S, Vatanpour V (2013) Preparation of superhydrophobic nanofiltration membrane by embedding multiwalled carbon nanotube and polydimethylsiloxane in pores of microfiltration membrane. Sep Purif Technol 111:98–107
63. Magrez A, Kasas S, Salicio V, Pasquier N, Seo JW, Celio M, Catsicas S, Schwaller B, Forro L (2006) Cellular toxicity of carbon-based nanomaterials. Nano Lett 6:1121–1125
64. Mah KH, Yussof KHY, Seman MNA, Mohammad AW (2016) Synthesis and characterization of polyester thin film composite membrane via interfacial polymerization: fouling behaviour of uncharged solute. Mater Sci Eng 162:012037
65. Majeed S, Fierro D, Buhr K, Wind J, Du B, Boschetti-de-Fierro A, Abetz V (2012) Multiwalled carbon nanotubes (MWCNTs) mixed polyacrylonitrile (PAN) ultrafiltration membranes. J Membr Sci 403–404:101–109
66. Majumder M, Chopra N, Hinds BJ (2005) Effect of tip functionalization on transport through vertically oriented carbon nanotube membranes. J Am Chem Soc 127:9062–9070
67. Majumder M, Keis K, Zhan X, Meadows C, Cole J, Hinds BJ (2008) Enhanced electrostatic modulation of ionic diffusion through carbon nanotube membranes by diazonium grafting chemistry. J Membr Sci 316:89–96
68. Majumder M, Stinchcomb A, Hinds BJ (2010) Towards mimicking natural protein channels with aligned carbon nano tube membranes for active drug delivery. Life Sci 86:563–568
69. Manawi Y, Kochkodan V, Hussein MA, Khaleel MA, Khraisheh M, Hilal N (2016) Can carbon-based nanomaterials revolutionize membrane fabrication for water treatment and desalination. Desalination 391:69–88
70. Manna SK, Sarkar S, Barr J, Wise K, Barrera EV, Jejelowo O, Rice-Ficht AC, Ramesh GT (2005) Single-walled carbon nanotube induces oxidative stress and activates nuclear transcription factor-κb in human keratinocytes. Nano Lett 5:1676–1684
71. Najjar A, Sabri S, Al-Gaashani R, Atieh MA, Kochkodan V (2019) Antibiofouling performance by polyethersulfone membranes cast with oxidized multiwalled carbon nanotubes and arabic gum. Membranes 9:1–32
72. Shen Jn, Yu Cc, Ruan Hm, Van Der Bruggen B (2013) Preparation and characterization of thin-film nanocomposite membranes embedded with poly (methyl methacrylate) hydrophobic modified multiwalled carbon nanotubes by interfacial polymerization. J Membr Sci 442:18–26
73. Nikolaev P, Broni kowski MJ, Bradley RK, Rohmund F, Colbert DT, Smith KA, Smalley RE (1999) Gas-phase catalytic growth of single-walled carbon nanotubes from carbon monoxide. Chem Phys Lett 313:91–97
74. Ong L-C, Chung FF-L, Tan Y-F, Leong C-O (2016) Toxicity of single-walled carbon nanotubes. Arch Toxicol 90:103–118

75. Park S-M, Jung J, Lee S, Baek Y, Yoon J, Seo DK, Kim YH (2014) Fouling and rejection behavior of carbon nanotube membranes. Desalination 343:180–186
76. Pendergast MM, Hoek EMV (2011) A review of water treatment membrane nanotechnologies. Environ Sci 4:1946–1971
77. Phao N, Nxumalo EN, Mamba BB, Mhlanga SD (2013) A nitrogen-doped carbon nanotube enhanced polyethersulfone membrane system for water treatment. Phys Chem Earth 66:148–156
78. Ping D, Wang C, Dong X, Dong Y (2016) Co-production of hydrogen and carbon nanotubes on nickel foam via methane catalytic decomposition. Appl Surf Sci 369:299–307
79. Prasek J, Drbohlavova J, Chomoucka J, Hubalek J, Jasek O, Adam V, Kizek R (2011) Methods for carbon nanotubes synthesis—review. J Mater Chem A 21:15872–15884
80. Prokudina NA, Shishchenko ER, Joo O-S, Hyung K-H, Han S-H (2005) A carbon nanotube film as a radio frequency filter. Carbon 43:1815–1819
81. Pulskamp K, Diabate S, Krug HF (2007) Carbon nanotubes show no sign of acute toxicity but induce intracellular reactive oxygen species in dependence on contaminants. Toxicol Lett 168:58–74
82. Qin S, Qin D, Ford WT, Resasco DE, Herrera JE (2004) Functionalization of single-walled carbon nanotubes with polystyrene via grafting to and grafting from methods. Macromolecules 373:752–757
83. Qu X, Alvarez PJJ, Li Q (2013) Applications of nanotechnology in water and wastewater treatment. Water Res 47:3931–3946
84. Rizzuto C, Pugliese G, Bahattab MA, Aljlil SA, Drioli E, Tocci E (2018) Multiwalled carbon nanotube membranes for water purification. Sep Purif Technol 193:378–385
85. Rodriguez-Yanez Y, Munoz B, Albores A (2013) A mechanisms of toxicity by carbon nanotubes. Toxicol Mech Methods 23:178–195
86. Saeed K, Park SY, Haider S, Baek JB (2009) In situ polymerization of multi-walled carbon nanotube/nylon-6 nanocomposites and their electrospun nanofibers. Nano Res Lett 4:1–39
87. Saito R, Hofmann M, Dresselhaus G, Jorio A, Dresselhaus MS (2011) Raman spectroscopy of graphene and carbon nanotubes. Adv Phys 60:413–550
88. Saththasivam J, Yiming W, Wang K, Jin J, Liu Z (2018) A novel architecture for carbon nanotube membranes towards fast and efficient oil/water separation. Sci Rep 8:7418
89. Shawky HA, Chae S, Lin S, Wiesner MR (2011) Synthesis and characterization of a carbon nanotube/polymer nanocomposite membrane for water treatment. Desalination 272:46–50
90. Shaban A, Ashraf AM, AbdAllah H, El-Salam HM (2018) Titanium dioxide nanoribbons/multi-walled carbon nanotube nanocomposite blended polyethersulfone membrane for brackish water desalination. Desalination 444:129–141
91. Shulga YM, Tien TC, Huang CC, Lo SC, Muradyan VE, Polyakova NV, Ling YC, Loutfy RO, Moravsky AP (2007) XPS study of fluorinated carbon multi-walled nanotubes. J Electron Spectrosc Rel Phenom 160:22–28
92. Spitalsky Z, Aggelopoulos C, Tsoukleri G, Tsakiroglou C, Parthenios J, Georga S, Krontiras C, Tasis D, Papagelis K, Galiotis C (2009) The effect of oxidation treatment on the properties of multi-walled carbon nanotube thin films. Mater Sci Eng B 165:135–138
93. Srivastava A, Srivastava ON, Talapatra S, Vajtai R, Ajayan PM (2004) Carbon nanotube filters. Nat Mater 3:610–614
94. Tanaka T (2010) Filtration characteristics of carbon nanotubes and preparation of buckypapers. Desalin Water Treat 17:193–198
95. Tankus KA, Issman L, Stolov M, Freger V (2018) Electrotreated carbon nanotube membranes for facile oil–water separations. ACS Appl Nano Mater 1:2057–2061
96. Thamaraiselvan C, Lerman S, Weinfeld-Cohen K, Dosoretz CG (2018) Characterization of a support-free carbon nanotube-microporous membrane for water and wastewater filtration. Sep Purif Technol 202:1–8
97. Tian E, Wang X, Wang X, Ren Y, Zhao Y, An X (2019) Characterization of thin-film nanocomposite membrane with high flux and antibacterial performance for forward osmosis. Ind Eng Chem Res 58:897–907

98. Vermisoglou EC, Pilatos G, Romanos GE, Karanikolos GN, Boukos N, Mertis K, Kakizis N, Kanellopoulos NK (2008) Synthesis and characterisation of carbon nanotube modified anodised alumina membranes. Microporous Mesoporous Mater 110:25–36
99. Vohrer U, Kolaric I, Haque MH, Roth S, Detlaff-Weglikowska U (2004) Carbon nanotube sheets for the use as artificial muscles. Carbon 42:1159–1164
100. Wang Y, Liu Y, Yu Y, Huang H (2018) Influence of CNT-rGO composite structures on their permeability and selectivity for membrane water treatment. J Membr Sci 551:326–332
101. Whitby RLD, Fukuda T, Maekawa T, James SL, Mikhalovsky SV (2008) Geometric control and tuneable pore size distribution of bucky paper and bucky discs. Carbon 46:949–956
102. Yu Z, ZengG Pan Y, Lv L, Hui M, ZhangL Yie H (2015) Effect of functionalized multi-walled carbon nanotubes on the microstructure and performances of PVDF membranes. RSC Adv 5:75998–76006
103. Zhao H, Qiu S, Wu L, Zhang L, Chen H, Gao C (2014) Improving the performance of polyamide reverse osmosis membrane by incorporation of modified multi-walled carbon nanotubes. J Membr Sci 450:249–256
104. Zirehpour A, Rahimpour A, Jahanshahi M, Peyravi M (2014) Mixed matrix membrane application for olive oil wastewater treatment: process optimization based on Taguchi design method. J Environ Manag 132:113–120

Efficient Carbon Nanocomposites as a Sustainable Adsorbents/Photocatalyst for Water Purification

Sheetal Sharma, Vishal Dutta, Pankaj Raizada, Vijay Kumar Thakur, and Pardeep Singh

Abstract Widespread pollution of water bodies by textile industries, agricultural wastes and organic contaminants has developed water contamination which has become one of the chief worldwide environmental disasters. The outcome of this uncivilized negligence is modeling solemn dangers to health of living beings thus today; sustaining water reserves for the spirit of life is of severe concern. Thus, there is a prerequisite for an effectual, cost-effective, steadfast, feasible and eco-friendly technology to remove pollutants and bacterium from wastewater. Advanced oxidation processes would be one of greatest favorable preferences for wastewater mitigation. This chapter reviews the preparation and characterization of carbon nanocomposites, and photocatalytic and anti-bacterial activity of carbon nanocomposites. Furthermore, the comprehension breakdowns and research confronts have been underlined, incorporating fabrication, optimization and critical concerns linked with carbon nanocomposites for its commercialization for the bacterial disinfection and contaminant degradation.

Keywords Carbon nanocomposites · Adsorption · Advanced oxidation process · Photocatalysis · Bacterial disinfection · Water purification

---

Sheetal Sharma and Vishal Dutta have contributed equally.

S. Sharma · V. Dutta · P. Raizada · P. Singh (✉)
Faculty of Basic Sciences, School of Chemistry, Shoolini University, Solan, Himachal Pradesh 173212, India
e-mail: pardeepchem@gmail.com

P. Raizada · P. Singh
Himalayan Centre for Excellence in Nanotechnology, Shoolini University, Solan, HP 173229, India

V. K. Thakur
Biorefining and Advanced Materials Research Centre, , , United Kingdom, Scotland's Rural College (SRUC), Edinburgh, United Kingdom

M. Jawaid et al. (eds.), *Environmental Remediation Through Carbon Based Nano Composites*, Green Energy and Technology,
https://doi.org/10.1007/978-981-15-6699-8_9

Abbreviations

BiOX	Bismuth oxyhalide
BPA	Bisphenol A
CNTs	Carbon nanotubes
eV	Electron volt
g-C_3N_4	Graphitic carbon nitride
MWCNT	Multi-walled carbon nanotube
NIR	Near infrared
RGO	Reduced graphite oxide
RhB	Rhodamine B
ROS	Reactive oxygen species
SEM	Scanning electron microscopy
SPR	Surface plasmon resonance
SWCNT	Single-walled carbon nanotube
TEM	Transmission electron microscopy
UCPL	Up-conversion photoluminescence
UV	Ultraviolet
XPS	X-ray photoelectron spectroscopy
XRD	X-ray diffraction

1 Introduction

With the gradual growth in the industrial world, environmental pollution has become a major concern [67, 112]. On daily basis, large numbers of toxic chemicals are disposed into the rivers, lakes and oceans. In the aquatic world, diverse toxic pollutants are identified as heavy metals, textile dyes, pesticides, surfactants and insecticides [54, 68]. Abundant consideration has been concentrated on the removal of these injurious and toxic pollutants from the water bodies in order to stop their precarious effect on the ecosystem [61, 77]. So far, various conventional techniques such as sedimentation, reverse osmosis, filtration, membrane filtration, and chemical and biological treatments have been deployed for the water mitigation [21, 56]. The treatment of decontaminated water using such conventional techniques is found quite unsatisfactory as the water contains various contaminants like pharmaceutical wastes, organic solvents and pesticides [17, 83]. As in case of adsorption process, activated carbon used as adsorbent loses its adsorbent capability after a number of repetitive cycles [78, 88]. The biological treatment for water mitigation is a slow technique and also abolishes only 75–85% of organic pollutants. The widespread wastewater mitigation techniques like coagulation and adsorption which are available currently just relocate the pollutants from one phase into another phase. Henceforth, there is a necessity to develop an alternative technique which can totally eliminate the water contaminants [38, 57]. To overcome such hurdles of the environmental water pollution, photocatalysis is a prominent chemical procedure as it is a promising technique for the elimination of toxic chemicals entirely [62, 80]. By means of photocatalysis, toxic water pollutants and toxic gas can be totally eliminated. In the current time, advance oxidation processes (AOPs) have gained very much attention in treatment of water mitigation due to their ability in degrading of organic chemicals in diverse range [55, 84]. Among all AOPs, visible light-supported oxidation processes have gained remarkable attraction because of their energy efficiency, eco-friendly nature, good stability and low cost [25, 73]. This efficient method has been studied widely for the indemnification of huge range of aqueous pollutants in water since last 30 years [64, 87]. In the basic principle of photocatalysis (Fig. 1a), under visible light absorption, molecules of pollutants pass off the photooxidative reactions which split pollutants into small molecular weight constituents and variation in their chemical, physical and mechanical properties helps to produce harmless by-products [7, 85]. The redox potentials and band end positions of semiconductor photocatalyst in the water (at pH 7) are effective to improve the reaction selectivity of reactant without contradicting the change (Fig. 1b). The photocatalytic mechanism is well clarified by Eqs. 1–21 [26].

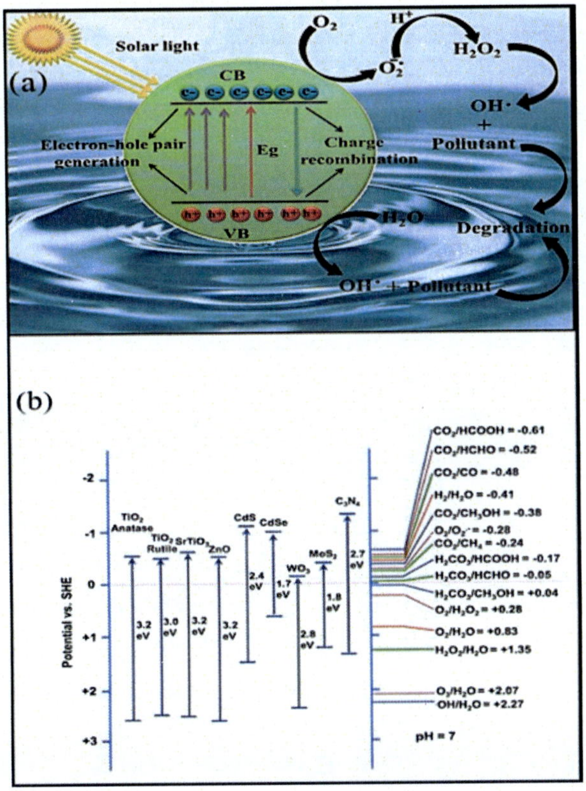

Fig. 1 **a** Fundamental photocatalytic mechanism of semiconductor photocatalyst under visible light illustrating degradation of organic contaminants present in the water and **b** potentials and band edge positions for redox couples (at pH 7) of semiconductor photocatalysts (copyright with license Id. 4653670201522)

$$AOP's + OH^{\cdot} \rightarrow \text{Pollutant molecule} \rightarrow CO_2 + H_2O + \text{Inorganic ions} \quad (1)$$

Route for charge separation for ROS production

$$\text{Photocatalyst} + h\upsilon \rightarrow e^- \text{ (conduction band)} + h^+ \text{ (valence band)} \quad (2)$$

$$O_2 + e^- \rightarrow O_2^{\cdot} \quad E_0' = -0.33\ V \quad (3)$$

$$HO_2^{\cdot} \rightarrow O_2^{\cdot} \quad pK_a = 4.8 \quad (4)$$

$$O_2^{\cdot} + 2H^+ + e^- \rightarrow H_2O_2 \quad E_0' = 0.89\ V \quad (5)$$

$$H_2O_2 \rightarrow HO_2^- + H^+ \quad pK_a = 11.7 \quad (6)$$

$$O_2 + 2H^+ + 2e^- \rightarrow H_2O_2 \quad E_0' = 0.28\ V \quad (7)$$

$$H_2O_2 + e^- \rightarrow OH^- + OH^{\cdot} \quad E_0' = 0.38\ V \quad (8)$$

$$H_2O + h^+ \rightarrow OH^{\cdot} + H^+ \quad E_0' = 2.32\ V \quad (9)$$

$$OH^{\cdot} \rightarrow O_2^- + H^+ \quad pK_a = 11.8 \quad (10)$$

$$2HO_2^{\cdot}\ {}^1O_2 \rightarrow H_2O_2 \quad (11)$$

$$O_2^{\cdot -} + h^+ \rightarrow {}^1O_2 \quad E_0' = 0.65\ V \quad (12)$$

Interaction among radicals

$$2HO_2^{\cdot} \rightarrow H_2O_2 + O_2 \quad (13)$$

$$2HO_2^{\cdot} + O_2^{\cdot -} + H_2O \rightarrow H_2O_2 + O_2 + OH^- \quad (14)$$

$$2HO_2^{\cdot} \rightarrow H_2O_2 \quad (15)$$

$$H_2O_2 + h\upsilon(UV) \rightarrow 2OH^{\cdot} \quad (16)$$

$$H_2O_2 + O_2^{-\cdot} \rightarrow OH^{\cdot} + O_2 + OH^- \quad (17)$$

$$3O_2 \rightarrow \text{Photocatalyst intersystem crossing} \rightarrow {}^1O_2 \quad (18)$$

Overall photocatalytic oxidation

$$\text{ROS},\ h^+ + \text{contaminants} \rightarrow \text{oxidised products } (H_2O,\ CO_2) \quad (19)$$

$$O_2 + \text{contaminants} + h\upsilon \rightarrow \text{Photocatalyst} \rightarrow H_2O, CO_2 \quad (20)$$

Recombination of generated hole and electron pairs:

$$e^- + h^+ \rightarrow \text{Heat} + \text{Radiationless decay} \quad (21)$$

Main obstacles which limit wide-scale recognition of AOPs are the inefficiency of commercially accessible visible absorbing photocatalytic materials and unsuccessful separation of generated holes and electron pairs [55, 92]. For the efficient degradation of water contaminants, the fabrication of economically accessible and visible light-absorbing photocatalyst is the most important step [69, 90]. From the past four decades, widely used semiconductor photocatalysts used for the elimination of contaminants of water are bismuth oxyhalide (BiOX, X = Cl, F, Br, I), Ag_3PO_4, $ZnFe_2O_4$, $CaFe_2O_4$, g-C_3N_4, ZrO_2, $BiFe_2O_4$, etc [25, 65].

To define the applicability of a photocatalytic semiconductor in a particular photocatalytic reaction, most important factor is band gap of the semiconductor [66, 86]. In a particular reaction, the elemental principle of photocatalyst depends on electron and hole pair excitation [89]. For the semiconductor owning the wide band gap ($E <$ 3 eV), the electrons and holes can only get excited by UV light because they require extra energy for the excitation of holes and electrons. Whereas the semiconductor owning the narrow band gap ($E >$ 3 eV) can easily go to their excited state by visible light, hence they do not require extra energy for the excitation of electron. Addition to this, electrons in semiconductor owning narrow band gap undergo quick combination of generated charge carriers [70]. However, the semiconductor photocatalysts also hold some drawbacks such as rapid recombination of hole pairs and electrons, and excitation happens only under UV and band gap properties. All these limitations affect the photocatalytic performance of the semiconductor photocatalyst. From the last decade, various strategies have been exploited to enhance the photocatalytic activity of semiconductor photocatalyst such as heterojunction formation, metal doping, noble metal doping and formation of semiconductor heterojunction composites [21]. Exclusively, various attempts have been made to combine carbon-based nanomaterials such as graphene, graphitic oxide, graphitic carbon nitride (g-C_3N_4), carbon nanotubes (CNTs) and carbon quantum dots (CQDs) with semiconductor photocatalyst to enhance the photocatalytic activity [44].

Photocatalysis by using metal oxides such as TiO_2 is most effective technique because it utilizes solar energy and leads toward the total mineralization of most of the organic contaminants which exist in aqueous medium as well as in air. However, metal oxides hold some drawbacks which limit their applications at mass production.

The photocatalytic region only absorbs a little fraction (<5%) of incident light, which represents its relativity with large band gap (<3.2 eV) [103]. Thus, insufficient utilization of visible light is a main factor for limitations in photocatalytic performance of the metal oxide. Also, because of poor affinity toward hydrophobic organic pollutants, the adsorption of pollutants on metal oxide surface becomes relatively low, hence resulting in poor photocatalytic degradation rate. Not restricted to this, metal oxide such as TiO_2, ZnO and $CuFe_2O_3$ may go through aggregation because of the instability of nano-size of particle. These aggregations of small particles hinder the incident light and hence reduce the photocatalytic activity [23]. In case of slurry system, one main challenge is to recover the nanoparticles from the treated water with concern to economic way. Limitations in the application of metal oxide particles for photocatalytic degradation of organic contaminants are listed in Fig. 2.

To enhance the photocatalytic performance of metal oxide materials, various methods have been exploited such as doping with metals and nonmetals [58], surface alteration with metal ions [42] and semiconductor nanoparticle modification. In recent times, researchers revealed that the addition of co-sorbent carbon-based nano-materials can enhance the photocatalytic performance of metal oxides [48]. Carbon nano-materials hold exceptional structural characters such as excellent thermal conductivity [14], mechanical strength [105], thermal stability [76] and unique electronic properties [14]. Hence, carbon nanocomposites can be used as an encouraging material for environmental purification.

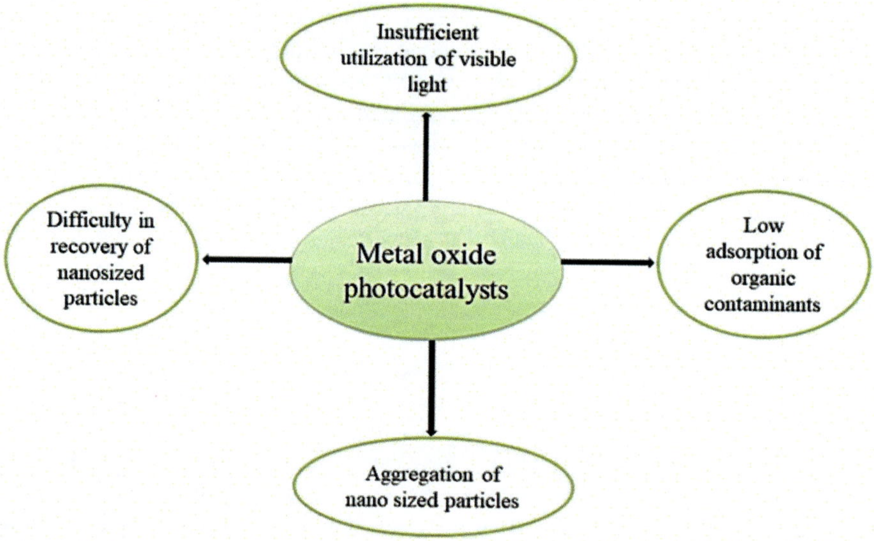

Fig. 2 Limitations of metal oxide particles during photodegradation of organic contaminants in water

2 Carbon-Based Nanocomposites

2.1 Graphene

Graphene consists of a specific layer of sp^2-bonded C atoms closely crammed into two-dimensional honeycomb-like structure (Fig. 3) [53]. This carbon material has gained a lot of attraction since after its discovery in 2004 because of its mechanical, optical, electrical and thermal properties. Graphene is one of all carbon-based nanomaterials which owns a high thermal conductivity (about 5000 W m^{-1} K^{-1}), displays extraordinary mobility of generated charge carriers even at room temperature (200,000 cm^2 V^{-1} s^{-1}) and also provides great surface area (about 2600 m^2 g^{-1}). At present, numerous techniques have been exploited for the fabrication of graphene which includes epitaxial growth, bottom-up organic synthesis, electrochemical and chemical reduction of graphite oxide and micromechanical exfoliation [74]. Out of all these fabrication techniques, the reduction of graphene oxide came out as a reliable and effective technique for the production of graphene nanosheets. This fabrication method is economic and results in huge scalability [59]. Fabrication of functionalized graphene-based nanocomposite can easily be done by just altering the surface properties via chemical modification [91]. In the present time, functionalized graphene-based semiconductor photocatalysts have gained a lot of attention because of their large specific surface area, high adsorption and good electron conductivity. The graphene-based nano-materials own unique optical and electronic properties and also hold good biocompatibility which represents their exploration in energy storage [95], biosensors [60], catalysis [96], drug delivery [13] and molecular imaging [2]. Lightcap et al. fabricated GO-TiO_2 nano-crystalline heterojunction by sonicating dispersed GO and TiO_2 nanoparticles in ethanol and revealed the practicability of using graphene as an electron transfer medium in the graphene/TiO_2 composite photocatalysts [44]. This work on graphene stimulated wide research on the modification, preparation and applications of graphene-based nano-material semiconductors. Zang et al. fabricated graphene-P25 TiO_2 under hydrothermal conditions

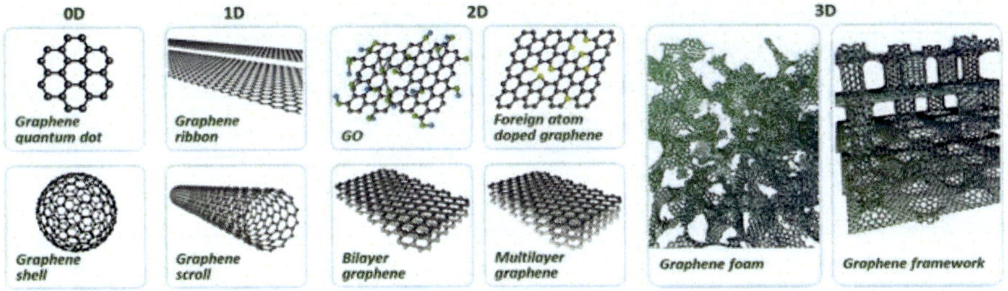

Fig. 3 Schematic illustrations of various configurations of graphene derivatives (copyright with license Id. 4653670484171)

for degradation of methylene blue in aqueous medium. The fabricated graphene-P25 TiO_2 displayed wide absorption range of extended light, efficient separation of charge carriers and great absorptivity of dyes. Therefore, in photodegradation of MB, enhancement in the photocatalytic performance was observed with graphene-P25 TiO_2 in comparison with pristine P25 and CNTs owning the same carbon content.

2.2 *Graphitic Carbon Nitride G-C_3N_4 (GCN)*

Presently, graphene family-based semiconductor photocatalysts are considered as promising photocatalysts because of their chemical stability, non-toxic nature, and economic and abundant nature [33]. Incidentally, GCN a π-conjugated semiconductor photocatalyst has gained much attention throughout the world [92]. GCN is a stable photocatalyst as it owns narrow band gap of 2.7 eV and displays two-dimensional configurations and easy fabrication route. The minimum conduction band of GCN (−1.12 eV vs. NHE) which is reliable for the high decline of generated electrons [111]. Despite having photocatalytic properties, GCN also holds minor limitations such as low adsorption area and rapid recombination rate of photo-generated electron–hole pairs. GCN used in bulk amasses the photocatalytic layers which results in overall decrement in photocatalytic behavior [93]. To date, various techniques have been exploited to overcome such limitations of photocatalyst. Out of all techniques, GCN nanosheets fabricated from decorticating bulk GCN have shown a good photocatalytic performance. These GCN nanosheets hold properties such as improved charge separation, beneficial alteration in band structure and wide exposure of active catalytic sites [100]. Recently, widely used fabrication method for synthesis of GCN is the thermal oxidation method. But due low yield percentage, it cannot be used for mass production of GCN [52]. Therefore, there is a need for development of new economic approaches which delaminate bulk GCN into nanosheets in large qualities. Moreover, it has been revealed that GCN materials when doped with heteroatoms (such as S, P, I and oxygen) own large charge carrier movement, enhanced light harvesting and changes in the band energy structure. Liu et al. fabricated sulfur-doped g-C_3N_4 for degradation of phenol under visible light. The fabricated composite displayed photocatalytic activity 7.2 and 8.9 times higher than pristine CGN under visible light. The complete oxidation of phenol under $\lambda > 400$ nm was done by sulfur-doped g-C_3N_4 and impossible for pristine GCN. This significant photocatalytic activity of sulfur-doped g-C_3N_4 is found to be a synergistic result of upshifting and widening of valence band, which is gained by homogeneous dispersal of sulfur dopant [46]. Zhang et al. fabricated iodine-doped GCN nano-material via in situ modification technique. The fabricated nano-material was obtained with enhanced optical absorption, accelerated charge transfers and also enlarged surface area. The iodine-doped GCN displayed excellent photocatalytic activity. The iodine-doped material displayed absorption extended to 600 nm, whereas pristine GCN is inactive at 500 nm. This result showed the advantage of nonmetal doping to enhance the band structure and texture of a photocatalyst [109].

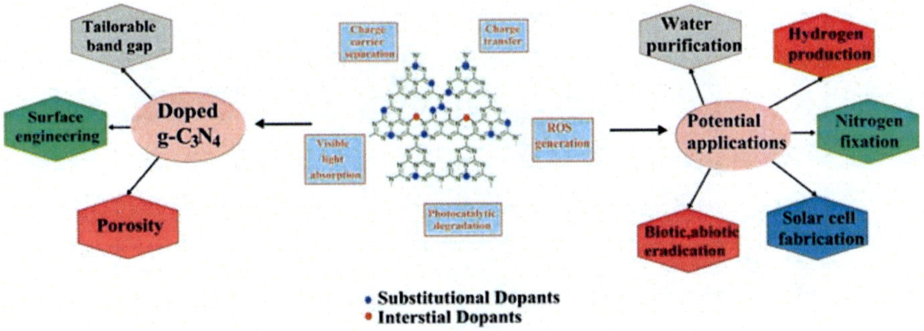

Fig. 4 Schematic representation of potential applications and modulations of g-C_3N_4 in various monarchies (copyright with license Id. 4654021257171)

Li et al. fabricated oxygen-doped GCN via hydrothermal technique. The oxygen doping effectively extended visible light response, enhanced the charge separation efficiency and also enlarged the surface area of the photocatalyst. Thus, such modifications consequently contributed to enhancement of the photocatalytic activity of pristine GCN [40]. In conclusion, sulfur-doped, oxygen-doped and iodine-doped GCN nano-materials have shown great potential as metal-free photocatalyst. For the development of green and cleaner environment, g-C_3N_4 has captivated researchers for protagonist function in wastewater mitigation, bacterial disinfection, organic contaminant degradation, water splitting, etc. Summary of review illustrating changes and probable applications of g-C_3N_4 in various monarchies is shown in Fig. 4.

2.3 *Carbon Quantum Dots (CQDs)*

Carbon quantum dots belong to the new family of carbon nanoparticles. These are biologically and environmentally workable materials in comparison with inorganic composites [34, 39]. For their fabrication, the carbon materials are utilized as adsorbents and dispersants and provide support to expand the surface area of the photocatalyst [19]. In general, the QDs are divided into two sub-types: carbon quantum dots (CQDs) and graphene quantum dots (GQDs) [22]. Graphene quantum dots are sp^2-hybridized and are crystal-like by nature, while carbon quantum dots are sp^3-hybridized and are amorphous in nature. The size of graphene quantum dots is 2–20 nm, and their fluorescence is because of quantum confinement (Fig. 5a) where carbon quantum dots are less than 10 nm and fluorescence is because of surface defects [114]. CQDs have gained much attention as an advanced family of nano-materials for the process of semiconductor visible light photocatalysis [106]. The photocatalytic process of CQD-based photocatalysts is represented in Fig. 5b. CQDs are synthesized by means of oxidation and carbonization, microwave technique, hydrothermal technique and electrochemical routes [11]. In the fabrication

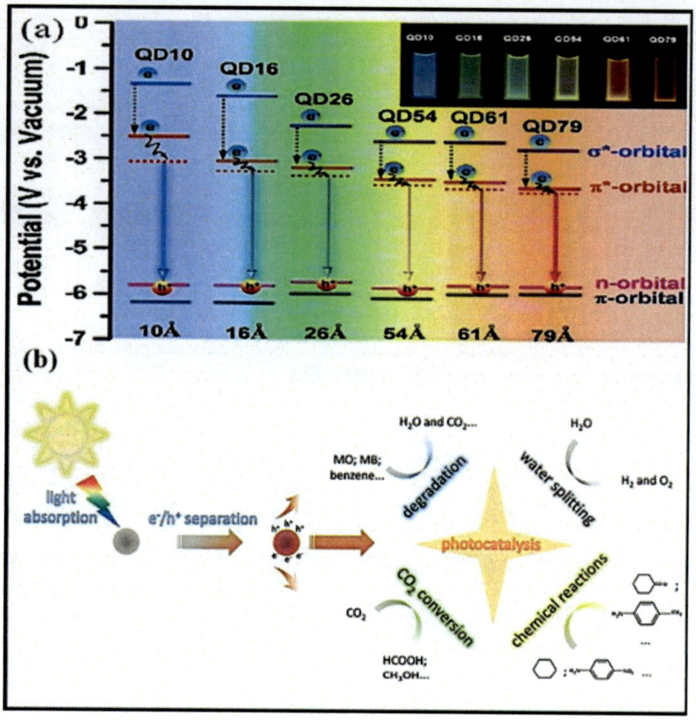

Fig. 5 **a** Quantum confinement effect of CQDs and their allied n-p* transition (copyright with license Id. 4653670201522) and **b** the photocatalytic process of CQD-mediated photocatalysts under visible light (copyright with license Id. 4654021433859)

procedure, the precursors are candle soot, citric acid, lampblack, papaya, grass, tulsi leaves, watermelon peels, lotus root, rice flour and potato [47, 49]. The importance of utilizing nature resources for fabrication of CQDs is that all the precursors are economic and eco-friendly in nature. Out of all, green synthesis techniques are highly acceptable than physical and chemical techniques. In recent times, researchers have encouraged to develop many new ways by using natural precursors. The CQDs are valuable resource because of their fine biocompatibility, abundant surface functional groups, low cost, small particle size, chemical inertness, wide varying optical properties, low toxicity and tuneable PL behavior [98, 101]. CQDs also have practical applications in electrocatalysis, light-emitting diodes, bio-sensing, nanomedicines, water treatment, drug/gene delivery, bio-imaging [24], disease detection, etc [113]. CQDs own extraordinary alteration capability to modify lower energy photons to higher energy photons. They are exploited as a spectral converter in order to use the overall spectra of the incident sunlight [16, 114]. In efficient, photocatalysis role of CQDs can be categorized as:

1. Mediator and acceptor for conduction band photo-generated electrons.

2. Enhancing visible light performance of wide-ranging band gap photocatalysts by the process of photosensitization.
3. Reducing agent during the fabrication of several metal nanoparticles with the help of surface plasmon resonance (SPR) phenomenon.
4. By using up-conversion photoluminescence (UCPL) phenomenon, efficient harvesting of wide solar spectrum is done in which the emission of shorter wavelength light is used for excitation of CQDs.

2.4 Carbon Nanotubes (CNTs)

Carbon nanotubes (CNTs) (Fig. 6) are classified as an illustrative kind of nanomaterials and acquire exceptional chemical and physical properties, which facilitate them to be favorably applied in several fields involving energy, medicine, environmental technology, etc., owing to porosity, high surface area, fast adsorption kinetics [35]. However, adsorption-based technologies are nondestructive and adsorbed contaminants are not mineralized. Thus, rejuvenation and organization of consumed adsorbents are critical to sustainability of adsorption-based procedures [94]. Lately, a number of reinforcement methods have been used to recover CNTs and lessen the treatment expenditure. The interaction between CNT functional groups under visible light produces reactive oxygen species (ROS) which promote the acceleration in the process of pollutant degradation and bacterial inactivation. Meanwhile, the possible environmental influences of CNTs have gradually attracted superior curiosities and concerns from global researchers [8]. CNTs were established to encourage DNA impairments and cytotoxic results toward eukaryotic cells and prokaryotic cells and adversely modify microbial multiplicity and community assemblies [4]. The physical interaction of CNTs with the microbial cells is the most feasible mechanisms for deactvation of toxic microbial cells.

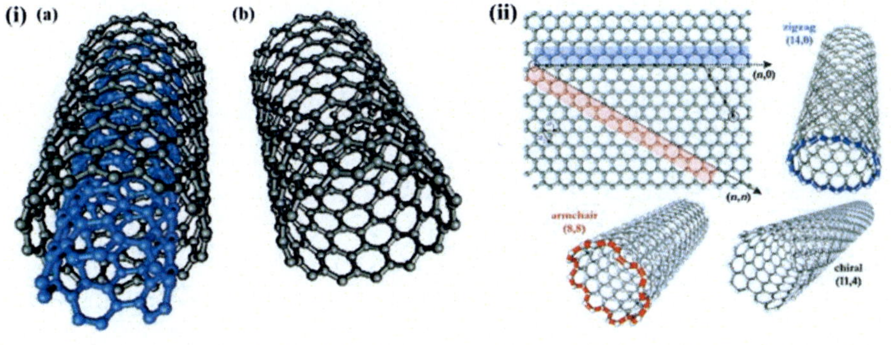

Fig. 6 (i) Structural representations of (a) multi-walled CNT and (b) single-walled CNT, and (ii) different configurations of CNTs (copyright with license Id. 4654030792512)

3 Preparation and Characterization of Carbon Nanocomposites

Recently, a range of techniques have been established to produce carbon-based nanocomposites with enhanced performance in water purification [6, 29]. The photocatalytic activity of these nanomaterials is related to their configuration and properties, which are administered by size of carbon material, number of walls or layers, density and type of defects and interfacial interaction among semiconductors and carbon nanocomposites [79]. Thus, to attain carbon nanocomposites with enhanced performance, their fabrication must be strategic and implemented in order to regulate their configurations and properties [80]. A greater surface area of carbon-based composites greatly impacts their photocatalytic power, the surface area can be estimated using nitrogen absorption, and chemical structures can be depicted by Raman spectroscopy. There are some substitute techniques to attain the enhanced performance of carbon nanocomposites, such as chemical surface functionalization, heteroatom doping and interface engineering. An et al. reported a method of hydrothermal aided by microwaves to provide compounds of Cu_2O coated with reduced graphene oxide (RGO) [5]. Examination of their XRD patterns revealed that composites were comprised of cubic phase of Cu_2O. The SEM and TEM studies showed the morphology of the Cu_2O/RGO materials. The configuration of these photocatalysts was also depicted by XPS. Surface chemical functionalization, e.g., chemical insertion of the chemical ligands and acid oxidation, upholds the core structure of material but can encourage development of surface defects and appropriate functional groups, which can be favorable for enlightening the photocatalytic activity of carbon nanocomposites [110]. Usually, functionalization can produce plentiful nucleation positions that encourage the development of unvarying nanoparticles, subsequently stimulate a well dispersal of combined semiconductor nanostructures and may additionally act as implementers for native photocatalytic reactions as they can act as supports for contributing reactants in reactions [97]. The electronic structure and electric conductivity of carbon nanocomposites can be the altered by heteroatom doping in demand to regulate their electron movement and charge transfer capacity [43].

4 Photocatalytic Activity of Carbon Nanocomposites

The method of photocatalysis is an effectual, cost-effective and green technique for wastewater treatment [3, 63]. Lately, improvement in carbon-based photocatalytic nanocomposites and nanotechnology has headed to invention of groundbreaking semiconductor photocatalysts via which deprivation of organic pollutants can be attained with higher effectiveness [6, 31].

Progresses in the photocatalytic properties of g-C_3N_4 for the wastewater mitigation have got remarkable attention in the field of research [30, 79]. Though, bare g-C_3N_4 experiences various bottlenecks such as fast recombination of generated electron–hole pairs, less surface area and inadequate absorption of light which lowers photocatalytic degradation activity [73, 92]. To advance the photocatalytic activity of g-C_3N_4, diverse approaches, e.g., tuning defects, fabricating heterojunctions, semiconductors have been attempted [71, 72]. Abdellatif et al. developed a simple method to enhance the oxidation capability of the electron–holes produced from the valence band of the g-C_3N_4 and elimination of NO [1]. Hu et al. fabricated K-doped g-C_3N_4 for the removal of Rhodamine B dye below visible light irradiation. K-doped g-C_3N_4 (0.5) showed 6.5 times more rate constant than bare g-C_3N_4, and also N-doped g-C_3N_4 amended the photocatalytic performance [33]. Xu and co-workers narrated fabrication of g-C_3N_4 via calcination and hydrothermal method for the photodegradation of Rhodamine B dye under the visible light radiation [97]. The g-C_3N_4-20 with super-cell structure fabricated by hydrothermal method exhibited higher separation of generated charge carriers, with larger surface area, thereby improving the photocatalytic activity for the mineralization of dye. g-C_3N_4 nanosheets were prepared using NH_4Cl as a precursor by Guo et al. for the elimination of cyanide [28]. The removal of cyanide using g-C_3N_4 0.18 mM nanosheets was 90% after 150 min. Also results revealed that the photocatalyst showed substantial recyclability after five successive cycles. The fabrication of carbon- and oxygen-doped g-C_3N_4 using malonic acid and urea as precursors via thermal polymerization was reported by Gu and co-researchers [27]. The fabricated photocatalyst degraded the 15 ppm bisphenol A (BPA) within 150 min, and photocatalytic activity of carbon- and oxygen-doped g-C_3N_4 was 4.8 times more than that of pure g-C_3N_4. The higher photocatalytic activity of the product was chiefly attributed to optical properties and electronic band structure and due to the positive charge density on the C atoms. On the further part, introduction of carbon atoms into g-C_3N_4 led to fabrication of π-bonds which favored the transfer of electrons [27]. The generation of hydroxyl radical plays a principal role in mineralization of BPA. The exclusive chemical and band configuration collected with porous morphology attributed to the greater photocatalytic action of the synthesized composite. The possible photocatalytic mechanism for the degradation of BPA using carbon- and oxygen-doped g-C_3N_4 is illustrated in Fig. 7a. A facile and easy approach for the synthesis of porous g-C_3N_4-covalent organic framework (COF) materials was conveyed by Yao and co-workers [102]. The synthesized hybrids exhibited greater photocatalytic activity, and results revealed 100% removal of orange II; bare g-C_3N_4 removed 10%, and COF removed 5% of dye. It was concluded that noble equilibrium among graphitization degree and N content helped in boosting photocatalytic activity of as-synthesized photocatalyst. The probable mechanism for the photocatalyst for dye mineralization is illustrated in Fig. 7b. The synthesized photocatalyst exhibited effective photocatalytic activity owing to exceptional porous structure, greater definite surface area, great concentration of N active sites and robust synergistic effects among COF and g-C_3N_4.

CNT-grounded strategies have drawn significant hold in an array of scientific arenas such as photocatalysts, adsorbents and membranes owed to its notable catalytic, electrical, chemical, structural and thermal properties [35]. Moreover, CNTs displayed tremendous adsorption ability with high adsorption competence

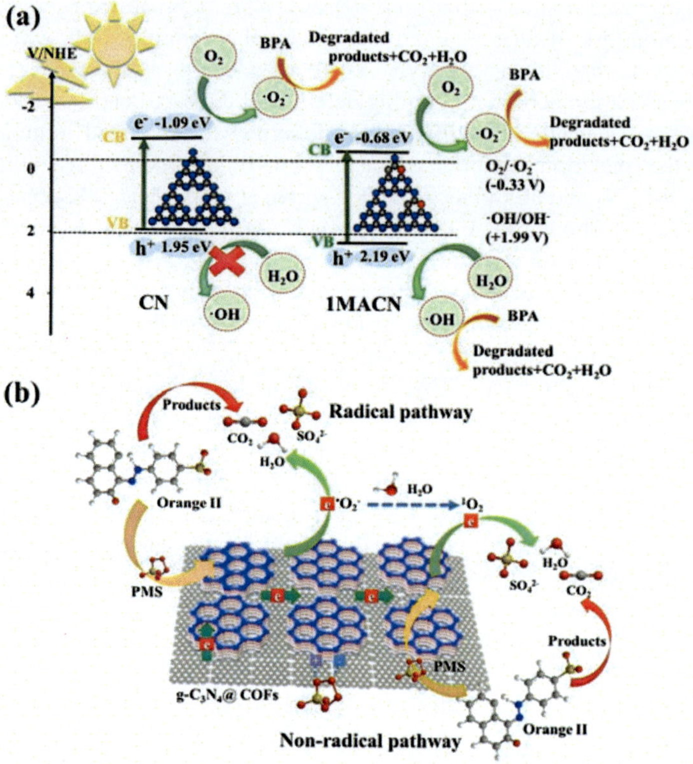

Fig. 7 **a** Pictorial representation of photocatalytic mechanism for the degradation of BPA using carbon- and oxygen-doped g-C_3N_4 (copyright with license Id. 4654031136570) and **b** the plausible mechanism for g-C_3N_4-covalent organic framework (COF) for the mineralization of dye orange II (copyright with license Id. 4654031289093)

than traditional granulated activated carbon, amorphous carbon and graphene [8]. CNTs emerge as an exceptional adsorbent due to its hefty surface area alongside with effective active locations [4]. Surface alteration of CNTs via adding acidic or basic solution hosted hydroxyl, carbonyl and carboxyl groups to the CNTs. The manifestation of these functional groups enhanced adsorption properties of the CNTs through eliminating contaminations on CNTs surface and besides delivered greater electrostatic force [18]. Zare and co-workers reported the removal of dye Congo red (CR) by operating MWCNTs [107]. The factors upsetting adsorption capability of MWCNTs were scrutinized, viz. pH and the initial dye concentration. Results revealed that dye CR adsorption was observed at pH 11 after 60 min. In other studies, MWCNTs were synthesized via photocatalytic chemical vapor deposition strategy for the degradation of dye reactive red 159, reactive yellow 81 and reactive blue 116 [94]. The photocatalytic activity of MWCNTs was significantly exaggerated by surface modification. It was found that the removal process followed Temkin model for all kind of dye used. A widespread research has been done by Sellaoui and associates for the degradation of crystal violet by di-functional MWCNTs [75]. Results showed that interaction among H-bonding is main active cause responsible for adsorption of dye. Lately, Banerjee et al. fabricated amorphous CNTs for removal of methyl orange (MO) and Rhodamine B (RhB) [10]. The fabricated CNTs degraded methyl orange in 30 min and Rhodamine B in 45 min, making it a potential unconventional adsorbent for the water mitigation. Bohdziewicz and co-worker reviewed the elimination of bisphenols (BPA) in the wastewater by single-walled CNTs (SWCNTs) and improved the functional groups CNTs [12]. It was revealed by the researchers that more pH value is not beneficial for the adsorption of BPA due to electrostatic repulsion among CNTs and bis-phenolate anions, which thereby reduced π–π interaction. Zhang and co-workers conveyed the photocatalytic activity of MWCNTs for the exclusion of bisphenol AP (BPAP) [110]. The results revealed that ionic potency is not an essential feature that led to the robust adsorption of molecules of BPAP on the MWCNTs. The MWCNTs displayed tremendous durability up to successive 8 cycles with the 95% retrieval.

Because of exceptional photosensitization, up-converted photoluminescence (UCPL) and charge carrier transfer, CQDs have been operated for amendment of a photocatalyst nanocomposite to nurture their photocatalytic performance [9, 104]. The mechanism of electron excitation and photocatalytic activity of CQDs is entirely depended on the band gap, conduction band and valence band [15]. The different functions of CQDs in effectual photocatalysis are acceptor and mediator for photo-generated electrons, enlightening the visible light photocatalytic activity of the photocatalyst via photosensitization methods, the reducing agent through construction of several metal nanoparticles with phenomenon of surface plasmon resonance (SPR), well-organized harvesting of solar energy via up-conversion photoluminescence (UCPL) [81]. In a stated work, Miao and co-workers exhibited the fabrication of CQDs/TiO_2 photocatalyst for the elimination of N-benzylideneaniline (NB) and methylene blue (MB) via sol-gel method, and ultrasonic and hydrothermal approaches below the visible light radiation [50]. The fabricated photocatalyst photodegraded 98% of dye MB after 1 h and 30% of dye NB after 2 h which was greater than bare CQDs and Ti-450. The existence of O_2 comprising groups and aromatic rings of the CQDs added the adsorption on MB and NB on the surface of nanocomposite. The electron reservoir and UCPL properties of the CQDs accelerated utilization of the visible light and obstructed the regrouping of e^-/h^+ pairs. Ali and associates narrated the fabrication of P25/CQDs nanocomposite for the degradation of 4-chlorophenol (4 CP), Rhodamine B (RhB) and methyl orange (MO) below visible light [3]. The nanocomposite degraded 49% of 4 CP, 80% of RhB and 40% of MO, whereas pristine P25 degraded 46% of 4 CP, 49% of RhB and 33% of MO. It was also stated that the formation of ROS played a vital function in the deprivation of dyes. The hydrothermal synthesis of the CQDs/TiO_2 nanocomposites was described using glucose (G) and citric acid (CA) as precursors for the deprivation of phenolic compounds under UV light radiation [82]. CQDs/G/TiO_2 photocatalyst revealed about 99% degradation ratio of the phenol which was greater than CQDs/CA/TiO_2 photocatalyst. CQDs assisted as electron reservoir and confined the photo-generated electrons from TiO_2 conduction band. The feasible mechanism

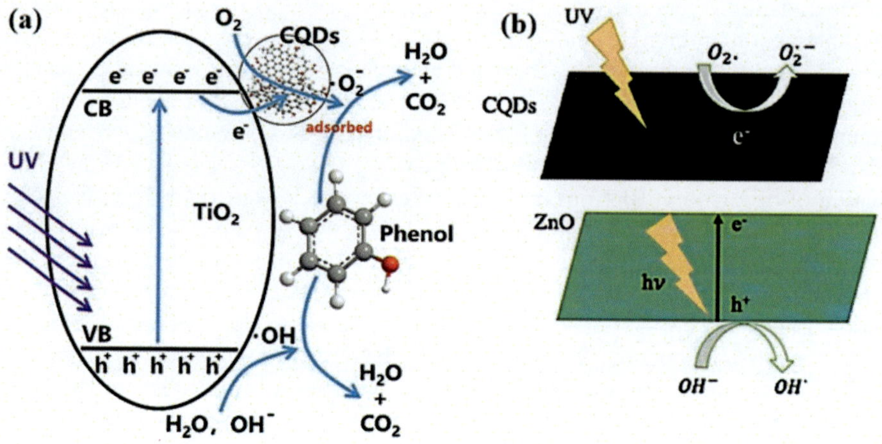

Fig. 8 **a** Feasible mechanism propositioned for elimination of phenol under visible light (copyright with license Id. 4654040037357) and **b** the photocatalytic mechanism using CQDs/ZnO photocatalysts for the degradation of the Rhodamine B dye where CQDs act as electron reservoir (copyright with license Id. 4654040444679)

proposed for the mineralization of phenol under visible light is shown in Fig. 8a. Li and co-researchers fabricated CQDs/ZnO photocatalyst composites via sol-gel technique besides spin-coating route for the degradation of the Rhodamine B dye [43]. CQDs/ZnO photocatalyst (4-layered) exhibited 3 times extra photocatalytic action than pure ZnO due to the electronic interaction among ZnO and CQDs molecules, improved the separation of photo-generated charge carriers and enriched the allocation of e^-–h^+ pairs (Fig. 8b). The upgraded photocatalytic performance of the photocatalyst was generally due to CQDs electron reservoir property and up-converted nature.

Muthulingam et al. delivered the groundwork of CQDs/N-ZnO photocatalyst for the deprivation of malachite blue (MB), methylene green (MG) and the fluorescein dyes under solar light [51]. The degradation percentage for the dye MG using CQDs/N-ZnO composite was 100%, and using N-ZnO composite was 60%, respectively, later 30 min. Photodegradation of the fluorescein dye using CQDs/N-ZnO nanocomposite was also 100% later 30 min, and using N-ZnO composite was 92% within 60 min. Ding and associates reported the fabrication of CQDs/ZnO foam for the elimination of Rhodamine B (RhB), methylene orange (MO) and methyl blue (MB) under the visible and UV light [20]. The rate constant (k) value for MO, MB and RhB was 0.0031, 0.0121 and 0.0092 min^{-1}, respectively. The mineralization of the dyes was in the order of MO < RhB < MB. Due to the up-conversion nature, CQDs enriched the photocatalytic action of the foam as deliberated by the excitation wavelengths in the range of 600–850 nm. The CQDs were represented as electron reservoir, photo-generated electrons were relocated from the ZnO surface to the CQDs surface, and the recombination of e^-–h^+ pairs inhibited well (Fig. 9).

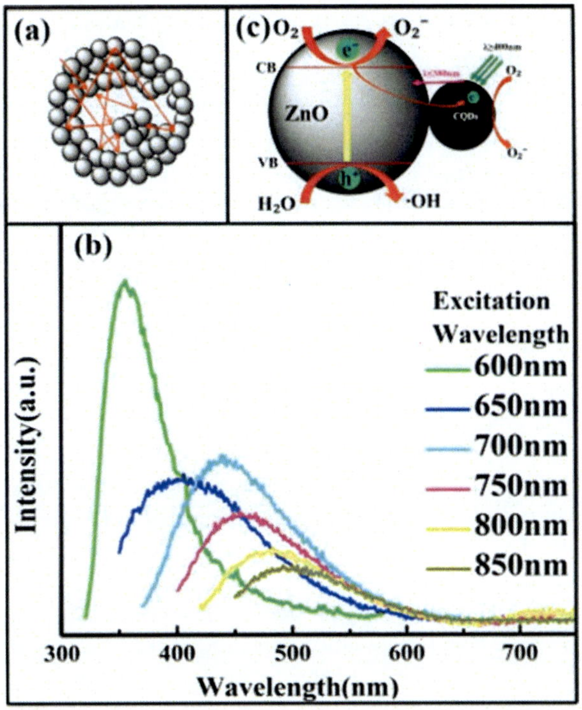

Fig. 9 **a** Schematic diagram of absorption and reflection of light in the ZnO foam, **b** CQDs up-converted spectra with excitation of visible and NIR wavelengths and **c** schematic representation of the photocatalytic process of ZnO foam CQDs nanocomposites (copyright with license Id. 4654040658534)

Kaur et al. stated a facile scheme for the preparation of CQDs revised ZnS nanocomposites via precipitation route for elimination of the dye alizarin red S (ARS) below visible light illumination [36]. The photocatalytic performance using CQDs/ZnS nanocomposites for the dye elimination was 89% within 250 min and that was greater than the catalytic action of ZnS (63%). After the introduction of CQDs into ZnS, electrons transferred from ZnS conduction band to the CQDs and caused an effectual separation of e^-–h^+ pairs. Liu and workfellows conveyed the fabrication of CQD-modified CdS nanocomposites via hydrothermal means for deprivation of Rhodamine B (RhB) below the visible light, fluctuating the CQDs concentration [47, 49]. The photodegradation productivity of dye RhB over CQDs/CdS photocatalyst composite was 90% and over CdS was 50%, respectively, within 1 h below visible light illumination. The photocatalytic performance of the CQDs/CdS composite was greater than that of pristine CdS as CQDs imprisoned the electrons and obstructed the generated e^-–h^+ pair recombination. Correspondingly, up-conversion property of CQDs made subsequent photocatalyst exploit the visible light extra efficiently and hence boosted the photocatalytic activity.

5 Photocatalytic Anti-bacterial Activity of Carbon Nanocomposites

The current confront in expenditure of the photocatalysis for the bacterial disinfection is the time needed for the disinfection process. The rate of bacterial inactivation can be augmented by varying rate of recombination of generated e^-–h^+ pairs and rate of production of reactive oxygen species (ROS) [32]. Boron and phenyl co-doped g-C_3N_4 was produced via thermal polycondensation of cyanamide to disinfect bacteria *Escherichia coli* (*E. coli*). The as-prepared photocatalysts displayed superior photocatalytic activity over UV to NIR light; 99.9% of bacteria were disinfected after 3 h with less concentration of the photocatalyst below sunlight. The bacterial disinfection mechanism was reviewed via scavenger experiments, demonstrating H_2O_2 was chief responsive species for the disinfection of bacteria. Conclusively, the photocatalyst was accumulated on solid material surface and also showed robust disinfection activity. The improved light absorption and effective charge separation capability of photocatalysts promoted their photocatalytic bacterial disinfection (Fig. 10a) [45]. The g-C_3N_4 quantum dots (QDs) were prepared via thermal polymerization and selective dialysis methodology for bacterial disinfection [99]. The fluorescent g-C_3N_4-QDs proficiently produce hydroxyl and superoxide radicals under visible light; ~99% of *E. coli* (Gram negative) and ~90% of S. aureus (Gram positive) were inactivated using synthesized photocatalyst. A widespread evaluation of g-C_3N_4-QDs with bulk g-C_3N_4, mesoporous g-C_3N_4, Ag-g-C_3N_4 and pristine Ag NPs signified them to be encouraging bactericidal material. The possible bacterial inactivation mechanism using as-fabricated composite is depicted in Fig. 10b.

The g-C_3N_4 nanosheet composite membranes were prepared via an effective and facile filtering method by aligning g-C_3N_4 membranes on PAN substrates for the disinfection of *E. coli*. The hydrophilic g-C_3N_4 membranes revealed superior antimicrobial capacity against *E. coli* than membranes without the g-C_3N_4 nanosheets. The nanosheets of g-C_3N_4 functionalized membranes with antibacterial performance and self-cleaning were support to the area of membrane separation [41]. Gram-negative bacteria *E. coli* is the highest studied bacteria to investigate bacterial inactivation procedures encouraged by g-C_3N_4-mediated photocatalysts under the visible light radiation (Fig. 11). Most of the analyses have showed microscopic evaluation for exploring the variations of bacterium morphology and approving demolition of cell reliability through photocatalytic disinfection [108].

ZnO- and TiO_2-conjugated CNTs and graphene oxide (GO) nanocomposites were examined for their antimicrobial effects on *E. coli*. Among four types of nanocomposites, ZnO-conjugated nanomaterials displayed greater antibacterial performance, subsequent in the antibacterial effect which was measured with progression inhibition of the cells in the order ZnO-GO > ZnO-CNTs > TiO_2-GO > TiO_2-CNTs. Among all the four probable antibacterial mechanisms, production of reactive oxygen species (ROS), chemical features and steric effect were part of causative mechanisms. The growing dispersion of TiO_2/ZnO on graphene oxide contributed to the antimicrobial effects due to increased surface areas. Likewise, noteworthy indemnities to cell

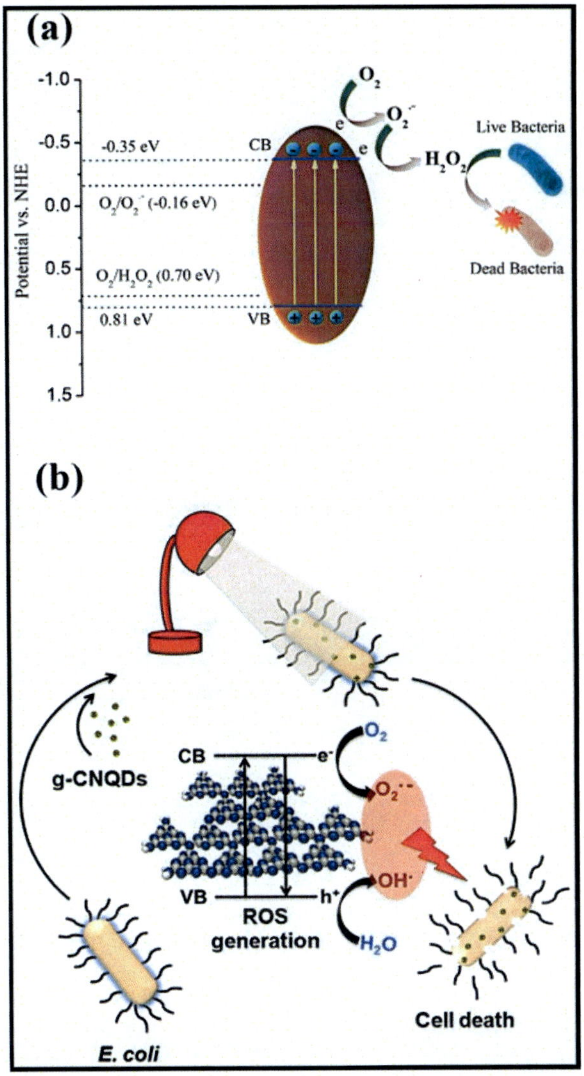

Fig. 10 **a** Mechanism of photocatalytic bacterial disinfection depicting improved light absorption and effective charge separation capability of photocatalysts (copyright with license Id. 4654040855898) and **b** the feasible bacterial inactivation mechanism using g-C_3N_4 quantum dots (QDs) (copyright with license Id. 4654041041902)

membranes of *E. coli* were found by GO nano-sheet with its shrill edges. The results proposed that applying GO-ZnO or TiO_2 was an effectual antibacterial mode, particularly for behavior of drug-resistant bacteria in water [8]. Improved photocatalysts were fabricated by coating TiO_2 on multi-walled CNTs for improving disinfection rate of the bacterial endospores. TiO_2-coated MWCNTs were verified for disinfection of *Bacillus cereus* endospores, and disinfection rate was double as equated to bare commercial TiO_2 nanoparticles. The inactivation of bacteria depended on half-life of reactive oxidative species (ROS), complexity and width of bacteria cell wall.

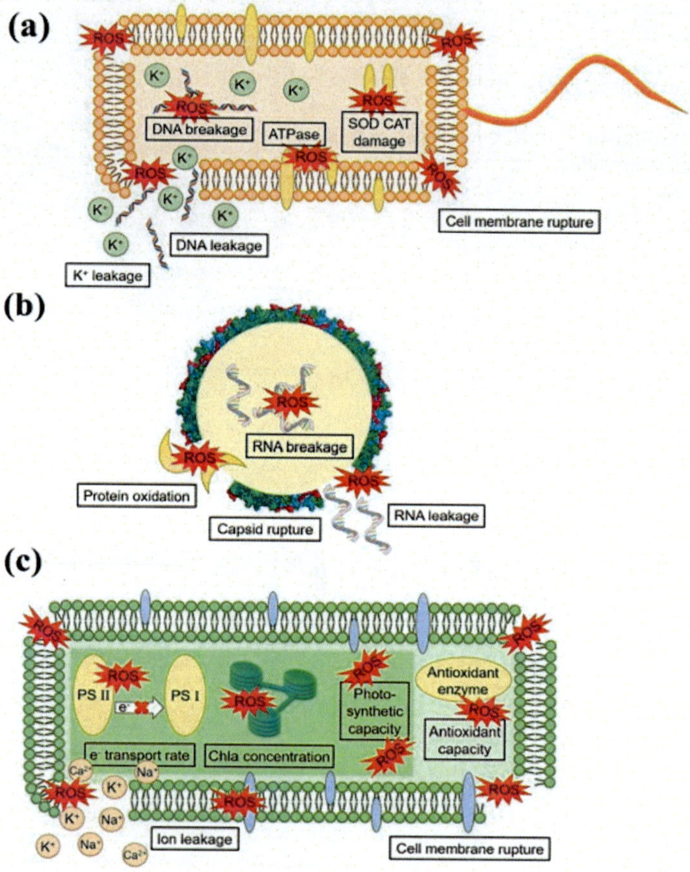

Fig. 11 g-C_3N_4-mediated photocatalytic bacterial disinfection: **a** bacterial death, **b** viral death and **c** micro-algal death (copyright with license Id. 4654050490577)

The present-day confront in the consumption of photocatalysts for the spore degradation is time needed for the bacterial inactivation. Improved photocatalyst composites have been established for improving the bacteria inactivation rate. The efficacy of TiO_2-coated MWCNTs was associated against commercial TiO_2 nanopowder for inactivation of *B. cereus* spores [37].

CQDs as a carbon-based substance are especially boosting because of their low toxicity, chemical inertness, noble biocompatibility and outstanding optical properties. Numerous semiconductor photocatalysts such as TiO_2, Bi_2WO_6, g-C_3N_4, ZnO and $BiVO_4$ can be united with CQDs to increase their photocatalytic efficacy. S-doped CQD-loaded hollow g-C_3N_4 semiconductor photocatalyst was successfully fabricated via ultrasonic assisted approach. In photocatalytic system, S-CQDs were acted as an electron reservoir and photosensitizer to enhance light absorption range. As related to bare hollow g-C_3N_4, CQD-loaded hollow g-C_3N_4 photocatalyst revealed better photocatalytic activity for TC photodegradation and *E. coli* disinfection.

6 Conclusion and Outlook

For the expansion of cleaner and green ecosystem, carbon-based materials have engrossed researchers for protagonist part in wastewater treatment, organic pollutant degradation, bacterial disinfection, etc. The abundant properties accompanying carbon nanocomposites have amazingly boosted their photocatalytic action. Carbon nanocomposites have acquired an admirable integrity during latest decades, owing to their high adsorption capacity, less-cost and bio-friendly nature. Though, literature analysis proposed that an excessive deal of conclusions on pollutant adsorption and bacterial disinfection has been restricted. In order to acquire a greater surface area and additional adsorption situates, many researchers did a lot of research for the reformation of carbon-based composites. Consequently, diverse nonmetals and metals were burdened on the carbon materials. Still, it would be probable to produce secondary pollution through adsorption process. Lately, carbon-based materials have been gradually used for the elimination of pollutants and bacterial inactivation. Carbon materials adsorb pollutants in their pores; conversely, contaminants persist in environment. Still, there are confronts forward how to incline adsorbed contaminants in carbon nanocomposites.

Nonetheless, broadened research is desirable to improve more capable materials which are cost-effective in great extents as well as check such materials in full-blown adsorption units. This entails, in turns, acquirement of more physical comprehensions into properties of carbon nanocomposites and their reformation to accomplish the anticipated properties for an assumed purpose, where molecular replications are playing a significant role. Furthermore, the massive advancement made in this field in previous 20 years, it is necessary to overawed some of existing limitations for an accurate application of carbonaceous materials for wastewater treatment. To discourse these deficits, there are definite predictable criteria to be accomplished, comprising (i) complete exploitation of solar energy from UV region to NIR region, (ii) accelerated separation of charge carriers and (iii) augmentation of the photocatalytic surface area to reinforce the anchoring capability of carbon-based nanomaterials. Hence, for the fabrication of effective carbon-based photocatalytic schemes, the following key guidelines must be taken into deliberation:

1. The researchers should emphasize on increasing the light garnering range from UV to NIR to gain widespread renewable solar energy to depreciate environmental insinuations via environment-friendly photocatalytic process. For this, carbon-based materials should be merged with metals/nonmetals.
2. The function of several morphologies, viz. nanotubes, nanofibers, nanosheets, nanorods structure of carbon-based materials, must be examined for attaining greater surface area.

3. The segregation and rapid recovery of carbon-based nanocomposites as photocatalysts are extremely desirable for advanced recyclability of the photocatalysts.
4. To elucidate the drawback of carbon-based composites enclosing pollutants, we should try to foster highly efficient and cheap reinforcement technology to increase the economic practicability.
5. There is an insistent necessity to elude metal ions (viz. copper, nickel and iron) leaching from carbon-based nanocomposites through adsorption method, consequently limiting the secondary pollution.

References

1. Abdellatif HR, Zhang G, Wang X, Xie D, Irvine JT, Ni J, Ni C (2019) Boosting photocatalytic oxidation on graphitic carbon nitride for efficient photocatalysis by heterojunction with graphitic carbon units. Chem Eng J 370:875–884
2. Akhavan O, Ghaderi E, Esfandiar A (2011) Wrapping bacteria by graphene nanosheets for isolation from environment, reactivation by sonication, and inactivation by near-infrared irradiation. J Phys Chem B 115:6279–6288
3. Ali MM, Nair JA, Sandhya KY (2019) Role of reactive oxygen species in the visible light photocatalytic mineralization of rhodamine B dye by P25 carbon dot photocatalyst. Dyes Pigment 163:274–284
4. Ali Q, Ahmed W, Lal S, Sen T (2017) Novel multifunctional carbon nanotube containing silver and iron oxide nanoparticles for antimicrobial applications in water treatment. Mater Today 4:57–64
5. An X, Li K, Tang J (2014) Cu_2O/reduced graphene oxide composites for the photocatalytic conversion of CO_2. Chemsuschem 7:1086–1093
6. Ani IJ, Akpan UG, Olutoye MA, Hameed BH (2018) Photocatalytic degradation of pollutants in petroleum refinery wastewater by TiO_2 and ZnO-based photocatalysts: recent development. J Clean Prod 205:930–954
7. Antoniadou M, Daskalaki VM, Balis N, Kondarides DI, Kordulis C, Lianos P (2011) Photocatalysis and photoelectrocatalysis using (CdS-ZnS)/TiO_2 combined photocatalysts. Appl Catal B 107:188–196
8. Baek S, Joo SH, Su C, Toborek M (2019) Antibacterial effects of graphene- and carbon-nanotube-based nanohybrids on *Escherichia coli*: Implications for treating multidrug-resistant bacteria. J Environ Manag 247:214–223
9. Bajorowicz B, Kobylański MP, Gołąbiewska A, Nadolna J, Zaleska-Medynska A, Malankowska A (2018) Quantum dot-decorated semiconductor micro-and nanoparticles: a review of their synthesis, characterization and application in photocatalysis. Adv Colloid Interface Sci 256:352–372
10. Banerjee D, Bhowmick P, Pahari D, Santra S, Sarkar S, Das B, Chattopadhyay KK (2017) Pseudo first ordered adsorption of noxious textile dyes by low-temperature synthesized amorphous carbon nanotubes. Physica E 87:68–76
11. Barati A, Shamsipur M, Arkan E, Hosseinzadeh L, Bdollahi H (2015) Synthesis of biocompatible and highly photoluminescent nitrogen doped carbon dots from lime: analytical applications and optimization using response surface methodology. Mater Sci Eng 47:325–332
12. Bohdziewicz J, Kamińska G (2013) Kinetics and equilibrium of the sorption of bisphenol A by carbon nanotubes from wastewater. Water Sci Technol 68:1306–1314
13. Cai W, Chen X (2007) Nanoplatforms for targeted molecular imaging in living subjects. Small 3:1840–1854

14. Cha SI, Kim KT, Lee KH, Mo CB, Jeng YJ, Hong SH (2008) Mechanical and electrical properties of cross-linked carbon nanotubes. Carbon 46:482–488
15. Chai YY, Qu DP, Ma DK, Chen W, Huang S (2018) Carbon quantum dots/Zn^{2+} ions doped-CdS nanowires with enhanced photocatalytic activity for reduction of 4-nitroaniline to p-phenylenediamine. Appl Surf Sci 450:1–8
16. Chan WC, Maxwell DJ, Gao X, Bailey RE, Han M, Nie S (2002) Luminescent quantum dots for multiplexed biological detection and imaging. Curr Opin Biotechnol 13:40–46
17. Chandel N, Sharma K, Sudhaik A, Raizada P, Hosseini-Bandegharaei A, Thakur VK, Singh P (2019) Magnetically separable ZnO/$ZnFe_2O_4$ and ZnO/$CoFe_2O_4$ photocatalysts supported onto nitrogen doped graphene for photocatalytic degradation of toxic dyes. Arab J Chem. https://doi.org/10.1016/j.arabjc.2019.08.005
18. Das R, Ali ME, Hamid SBA, Ramakrishna S, Chowdhury ZZ (2014) Carbon nanotube membranes for water purification: a bright future in water desalination. Desalination 336:97–109
19. Das R, Bandyopadhyay R, Pramanik P (2018) Carbon quantum dots from natural resource: a review. Mater Today Chem 8:96–109
20. Ding D, Lan W, Yang Z, Zhao X, Chen Y, Wang J, Zhang X, Zhang Y, Su Q, Xie E (2016) A simple method for preparing ZnO foam/carbon quantum dots nanocomposite and their photocatalytic applications. Mater Sci Semicond Process 47:25–31
21. Dutta V, Singh P, Shandilya P, Sharma S, Raizada P, Saini AK, Gupta VK, Hosseini-Bandegharaei A, Agarwal S, Rahmani-Sani A (2019) Review on advances in photocatalytic water disinfection utilizing graphene and graphene derivatives-based nanocomposites. J Environ Chem Eng 7:103132
22. Feng H, Qian Z (2018) Functional carbon quantum dots: a versatile platform for chemosensing and biosensing. Chem Rec 18:491–505
23. Gao B, Yap PS, Lim TM, Lim TT (2011) Adsorption photocatalytic degradation of Acid Red 88 by supported TiO_2: effect of activated carbon support and aqueous anions. Chem Eng J 171:1098–1107
24. Gao X, Cui Y, Levenson RM, Chung LW, Nie S (2004) In vivo cancer targeting and imaging with semiconductor quantum dots. Nat Biotechnol 22:969
25. Gautam S, Shandilya P, Priya B, Singh VP, Raizada P, Rai R, Valente MA, Singh P (2017) Superparamagnetic $MnFe_2O_4$ dispersed over graphitic carbon sand composite and bentonite as magnetically recoverable photocatalyst for antibiotic mineralization. Sep Purific Technol 172:498–511
26. Gautam S, Shandilya P, Singh VP, Raizada P, Singh P (2016) Solar photocatalytic mineralization of antibiotics using magnetically separable $NiFe_2O_4$ supported onto graphene sand composite and bentonite. J Water Process Eng 14:86–100
27. Gu J, Chen H, Jiang F, Wang X (2019) Visible light photocatalytic mineralization of bisphenol A by carbon and oxygen dual-doped graphitic carbon nitride. J Colloid Interface Sci 540:97–106
28. Guo Y, Wang Y, Zhao S, Liu Z, Chang H, Zhao X (2019) Photocatalytic oxidation of free cyanide over graphitic carbon nitride nanosheets under visible light. Chem Eng J 369:553–562
29. Hasija V, Raizada P, Sudhaik A, Sharma K, Kumar A, Singh A, Jonnalagadda SB, Thakur VK (2019) Recent advances in noble metal free doped graphitic carbon nitride based nanohybrids for photocatalysis of organic contaminants in water: a review. Appl Mater Today 15:494–524
30. Hasija V, Sudhaik A, Raizada P, Hosseini-Bandegharaei A, Singh P (2019) Carbon quantum dots supported AgI/ZnO/phosphorus doped graphitic carbon nitride as Z-scheme photocatalyst for efficient photodegradation of 2,4-dinitrophenol. J Environ Chem Eng 7:103272
31. Hasija V, Raizada P, Sudhaik A, Singh P, Thakur VK, Khan AAP (2020) Fabrication of Ag/AgI/WO3 heterojunction anchored P and S co-doped graphitic carbon nitride as a dual Z scheme photocatalyst for efficient dye degradation. Solid State Sci 100:106095
32. Heo NS, Shukla S, Oh SY, Bajpai VK, Lee SU, Cho HJ, Kim S, Kim Y, Kim HJ, Lee SY, Jun YS (2019) Shape-controlled assemblies of graphitic carbon nitride polymer for efficient sterilization therapies of water microbial contamination via 2D g-C_3N_4 under visible light illumination. Mater Sci Eng 104:109846

33. Hu S, Li F, Fan Z, Wang F, Zhao Y, Lv Z (2015) Band gap-tunable potassium doped graphitic carbon nitride with enhanced mineralization ability. Dalton Trans 44:1084–1092
34. Jamwal D, Kaur G, Raizada P, Singh P, Pathak D, Thakur P (2015) Twin-tail surfactant peculiarity in superficial fabrication of semiconductor quantum dots: toward structural, optical, and electrical features. J Phys Chem 119:5062–5073
35. Kassem A, Ayoub GM, Malaeb L (2019) Antibacterial activity of chitosan nano-composites and carbon nanotubes: a review. Sci Total Environ 668:566–576
36. Kaur S, Sharma S, Kansal SK (2016) Synthesis of ZnS/CQDs nanocomposite and its application as a photocatalyst for the degradation of an anionic dye. Superlattices Microstruct 98:86–95
37. Krishna V, Pumprueg S, Lee SH, Zhao J, Sigmund W, Koopman B, Moudgil BM (2005) Photocatalytic disinfection with titanium dioxide coated multi-wall carbon nanotubes. Process Saf Environ Prot 83:393–397
38. Kumar A, Raizada P, Singh P, Saini R, Saini A, Hosseini-Bandegharaei A (2019) Perspective and status of polymeric graphitic carbon nitride based Z-scheme photocatalytic systems for sustainable photocatalytic water purification. Chem Eng J 123496. https://doi.org/10.1016/j.cej.2019.123496
39. Li H, He X, Kang Z, Huang H, Liu Y, Liu J, Lian S, Tsang CHA, Yang X, Lee ST (2010) Water-soluble fluorescent carbon quantum dots and photocatalyst design. Angew Chem 49:4430–4434
40. Li J, Shen B, Hong Z, Lin B, Gao B, Chen Y (2012) A facile approach to synthesize novel oxygen-doped g-C_3N_4 with superior visible-light photoreactivity. Chem Commun 48:12017–12019
41. Li R, Ren Y, Zhao P, Wang J, Liu J, Zhang Y (2019) Graphitic carbon nitride (g-C_3N_4) nanosheets functionalized composite membrane with self-cleaning and antibacterial performance. J Hazard Mater 365:606–614
42. Li XY, Zou XJ, Qu ZP, Zhao QD, Wang LZ (2011) Photocatalytic degradation of gaseous toluene over Ag-doping TiO_2 nanotube powder prepared by anodization coupled with impregnation method. Chemosphere 83:674–679
43. Li Y, Zhang BP, Zhao JX, Ge ZH, Zhao XK, Zou L (2013) ZnO/carbon quantum dots heterostructure with enhanced photocatalytic properties. Appl Surf Sci 279:367–373
44. Lightcap IV, Kosel TH, Kamat PV (2010) Anchoring semiconductor and metal nanoparticles on a two-dimensional catalyst mat. Storing and shuttling electrons with reduced graphene oxide. Nano Lett 10:577–583
45. Lin T, Song Z, Wu Y, Chen L, Wang S, Fu F, Guo L (2018) Boron- and phenyl-codoped graphitic carbon nitride with greatly enhanced light responsive range for photocatalytic disinfection. J Hazard Mater 358:62–68
46. Liu G, Niu P, Sun CH, Smith SC, Chen SG, Lu GQ, Cheng HM (2010) Novelty in designing of photocatalysts for water splitting and CO_2 reduction. J Am Chem Soc 132:11642
47. Liu R, Huang H, Li H, Liu Y, Zhong J, Li Y, Zhang S, Kang Z (2013) Metal nanoparticle/carbon quantum dot composite as a photocatalyst for high efficiency cyclohexane oxidation. ACS Catal 4:328–336
48. Liu SX, Chen XY, Chen X (2007) A TiO_2/AC composite photocatalyst with high activity and easy separation prepared by a hydrothermal method. J Hazard Mater 143:257–263
49. Liu Y, Yu YX, Zhang WD (2013) Carbon quantum dots-doped CdS microspheres with enhanced photocatalytic performance. J Alloys Compd 569:102–110
50. Miao R, Luo Z, Zhong W, Chen SY, Jiang T, Dutta B, Nasr Y, Zhang Y, Suib SL (2016) Mesoporous TiO_2 modified with carbon quantum dots as a high-performance visible light photocatalyst. Appl Catal B 189:26–38
51. Muthulingam S, Lee IH, Uthirakumar P (2015) Highly efficient degradation of dyes by carbon quantum dots/N-doped zinc oxide (CQD/N-ZnO) photocatalyst and its compatibility on three different commercial dyes under daylight. J Colloid Interface Sci 455:101–109
52. Niu P, Liu G, Cheng HM (2012) Nitrogen vacancy-promoted photocatalytic activity of graphitic carbon nitride. J Phys Chem C 116:11013–11018

53. Novoselov KS, Geim AK, Morozov S, Jiang D, Katsnelson MI, Grigorieva I, Firsov AA (2005) Two-dimensional gas of massless Dirac fermions in graphene. Nature 438:197
54. Oliveira HG, Nery DC, Longo C (2010) Effect of applied potential on photocatalytic phenol degradation using nanocrystalline TiO_2 electrodes. Appl Catal B 93:205–211
55. Pare B, Singh P, Jonnalagadda SB (2008) Visible light induced heterogeneous advanced oxidation process to degrade pararosanilin dye in aqueous suspension of ZnO. Ind J Chem 47A:830–835. http://nopr.niscair.res.in/handle/123456789/2107
56. Pare B, Singh P, Jonnalgadda SB (2009) Artificial light assisted photocatalytic degradation of lissamine fast yellow dye in ZnO suspension in a slurry batch reactor. Ind J Chem 48A(10):1364–1369. http://nopr.niscair.res.in/handle/12345689/6122
57. Pare B, Singh P, Jonnalgadda SB (2009) Degradation and mineralization of victoria blue B dye in a slurry photoreactor using advanced oxidation process. J Sci Ind Res 68:724–729
58. Park JH, Kim S, Bard AJ (2006) Novel carbon-doped TiO_2 nanotube arrays with high aspect ratios for efficient solar water splitting. Nano Lett 6:24–28
59. Park S, Ruoff RS (2009) Chemical methods for the production of graphenes. Nat Nanotechnol 4:217
60. Park S, Mohanty N, Suk JW, Nagaraja A, An J, Piner RD, Ruoff RS (2010) Biocompatible, robust free-standing paper composed of a TWEEN/graphene composite. Adv Mater 22:1736–1740
61. Priya B, Shandilya P, Raizada P, Thakur P, Singh N, Singh P (2016) Photocatalytic mineralization and degradation kinetics of ampicillin and oxytetracycline antibiotics using graphene sand composite and chitosan supported BiOCl. J Mol Catal A: Chem 423:400–413
62. Priya B, Raizada P, Singh N, Thakur P, Singh P (2016) Adsorptional photocatalytic mineralization of oxytetracycline and ampicillin antibiotics using Bi_2O_3/BiOCl supported on graphene sand composite and chitosan. J Colloid Interface Sci 479:271–283
63. Raizada P, Priya B, Thakur P, Singh P (2016) Solar light induced photodegradation of oxytetracyline using Zr doped TiO_2/CaO based nanocomposite. Ind J Chem 55A(07):803–809
64. Raizada P, Kumari J, Shandilya P, Dhiman R, Singh VP, Singh P (2017) Magnetically retrievable Bi_2WO_6/Fe_3O_4 immobilized on graphene sand composite for investigation of photocatalytic mineralization of oxytetracycline and ampicillin. Process Saf Environ Prot 106:104–116
65. Raizada P, Kumari J, Shandilya P, Singh P (2017) Kinetics of photocatalytic mineralization of oxytetracycline and ampicillin using activated carbon supported ZnO/$ZnWO_4$. Desalin Water Treat 79:204–213
66. Raizada P, Shandilya P, Singh P, Thakur P (2017) Solar light-facilitated oxytetracycline removal from the aqueous phase utilizing a H_2O_2/$ZnWO_4$/CaO catalytic system. J Taibah Univ Sci 11:689–699
67. Raizada P, Singh P, Kumar A, Sharma G, Pare B, Jonnalagadda SB, Thakur P (2014) Solar photocatalytic activity of nano-ZnO supported on activated carbon or brick grain particles: role of adsorption in dye degradation. Appl Catal A 486:159–169
68. Raizada P, Singh P, Kumar A, Pare B, Jonnalagadda SB (2014) Zero valent iron-brick grain nanocomposite for enhanced Solar-Fenton removal of malachite green. Sep Purif Technol 133:429–437
69. Raizada P, Sudhaik A, Singh P (2019) Photocatalytic water decontamination using graphene and ZnO coupled photocatalyst: a review. Mater Sci Energy Technol 2:509–525
70. Raizada P, Sudhaik A, Singh P, Shandilya P, Gupta VK, Hosseini Bandegharaei A, Agrawal S (2019) Ag_3PO_4 modified phosphorus and sulphur co-doped graphitic carbon nitride as a direct Z-scheme photocatalyst for 2,4-dimethyl phenol degradation. J Photochem Photobiol A Chem 374:22–35
71. Raizada P, Sudhaik A, Singh P, Shandilya P, Saini AK, Gupta VK, Lim JH, Jung H, Hosseini-Bandegharaei A (2019) Fabrication of Ag_3VO_4 decorated phosphorus and sulphur co-doped graphitic carbon nitride as a high-dispersed photocatalyst for phenol mineralization and *E. coli* disinfection. Sep Purif Technol 212:887–900

72. Raizada P, Sudhaik A, Singh P, Hosseini-Bandegharaei A, Thakur P (2019) Converting type II AgBr/VO into ternary Z scheme photocatalyst via coupling with phosphorus doped g-C_3N_4 for enhanced photocatalytic activity. Sep Purif Technol 12:115692
73. Raizada P, Sudhaik A, Singh P, Shandilya P, Thakur P, Jung H (2018) Visible light assisted photodegradation of 2,4-dinitrophenol using Ag_2CO_3 loaded phosphorus and sulphur co-doped graphitic carbon nitride nanosheets in simulated wastewater. Arab J Chem. https://doi.org/10.1016/j.arabjc.2018.10.004
74. Rao CNR, Sood AK, Subrahmanyam KS, Govindaraj A (2009) Graphene: the new two-dimensional nanomaterial. Angew Chem 48:7752
75. Sellaoui L, Dotto GL, Peres EC, Benguerba Y, Lima ÉC, Lamine AB, Erto A (2017) New insights into the adsorption of crystal violet dye on functionalized multi-walled carbon nanotubes: experiments, statistical physics and COSMO–RS models application. J Mol Liq 248:890–897
76. Serp P, Corrias M, Kalck P (2003) Carbon nanotubes and nanofibers in catalysis. Appl Catal A 253:337–358
77. Shandilya P, Mittal D, Soni M, Raizada P, Hosseini-Bandegharaei A, Saini AK, Singh P (2018) Fabrication of fluorine doped graphene and $SmVO_4$ based dispersed and adsorptive photocatalyst for abatement of phenolic compounds from water and bacterial disinfection. J Clean Prod 203:386–399
78. Shandilya P, Mittal D, Soni M, Raizada P, Lim JH, Jeong DY, Dewedi RP, Saini AK, Singh P (2018) Islanding of $EuVO_4$ on high-dispersed fluorine doped few layered graphene sheets for efficient photocatalytic mineralization of phenolic compounds and bacterial disinfection. J Taiwan Inst Chem Eng 93:528–542
79. Shandilya P, Mittal D, Sudhaik A, Soni M, Raizada P, Saini AK, Singh P (2019) $GdVO_4$ modified fluorine doped graphene nanosheets as dispersed photocatalyst for mitigation of phenolic compounds in aqueous environment and bacterial disinfection. Sep Purif Technol 210:804–816
80. Sharma K, Dutta V, Sharma S, Raizada P, Hosseini-Bandegharaei A, Thakur P, Singh P (2019) Recent advances in enhanced photocatalytic activity of bismuth oxyhalides for efficient photocatalysis of organic pollutants in water: a review. J Ind Eng Chem 78:1–20
81. Sharma S, Dutta V, Singh P, Raizada P, Rahmani-Sani A, Hosseini-Bandegharaei A, Thakur VK (2019) Carbon quantum dot supported semiconductor photocatalysts for efficient degradation of organic pollutants in water: a review. J Clean Prod 228:755–769
82. Shen T, Wang Q, Guo Z, Kuang J, Cao W (2018) Hydrothermal synthesis of carbon quantum dots using different precursors and their combination with TiO_2 for enhanced photocatalytic activity. Ceram Int 44:11828–11834
83. Singh P, Gautam S, Shandilya P, Priya B, Singh VP, Raizada P (2017) Graphene bentonite supported $ZnFe_2O_4$ as superparamagnetic photocatalyst for antibiotic degradation. Adv Mater Lett 8:229–238
84. Singh P, Priya B, Shandilya P, Raizada P, Singh N, Pare B, Jonnalagadda SB (2016) Photocatalytic mineralization of antibiotics using 60% WO_3/BiOCl stacked to graphene sand composite and chitosan. Arab J Chem. https://doi.org/10.1016/j.arabjc.2016.08.005
85. Singh P, Raizada P, Pathania D, Sharma G, Sharma P (2013) Microwave induced KOH activation of guava peel carbon as an adsorbent for congo red dye removal from aqueous phase. Ind J Chem Technol 20:305–311
86. Singh P, Raizada P, Pathania D, Kumar A, Thakur P (2013) Preparation of BSA-$ZnWO_4$ nanocomposites with enhanced adsorptional photocatalytic activity for methylene blue degradation. Int J Photoenergy. https://doi.org/10.1155/2013/726250
87. Singh P, Raizada P, Kumari S, Kumar A, Pathania D, Thakur P (2014) Solar-Fenton removal of malachite green with novel Fe0-activated carbon nanocomposite. Appl Catal A 476:9–18
88. Singh P, Shandilya P, Raizada P, Sudhaik A, Rahmani-Sani A, Hosseini-Bandegharaei A (2018) Review on various strategies for enhancing photocatalytic activity of graphene based nanocomposites for water purification. Arab J Chem. https://doi.org/10.1016/j.arabjc.2018.12.001

89. Singh P, Raizada P, Sudhaik A, Shandilya P, Thakur P, Agarwal S, Gupta VK (2019) Enhanced photocatalytic activity and stability of AgBr/BiOBr/graphene heterojunction for phenol degradation under visible light. J Saudi Chem Soc 23:586–599
90. Singh P, Sharma K, Hasija V, Sharma V, Sharma S, Raizada P, Singh M, Saini AK, Hosseini-Bandegharaei A, Thakur VK (2019) Systematic review on applicability of magnetic iron oxides-integrated photocatalysts for degradation of organic pollutants in water. Mater Today Chem 14:100186
91. Stankovich S, Dikin DA, Dommett GH, Kohlhaas KM, Zimney EJ, Stach EA, Ruoff RS (2006) Graphene-based composite materials. Nature 442:282
92. Sudhaik A, Raizada P, Shandilya P, Singh P (2018) Magnetically recoverable graphitic carbon nitride and $NiFe_2O_4$ based magnetic photocatalyst for degradation of oxytetracycline antibiotic in simulated wastewater under solar light. J Environ Chem Eng 6:3874–3883
93. Sudhaik A, Raizada P, Shandilya P, Jeong DY, Lim JH, Singh P (2018) Review on fabrication of graphitic carbon nitride based efficient nanocomposites photodegradation of aqueous phase organic pollutants. J Ind Eng Chem 67:28–51
94. Vuono D, Catizzone E, Aloise A, Policicchio A, Agostino RG, Migliori M, Giordano G (2017) Modelling of adsorption of textile dyes over multi-walled carbon nanotubes: equilibrium and kinetic. Chin J Chem Eng 25:523–532
95. Wang K, Ruan J, Song H, Zhang J, Wo Y, Guo S, Cui D (2011) Biocompatibility of graphene oxide. Nanoscale Res Lett 6:8
96. Wang Q, Guo X, Cai L, Cao Y, Gan L, Liu S, Wang Z, Zhang H, Li L (2011) TiO_2-decorated graphenes as efficient photoswitches with high oxygen sensitivity. Chem Sci 2:1860–1864
97. Xu G, Xu Y, Zhou Z, Bai Y (2019) Facile hydrothermal preparation of graphitic carbon nitride supercell structures with enhanced photodegradation activity. Diamond Relat Mater 97:107461
98. Xu X, Ray R, Gu Y, Ploehn HJ, Gearheart L, Raker K, Scrivens WA (2004) Electrophoretic analysis and purification of fluorescent single-walled carbon nanotube fragments. J Am Chem Soc 126:12736–12737
99. Yadav P, Nishanthi ST, Purohit B, Shanavas A, Kailasam K (2019) Metal-free visible light photocatalytic carbon nitride quantum dots as efficient antibacterial agents: an insight study. Carbon 152:587–597
100. Yang S, Gong Y, Zhang J, Zhan L, Ma L, Fang Z, Ajayan PM (2013) Exfoliated graphitic carbon nitride nanosheets as efficient catalysts for hydrogen evolution under visible light. Adv Mater 25:2452–2456
101. Yang ZC, Wang M, Yong AM, Wong SY, Zhang XH, Tan H, Chang AY, Li X, Wang J (2011) Intrinsically fluorescent carbon dots with tunable emission derived from hydrothermal treatment of glucose in the presence of monopotassium phosphate. Chem Commun 47:11615–11617
102. Yao Y, Hu Y, Hu H, Chen L, Yu M, Gao M, Wang S (2019) Metal-free catalysts of graphitic carbon nitride-covalent organic frameworks for efficient pollutant destruction in water. J Colloid Interface Sci 554:376–387
103. Yin S, Zhang Q, Saito F, Sato T (2003) Preparation of visible light-activated titania photocatalyst by mechanochemical method. Chem Lett 32:358–359
104. Yu H, Zhang H, Huang H, Liu Y, Li H, Ming H, Kang Z (2012) ZnO/carbon quantum dots nanocomposites: one-step fabrication and superior photo-catalytic ability for toxic gas degradation under visible light at room temperature. New J Chem 36:1031–1035
105. Yu MF, Files BS, Arepalli S, Ruoff RS (2000) Tensile loading of ropes of single wall carbon nanotubes and their mechanical properties. Phys Rev Lett 84:5552–5553
106. Yu X, Liu J, Yu Y, Zuo S, Li B (2014) Preparation and visible light photocatalytic activity of carbon quantum dots/TiO_2 nanosheet composites. Carbon 68:718–724
107. Zare K, Sadegh H, Shahryari-Ghoshekandi R, Maazinejad B, Ali V, Tyagi I, Gupta VK (2015) Enhanced removal of toxic Congo red dye using multi walled carbon nanotubes: kinetic, equilibrium studies and its comparison with other adsorbents. J Mol Liq 212:266–271

108. Zhang C, Li Y, Shuai D, Shen Y, Xiong W, Wang L (2018) Graphitic carbon nitride ($g\text{-}C_3N_4$)-based photocatalysts for water disinfection and microbial control: a review. Chemosphere 214:462–479
109. Zhang GG, Zhang MW, Ye XX, Qiu XQ, Lin S, Wang XC (2014) Iodine modified carbon nitride semiconductors as visible light photocatalysts for hydrogen evolution. Adv Mater 26:805
110. Zhang L, Fang P, Yang L, Zhang J, Wang X (2013) Rapid method for the separation and recovery of endocrine-disrupting compound bisphenol AP from wastewater. Langmuir 29:3968–3975
111. Zhang Y, Pan Q, Chai G, Liang M, Dong G, Zhang Q, Qiu J (2013) Synthesis and luminescence mechanism of multicolor-emitting $g\text{-}C_3N_4$ nanopowders by low temperature thermal condensation of melamine. Sci Rep 3:1943
112. Zhang X, Wang Y, Liu B, Sang Y, Liu H (2017) Heterostructures construction on TiO_2 nanobelts: a powerful tool for building high-performance photocatalysts. Appl Catal B 202:620–641
113. Zhou J, Zhou H, Tang J, Deng S, Yan F, Li W, Qu M (2017) Carbon dots doped with heteroatoms for fluorescent bioimaging: a review. Mikrochim Acta 184:343–368
114. Zhu S, Meng Q, Wang L, Zhang J, Song Y, Jin H, Zhang K, Sun H, Wang H, Yang B (2013) Highly photoluminescent carbon dots for multicolor patterning, sensors, and bioimaging. Angew Chem 125:4045–4049

Carbon-Based Nanocomposites: Preparation and Application in Environmental Pollutants Removal

Ambika and Pradeep Pratap Singh

Abstract Environmental pollution is problem of enormous public concern worldwide. With the increasing population, the demand for fresh water is also increasing while per capita annual availability of water has reduced. Due to the rapid industrial growth, the water is getting polluted. Various pollutants, such as heavy metals, dyes, pesticides, insecticides, herbicides, antibiotics, oil spills, plant nutrients, bacteria, viruses, etc., pose serious risks to the environment. Thus, there is an urgent need to develop new methodologies and material for the removal of pollutants from environment. Carbon-based nanocomposites have drawn the attention of scientists because of their unique chemical and physical properties. These nanocomposites pose a great potential for application in various environmental fields including, air pollution biotechnologies, monitoring, wastewater treatment, etc. The present article describes preparation of carbon-based nanocomposites and their application in environmental pollutants removal.

Keywords Carbon-based nanocomposites · Pollutants · Carbon nanotubes · Graphene

1 Introduction

One of the major global concerns in the twenty-first century is problems related to the environment. Environmental pollutants can comfortably spread into the surroundings via different pathways [95]. Anthropogenic activities are treated as imperative part which could contaminate the different ecosystems [93]. Various types of pollutants, such as heavy metals [59, 69, 91], organic dyes [15, 52, 60, 79, 82], bacteria/viruses [125, 140], can pollute the different ecosystems. The polluted environment not only

Ambika
Department of Chemistry, Hansraj College, University of Delhi, Delhi 110007, India

P. P. Singh (✉)
Department of Chemistry, Swami Shraddhanand College, University of Delhi, Delhi 110036, India
e-mail: ppsingh@ss.du.ac.in

M. Jawaid et al. (eds.), *Environmental Remediation Through Carbon Based Nano Composites*, Green Energy and Technology,
https://doi.org/10.1007/978-981-15-6699-8_10

have adverse effect on the human health but also affects the natural balance of the ecosystems. The contaminants like heavy metals, pesticides, etc. can easily enter and influence the different trophic levels [27]. These pollutants can be removed by employing different nanomaterials (NMs) with improved properties.

Nanotechnology is the science in which the size of the materials can be manipulated by one-billionth of a meter. NMs are the materials which have nanosized particles or constituents. Generally, these materials have marked thermal, optical, mechanical, electronic properties as compared to their bulk materials [55]. Nowadays, there is an intense research is taking place in the field of nanocomposites (NCs). The NCs offer a huge potential with a wide variety of applications in different fields and thus can be employed as an alternative to conventional composite materials. These NCs possess high surface area to volume ratio of the reinforcing nanoparticles as compared to conventional composites. Therefore, even a minute quantity of reinforcement material can significantly influence the macroscopic properties of NCs due to the nanosize of their components. The reinforcing material could be prepared by using particles (e.g., minerals, metallic nanoparticles, carbon-based NMs), sheets (e.g., graphene), or fibers (electrospun nanofibres) [58].

Carbon is an important element to the researchers and scientists due to its unique properties. Carbon-based NMs possess various physical and chemical properties, due to which these NMs have been utilized for the preparation of carbon-based nanocomposites (CNCs). Recently, the CNCs have attracted the curiosity of scientists owing to their potential for the discovery of new materials as well as for the development of new technologies. CNCs possess high surface area, high mechanical strength, good chemical stability, high temperature stability, etc. Owing to the unique properties, they offer potential advances in energy and environmental systems toward energy efficiency, pollutant transformation, and toxicity control. The present article describes the preparation of CNCs and their application in environmental pollutants removal.

2 Types of Carbon-Based Nanocomposites Used in Pollutants Removal from Environment

Carbon NMs (CNMs) such as carbon nanotubes (CNTs), fullerenes, and graphene have been developed for various environmental applications [61, 92] (Scheme 1). CNCs have been prepared by using different types CNMs, few of them are discussed below.

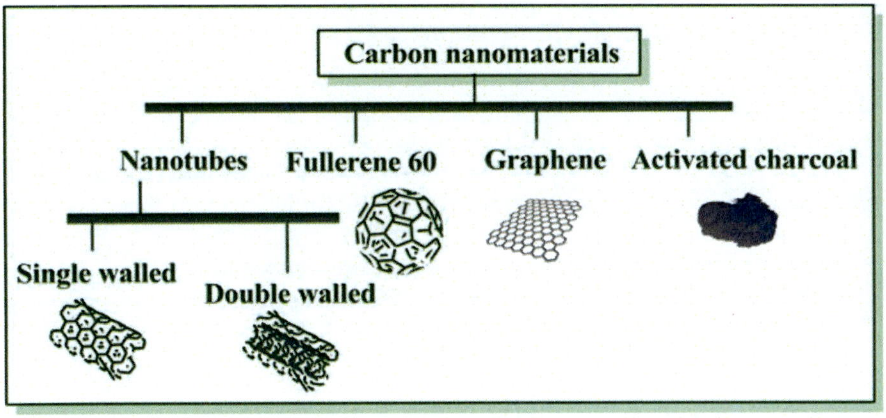

Scheme 1 Different types of carbon-based nanomaterials

2.1 Carbon Nanotubes-Based Nanocomposites

Carbon nanotubes (CNTs) are the NMs of carbon with diameters measured in nanometers. They can be of two types: (a) single-walled (SWCNT) (b) multi-walled (MWCNT). SWNTs possess diameter in the range of 0.5–3.0 nm, whereas the diameter of MWNTs lies in the range of 1.5–100 nm. CNTs possess high aspect ratio, excellent electrical conductivity, chemical stability, and mechanical robustness [13]. Different types of NCs can be prepared by utilizing CNTs, due to their versatile properties. CNTs-based NCs possess significantly enhanced electrical and mechanical properties. These NCs are utilized a variety of applications such as electronic, aerospace, military applications, and environmental remediation [13].

2.2 Graphene-Based Nanocomposites

Graphene is an allotrope of carbon with honey comb-like structure and a zero band gap. It possesses large surface area, high charge carrier mobility, optical transparency, excellent mechanical stiffness, high electrical conductivity [80, 81]. Recently, Graphene have been employed as support NM for the preparation of NCs. Graphene-based NCs posses biocompatibility, high water dispersion, easy modification, etc. [42, 130]. These NCs have been utilized in photocatalysis and adsorption of contaminants from environment [62, 119].

2.3 Fullerenes-Based Nanocomposites

Fullerene is another allotrope of carbon which consists of hollow molecular cages, balls, or tubes of strongly bonded carbon atoms connected by single and double bonds [85]. Recently, fullerenes have attracted the attention of materials scientists. Fullerenes and their derivatives have been employed for fabrication of various types of NCs. These NCs have been employed for important applications in various fields such as electronics, energy, organocatalysis, sensing, biomedical sciences, photoelectrochemistry, environmental remediation [26, 31, 49, 67, 107].

3 Different Processing Methods for Carbon-Based Nanocomposites

The CNCs can be processed employing the following methods:

3.1 Melt Blending

Melt blending is a cost-effective method used for the fabrication of composites in the absence of organic solvents. In this method, the polymer pellets are melted to form the viscous liquid and the nanofillers such as CNTs, graphene, etc. are dispersed into it by using high shear force. However, there are limited studies on the melt blending of graphene and polymers due to the low thermal stability of most chemically modified graphene and its low bulk density [48, 78]. The high shear force and high temperature can degrade the properties of NCs; therefore, the process must be fine-tuned in order to obtain optimum conditions.

3.2 Solvent Processing

This method can be used for the production of graphene or CNTs-based polymer composites. The nanoparticles in a polymer dissolved in a solvent are agitated for the preparation. It involves the evaporation of solvent before casting in a mold and evaporation of the solvent, of both thermoplastic and thermoset materials. Different solvents ranging from aqueous to organic can be employed. However, the removal of organic solvent after casting can affect the environment.

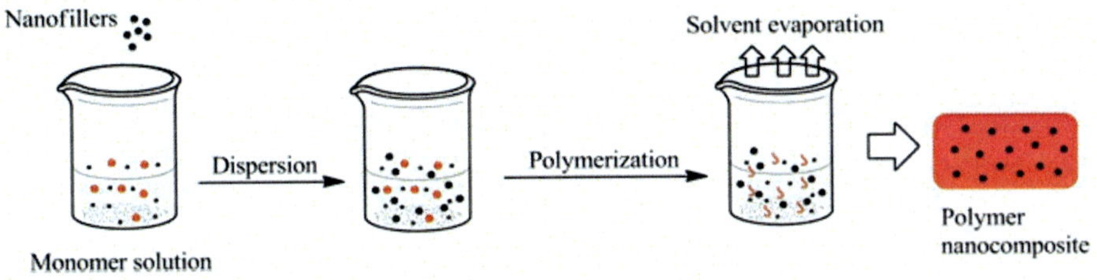

Scheme 2 In situ polymerization method of carbon-based NCs

3.3 In Situ Polymerization

It involves an initiation step followed by a series of polymerization steps for the formation of a hybrid between polymer molecules and nanoparticles. Intercalated or exfoliated NCs are produced by spreading the nanoparticles in a liquid monomer of relatively low molecular weight which percolates in between the interlayers followed by the polymerization of monomers [88]. The properties of final composite can be improved by the grafting of polymer on filler surface (Scheme 2).

3.4 Template Synthesis (Sol-Gel Technology)

In this method, a colloidal suspension of solid nanoparticles (sol) is formed by the dispersion of solid nanoparticles in the monomer solution, which serves as the precursor for an integrated network (gel) of discrete particles. The polymer acts as the nucleating agent as well as assists the growth of the filler crystals. The NCs formation takes place as the crystal grows with polymers trapped within the layers [2].

4 Applications of Carbon-Based Nanocomposites

Due to the extensive growth of industries, the different ecosystems are getting contaminated with hazardous contaminants like, dyes, polychlorinated compounds, polycyclic aromatic hydrocarbons (PAHs), insecticides, etc. Thus, new materials and treatment approaches have been developed for the remediation of the environment.

Recently, CNCs have received particular interest for the removal of the contaminants due to their excellent physicochemical properties, biocompatibility, and unlimited probabilty of functionalization. CNCs involve photocatalysis, adsorption, and disinfection methods for environmental remediation (Scheme 3). Herein, we have discussed the removal of contaminants using CNCs mainly by two process (a) adsortion and (b) photocatalysis.

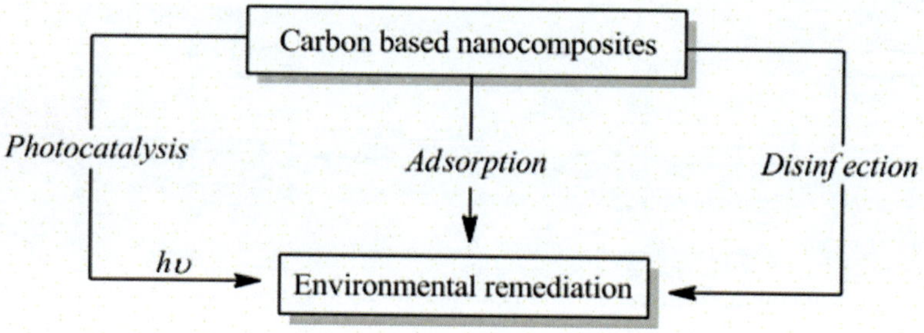

Scheme 3 Applications of carbon-based nanocomposites for environmental remediation

4.1 Adsorption of Contaminants Using Carbon-Based Nanocomposites

Adsorption is the process in which a film of the adsorbate is formed at the surface of the solid (adsorbent). Heavy metals, dyes, PAHs, pesticides, etc. are few of the hazardous pollutants which are deleterious to human health and environment [25, 89, 94, 102]. Various anthropogenic activities such as mining, industrial discharge, agricultural run offs etc. has led to an increase in the environmental pollution [6, 86, 120]. These contaminants can bioaccumulate into food chains resulting in serious life threatening effects on living beings [29, 41, 115]. Different types of CNCs have been employed for the adsorption of these contaminants and are discussed in Table 1.

4.1.1 Adsorption of Contaminants by Activated Carbon-Based Nanocomposites

Activated carbon (AC) is an important form of carbon which has been employed for the development of NCs and their utilization in environmental applications. Different magnetic NCs have been employed for the removal of contaminants from the environment. Magnetic NCs alginate beads were fabricated for the simultaneous removal of cationic (e.g., copper), anionic (e.g., phosphate), and organic (e.g., toluene) pollutants in highly acidic water. These alginate beads were impregnated with a NC material composed of zeolites, AC, layered double hydroxides, and magnetic nanoparticles bound together by xanthan gum with an aspect ratio of 3:4:1 (alginate: NCs: xanthan gum). The beads exhibited high adsorption capacities [91]. Highly porous N/S doped magnetic carbon aerogel (N/S-MCA) was utilized for the removal of bisphenol–A (BPA) from aqueous solution. The maximum removal of the BPA by the above NC depends on pH, temperature [5]. The Fe_3O_4/Ag/C magnetic NC was employed for the adsorption of Methylene Blue (MB), Acid Orange 7 (AO7), and Rhodamine 6G (Rh 6G) in single and multi-component system from aqueous solutions [82]. C/$ZnFe_2O_4$ NC was utilized as an adsorbent for the removal of PAHs such as naphthalene and

Table 1 Removal of different pollutants by carbon-based nanocomposites

S. No.	Class of pollutants	Type of pollutants	Types of carbon-based nanocomposites	Reference
1.	Heavy metal ions	Copper	Magnetic alginate beads, MWCNT-PEI/PAN, CS-CNTs, CNT-PDA-CS, SO_3H-Fe_3O_4-GO, G/Fe_3O_4, CNTs-G, GO-Ca-alginate, GO-gelatin-CS, PFSP, MnO_2/GNS	Phiri et al. [91], Deng et al. [30], Dou et al. [32], Zeng et al. [135], Hu et al. [46], Wu et al. [128], Sui et al. [116], Algothmi et al. [10], Zhang et al. [139]; Rikame et al. [96, 99]
		Lead	MWCNT-PEI/PAN, L-CNTs, CS/GO GO-Fe_3O_4, CNTs-G, PVC-GO GO-TiO_2, MnO_2/GNS, G/Fe_3O_4, Fe(0)-Fe_3O_4, GO-gelatin-CS, G/SiO_2	Hao et al. [45], Deng et al. [30], Li et al. [69], Debnath et al. [28], Li et al. [65], Fan et al. [37, 38]; Musico et al. [83], Bhunia et al. [19]; Liu et al. [70, 71], Sui et al. [116], Rikame et al. [99], Ren et al. [96], Lee and Yang [63]
		Zinc	GO-TiO_2	Li et al. [69]
		Ni(II)	δ-MnO_2-G	Ren et al. [97]
		Gold	CS-GO	Liu et al. [70, 71]
		Mercury	CNTs-G, RGO-polypyrrole, MnO_2-RGO and Ag-RGO, Fe(0)-Fe_3O_4	Sui et al. [116], Chandra and Kim [20]; Sreeprasad et al. [112]; Bhunia et al. [19]
		Arsenic	GO/$Fe(OH)_3$, GO-Fe_3O_4, Mn-Fe_3O_4-G, GO-ZrO	Zhang et al. [137]; Zhang et al. [136]; Nandi et al. [84]; Luo et al. [75]

(continued)

Table 1 (continued)

S. No.	Class of pollutants	Type of pollutants	Types of carbon-based nanocomposites	Reference
		Cadmium	CS/AC/Fe, ZnO/AC, Fe(0)-Fe_3O_4 GO-TiO_2	Srivastava et al. [114]; Bhunia et al. [19]; Li et al. [69]
		Chromium	GnZVI/PAC, MNP/MWCNTs, Chitin/Fe_3O_4/MWCNTs, Branched PEI-MWCNT G-Fe(0), MgAl-G double-layered hybrid, NiO/rGO, Fe(0)-Fe_3O_4	Khosravi et al. [59]; Lu et al. [72]; Salam et al. [103]; Phiri et al. [90]; Li et al. [67]; Jabeen et al. [51]; Zhang et al. [138, 140]
		Uranium (VI)	PECQDs/$MnFe_2O_4$	Huang et al. [47]
2.	Dyes	Rhodamine 6G	Fe_3O_4/Ag/C, G-CNT, G-asphalt	Sreeprasad et al. [113]; Muntean et al. [82]
		Malachite Green	NH_2-CNT/Fe_2O_3/ZIF-8, Fe_3O_4-RGO	Sun et al. [117]
		Rose Bengal	MWCNTs/Chitin/Fe_3O_4	Salam et al. [105]
		Crystal Violet	(OMWCNT)-κ-carrageenan-Fe_3O_4	Duman et al. [33]
		Reactive Orange 84	CuNPs-MWCNT	Jafari et al. [52]
		Congo Red	ZnO/MWCNTs, G/Fe_3O_4	Kirti et al. [60]; Arabi et al. [15]
		Fuchsine dye	G/Fe_3O_4, G-CNT	Wang et al. [122]; Li et al. [68]
		Rhodamine-B	Fe_3O_4-RGO	Mohammad et al. [79], Wang et al. [123, 126]
		Methyl Blue	Magnetic CS-GO	Fan et al. [36]
		Reactive Black 5	CS/RGO,CS-G	Duman et al. [33], Cheng et al. [24]
		Pararosaniline	G/Fe_3O_4	Wang et al. [122]
		Victoria blue	G/Fe_3O_4	Wang et al. [124]

(continued)

Table 1 (continued)

S. No.	Class of pollutants	Type of pollutants	Types of carbon-based nanocomposites	Reference
		Methyl Orange	G/Fe_3O_4, Co-Fe_2O_4/G	Wang et al. [124]; Li et al. [64]
		Brilliant yellow	G/Fe_3O_4	Wang et al. [124]
		Neutral red	G/Fe_3O_4	Wang et al. [124]
		Alizarin red	G/Fe_3O_4	Wang et al. [124]
		Eosin Y	GO-CS	Chen et al. [23]
		Safranine T	G/Fe_3O_4	Wang et al. [124]
		Methyl green	$CoFe_2O_4$-FGS	Farghali et al. [39]
		Bisphenol-A	Cu-BDC@GrO, Cu-BDC@CNT	Ahmad et al. [5]
		Acid yellow 36	GO-Cs	Mirzaee et al. [77]
		Acid blue 74	GO-Cs	Banerjee et al. [17]
		Ciprofloxacin	SA/GO	Wu et al. [128]
3.	Other organic species	Toluene	Fe_3O_4-alginate beads Ncs	Phiri et al. [91], Srivastava et al. [114]
		Aniline	Magnetic MWCNT/ferrite ($NiFe_2O_4$)	Salam [103]
		4-Chlorophenol	CLDH/SWCNT	Zhang et al. [141]
		Phenol	CLDH/SWCNT, CS/MWCNT	Guo et al. [43]
		Picric acid	MWCNT-CS	Khakpour and Tahermansouri [57]
		Humic Acid	M-PAC-MWCNT	Shaoxiu
		PAHs such as naphthalene and s2-naphthol	C/$ZnFe_2O_4$	Sharma et al. [109]
		Methyl-ethyl ketone	Mesoporous C/SiO_2	Janus et al. [53]

2-naphthol from aquatic system. The above NC can be efficiently regenerated using NaOH-ethanol for four cycles using different desorbing agents [109].

Different bionanocomposites have been employed for the environmental remediation. For example, biomaterials such as unripened fruit of *Cassia fistula* (Golden shower) and *Aloe vera* were utilized for the preparation of multifunctional super paramagnetic NC for dye removal [Methyl Blue (MTB) and Congo Red (CR)] and disinfection. Moreover, even at very low nanoparticle content, the above NC manifested excellent pollutant removal and disinfection properties while the *Aloe vera*-based bionanocomposites have potential for cost reduction to the extent of ten times as compared to only magnetite nanoparticles [60]. Powdered activated carbon (PAC) was prepared from *Peganum harmala* seed and its extract was used for the synthesis of zero-valent iron nanoparticles (GnZVI). Then GnZVI was loaded on PAC and used as a green NC to remove Cr(VI) from aqueous solutions. The adsorption capacity of Cr(VI) increases with the loading of GnZVI and increasing temperature [59]. Two types of NCs, (HAP/TE/GAC) and (HAP/GAC), were synthesized one using granular activated carbon (GAC) coated with both hydroxyapatite nanoflakes and turmeric extract while the other composite with only HAP nanoflakes coating on GAC for the adsorption of heavy metal ions. HAP/TE/GAC NC displayed better activity as compared with HAP/GAC [54].

Carbon--silica materials with hierarchical pores consisting of micropores and mesopores demonstrated excellent adsorption and desorption capacity for different volatile organic compounds (VOCs) and organic waste gases, and it could also be regenerated for further use [73]. Similarly, mesoporous carbon/silica NC has utilized as a highly stable and reusable materials for the adsorption of methyl--ethyl ketone from gas phase. The MCM-41-based composites containing highly dispersed carbon layers on the surface were found to be the most promising adsorbent for a commercial application [53]. ZnO/AC NC exhibited enhanced electrostatic interactions for the effective adsorption of Cd^{2+} with a maximum adsorption capacity of 96.2 mg/g for Cd^{2+} ions [11]. Functionalized carbon-micro NCs were developed for the adsorption of hydrocarbons (e.g., toluene, ethyl benzene, o-xylene). The above NC can be reused for five cycles without any decrease in the sorption capacity [114].

4.1.2 Adsorption of Contaminants by Carbon Nanotubes-Based Nanocomposites

Carbon nanotubes (CNTs) are one of most researched carbon-based materials with unique physical and chemical properties which makes them a material of choice for environmental remediation. Various magnetic CNTs-based NCs have been utilized for the removal of contaminant from the environment. Magnetic Fe_3O_4 nanoparticles (MNP) coated with different types of CNTs have been employed for magnetic solid phase extraction (MSPE) for mercury speciation analysis. SWCNT-MNP showed higher adsorption capacity than MWCNTs. Also, above magnetic NCs can be reused at least seven times without any loss in efficiency [98]. Similarly, MNPs and MWCNT (MNP/MWCNTs)-based NC has been employed for the adsorption of Cr(VI) from aqueous solution. The removal efficiency of above NC depends on pH and temperature of the solution. The NC could be regenerated by using an external magnetic field and can be reused for several cycles [72]. Chitin/magnetite/MWCNTs (CMM) magnetic NC has been employed as a potential and promising adsorbent for the efficient removal of Cr(VI) as compared with natural chitin [103]. Magnetic amine functionalized CNT (NH_2-CNT/Fe_2O_3)-zeolitic imidazolate framework-8 (ZIF-8) [NH_2-CNT/Fe_2O_3/ZIF-8: NCFZ] NCs with different amounts of NH_2-CNT/Fe_2O_3 (5, 10, and 15 wt% denoted as NCFZ-5, NCFZ-10, and NCFZ-15) were employed for the selective removal of cationic dyes [Malachite Green (MCG) and Rhodamine B (RhB)] from a binary system [79]. The MWCNTs/chitin/magnetite (MCM) NC has been employed for the adsorption of Rose Bengal (RB) [105]. Magnetic oxidized MWCNT [(OMWCNT)-κ-carrageenan-Fe_3O_4] NC has been employed as an adsorbent for the removal of cationic Crystal Violet (CV) and anionic reactive black 5 (RB5) dyes. An increase in the initial dye concentrations and the temperature of dye solutions led to an increase in the adsorption amounts of magnetic adsorbents. A decrease in the adsorption amount of CV dye was observed at low pH values, on the contrary to RB5 dye. On the contrary to anionic RB5 dye, the adsorption capacity of magnetic OMWCNT-κ-carrageenan-Fe_3O_4 NC for cationic CV dye is higher than that of magnetic OMWCNT-Fe_3O_4 NC. Therefore, magnetic OMWCNT-κ-carrageenan-Fe_3O_4 NC may be used as a potential adsorbent to remove the cationic dyes from aqueous solution [33]. A pH-dependent adsorption of the dyes such as Direct Blue 71 (DB71) and Reactive Blue 19 (RB19) from aqueous solution has been demonstrated involving CS/SiO_2/CNTs magnetic NCs. The maximum adsorption of DB71 occurred at pH 6.8, whereas RB19 adsorb maximally at pH 2.0 [1]. A magnetic titanium nanotube/CNT, (magnetiteTNT@CNT) NC was employed for the oxidative degradation of BPA from high saline polycarbonate plant wastewater (PCW) using catalytic wet peroxide oxidation [77]. A magnetic polyaluminium chloride (M-PAC)-MWCNT NC was utilized for removal of humic acid (HA) from aqueous solution. The adsorption of MWCNTs increases after the magnetization and modification by PAC of MWCNTs, which may be attributed to the interaction between PAC and HA through hydrogen bond and electrostatic attraction. Magnetic MWCNT/ferrite ($NiFe_2O_4$) NC was used for the removal of organic pollutants (e.g., aniline) from aqueous solution. The magnetic NC displayed high efficiency for the removal of aniline with the ease of separation of the nanoparticles from the aqueous solution using an ordinary magnet [104].

CNTs-based bionanocomposites have been employed efficiently for the environmental remediation. For example, CNTs filled biopolymer composites such as CS have a promising adsorption properties [106]. These composite systems could be employed for the removal of heavy metal ions and treatment of wastewater [74]. A CS-coated CNTs composites with high affinity and fast kinetics have been developed for the adsorption of Cu^{2+} ions from aqueous solution. The adsorption capacity of the composites was found to be two times that of pristine CNTs [32]. CNT-PDA-CS have been utilized for the adsorption of Cu^{2+} from aqueous solution. The above NC exhibited enhanced Cu^{2+} removal capability as compared with the unmodified CNT [135]. A low-cost and eco-friendly NC based on lignin grafted CNTs (L-CNTs) with good water-dispersibility and excellent adsorption capability for lead ion and oil droplet has been utilized as an adsorbent for water remediation [69]. Calcium-alginate (CA) and MWCNT-COOH beads (CA-MWCNT-COOH) NCs have been employed for the removal of MB from aqueous solution. The above NC exhibited high adsorption capacity as compared to both CA beads alone and the undispersed MWCNT-COOH. Furthermore, the impregnation also enhances the adsorption of MB onto other types of MWCNTs, indicating CA beads are an excellent supporting material to disperse and stabilize CNTs for their optimal application as a high-capacity adsorbent [121].

Different CNTs-based NCs have been employed in contaminants removal from the environment. For example, MWCNTs and polyethylenimine, polyacrylonitrile (PEI/PAN) were utilized to prepare a NC membrane (MWCNT-PEI/PAN) which possess higher mechanical strength, improved hydrophilicity, and excellent removal efficiency for metal (e.g., Pb^{2+} and Cu^{2+}) ions as compared to plain PAN membrane [30]. A cross-linked NC film of polyvinyl alcohol (PVA) incorporated with functionalized(f)-MWCNTs at different concentration with good recyclability has been employed for the removal of heavy metals, pesticides, bacteria, and fungi from wastewater [133]. Carbon ceramic electrode consisting of CuNPs and MWCNT was developed to treat reactive orange 84 (RO84) using ultrasound-assisted electrochemical degradation. The carbon ceramic electrode made with 4.0 wt% CuNPs and 4.0 wt% MWCNT exhibited high removal efficiency in a phosphate buffer with pH 8.0 [52]. ZnO/MWCNTs NCs can act as a promising, environment-friendly, and efficient adsorbent for the removal of CR dye from wastewater [15]. Calcined products of layered double hydroxides/SWCNT were developed for the removal of phenolic pollutants such as phenol and 4-chlorophenol with the removal rates of 91.7% and 99.5%, respectively. The above NC possess high adsorption capacity, wide range of effective pH for adsorption, fast adsorption speed, and excellent recyclability in comparison to most of the existing adsorbents [141]. Biomaterials functionalized CNTs possess strong affinity for dyes. Impregnation of carbon tubes with CS hydrogel beads has been demonstrated to be an efficient biosorbent material for the removal of CR [22]. CS/MWCNT NC has been employed for the phenol adsorption from water. The adsorption capacity (86.96 mg/g) of the above NC was improved compared to the original CS (61.69 mg/g) [43]. A NC material (MWCNT-Cs) has been developed by the modification of MWCNT-COOH with CS for the removal of picric acid from aqueous solutions. The picric acid molecules can be desorbed from MWCNT-CS up to 90% at pH = 9 and the NC can be recycled for five times after regeneration [57].

4.1.3 Adsorption of Contaminants by Carbon Quantum Dots-Based Nanocomposites

Uranium(VI) can be removed from aqueous solution by the application of a magnetic polyethyleneimine-functionalized carbon quantum dots/$MnFe_2O_4$ (PECQDs/$MnFe_2O_4$) NC. The enhanced adsorption of U(VI) may be attributed to the cation exchange and interaction between uranyl ions and abundant functional groups on PECQDs/$MnFe_2O_4$ [47].

4.1.4 Adsorption of Contaminants by Graphene and Its Derivatives-Based Nanocomposites

Graphene and its derivatives-based nanocomposites are important materials having unique properties which make them a good candidate for their utilization as an adsorbent for the removal of environmental contaminants. A variety of magnetically modified graphene-based NCs have been designed and employed for the removal of contaminants such as heavy metals, dyes, etc. from the environment. For example, a magnetically separable Fe_3O_4/porous graphene NC was employed for the adsorption of dyes and heavy metal ions from wastewater. Due to the high specific surface area and porous nature of graphene and high magnetic property of Fe_3O_4 nanoparticles, above NC demonstrates rapid adsorption with high adsorption capacity, easy separation, and reusability [18]. GO/$Fe(OH)_3$ material was utilized for the sorption of As(V) in polluted drinking water [136, 137]. GO-Fe_3O_4 has been employed for the removal of As(III) and As(V) in water [21]. High sorption could be attributed to the formation of surface complex. Mn-magnetite-graphene hybrid magnetic material has been utilized for 99.9% arsenite removal at pH = 7. The used sorbent could be separated by magnetic field [84]. Graphene sheets decorated with zero valent iron nanoparticles were utilized for pH-dependent Cr(VI) uptake [51]. Magnetic graphene NC and pristine graphene were compared for Cr(VI) removal for water treatment. The former NC exhibited good sorption ability as compared to the later [142]. Graphene nanohybrid (magnetic) anchored via core@double shell nanoparticles of crystalline iron oxide and inside shell of amorphous Si-S-O has been employed for Cr(VI) uptake under acidic pH (1–3) [143]. Cr(VI) has been adsorbed by the application of magnetically modified GO/CS/ferrite (GCF) NCs at pH 2.0 with high adsorption capacity [90]. Similarly, reusable, magnetic CS/GO NC can also be employed for the removal of contaminants such as heavy metal ions [Cr(VI), Pb(II)] and dyes [AO 7 and MB] from wastewater. The adsorption of the metal ions and dyes is pH dependent [28, 36–38, 65, 110]. GO anchored on magnetite was utilized for Pb(II) uptake with a sorbtion capacity of 588.24 mg/g [131]. Sulfonated Fe_3O_4-GO composite has been used for the removal of Cu(II) from water with a sorption capacity of 62.73 mg/g at pH 4.68 and 323 K temperature [46]. Magnetite-RGO (MRGO) hybrid material has demonstrated the uptake of 94 and 91% of malachite green (MG) and Rh B dyes in water with sorption capacities of 22.0 and 13.15 mg/g for MG and Rh B. The hybrid material can be recycled for several cycles [117]. Magnetic CS-GO (MCGO) has been employed for the uptake of dyes like MTB, MB in water with sorption capacity of 95.31 mg/g and 180.83 mg/g, respectively. Moreover, the sorbent can be regenerated with 0.5 M NaOH without any significant loss in sorption capacity after four consecutive cycles [35, 36]. Magnetic β-cyclodextrin-chitosan-GO sorbent demonstrated the uptake of MB with 84.32 mg/g as sorption capacity. The sorbent posses several advantages such as cheap, quick generation, and easy operation for water treatment [37, 38]. Magnetic cellulose/GO NC was employed as an adsorbent for the separation of MB from wastewater effectively under alkaline conditions. The removal of dye is proportionally affected by dose of NC and the initial concentration of dye. The maximum capacity of adsorption was 70.03 mg/g, and the desorption can be effectivity performed using 0.1 M NaOH [111].

Various NCs involving the use of different biomaterial have been reported in the recent past. GO-calcium alginate composite material was employed for the removal of Cu(II) metal ion in water with a maximum sorption capacity of 60.2 mg/g [10]. Various studies involving CS-GO hybrid material were reported for the removal of heavy metals such as Pb, Au and have high sorption capacity under different conditions [70, 71]. Spongy biodegradable GO-gelatin-CS has been utilized for Pb(II) and Cu(II) removal. The material possess good metal removing capacities and can be recycled several times without any loss in sorption capacity [139]. GO-calcium alginate (GOCA) hybrid sorbent has been exploited for uptake of MB from water with sorption capacity of 163.93 mg/g [66]. Glycerol plasticized-starch/ascorbic acid-MWCNTs (GPS/AA-MWCNTs) NC was used as a good adsorbent for removal of MB dye from aqueous solutions [76]. Sodium alginate/graphene oxide NC (SA/GO) in different forms (fibers, beads, and hydrogels) was employed for the adsorption of dyes (MB) and drugs like (Ciprofloxacin) from wastewater under different reaction conditions. The adsorption process of MB is not affected by the pH of the solution, whereas the adsorption capacity increases on decreasing the temperature. However, the adsorption of dye is pH dependent and the maximum adsorption of the dye can occur at pH 5.9. The optimum desorption of dye can be obtained at acidic pH, which may be attributed to the competition over the adsorption sites of H^+ with the positively charged molecules of MB [128]. GO-chitosan (GO-CS) hybrid material has been developed for the uptake of eosin Y (EY) (acidic dye), reactive black 5 (RB 5), and MB (basic dye) in water with the corresponding uptake capacities of 326 and 390 mg/g [23]. Graphene-asphalt composite has been utilized for removal of Rh 6G in water with high sorption capacity. The sorption depends on carbon loading and particle size of sand particles [23, 113]. Ultrasound assisted GO nanoplatelets embedded in CS matrix (GO-Cs–Nc) was employed for the simultaneous adsorption of acid yellow 36 (AY) and acid blue 74 (AB) from their aqueous solutions. The above NC offers several advantages such as inexpensive, effective in low dosage, and sustainable adsorbents, with reduced time and can be used for the treatment of industrial effluents rich in mixed dyes [17].

GO-ZrO_2 NC exhibited very fast and high adsorption of As(III) and As(V) from water [75]. Arsenite uptake with sorption capacity of 44.4 mg/g can be achieved by utilizing RGO-Fe(0)-Fe_3O_4 hybrid material. The above NC can also be employed for other metal ions such as Cd(II), Hg (II), Cr(VI), and Pb(II) with 1.91, 22.0, 31.1, and 19.7 mg/g, respectively [19]. Graphene/SiO_2 material was employed for Pb(II) removal from water with an uptake capacity of 113.6 mg/g [45]. CNTs-graphene composite aerogels were employed for the efficient removal of metal ions like Cu(II), Ag(II), Pb(II), and Hg(II). The good sorption of these metal ions can be attributed to the presence of oxygen containing groups in the composite [116]. Poly(N-vinylcarbazole) (PVC)-GO hydrid material was employed for Pb(II) removal with 887.98 mg/g sorption capacity at high pH. The removal of the metal ions is directly proportional to the amount of GO in hybrid due to enhanced oxygen functional groups [83]. GO-TiO_2 composite has been employed for the removal of Cd(II), Pb(II), and Zn(II) from water. The sorption capacities for Cd(II), Pb(II), and Zn(II) were 72.8, 65.6 and 88.9 mg/g, respectively [63]. $MnO_{2/}$GNS has been utilized for Pb(II) and Cu(II) adsorption with sorption capacities of 793.65 and 1637.9 μmol/g, respectively. The high sorption of the metal ions can be attributed to tetradentate surface complexes formation of bidentate mononuclear, monodentate, multidentate configurations, and bidentate binuclear [96]. MgAl-graphene double-layered hybrid nanomaterial was used for Cr(VI) uptake with 183.82 mg/g sorption capacity [134]. A pH controlled sorption of Cr(VI) has been demonstrated using NiO/rGO nano hybrid. δ-MnO_2-graphene nanosheets were exploited for the uptake of Ni(II) in aqueous solution. The maximum sorption capacity was 46.55 mg/g and the desorption process can be achieved by 0.1 M HCl with only 9% loss [97]. RGO-polypyrrole was used for uptake Hg(II) in water with a sorption capacity of 979.54 mg/g [20]. Similarly MnO_2-RGO and Ag-RGO sorbents were utilized for uptake of Hg(II) in water [112]. $CoFe_2O_4$-FGS has been exploited for methyl green (MTG) uptake in water [39]. Cu-BDC-based absorbents decorated over GrO and CNTs hybrid NC, such as Cu-BDC@GrO and Cu-BDC@CNT, were utilized for water remediation using BPA as a model organic pollutant. The hybrid NMs exhibits great adsorption capacity (182.2 and 164.1 mg/g) toward the removal of BPA, as compared to Cu-BDC MOF (60.2 mg/g) [7]. Reduced graphene oxide/ZnO composite has been exploited for uptake of Rh B and 99% of the sorbents could be recovered after four cycles [123, 126]. Also, graphene-CNT composite was employed for the removal of Fuschine (FS), MB and Rh B (basic dyes) in water, with the corresponding sorption capacities of 180.8, 191.0, and 150.2 mg/g for FS, MB, and Rh B, respectively [116]. Cylindrical graphene-g-CNT composite (G-CNT) with multilayered sorption capacity has been employed for the removal of MB in water with 81.97 mg/g sorption capacity [8]. $Fe_3O_{4/}SiO_2$-GO nanohybrid material was utilized for MB uptake with 111.1 mg/g as sorption capacity [132]. GO-TiO_2 nanoparticles involving chemisorption process have been employed for the uptake of the dye MB with sorption capacity 83.26 mg/g [87]. Cationic (basic) FS can be adsorbed effectively on magnetic CS/GO NC at pH 5.5. This may be attributed to the fact that under acidic conditions the protonation of its amido groups takes place which increases its solubility in water [68]. CS/RGO mesoporous NC was employed for the adsorption of the anionic azo dye RB5. The decolorization of dye solutions can be attributed to the strong electrostatic interactions, hydrogen bonding, and van der Waals forces [24]. Graphene-magnetite nano-hybrid (G/Fe_3O_4) was employed for removal of various dyes like CR, MB, pararosaniline (PR), FS uptake in water with up to 99% sorption capacity within 30 min. Moreover, the NC can be reused for five cycles without any loss in sorption capacity [9, 122]. G/Fe_3O_4 was modified to graphene sulfonic-magnetite and utilized for the victoria blue (VB), MO, brilliant yellow (BY), neutral red (NR), alizarin red (AR), and safranine T (ST) removal. The above NC exhibits excellent sorption capacity of 93% for these dyes within 10 min. The sorption of dyes could be attributed to the electron-donating effect of amino group of cationic dyes. The sorbent could be recycled for six cycles without any loss in sorption capacities [124]. Magnetic Co-Fe_2O_4 graphene sheets have been utilized for the removal of MO in water with 71.54 mg/g sorption capacity [64].

4.1.5 Adsorption of Contaminants by Fullerenes-Based Nanocomposites

The phosphorylated fullerene/sulfonated polyvinyl alcohol (PFSP) cation exchange membrane was prepared for Cu(II) removal along with the power generation. The maximum Cu(II) removal was 73.2% as compared to 63.2% for Ultrex CMI 7000; a commercial membrane [99].

4.2 Photocatalysis

Photocatalytic processes have shown a great potential as a low-cost environment-friendly and sustainable treatment technology for water purification. This has been spurred by the need to reduce the use of large quantity of chemical additives and disinfectants. Due to the unique properties of carbon-based NCs, they offer a huge potential for the photocatalytic oxidation or degradation of various pollutants in industrial wastewaters (Scheme 4).

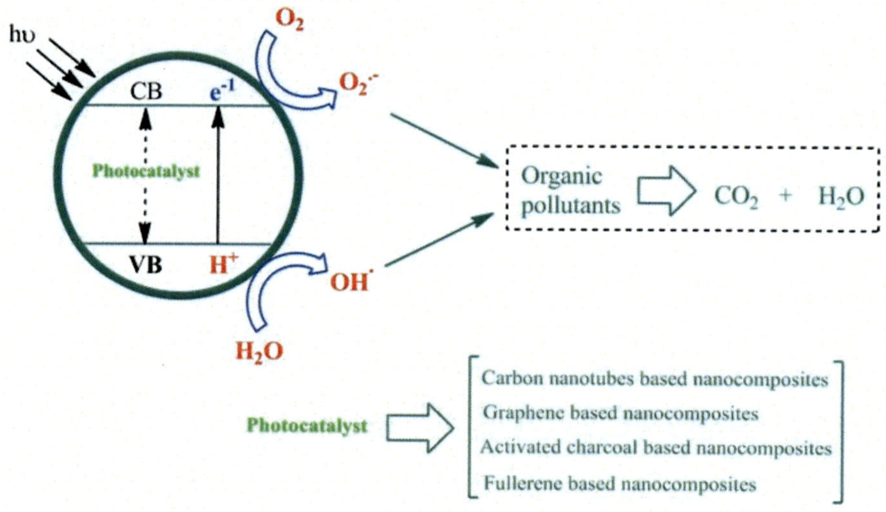

Scheme 4 Photocatalytic degradation of pollutants using carbon-based nanocomposites

4.2.1 Photocatalytic Activity of Activated Carbon-Based Nanocomposites

A reusable TiO_2-based nano-photocatalysts coupled with AC and $SrTiO_3$ exhibited exceptional photoactivities for the degradation of pollutants like 2,4-dichlorophenol (2,4-DCP) and BPA, as compared to commercial P25 TiO_2 [12]. TiO_2 nanofiber/activated carbon fiber (TiNF/ACF) porous composites were employed for the photoctalytic degradation of VOCs such as toluene. The photodegradation activity of the composites can be enhanced due to the synergetic effect between the TiNFs and ACFs. The ACFs is responsible for the enhanced quantum efficiency and the light adsorption ability of TiO_2 [118].

4.2.2 Photocatalytic Activity of Carbon Nanotube-Based Nanocomposites

Carbon nanotube-based NCs posses excellent photocatalytic activity for the degradation of environmental pollutants. Carbon/SnS_2 NC with 3D hierarchical macroporous (MCS) has been employed for rapid photocatalytic treatment of Cr(VI). The MCS displayed excellent adsorption properties for Cr(VI) under mild visible light irradiation, which may be attributed to large specific surface area [44]. ZnS:Mn/MWCNTs NC demonstrated good photodegradation of organic pollutants. The photocatalytic activity of the ZnS nanoparticles was effectively enhanced by the MWCNTs and the composites with the carboxylic functional group exhibited greater photocatalytic activity [108]. CNT/TiO_2 nanofibers have been employed as a visible light active photocatalysts for the decolorization of MB dye and degradation of gaseous benzene under visible light irradiation. The 50-CNT/TiO_2 nanofibers (calcined CNT/TiO_2 nanofibers fabricated from a spinning solution of 50% wt CNT/TiO_2 based on PVP) exhibited higher degradation efficiency for the above pollutants than other CNT/TiO_2 nanofibers and pristine TiO_2 nanofibers under visible light irradiation. The strong adsorption ability and greater visible light adsorption can be attributed to the synergistic effects of the larger surface area and lower band gap energy of CNT/TiO_2 nanofibers [127]. ZnO/MWCNTs NC has been employed for the removal of CR dye from aqueous solutions [15]. A stable and reusable mesoporous anatase TiO_2/MWCNTs NC has been developed with improved pollutant adsorption capacity, electron--hole pair lifetime, light absorption capability, and absorbance of visible light. The above NC exhibits enhanced photocatalytic activity due to the large specific surface areas, presence of TiO_2 in the anatase phase, and the reduced band gap energy. Also, the NC consisting of 20 wt% MWCNTs exhibited the best photocatalytic efficiency and degradation rate [129].

Magnetic MWCNTs-CeO_2 NC has been utilized for the photocatalytic degradation of organic pollutants such as MB. The above NC displayed relatively high degradation efficiency (97.5%) of MB in the presence of H_2O_2. The incorporation of magnetic nanoparticles could not only facilitate the separation of photocatalyst from the solution after treatment, but also enhance the photocatalytic degradation efficiency [40].

4.2.3 Photocatalytic Activity of Graphene-Based Nanocomposite

Graphene-based NC posses a huge potential as heterogeneous photocatalyst in the environmental purification. Graphene nanosheets on introduction to TiO_2 enhance the photocatalytic decomposition activity under different light sources as compared to those of corresponding pure TiO_2 structures [14, 34, 56, 100, 101, 123, 126]. The enhancement of the activity can be attributed to effective charge transfer from TiO_2 to the adsorbed VOCs molecules via graphene due to the narrowing of the bandgap. Also, TiO_2 and graphene nanosheets interaction assists the acceptance of photoelectron to target VOCs molecule. RP-MoS_2/rGO NC has been employed for photocatalytic activity, including the photoreduction of Cr(VI) and photooxidation of different organic pollutants which may be attributed to the increased number of excited electrons/holes and enhanced separation efficiency of charge carriers [16]. P25-graphene composite with different graphene content has been utilized for the photocatalytic degradation of dye such as MB. The increment in graphene content results in the enhancement of the absorption intensity of visible light as well as specific surface area [67]. Fe-doped TiO_2/rGO NC has been developed for the photocatalytic degradation of dyes such as RhB. Addition of H_2O_2 results in the enhancement of photocatalytic efficiency for complete degradation. The raw wastewater biodegradability increased after the photodegradation [50]. CdO/GO NC has been utilized for the photocatalytic degradation of organic pollutants such as MB, MO, and RhB dyes. The above NC exhibits much higher photocatalytic activity as compared to pure CdO nanoparticles on visible light irradiation and the NC having 3.3% GO possess the highest photocatalytic activity [3]. Similarly, NiO/GO NC has been utilized for the photocatalytic degradation of MB. The above NC possess higher degradation efficiency as compared to the cubic NiO nanopowder, which may be due to the formation of p-n heterojunction. However, the enhancement of photocatalytic activity can be attributed to the high separation efficiency of photogenerated electrons and holes resulting from the interaction between NiO and GO [4].

4.2.4 Photocatalytic Activity of Fullerene Nanocomposites

Different NC of fullerene C60 with tetrahydrofuran (THF-nC60), as well as fullerenol C-60(OH)24 nanoparticles (FNP) such as TiO_2/THF-nC60, and TiO_2/FNP have been developed for the photocatalytic degradation of mesotrione under sunlight. TiO_2/FNP system demonstrated the highest photoactivity. Also, the degradation efficiency of mesotrione increases in the presence of different electron acceptors such as H_2O_2 and $KBrO_3$ as compared to O_2 alone [31]. Fullerene-modified lead molybdate (C60-$PbMoO_4$) was developed for the photocatalytic degradation of RhB under UV and visible light irradiation. Increase of C60 weight ratio in the NC also increases the photocatalytic activity under visible light irradiation. Excellent light absorption and charge separation on the interfaces between C60 and $PbMoO_4$ can be responsible for the significant photocatalytic activity of C60-$PbMoO_4$ [26]. TiO_2 nanotubes/Polyhydroxyfullerene (PHF-HNT) has been utilized for the photocatalytic degradation of formic acid. Higher photocatalytic activity can be achieved for monolayer PHF as compared to TiO_2 nanotubes [66].

5 Conclusions and Future Perspectives

Nanocomposites offer a further degree of tunability to the properties of carbon and its allotropes through their inclusion with other materials. The unique properties of carbon-based NCs offer new candidates to construct better materials for environmental monitoring studies. CNCs can be used for the elimination of all type of noxious organic and inorganic contaminants such as dyes, PAHs and heavy metal ions, etc. released by industries, combustion of fossil fuels, agricultural runoffs, etc. CNCs possess significant potential to be utilized as a photocatalyst for water treatment. These NCs can be used to adsorb and photodegrade the pollutants from the environment effectively. Hence, they can be used in industries for the purification of wastewater. CNCs have many advantages as well as limitations in wastewater treatment; it is indeed potential NMs for solving diverse environmental problems. Thus, these materials pose a great potential for application in various environmental fields.

CNCs possess high specific surface areas, excellent electrical, optical, thermal, and chemical activity. These NCs are among one of the most prospective materials for the removal of chemical and biological contaminants from environment. However, the different synthetic methodologies employed for efficient fabrication of various CNCs need to be simplified. Moreover, various strategies should be optimized for the synthesis of these NCs in order to achieve finer quality NCs with improved adsorption and photocatalytic properties. Also, commercial large-scale production of these NCs is a huge challenge and methods should be developed for broad range applications. CNTs and graphene-based NCs in aqueous phase undergoes agglomeration which may lead to reduction in the surface area as well as active sites and can affect their efficiency for the removal of pollutants. Further, research should be focused on the targeted modification of the various NCs for the enhanced removal efficiency as well as selectivity and affinity toward specific contaminants. For the evaluation of the efficiency and applicability of diverse carbon-based NCs, various research studies should be carried out by treating the samples collected directly from real world polluting sources rather than testing them on laboratory samples. The cytotoxicity and risk assessment for various CNCsshould be evaluated for understanding their effect on health of human beings, living organisms and ecosystems. Furthermore, more research needs to be carried out to explore the use of low-cost, efficient, and biodegradable NCs.

References

1. Abbasi M (2017) Synthesis and characterization of magnetic nanocomposite of chitosan/SiO_2/carbon nanotubes and its application for dyes removal. J Clean Prod 145:105–113
2. Abedi S, Abdouss MA (2014) Review of clay-supported Ziegler-Natta catalysts for production of polyolefin/clay nanocomposites through in situ polymerization. Appl Catal A Gen 475:386–409
3. Ahmad J, Majid K (2018) Enhanced visible light driven photocatalytic activity of CdO-graphene oxide heterostructures for the degradation of organic pollutants. New J Chem 42:3246–3257
4. Ahmad J, Majid K, Dar MA (2018) Controlled synthesis of p-type NiO/n-type GO nanocomposite with enhanced photocatalytic activity and study of temperature effect on the photocatalytic activity of the nanocomposite. Appl Surf Sci 457:417–426
5. Ahmad T, Naushad M, Ruksana AA, Alshehri SM (2019) N/S doped highly porous magnetic carbon aerogel derived from sugarcane bagasse cellulose for the removal of bisphenol–A. Int J Biol Macromol 132:1031–1038
6. Ahmaruzzaman M, Gupta VK (2011) Rice husk and its ash as low-cost adsorbents in water and wastewater treatment. Ind Eng Chem Res 50:13589–13613
7. Ahsan MA, Jabbari V, Islam MT, Turley RS, Dominguez N, Kim H, Castro E, Hernandez-Viezcas JA, Curry ML, Lopez J, Gardea-Torresdey JL, Noveron JC (2019) Sustainable synthesis and remarkable adsorption capacity of MOF/graphene oxide and MOF/CNT based hybrid nanocomposites for the removal of Bisphenol A from water. Sci Total Environ 673:306–317
8. Ai L, Jiang J (2012) Removal of methylene blue from aqueous solution with self-assembled cylindrical graphene-carbon nanotube hybrid. J Chem Eng 192:156–163
9. Ai L, Zhang C, Chen Z (2011) Removal of methylene blue from aqueous solution by a solvothermal-synthesized graphene/magnetite composite. J Hazard Mater 192:1515–1524
10. Algothmi WM, Bandaru NM, Yu Y, Shapter JG, Ellis AV (2013) Alginate-graphene oxide hybrid gel beads, an efficient copper adsorbent material. J Colloid Interface Sci 397:32–38
11. Alhana S, Nehraab M, Dilbaghia N, Kumar N, Kimd SKH, Kumarad S (2019) Potential use of ZnO@activated carbon nanocomposites for the adsorptive removal of Cd^{2+} ions in aqueous solutions. Environ Res 173:411–418
12. Ali S, Li Z, Chen S, Zada A, Khan I, Khan I, Ali W, Shaheen S, Qu Y, Jing L (2019) Synthesis of activated carbon-supported TiO_2-based nano-photocatalysts with well recycling for efficiently degrading high-concentration pollutants. Catal Today 335:557–564
13. Ambika, Singh PP (2018) Nanotechnology: greener approach for sustainable environment, In: Hussain CM, Mishra AK (ed) Nanotechnology in environmental Sciences, Wiley VCH Verlag GmbH & Co. KGaA, Boschstr. 12, 69469 Weinheim, Germany, pp 805–824
14. Andryushina NS, Stroyuk OL (2014) Influence of colloidal graphene oxide on photocatalytic activity of nanocrystalline TiO_2 in gas-phase ethanol and benzene oxidation. Appl Catal B Environ 148–149:543–549
15. Arabi SMS, Lalehloo RS, Olyai MRTB, Ali GAM, Sadegh H (2019) Removal of congo red azo dye from aqueous solution by ZnO nanoparticles loaded on multiwall carbon nanotubes. Physica E 106:150–155
16. Bai X, Du Y, Hu X, He Y, He C, Liu E, Fan J (2018) Synergy removal of Cr (VI) and organic pollutants over RP-MoS_2/rGO photocatalyst. Appl Catal B 239:204–213
17. Banerjee P, Barmana SR, Mukhopadhayaya A (2017) A Ultrasound assisted mixed azo dye adsorption by chitosan-graphene oxide nanocomposite. Chem Eng Res Des 117:43–56
18. Bharath G, Alhseinat E, Ponpadian N, Khan MA, Siddiqui MR, Ahmed F, Alsharaeh EH (2017) Development of adsorption and electrosorption techniques for removal of organic and inorganic pollutants from waste waterusing novel magnetite/porous graphene-based nanocomposites. Sep Purif Technol 188:206–218

19. Bhunia P, Kim G, Baik C, Lee H (2012) A strategically designed porous iron-iron oxide matrix on graphene for heavy metal adsorption. Chem Commun 48:9888–9890
20. Chandra V, Kim KS (2011) Highly selective adsorption of Hg^{2+} by a polypyrrole-reduced graphene oxide composite. Chem Commun 47:3942–3944
21. Chandra V, Park J, Chun Y, Lee JW, Hwang IC, Kim KS (2010) Water-dispersible magnetite-reduced graphene oxide composites for arsenic removal. ACS Nano 4:3979–3986
22. Chatterjee S, Lee MW, Woo SH (2010) Adsorption of congo red by chitosan hydrogel beads impregnated with carbon nanotubes. Bioresour Technol 101:1800–1806
23. Chen Y, Chen L, Bai H, Li L (2013) Graphene oxide-chitosan composite hydrogels as broad-spectrum adsorbents for water purification. J Mater Chem A 1:1992–2001
24. Cheng JS, Du J, Zhu W (2012) Facile synthesis of three-dimensional chitosan-graphene mesostructures for reactive black 5 removal. Carbohydr Polym 88:61–67
25. Crini G, Badot PM (2008) Application of chitosan, a natural amino polysaccharide, for dye removal from aqueous solutions by adsorption processes using batch studies, a review of recent literature. Prog Polym Sci 33:399–447
26. Dai K, Yao Y, Liu H, Mohamed I, Chen H, Huang Q (2013) Enhancing the photocatalytic activity of lead molybdate by modifying with fullerene. J Mol Catal A Chem 374–375:111–117
27. Das D, Charumathi D, Das N (2011) Bioaccumulation of the synthetic dye Basic Violet 3 and heavy metals in single and binary systems by Candida tropicalis grown in a sugarcane bagasse extract medium: modelling optimal conditions using response surface methodology (RSM) and inhibition kinetics. J Hazard Mater 186(2–3):1541–1552
28. Debnath S, Maity A, Pillay K (2014) Magnetic chitosan-GO nanocomposite: Synthesis, characterization and batch adsorber design for Cr(VI) removal. J Environ Chem Eng 2:963–973
29. Demirbas A (2008) Heavy metal adsorption onto agro-based waste materials: a review. J Hazard Mater 157:220–229
30. Deng S, Liu X, Liao J, Lin H, Liu F (2019) PEI modified multiwalled carbon nanotube as a novel additive in PAN nanofiber membrane for enhanced removal of heavy metal ions. Chem Eng J 375:122086
31. Djordjevic A, Merkulov DS, Lazarevic M, Borisev I, Medic I, Pavlovic V, Miljevic B, Abramovic B (2018) Enhancement of nano titanium dioxide coatings by fullerene and polyhydroxy fullerene in the photocatalytic degradation of the herbicide mesotrione. Chemosphere 196:145–152
32. Dou J, Gan D, Huang Q, Liu M, Deng F, Zhu X, Wen Y, Zhang X, Wei Y (2019) Functionalization of carbon nanotubes with chitosan based on MALI multicomponent reaction for Cu^{2+} removal. Int J Biol Macromol 136:476–485
33. Duman O, Tunc S, Kanc B, Kanc B, Tulin B, Polat G (2016) Removal of triphenylmethane and reactive azo dyes from aqueous solution by magnetic carbon nanotube-κ-carrageenan-Fe_3O_4 nanocomposite. J Alloy Compd 687:370–383
34. Ebrahimi A, Fatemi S (2017) Titania-reduced graphene oxide nanocomposite as a promising visible light-active photocatalyst for continuous degradation of VOC in air purification process. Clean Technol Environ Policy 19:2089–2098
35. Fan L, Luo C, Li X, Lu F, Qiu H, Sun M (2012) Fabrication of novel magnetic chitosan grafted with graphene oxide to enhance adsorption properties for methyl blue. J Hazard Mater 215–216:272–279
36. Fan L, Luo C, Sun M, Li X, Lu F, Qiu H (2012) Preparation of novel magnetic chitosan/graphene oxide composite as effective adsorbents toward methylene blue. Bioresour Technol 114:703–706
37. Fan L, Luo C, Sun M, Li X, Qiu H (2013) Highly selective adsorption of lead ions by water-dispersible magnetic chitosan/graphene oxide composites. Colloids Surf B Biointerfaces 103:523–529
38. Fan L, Luo C, Sun M, Qiu H, Li X (2013) Synthesis of magnetic β-cyclodextrinchitosan/graphene oxide as nanoadsorbent and its application in dye adsorption and removal. Colloids Surf B Biointerfaces 103:601–607

39. Farghali AA, Bahgat M, Khedr MH (2013) Preparation, decoration and characterization of graphene sheets for methyl green adsorption. J Alloys Compd 555:193–200
40. Feng K, Song B, Li XF, Liao F, Gong J (2019) Enhanced photocatalytic performance of magnetic multi-walled carbon nanotubes/cerium dioxide nanocomposite. Ecotoxicol Environ Safe 171:587–593
41. Fu F, Wang Q (2011) Removal of heavy metal ions from wastewaters: a review. J Environ Manage 92:407–418
42. Georgakilas V, Otyepka M, Bourlinos AB, Chandra V, Kim N, Kemp KC, Hobza P, Zboril R, Kim KS (2012) Functionalization of graphene: covalent and non-covalent approaches, derivatives and applications. Chem Rev 112:6156–6214
43. Guo M, Wang J, Wang C, Strong PJ, Jiang P, Ok YS, Wang H (2019) Carbon nanotube-grafted chitosan and its adsorption capacity for phenol in aqueous solution. Sci Total Environ 82:340–347
44. Han L, Zhong YL, Su Y, Wang L, Zhu L, Fei X, Dong Y, Hong G, Zhou Y, Fang D (2019) Nanocomposites based on 3D macroporous biomass carbon with SnS_2 nanosheets hierarchical structure for efficient removal of hexavalent chromium. Chem Eng J 369:1138–1149
45. Hao L, Song H, Zhang L, Wan X, Tang Y, Lv Y (2012) SiO_2/graphene composite for highly selective adsorption of Pb(II) ion. J Colloid Interface Sci 369:381–387
46. Hu XJ, Liu YG, Wang H, Chen A, Zeng G, Liu S, Guo Y, Hu X, Li T, Wang Y, Zhou L, Liu S (2013) Removal of Cu(II) ions from aqueous solution using sulfonated magnetic graphene oxide composite. Sep Purif Technol 108:189–195
47. Huang S, Jiang S, Pang H, Wen T, Asiri AM, Alamry KA, Alsaedi A, Wang X, Wang S (2019) Dual functional nanocomposites of magnetic $MnFe_2O_4$ and fluorescent carbon dots for efficient U(VI) removal. Chem Eng J 368:941–950
48. Hyunwoo K, Abdala AA, Macosko CW (2010) Graphene/polymer nanocomposites. Macromolecules 43:6515–6530
49. Innocenzi P, Brusatin G (2001) Fullerene-based organic-inorganic nanocomposites and their applications. Chem Mater 13:3126–3139
50. Isari AA, Payan A, Fattahi M, Jorfi S, Kakavandi B (2018) Photocatalytic degradation of rhodamine B and real textile wastewater using Fe-doped TiO_2 anchored on reduced graphene oxide (Fe-TiO2/rGO): characterization and feasibility, mechanism and pathway studies. Appl Surf Sci 462:549–564
51. Jabeen H, Chandra V, Jung S, Lee JW, Kim KS, Kim SB (2011) Enhanced Cr(VI) removal using iron nanoparticle decorated graphene. Nanoscale 3:3583–3585
52. Jafari F, Nasirizadeh N, Mirjalili M (2019) Enhanced degradation of reactive dyes using a novel carbon ceramic electrode based on copper nanoparticles and multiwall carbon nanotubes. Chin J Chem Eng (In press)
53. Janus R, Kustrowski P, Dudek B, Piwowarska Z, Kochanowski A, Michalik M, Cool P (2011) Removal of methyl-ethyl ketone vapour on polyacrylonitrile-derived carbon/mesoporous silica nanocomposite adsorbents. Microporous Mesoporous Mater 145:65–73
54. Jayaweera HDAC, Siriwardane I, De Silva KMN, De Silva RM (2018) Synthesis of multifunctional activated carbon nanocomposite comprising biocompatible flake nano hydroxyapatite and natural turmeric extract for the removal of bacteria and lead ions from aqueous solution. Chem Cent J 12:18
55. Jeevanandam J, Barhoum A, Chan YS, Dufresne A, Danquah MK (2018) Review on nanoparticles and nanostructured materials: history, sources, toxicity and regulations. Beilstein J Nanotechnol 9:1050–1074
56. Jo WK, Kang HJ (2013) Titanium dioxide-graphene oxide composites with different ratios supported by Pyrex tube for photocatalysis of toxic aromatic vapors. Powder Technol 250:115–121
57. Khakpour R, Tahermansouri H (2018) Synthesis, characterization and study of sorption parameters of multi-walled carbon nanotubes/chitosan nanocomposite for the removal of picric acid from aqueous solutions. Int J Biol Macromol 109:598–610

58. Khan WS, Hamadneh NN, Khan WA (2016) Polymer nanocomposites-synthesis techniques, classification and properties. In: Sia PD (ed) Science and applications of tailored nanostructures. One Central Press, UK, pp 50–66
59. Khosravi R, Moussavi G, Ghaneian MT, Ehrampoush MH, Barikbin B, Ebrahimi AA, Sharifzadeh G (2018) Chromium adsorption from aqueous solution using novel green nanocomposite: adsorbent characterization, isotherm, kinetic and thermodynamic investigation. J Mol Liq 256:163–174
60. Kirti S, Bhandari VM, Jena J, Bhattacharyya AS (2018) Elucidating efficacy of biomass derived nanocomposites in water and wastewater treatment. J Environ Manage 226:95–105
61. Krueger A (ed) (2010) Carbon-element of many faces, In: Carbon materials and nanotechnology. Wiley-VCH, Weinheim, Germany, pp 1–32
62. Kubacka A, Fernandez-Garcia M, Colon G (2011) Advanced nano-architectures for solar photocatalytic applications. Chem Rev 112:1555–1614
63. Lee YC, Yang JW (2012) Self-assembled flower-like TiO_2 on exfoliated graphite oxide for heavy metal removal. J Ind Eng Chem 18:1178–1185
64. Li B, Cao H, Yin G, Lu Y, Yin J (2011) Cu_2O@reduced graphene oxide composite for removal of contaminants from water and super capacitor. J Mater Chem 21:10645–10648
65. Li J, Sl Z, Hong GB, Chang CT (2013) Hydrothermal preparation of P25-graphene composite with enhanced adsorption and photocatalytic degradation of dyes. Chem Eng J 219:486–491
66. Li Y, Du Q, Liu TJ, Wang Y, Wu S, Wang Z, Xia Y, Xia L (2013) Methylene blue adsorption on graphene oxide/calcium alginate composites. Carbohydr Polym 95:501–507
67. Li Y, Du Q, Liu T, Qiu H, Li X, Duan H, Luo C (2013) Comparative study of methylene blue dye adsorption onto activated carbon, graphene oxide and carbon nanotubes. Chem Eng Res Des 91:361–368
68. Li L, Fan L, Luo C, Duan H, Wang X (2014) Study of fuchsine adsorption on magnetic chitosan/graphene oxide. RSC Adv 4:24679–24685
69. Li Z, Chen J, Ge Y (2017) Removal of lead ion and oil droplet from aqueous solution by lignin-grafted carbon nanotubes. Chem Eng J 308:809–817
70. Liu T, Li Y, Du Q, Jiao Y, Yang G, Wang Z, Xia Y, Zhang W, Wang K, Zhu H, Wu D (2012) Adsorption of methylene blue from aqueous solution by grapheme. Colloids Surf B Biointerfaces 90:197–203
71. Liu L, Li C, Bao C, Jia Q, Xiao P, Liu X, Zhang Q (2012) Preparation and characterization of chitosan/graphene oxide composites for the adsorption of Au(III) and Pd(II). Talanta 93:350–357
72. Lu W, Li J, Sheng Y, Zhang X, You J, Chen L (2017) One-pot synthesis of magnetic iron oxide nanoparticle-multiwalled carbon nanotube composites for enhanced removal of Cr(VI) from aqueous solution. J Colloids Interface Sci 505:1134–1146
73. Lu X, He J, Xie J, Zhou Y, Liu S, Zhu Q, Lu H (2020) Preparation of hydrophobic hierarchical pore carbon-silica composite and its adsorption performance toward volatile organic compounds. J Environ Sci 87:39–48
74. Lu YM, Gong QM, Liang J (2009) Preparation of carbon nanotubes/activated carbon composite microspheres and their application to adsorption of VB_{12}. Acta Phys Chim Sin 25:1697–1702
75. Luo X, Wang C, Wang L, Deng F, Luo S, Tu X, Au C (2013) Nanocomposites of graphene oxide-hydrated zirconium oxide for simultaneous removal of As(III) and As (V) from water. Chem Eng J 220:98–106
76. Mallakpour S, Rashidimoghadam S (2018) Application of ultrasonic irradiation as a benign method for production of glycerol plasticized-starch/ascorbic acid functionalized MWCNTs nanocomposites: Investigation of methylene blue adsorption and electrical properties. Ultrason Sonochem 40:419–432
77. Mirzaee SA, Jaafarzadeh N, Gomes HT, Jorfi S, Magnetic AM (2019) Magnetic titanium/carbon nanotube nanocomposite catalyst for oxidative degradation of Bisphenol A from high saline polycarbonate plant effluent using catalytic wet peroxide oxidation. Chem Eng J 370:372–386

78. Mittal V (ed) (2010) Optimization of polymer nanocomposite properties. Wiley VCH Verlag GmbH & Co. KGaA, Weinheim, pp 1–19
79. Mohammad N, Oveisi M, Taghizadeh A, Taghizadeh M (2019) Novel magnetic amine functionalized carbon nanotube/metal-organic framework nanocomposites: from green ultrasound-assisted synthesis to detailed selective pollutant removal modelling from binary systems. J Hazard Mat 368:746–775
80. Mondal A, Jana NR (2014) Graphene-nanoparticle composites and their applications in energy, environmental and biomedical science. Rev Nanosci Nanotechnol 3:77–192
81. Moser J, Barreiro A, Bachtold A (2007) Current-induced cleaning of graphene. Appl Phys Lett 91:163513
82. Muntean SG, Nistor MA, Ianos R, Pacurariu C, Capraru A, Surdu VA (2019) Combustion synthesis of Fe_3O_4/Ag/C nanocomposite and application for dyes removal from multicomponent systems. Appl Surf Sci 481:825–837
83. Musico YLF, Santos CM, Dalida MLP, Rodrigues DF (2013) Improved removal of lead(II) from water using a polymer-based graphene oxide nanocomposite. J Mater Chem 1:3789–3796
84. Nandi D, Gupta K, Ghosh AK, De A, Banerjee S, Ghosh UC (2012) Manganese incorporated iron(III) oxide-graphene magnetic nanocomposite, synthesis, characterization, and application for the arsenic(III)-sorption from aqueous solution. Nanotechnol Sustain Dev 14:149–162
85. Nasir S, Hussein MZ, Zainal Z, Yusof NA (2018) Carbon-based nanomaterials/allotropes: a glimpse of their synthesis, properties and some applications. Materials 11:295
86. Ngah WWS, Hanafiah MA (2008) Removal of heavy metal ions from wastewater by chemically modified plant wastes as adsorbents: a review. Bioresour Technol 99:3935–3948
87. Nguyen-Phan TD, Pham VH, Kim EJ, Oh ES, Hur SH, Chung JS, Lee B, Shin EW (2012) Reduced graphene oxide–titanate hybrids, morphologic evolution by alkali solvothermal treatment and applications in water purification. Appl Surf Sci 258:4551–4557
88. Pavlidou S, Papaspyrides CD (2008) A review on polymer-layered silicate nanocomposites. Prog Polym Sci 33:1119–1198
89. Pekakis PA, Xekoukoulotakis NP, Mantzavinos D (2006) Treatment of textile dyehouse wastewater by TiO_2 photocatalysis. Water Res 40:1276–1286
90. Phiri J, Johansson LS, Gane P, Thad M (2018) A comparative study of mechanical, thermal and electrical properties of graphene, graphene oxide- and reduced graphene oxide-doped microfibrillated cellulose nanocomposites. Compos B Eng 147:104–113
91. Phiri I, Ko JM, Mushonga P, Onani MO, Msamadya S, Kim SJ, Bon CY, Mugobera S, Choto KS, Madzvamuse A (2019) Simultaneous removal of cationic, anionic and organic pollutants in highly acidic water using magnetic nanocomposite alginate beads. J Water Process Eng 31:100884
92. Pierson HO (1993) Handbook of carbon, graphite, diamond and fullerenes: Properties, processing and applications. Noyes Publications, Park Ridge, NJ
93. Pradhan D, Sukla LB, Sawyer M, Rahman PK (2017) Recent bioreduction of hexavalent chromium in wastewater treatment: a review. J Ind Eng Chem 55:1–20
94. Rao GP, Lu C, Su F (2007) Sorption of divalent metal ions from aqueous solution by carbon nanotubes: a review. Sep Purif Technol 58:224–231
95. Reddy KR, Cameselle C (2009) Electrochemical remediation technologies for polluted soils, sediments and groundwater. Wiley, pp 1–28
96. Ren Y, Yan N, Feng J, Ma J, Wen Q, Li N, Dong Q (2012) Adsorption mechanism of copper and lead ions onto graphene nanosheet/δ-MnO_2. Mater Chem Phys 136:538–544
97. Ren Y, Yan N, Wen Q, Fan Z, Wei T, Zhang M, Ma J (2011) Graphene/delta-MnO_2 composite as adsorbent for the removal of nickel ions from wastewater. Chem Eng J 175:1–7
98. Ricardo AIC, Cachero AS, Moreno MJ, Bernardo FJG, Doimeadios RCRM, Rios A (2018) Carbon nanotubes magnetic hybrid nanocomposites for a rapid and selective preconcentration and clean-up of mercury species in water samples. Talanta 179:442–447

99. Rikame SS, Mungray AA, Mungray AK (2017) Synthesis, characterization and application of phosphorylated fullerene/sulfonated polyvinyl alcohol (PFSP) composite cation exchange membrane for copper removal. Sep Purif Technol 177:29–39
100. Roso M, Boaretti C, Pelizzo MG, Lauria A, Modesti M (2017) Nanostructured photocatalysts based on different oxidized graphenes for VOCs removal. Ind Eng Chem Res 56:9980–9992
101. Roso M, Lorenzetti A, Boaretti C, Bernardo FJG, Doimeadios RCRM, Rios A (2015) Graphene/TiO_2 based photo-catalysts on nanostructured membranes as a potential active filter media for methanol gas-phase degradation. Appl Catal B Environ 176–177:225–232
102. Saha P, Chowdhury S, Gupta S, Kumar I (2010) Insight into adsorption equilibrium, kinetics and thermodynamics of Malachite Green onto clayey soil of Indian origin. Chem Eng J 165:874–882
103. Salam MA (2017) Preparation and characterization of chitin/magnetite/multiwalled carbon nanotubes magnetic nanocomposite for toxic hexavalent chromium removal from solution. J Mol Liq 233:197–202
104. Salam MA, Gabal MA, Obaid AY (2012) Preparation and characterization of magnetic multi-walled carbon nanotubes/ferrite nanocomposite and its application for the removal of aniline from aqueous solution. Synth Metals 161:2651–2658
105. Salam MA, Shishtawy RME, Obaid AY (2014) Synthesis of magnetic multi-walled carbon nanotubes/magnetite/chitin magnetic nanocomposite for the removal of Rose Bengal from real and model solution. J Ind Eng Chem 20:3559–3567
106. Salehi E, Madaeni SS, Rajabi L, Vatanpour V, Derakhshan AA, Zinadin S, Ghorabi S, Monfared HA (2012) Novel chitosan/poly(vinyl) alcohol thin adsorptive membranes modified with amino functionalized multi-walled carbon nanotubes for Cu(II) removal from water: preparation, characterization, adsorption kinetics and thermodynamics. Sep Purif Technol 89:309–319
107. Shan SJ, Zhao Y, Tang H, Cui FY (2017) A mini-review of carbonaceous nanomaterials for removal of contaminants from wastewater. In: IOP conference series: earth environmental science, vol 68, p 012003
108. Sharifi A, Montazerghaem L, Naeimi A, Abhari AR, Vafaee M, Ali GAM, Sadegh H (2019) Investigation of photocatalytic behavior of modified ZnS:Mn/MWCNTs nanocomposite for organic pollutants effective photodegradation. J Environ Manage 247:624–632
109. Sharma A, Siddiqi ZM, Pathania D (2017) Adsorption of polyaromatic pollutants from water system using carbon/$ZnFe_2O_4$ nanocomposite: equilibrium, kinetic and thermodynamic mechanism. J Mol Liq 240:361–371
110. Sheshmani S, Ashori A, Hasanzadeh S (2014) Removal of acid orange 7 from aqueous solution using magnetic graphene/chitosan: a promising nano-adsorbent. Int J Biol Macromol 68:218–224
111. Shi H, Li W, Zhong L, Xu C (2014) Methylene blue adsorption from aqueous solution by magnetic cellulose/graphene oxide composite: equilibrium, kinetics, and thermodynamics. Ind Eng Chem Res 53:1108–1118
112. Sreeprasad TS, Maliyekkal SM, Lisha KP, Pradeep T (2011) Reduced graphene oxide-metal/metal oxide composites, facile synthesis and application in water purification. J Hazard Mater 186:921–931
113. Sreeprasad TS, Sen Gupta S, Maliyekkal SM, Lisha KP, Pradeep T (2013) Immobilized graphene-based composite from asphalt, facile synthesis and application in water purification. J Hazard Mater 246–247:213–220
114. Srivastava I, Mishra S, Singh PK, Gupta T, Sankararamakrishnan N (2018) Fast and efficient removal of Toluene, Ethylbenzene and o-Xylene from aqueous phase by functionalized carbon micro/nano composites. J Environ Chem Eng 6:4917–4926
115. Sud D, Mahajan G, Kaur MP (2008) Agricultural waste material as potential adsorbent for sequestering heavy metal ions from aqueous solutions—a review. Bioresour Technol 99:6017–6027
116. Sui Z, Meng Q, Zhang X, Ma R, Cao B (2012) Green synthesis of carbon nanotube-graphene hybrid aerogels and their use as versatile agents for water purification. J Mater Chem 22:8767–8771

117. Sun HM, Cao LY, Lu LH (2011) Magnetite/reduced graphene oxide nanocomposites, one step solvothermal synthesis and use as a novel platform for removal of dye pollutants. Nano Res 4:550–562
118. Tian MJ, Liao F, Ke QF, Guo YJ, Guo YP (2017) Synergetic effect of titanium dioxide ultralong nanofibers and activated carbon fibers on adsorption and photodegradation of toluene. Chem Eng J 328:962–976
119. Tong H, Ouyang S, Bi Y, Guo YJ, Guo YP (2012) Nano-photocatalytic materials: possibilities and challenges. Adv Mater 24:229–251
120. Vijayaraghavan K, Winnie HYN, Balasubramanian R (2011) Biosorption characteristics of crab shell particles for the removal of manganese(II) and zinc(II) rom aqueous solutions. Desalination 266:195–200
121. Wang B, Gao B, Zimmerman AR, Lee X (2018) Impregnation of multiwall carbon nanotubes in alginate beads dramatically enhances their adsorptive ability to aqueous methylene blue. Chem Eng Res Des 133:235–242
122. Wang C, Feng C, Gao Y, Ma X, Wu Q, Zhi W (2011) Preparation of a graphene based magnetic nanocomposite for the removal of an organic dye from aqueous solution. Chem Eng J 173:92–97
123. Wang J, Tsuzuki T, Tang B, Hou X, Sun L, Wang X (2012) Reduced graphene oxide/ZnO composite, reusable adsorbent for pollutant management. ACS Appl Mater Interfaces 4:3084–3090
124. Wang S, Wei J, Lv S, Guo Z, Jiang F (2013) Removal of organic dyes in environmental water onto magnetic-sulfonic graphene nanocomposite. Clean-Soil Air Water 41:992–1001
125. Wang W, Hao X, Chen S, Yanga Z, Wanga C, Yana R, Zhang X, Liu H, Shaod Q, Guo Z (2018) pH-responsive Capsaicin@chitosan nanocapsules for antibiofouling in marine applications. Polymer 158:223–230
126. Wang WG, Yu JG, Xiang QJ, Cheng B (2012) Enhanced photocatalytic activity of hierarchical macro/mesoporous TiO_2-graphene composites for photodegradation of acetone in air. Appl Catal B Environ 119–120:109–116
127. Wongaree M, Chiarakorn S, Chuangchote S, Sagawa T (2016) Photocatalytic performanceof electrospun CNT/TiO_2 nanofibers in a simulated air purifier under visible light irradiation. Environ Sci Pollut Res 23:21395–21406
128. Wu S, Zhao X, Li Y, Zhao C, Du Q, Sun J, Wang Y, Peng X, Xia Y, Wang Z, Xia L (2013) Adsorption of ciprofloxacin onto biocomposite fibers of graphene oxide/calcium alginate. Chem Eng J 230:389–395
129. Yang HM, Park SJ (2017) Effect of incorporation of multiwalled carbon nanotubes on photodegradation efficiency of mesoporous anatase TiO_2 spheres. Mater Chem Phys 186:261–270
130. Yang MQ, Zhang N, Pagliaro M, Xu YJ (2014) Artificial photosynthesis over graphene-semiconductor composites. Are we getting better? Chem Soc Rev 43:8240–8254
131. Yang X, Chen C, Li J, Zhao G, Ren X, Wang X (2012) Graphene oxide-iron oxide and reduced graphene oxide-iron oxide hybrid materials for the removal of organic and inorganic pollutants. RSC Adv 2:8821–8826
132. Yao Y, Miao S, Yu S, Ma LP, Sun H, Wang S (2012) Fabrication of Fe_3O_4/SiO_2 core/shell nanoparticles attached to graphene oxide and its use as an adsorbent. J Colloid Interace Sci 379:20–26
133. Youssef AM, El-Naggar ME, Malhat FM, Sharkawi HME (2019) Efficient removal of pesticides and heavy metals from wastewater and the antimicrobial activity of *f*-MWCNTs/PVA nanocomposite film. J Clean Prod 206:315–325
134. Yuan X, Wang Y, Wang J, Zhaou C, Tang Q, Rao X (2013) Calcined graphene/MgAl-layered double hydroxides for enhanced Cr(VI) removal. Chem Eng J 221:204–213
135. Zeng G, Liu X, Liu M, Huang Q, Xu D, Wan Q, Huang H, Deng F, Zhang X, Wei Y (2016) Facile preparation of carbon nanotubes based carboxymethyl chitosan nanocomposites through combination of mussel inspired chemistry and Michael addition reaction: Characterization and improved Cu^{2+} removal capability. J Taiwan Inst Chem Eng 68:446–454

136. Zhang HB, Zheng WG, Yan Q, Yang Y, Wang JW, Lu ZH, Ji GH, Yu ZZ (2010) Electrically conductive polyethylene terephthalate/graphene nanocomposites prepared by melt compounding. Polymer 51:1191–1196
137. Zhang K, Dwivedi V, Chi C, Wu J (2010) Graphene oxide/ferric hydroxide composites for efficient arsenate removal from drinking water. J Hazard Mater 182:162–168
138. Zhang K, Li H, Xu X, Yu H (2018) Synthesis of reduced graphene oxide/NiO nanocomposite for the removal of Cr(VI) from aqueous water by adsorption. Microporous Mater 255:7–14
139. Zhang N, Qiu H, Si Y, Wang W, Gao J (2011) Fabrication of highly porous biodegradable monoliths strengthened by graphene oxide and their adsorption of metal ions. Carbon 49:827–837
140. Zhang P, Ge T, Yang H, Shiwei L, Yang C, Xinxin ZC, Hu L, Ahmad U, Zhanhu G (2018) Antifouling of titania nanostructures in real maritime conditions. Sci Adv Mater 10:1216–1223
141. Zhangf ZY, Sun D, Li G, Zhang B, Zhang B, Qiu S, Li Y, Wu T (2019) Calcined products of Mg–Al layered double hydroxides/single-walled carbon nanotubes nanocomposites for expeditious removal of phenol and 4-chlorophenol from aqueous solutions. Colloids Surf A Phys Eng Aspects 565:143–153
142. Zhu J, Wei S, Chen M, Gu H, Rapole SB, Pallavkar S, Ho TC, Hopper J, Guo Z (2013) Magnetic nanocomposites for environmental remediation. Adv Powder Technol 24:459–467
143. Zhu J, Wei S, Gu H, Rapole SB, Wang Q, Luo Z, Haldolaarachchige N, Young DP, Guo Z (2012) One-pot synthesis of magnetic graphene nanocomposites decorated with core@double-shell nanoparticles for fast chromium removal. Environ Sci Technol 46:977–985

Use of Carbon Nanomaterials as Potential Ion-Exchange

Gunjan Bhalla, Anupamdeep Sharma, Vaneet Kumar, Barjinder Bhalla, Saruchi, and Harsh Kumar

Abstract This chapter deals with the use of different varieties of carbon nanomaterial's (CNMs) as a potential ion-exchange material. Potential modifications of CNMs to enhance their ion-exchange properties such as functionalization of the surfaces of these materials and heteroatom doping have been discussed. Primarily, it is based on functional groups addition which modifies the surface of CNMs thus increasing ion exchange capabilities. The specificity of CNMs for particular ions needs improvement. Heteroatom-doped carbon nanotubes (CNTs) depict better properties over common CNTs.

Keywords Ion-exchange · Carbon nanomaterial's · Functionalization · Heteroatom doping · Carbon nanotubes

---

G. Bhalla
Department of Civil Engineering, CT Group of Institutions, Shahpur Campus Jalandhar, Punjab 144020, India

A. Sharma
CT Group of Institutions, Shahpur Campus, Jalandhar, Punjab 144020, India

V. Kumar (✉)
Department of Applied Sciences, CT Group of Institutions, Shahpur Campus, Jalandhar, Punjab 144020, India
e-mail: vaneet2106@gmail.com

B. Bhalla
Pushpa Gujral Science City, Kapurthala, Punjab 144601, India

Saruchi
Department of Biotechnology, CT Group of Institutions, Shahpur Campus Jalandhar, Punjab 144020, India

H. Kumar
Department of Chemistry, Dr. B. R. Ambedkar National Institute of Technology, Jalandhar, Punjab 144011, India

M. Jawaid et al. (eds.), *Environmental Remediation Through Carbon Based Nano Composites*, Green Energy and Technology,
https://doi.org/10.1007/978-981-15-6699-8_11

1 Introduction

Nanotechnology grows quickly and encourages innovations in different spheres of science and technology. Different nanomaterials comprised of carbon atoms are called as CNMs [1]. Categorization of CNMs is mainly done on the basis of their structure. CNMs can be horn-molded, tube-formed, ellipsoidal, or spherical. CNMs having tube-formed are known as CNTs; particles having horn-shaped are called nano-horns and ellipsoids or spheres are related to fullerene group. Meanwhile, CNMs comprises of different technological utilizations [10, 17]. Surrounded by different carbon-based nanomaterials; CNTs are having outstanding properties appropriate for technological utilization. It was first revealed in 1991 by a Japanese investigator Iijima [15]. CNTs are cylindrical in shape with diameter of few nanometers, comprising of trolled sheets of graphene (Fig. 1).

CNTs differ in diameter, length, number of layers, and chirality. Based on structure, CNTs can be divided into two major categories: single-walled CNTs and multi-walled CNTs. In general, single-walled CNTs have about 1–3 nm diameter and length of few micrometers. Multi-walled CNTs have 5–40 nm diameter and a length of about 10 micrometers [39]. The CNTs structures have outstanding properties with a blend of elasticity, strength, and rigidity in contrast to various other fibrous materials. For instance, CNTs have larger length to diameter ratios in comparison

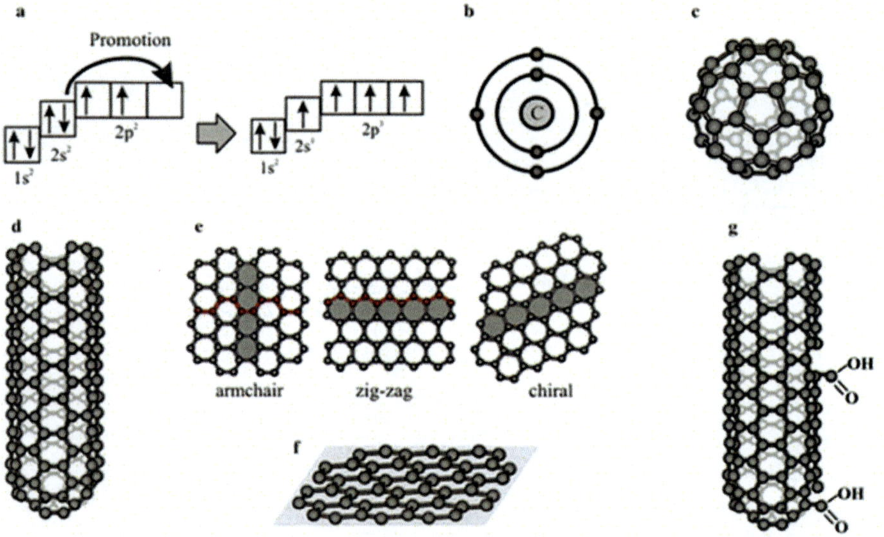

Fig. 1 Schematic illustration of carbon atomic structure and of CNMs, **a** electron arrangement of a carbon atom before and after promotion of one s-electron, **b** demonstration of a carbon atomic structure with two-electron orbitals surrounding the nucleus with six electrons distribution, **c** fullerene structure, **d** structure of a single-walled nanotube **e** various forms of single-walled nanotubes, **f** graphene sheet structure, **g** oxidized single-walled nanotubes structure [37]

to others and SWCNTs have greater aspect ratios then MWCNTs because of its lesser diameter. Furthermore, CNTs have greater electrical and thermal conductivity in contrast to various other materials. SWCNTs electrical properties depend on their chirality concerning the tube axis. SWCNTs are divided into three main categories: (i) armchair, (ii) zigzag and (iii) chiral (Fig. 1). Whereas, MWCNTs contains multiple layers and with changeable chirality which provides amazing mechanical properties [37].

2 CNMs: An Effective Ion-Exchanger

There are several techniques to induct carbon nanoparticles into Ion Exchange Membranes (IEMs), for example, in situ polymerization, plasma treatment, additive blending sol-gel process, etc. Different mechanisms have been there to illustrate the techniques by which CNMs increases the electrochemical characteristics of IEMs. Numerous studies have been attributed to improve the IEMs properties due to the existence of functionalized CNMs which gave extra ionic groups for better exchange of ions [4, 18, 35, 41]. This phenomenon of the stipulation of added functional groups for better ion exchange could be applicable for a precise case, i.e., when the CNMs functionalization grade is higher and the amount of CNMs functionalized in the nano-composite IEMs is significantly higher to generate a noteworthy augment in the IEC by means of added application of the ionic groups. This clarification is not agreeable for cases where the CNMs are not functionalized. There are abundant cases where adding of non-functionalized CNMs leads to enhancements in the IEC of IEMs [18, 22, 40]. Consequently, there must be a mechanism accountable for enhanced characteristics of carbon nano-composite IEMs, which is called as ionic collection dispersal. It explains that the amalgamation of CNMs facilitates the generation of interconnected ion-conducting pathways within the membrane matrix of carbon nano-composite IEMs [9, 19, 21, 36]. The addition of CNMs improved the allocation of ionic masses in carbon nano-composite IEMs [13, 20]. This enhanced dispersal of ionic masses leads to development of added ion-conducting passages which give added pathways for ion transport. A detailed illustration of the Ionic Cluster Dispersion Mechanism (ICDM) is depicted in Fig. 2. ICDM is a significant method for enhancing the characteristics of IEMs by CNMs. Moreover, it gives a rational justification for additions in IEC wherever non-functionalized CNMs are engaged in the nano-composite IEMs. It brings enhancement in IEC where CNMs functionalized are utilized in the nano-composite IEMs. The enhancement in IEC is because of a better acquaintance of ionic functional groups for the exchange of ions because of improved dispersal of the ionic masses in the matrix. Functionalized CNMs, it can be understood that interaction amongst both of the mechanisms outcomes in the pragmatic enhancement in nano-composite IEMs, while the CNMs, which are non-functionalized, only the later method leads. The exterior changes to CNTs have made a noteworthy job in affecting their IE characteristics. Oxidized CNTs showed an enhanced prospective for captions uptake than unoxidized CNTs

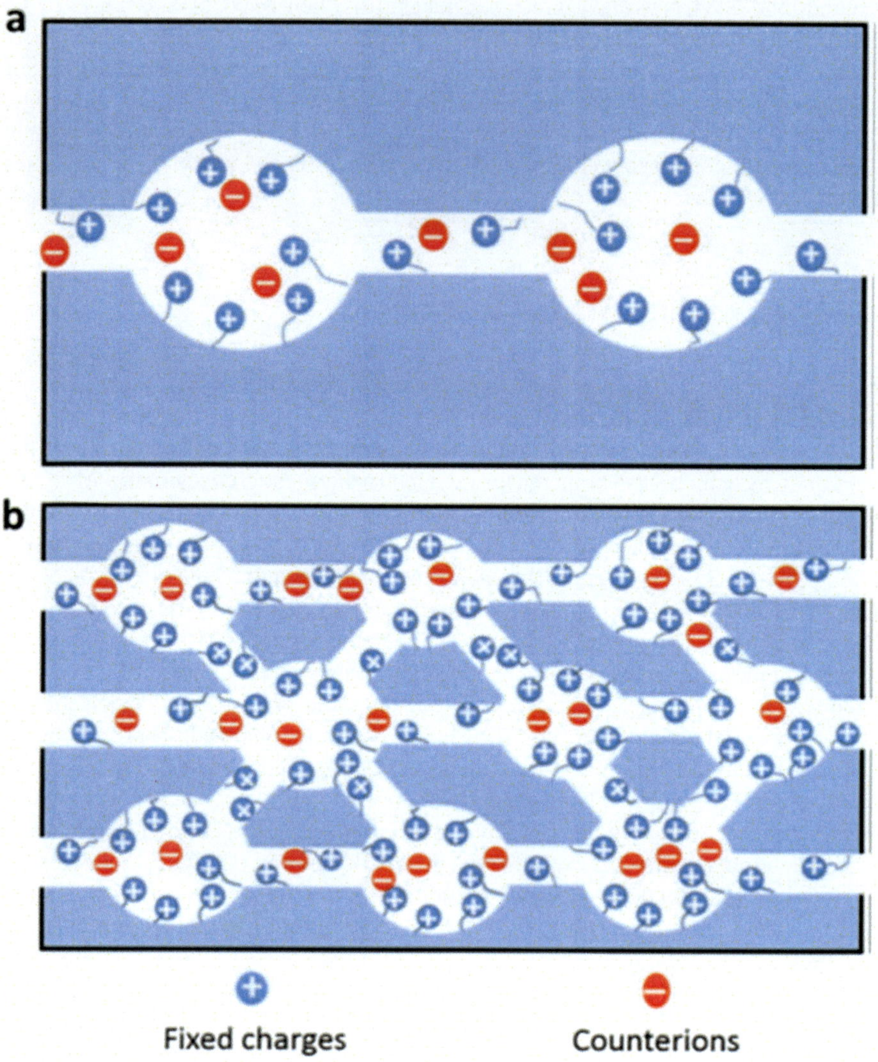

Fig. 2 Details of ionic collection dispersal mechanism in Anion Exchange Membranes (AEMs). **a** Virgin AEMs. **b** AEMs incorporated with NMs. The clusters (shown in **a**, **b**) are connected by ion-conducting channels which creates a network for the migration of the oppositely charged ionic groups (counter ions). In b the incorporation of NMs facilitates the dispersion of the ionic clusters and consequently creates more interconnected ion-conducting channels. Moreover, the distance between the ionic clusters is reduced. These changes ensure higher exposure of the fixed charges for ion exchange and also promote the transport of counter ions through the additional and shorter conducting channels of the AEMs [2]

[31]. While, unoxidized multi-walled CNTs are more efficient for the anions uptake, for example, dichromate than oxidized multi-walled CNTs [28]. It has been noticed that oxidation regulates the surface charge of these materials [28, 31]. In general, oxidation decreases the pH PZC (pH value at zero charges) thus resulting in a negative charged surface which is extra capable for cations uptake. Similarly, unoxidized CNTs have more pH PZC which is extra promising for anion uptake.

Rao et al. [31] noticed that oxidized single-walled CNTs depicted more uptake for Ni^{2+} and Zn^{2+} ions than unchanged multi-walled CNTs. Pillay et al. [28] said that unchanged multi-walled CNTs are extra successful for the exclusion of Cr^{6+} in comparison to oxidized multi-walled CNTs because of greater pHPZC. The chief success in carbon nano-composite ion exchange membranes study focused on utilizing CNTs in increasing properties like mechanical strength, thermal stability, ion exchange capacity, and ionic conductivity at a laboratory scale [2]. To regulate the ion-exchange properties the exterior amendments be done by two methods. It can be through the adding up of functional groups to surface and the other technique is by doping with heteroatom.

2.1 CNTs Functionalization

Functionalization is a crucial method to make CNTs receptive to mechanical and electro-magnetic forces. Magnetic CNTs, for instance, are attractive for use in polymer composites as stirrers in nano-fluidic devices. Functionalization of the outer surface of CNTs supplements has made the CNTs with extra characteristics, like compatibility and solubility with various substances. CNTs elite properties make it attractive for varied uses. CNTs required functionalization for the bulk of these uses, for example, changing the properties of graphite to create CNTs solvable diversity, or combining various clusters or inorganic substances for the prospective use of changed CNTs. Different techniques of CNTs functionalization can be grouped into two main categories:

i. Functionalization with chemical from external (exohedral) [14] Fig. 3a–d. It is further divided into three groups on the basis of methods of add-on of various groups to CNTs side membrane:

   - Functionalization (Covalent) by connecting functional groups to CNTs ends or defects [5, 14].
   - Functionalization (Covalent) by sidewall functionalization [5, 14].
   - Exohedral functionalization(Non-covalent), for instance, polymers wrapping of CNTs [14].

ii. Functionalization from the interior (endohedral) [14] depicted in Fig. 3e, that CNTs are functionalized by adding them with various nanoparticles, this can be accomplished either by

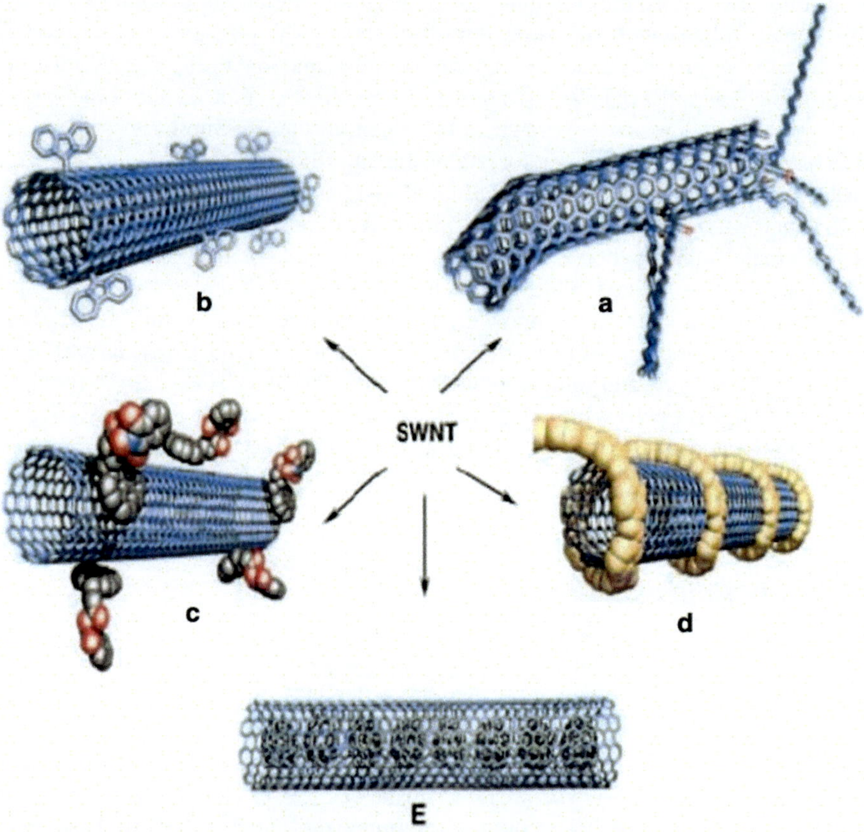

Fig. 3 Functionalization prospective for single-walled CNTs. **a** Functionalization with defects, **b** functionalization with covalent sidewall, **c** non-covalent exohedral functionalization, **d** non-covalent exohedral functionalization with polymers and **e** functionalization from inside (endohedral)

- Using the technique of continues invasion when CNTs are packed with suspension of colloids and further liquid carrier evaporation, or by
- In this, CNTs are packed with different substances, which respond with specific chemical or thermal situations and generate CNTs. These CNMs are then confined in the CNTs.

A large amount of functional groups can be combined with the exterior of CNTs. These are generally functional groups in which oxygen like –OH and –COOH groups. This functionalization category has been chiefly by means of CNTs oxidation in acid which leads to the addition of carboxyl group. As a result, this CNTs category has been utilized in cations uptake [27]. The above reference studies supported the fact that collaboration in between the cations and the functional groups are chiefly responsible to retain cations. Ion exchange mechanism is dependent on pH. At low pH, uptake of cations is less and high at higher pH. In low pH, the electrostatic repulsions obstruct the cations uptake while electrostatic attractions at higher pH the uptake of the cations increases. Another captivating feature of the above results revealed that surface amended CNTs gave better results than traditional adsorbents. Cech et al. [7] revealed that lateral-wall separation of multi-walled CNTs can be attained by treatment with P_2S_5. Pillay et al. [29] amended this technique to generate Sulphur comprising multi-walled CNTs which revealed selective and enhanced receiving of Hg^{2+}. Therefore, the chemical treatment of CNTs can be controlled both by functional groups and heteroatom. Pillay et al. [27] studied the uptake of both anions and cations by weak and strong acid mixtures. It was noticed that the preamble of functional groups containing oxygen privileged uptake of cation by decreasing the pHPZC of materials. On the other hand, treatment with a weak base like NH_3 favored uptake of anions by raising pHPZC. Consequently, acid and base treatment is dependent on acid or base strength and heteroatoms.

2.1.1 CNMs as Cation Exchanger

Generally, a large movement of lattice ions is required to help solid-state ion-exchange dispersal. Cations dispersion is more conveniently occur as their ionic radii are commonly lesser than anions. Thus, cation exchange reactions (CERs) are more probable to occur in comparative to anion exchange reactions (AERs). CERs commonly lead to new products with compound structures like heterostructures and metastable phases which are not manageable by common artificial methods. The crystalline structure of nanoparticles (NPs) can be created by NPs template. For instance, the rock salt configuration of PbSe NRs regulates the crystalline structure of the ultimate product, CdSe [6]; this can be zinc-blended or wurtzite, though the previous is less stable thermodynamically. Zinc-blended CdSe with an alike crystalline structure to the PbSe template is made by CERs. Likewise, roxbyite $Cu_{2-x}S$ NPs are utilized as a template, wurtzite CoS, and Mn Scan is attained as products because of their analogous crystalline structures [30]. Figure 4 depicts the framework of the crystalline structure of cation and anion. Anion sublattice of roxbyite is a disturbed hexagonal close-packed (HCP) arrangement, whereas the cation sublattice is of trigonal and tetrahedral arrangements. The structure of the NP product i.e. wurtzite is metastable in bulk, whereas the anion and cation frameworks are similar to roxbyite. The crystalline structure resemblance among the NPs template and their products is a significant measure to determine the crystalline structure of NPs products.

In NP crystalline structure the interstitial sites are efficient means for ions diffusion. Figure 5 depicted activation energy analysis by CERs among PdS and CdS [11]. Cd ions disperse through interstitial sites (Fig. 5b) and not by the vacancy sites (Fig. 5a), which is beneficial by dropping the energy of activation. Similarly, the CERs arbitrated by the interstitial sites and vacancy can be encouraged because of the lesser energy of activation for ions diffusion.

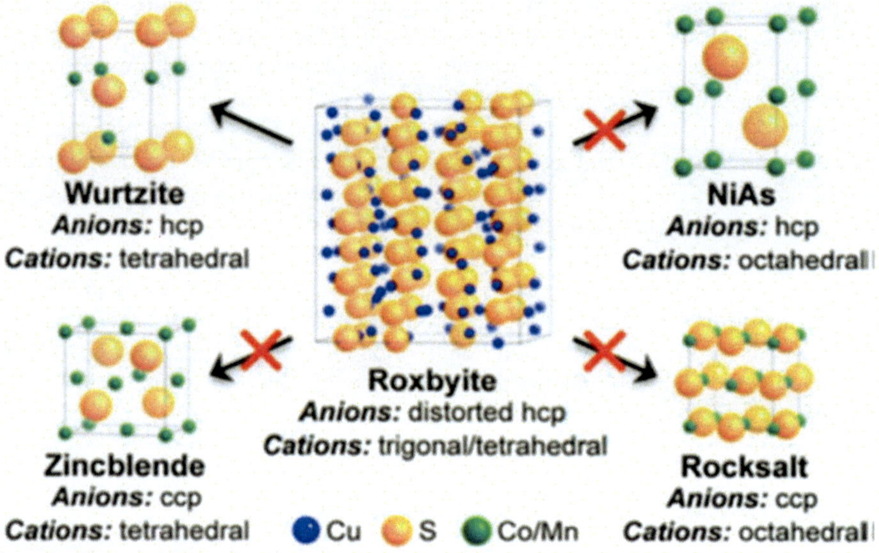

Fig. 4 Wurtzite (CoS/MnS) and Roxbyite ($Cu_{2-x}S$) crystalline structures. The transformation of $Cu_{2-x}S$ NPs into CoS and MnS NPs is linked oHCP anion lattice during cation exchange. Reprinted with permission from American Chemical Society [8, 30]

The generation of NPs through CERs delivers a novel passageway for designing compound NPs structure, for example, NRs segmented, which are problematic to get by common methods. Sadtler et al. [33] and Robinson et al. [32] generated CdS–Ag_2S and CdS–Cu_2S NRs by incomplete CERs of CdS NRs and depicted that NRs configuration got changed reliant on choice of cation (Fig. 6). When Cd^{2+} is replaced by Ag^+ ions, Ag_2S with steady space encouraged by strain is generated, subsequently superlattice CdS–Ag_2S NRs. When Cd^{2+} is replaced with Cu^+, heterostructured CdS–Cu_2S NRs are made. This structural alteration is formed by favorability of chemical and elastic alteration.

Fenton et al. [12] created different NPs from 1stgeneration nanostructures to 3rd generation nanostructures by means of CERs with $Cu_{1.8}S$. The $Cu_{1.8}S$ NPs are converted to CdS and ZnS by CER. An alteration in the structure was perceived with reaction time. Transition metals like Co, Mn, and Ni along with group II elements such as Zn and Cd have been used for CER to generate heterostructured NPs. Figure 7 represents CdS–ZnS–$Cu_{1.8}S$–ZnS NPs which is fabricated by successive CER practices. Justo et al. [16] has made compound-structured NPs (dot-in-rod PbS/CdS NRs) by incomplete CERs from PbS to CdS (Fig. 8). The photoluminescence (PL) spectra of these NRs are organized by the CERs situations for example, reaction temperature and time, leads to the PL yield of 55%. The amount of PbS dots in an NR was attained by the length of the preliminary substance, CdS NRs. Zhang et al. [38] witnessed to generate CdS–PbS Janus-like NPs structure by a well-controlled CERs in CdS NPs (Fig. 9). The CERs are generated alongside the <111> direction, thus

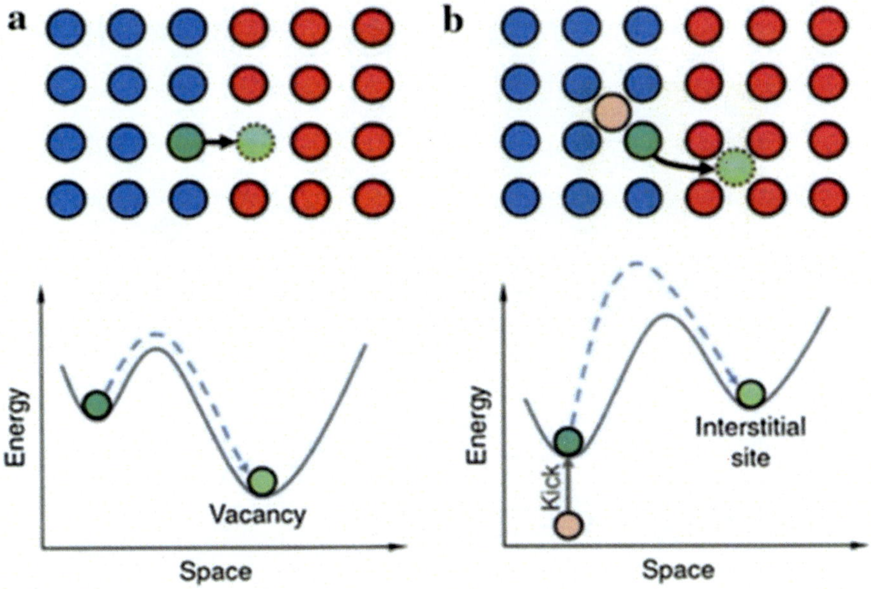

Fig. 5 Depiction of ion dispersion through vacancies: **a** interstitial sites, **b** PbS/CdS interfacial site. The blue circles depict Pd, red circles depict Cd and pink circle is an interstitial site. They have different activation energies. In this case, Cd ions diffuse through the interstitial site (**b**) is advantageous in terms of the activation energy. Reprinted with permission from Springer Nature Publishing [8, 11]

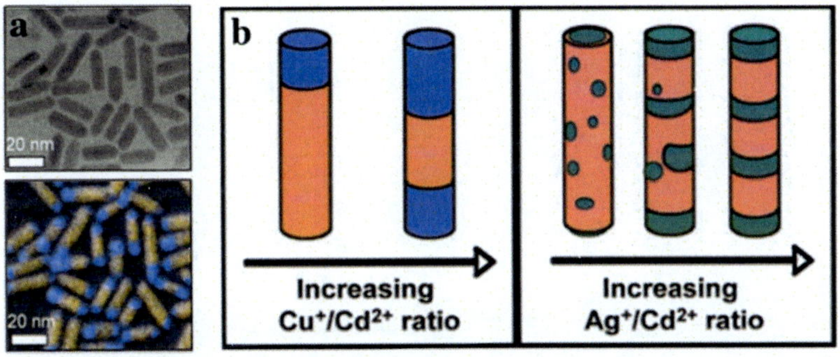

Fig. 6 **a** An electron microscopy bright-field transmission picture (top) and a color-composite energy-filtered transmission electron microscopy picture (bottom) of CdS–Cu_2S binary NRs The orange regions correspond to the Cd energy-filtered mapping and the blue regions correspond to the Cu-mapping. **b** A schematic of the structural transformation of the CdS–Ag_2S and CdS–CuS NRs. Reprinted with consent from American Chemical Society [8, 32, 33]

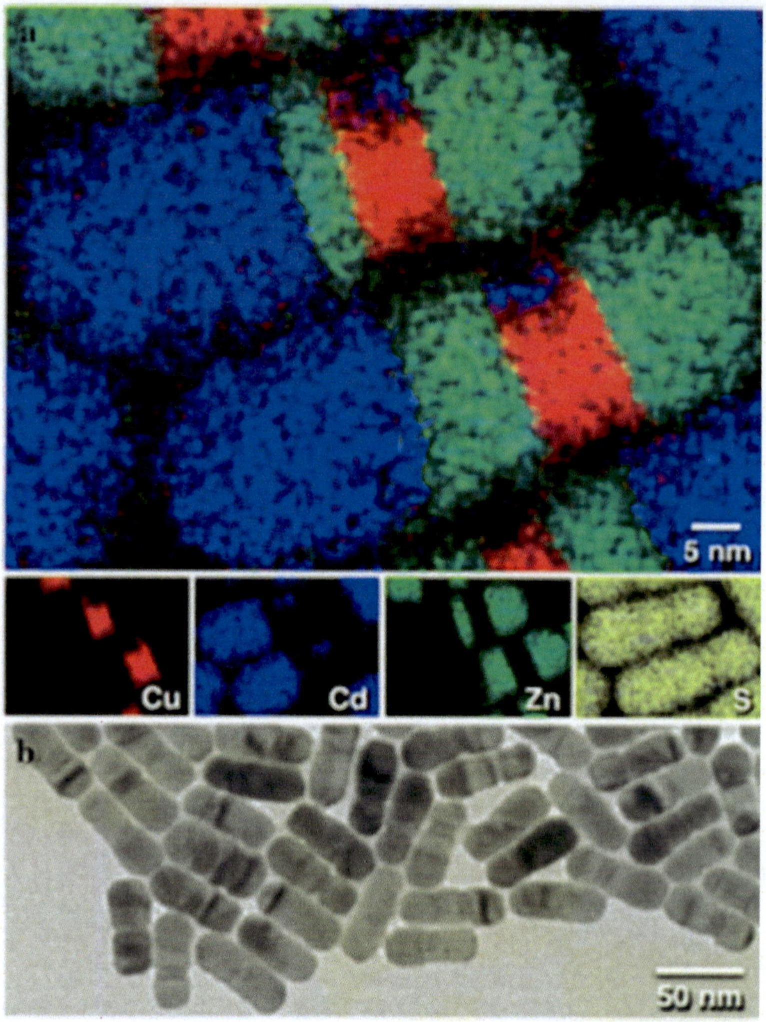

Fig. 7 **a** Electron microscopic energy scattered X-ray spectroscopy image and **b** transmission electron microscopy image of CdS–ZnS–$Cu_{1.8}$S–ZnS nanorod obtained from sequential cation exchange from $Cu_{1.8}$S nanorod. Reprinted with permission from American Chemical Society [8, 12]

generating Janus-like NPs structure that can be utilized to surge the proficiency of solar cells because of flexible PL features by CERs characteristics such as temperature and process time. Park et al. [25] made (Au_2S–$Cu_{1.81}$S)@IrxSy nanoplates and (PdS–$Cu_{1.81}$S)@IrxSy nanoplates with Janus-like and hexagonal structures by CERs of Au and Pd in $Cu_{1.81}$S@IrxSy (Fig. 10). When the CERs occurred, the six bends of hexagonal $Cu_{1.81}$S nanoplates executed as CERs sites, and the course of the CERs are obtained.

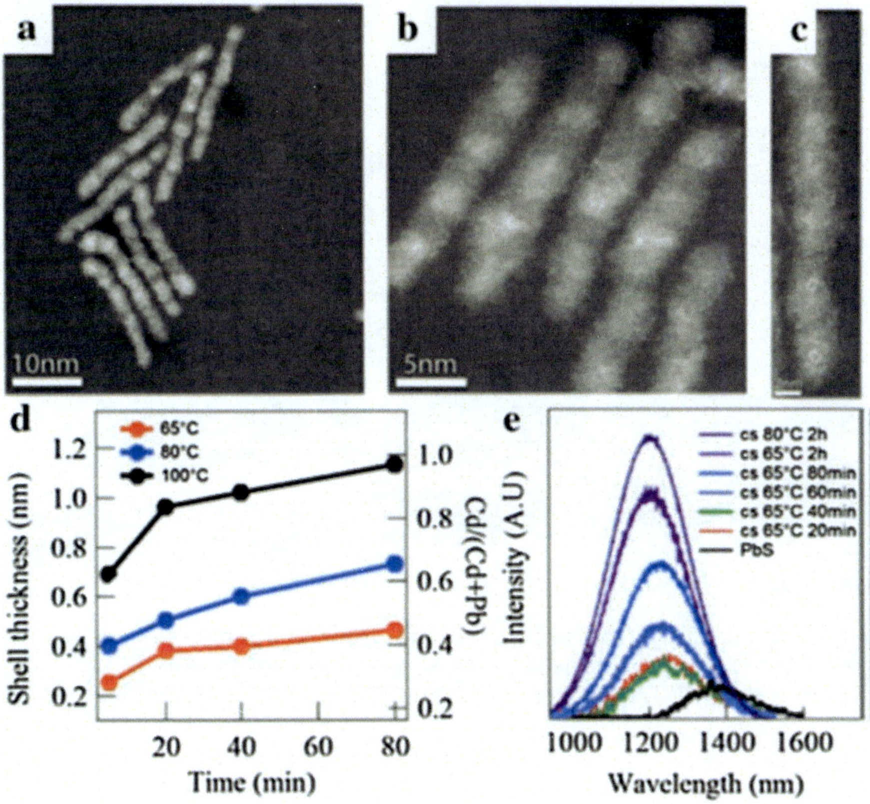

Fig. 8 **a–c** Electron microscopic large angle annul are pictured in dark area of PbS/CdS rods revealing various PbS dots in between the rods. **d** Ratio Cd/(Cd + Pb) and CdS shell thickness as a function of reaction time for various reaction temperatures. **e** Photoluminescence spectra with various reaction times at 65 and 80 °C. Reprinted with consent from American Chemical Society [8, 12]

2.1.2 CNMs as Anion Exchanger

Usually, AERs are leisure lier than CERs because of less movement and hugeionic radii of anions [3]. Consequently, slothful AERs always needed a lengthier time and temperature during reaction in contrast to CERs. The advantages of the slowness can be used in incomplete AERs by regulating the sluggish kinetics [34]. The process of AERs can be described by the thermodynamic energy and the theories of mass action. The AERs described by mass action theory [3], growing the reagents concentration in solution stimulates the reaction kinetics alike to CERs. The thermodynamic theory developed for AERs is similar to CERs [3]. The thermodynamic extemporaneity of AERs is due to incoming anions precursors and further by outgoing ions reactions [3]. Nedelcu et al. [23] described rapid, low-temperature incomplete or complete AERs that can be regulated in tremendously luminescent semiconductor NPs of

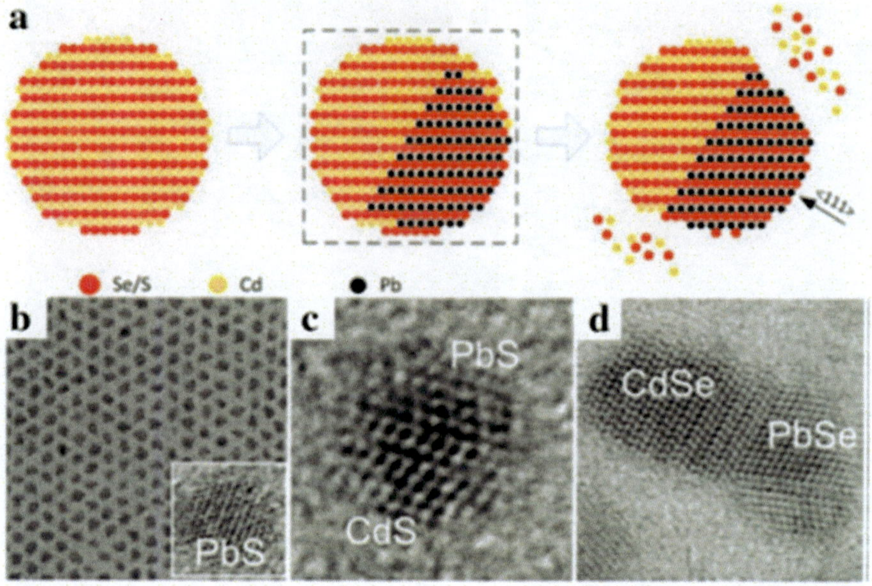

Fig. 9 **a** A representation of the CERs in CdS or CdSe with Pb^{2+}, **b** partially replaced CdS NPs, **c** spherical in nature, and heterostructure can be detected, **d** partially exchanged CdSe NPs. The PbSe and CdSe domains form two (111) interfaces. Reprinted with consent from American Chemical Society [8, 38]

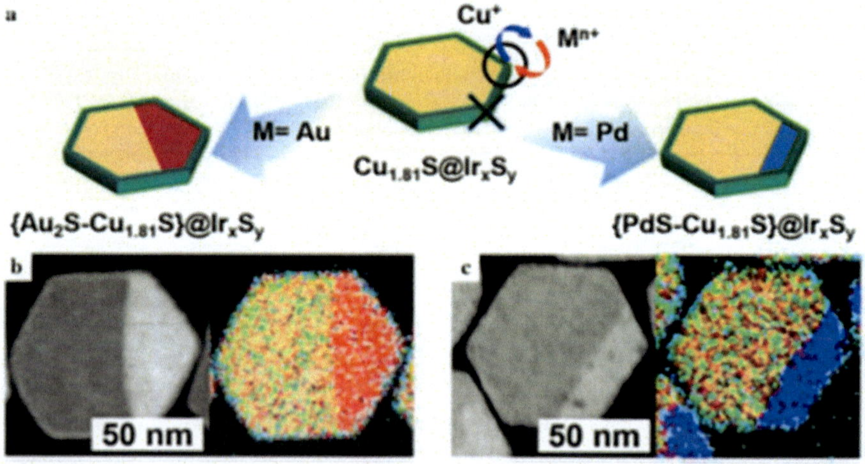

Fig. 10 **a** A representation of the CERs in $Cu_{1.81}S$@IrxSy concerning Au and Pd ions. Scanning transmission electron microscopy pictures of the Janus-like (Au_2S–$Cu_{1.81}S$)@IrxSy NPs. **b** The Janus-like (PdS–$Cu_{1.81}S$)@Ir$_x$S$_y$ NPs. Reprinted with consent from American Chemical Society [8, 25]

$CsPbY_3$(Y = Cl, Br, or I). By amending the shares of halide in colloidal, the photoluminescence attuned over the complete visible spectral region while a quantum gain of 20–80% was sustained. NPs heterostructured can be attained by AERs followed by sluggish reaction kinetics. Park et al. [26] specified the anion exchange from ZnO to solid, hollow structures of monocrystalline ZnS NPs. The study of AERs from ZnO to ZnS indicated that the chemical transformation of NPs followed by the "Kirkendall effect" usually produces NP polycrystalline products, whereas monocrystalline products are barely attained (Fig. 11).

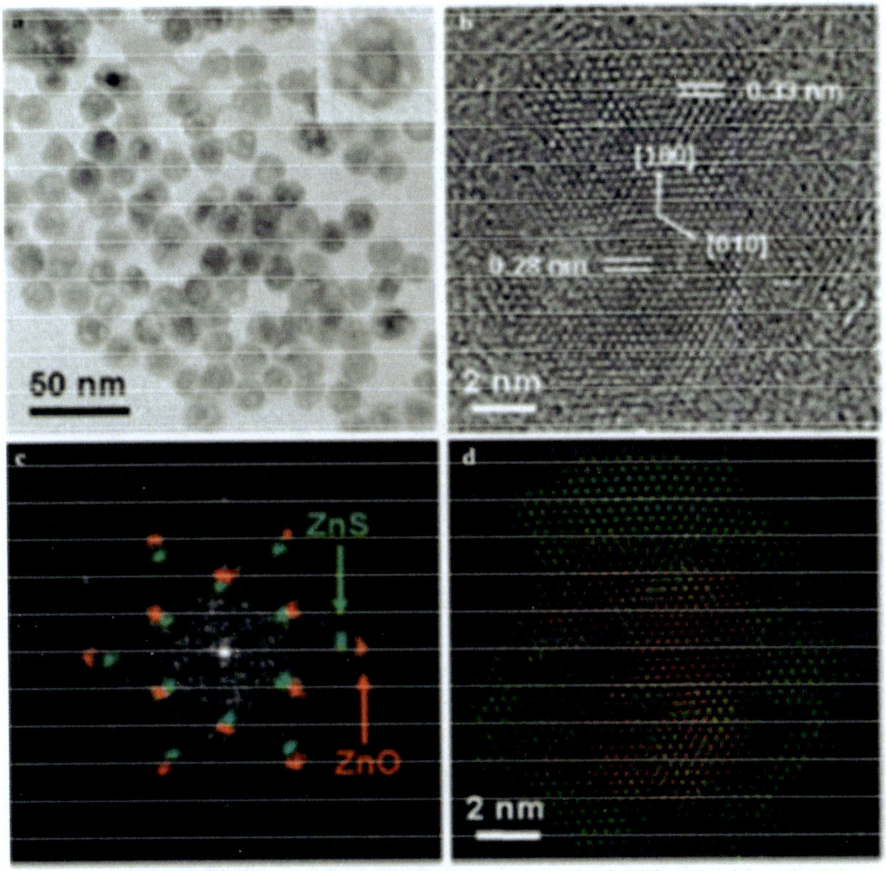

Fig. 11 **a** High-resolution transmission electron microscopy image of partly wrapped yolk-shell NPs and **b** a high-resolution transmission electron microscopy image and fast Fourier transform (**c**) of a single yolk-shell NP, **d** the reconstructed crystal structure from (**c**). Reprinted with consent from American Chemical Society [8, 26]

2.2 Effect of Heteroatom Dopants on CNTs

CNMs have extraordinary intrinsic properties, which can be used for many applications. One possible application is the introduction of different elements (for example Si, P, N, and B) as heteroatom dopants in the lattice structure. Due to hetero atoms doping, the electronic properties of CNTs are notably tuned. Heteroatom-doped CNTs showed superior properties over general CNTs. Figure 12 represents SWCNTs doped with nitrogen heteroatoms, which is highlighted in green color. The adding of functional groups containing oxygen and oxygen as heteroatom favors uptake of cations by decreasing pHP ZC. Likewise, functional groups containing Sulphur and Sulphur depicting as hetero atom have revealed analogous results for the acceptance of Hg^{2+} [29]. Nevertheless, the effect of other heteroatoms such as nitrogen has been poorly known. Nitrogen-doped carbon nanotubes (NCNTs) have been receiving much attention now. Perez-Aguilar et al. [24] revealed that NCNTs which are oxidized is more dynamic for the uptake of Cd^{2+} and Pb^{2+} than undoped unoxidized multi-walled CNTs.

Pillay et al. [27] investigated that the nitrogen atom primarily dependent on nitrogen forms present on CNTs. For example, if quaternary nitrogen exists it leads

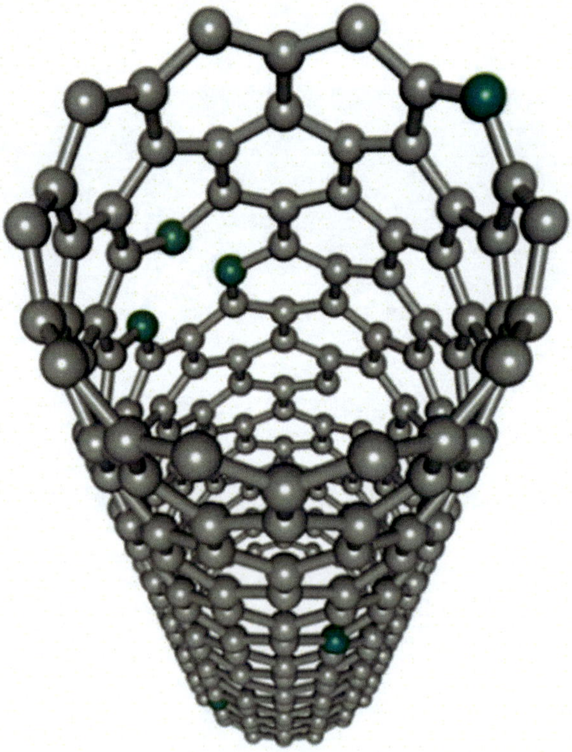

Fig. 12 SWCNTs doped with Nitrogen hetero atoms [24]

to positively charged nitrogen which raises the pHP ZC and favors anion uptake. Otherwise, if nitrogen is present in aggregation with oxygen this decreases the pHP ZC, thus, favoring cations uptake. Stress on the use of heteroatoms such as O_2, S, and N_2 atoms has thus been made. The concern about the influence of addition of heteroatom on selectivity of CNTs for particular pollutants is required to be explored in detail.

3 Conclusion and Future Perspective

CNTs ion-exchangers have shown its prospective to remove a large number of anions and cations on the basis of functional groups and heteroatoms. Though, the choosiness of CNTs ion-exchangers for particular pollutants is quiet ambiguous. Insufficient studies have been done in this zone. The past literature review strongly depicted that CNMs are exceptional ion-exchangers. CNTs can be amended to work both as anion and cation exchangers by the addition of functional groups and heteroatoms which eventually influence the surface charge. Though, the choosiness of CNTs ion-exchangers for particular ions still required enhancement.

References

1. Aguiar AL, Fagan SB, da Silva LB, Mendez Filho J, Souza Filho AG (2010) Benzonitrile adsorption on Fe-doped carbon nanostructures. J Phys Chem C 114:10790–10795
2. Alabi A, Alhajaj A, Cseri L, Szekely G, Budd P, Zou L (2018) Review of nanomaterials-assisted ion exchange membranes for electro membrane desalination. Nat Res J Clean Water 1(1):1–22
3. Anderson BD, Tracy JB (2014) Nanoscale 6(21):12195
4. Bai H et al (2015) Anhydrous proton exchange membranes comprising of chitosan and phosphorylated graphene oxide for elevated temperature fuel cells. J Membr Sci 495:48–60
5. Banerjee S, Hemraj-Benny T, Wong SS (2005) Covalent surface chemistry of single-walled carbon nanotubes. Adv Mater 17(1):17–29
6. Casavola M, van Huis MA, Bals S, Lambert K, Hens Z, Vanmaekelbergh D (2012) Chem Mater 24(2):294
7. Cech J, Curran SA, Zhang D, Dewald JL, Avandhandula A, Kandadai M, Roth S (2006) Functionalisation of multi-walled carbon nanotubes: direct proof of sidewall thiolation. Phys Stat 2, 243(13):3221–3225
8. Cho G, Park Y, Hong Y-K, Ha D-H (2019) Ion exchange: an advanced synthetic method for complex nanoparticles. Nano Convergence 6:1–17
9. Choi BG et al (2011) Innovative polymer nanocomposite electrolytes: nanoscale manipulation of ion channels by functionalized graphenes. ACS Nano 5:5167–5174
10. De Volder MF, Tawfick SH, Baughman RH, Hart AJ (2013) Carbon nanotubes: present and future commercial applications. Science 339:535–539
11. Fan Z, Lin LC, Buijs W, Vlugt TJH, van Huis MA (2016) Nat Commun 7:11503
12. Fenton JL, Steimle BC, Schaak RE (2018) J Am Chem Soc 140(22):6771
13. He Y, Tong C, Geng L, Liu L, Lü C (2014) Enhanced performance of the sulfonated polyimide proton exchange membranes by graphene oxide: size effect of graphene oxide. J Membr Sci 458:36–46

14. Hirsch A (2002) Functionalization of single-walled carbon nanotubes. Angew Chem Int Ed 41(11):1853–1859
15. Iijima S (1991) Synthesis of carbon nanotubes. Nature 354:56–58
16. Justo Y, Goris B, Kamal JS, Geiregat P, Bals S, Hens Z (2012) J Am Chem Soc 134(12):5484
17. Khare R, Bose S (2005) Carbon nanotube based composites—a review. J Miner Mater Charact Eng 4:31–46
18. Klaysom C, Marschall R, Wang L, Ladewig BP, Lu GQM (2010) Synthesis of composite ion-exchange membranes and their electrochemical properties for desalination applications. J Mater Chem 20:4669–4674
19. Lee DC, Yang HN, Park SH, Kim WJ (2014) Nafion/graphene oxide composite membranes for low humidifying polymer electrolyte membrane fuel cell. J Membr Sci 452:20–28
20. Li N, Zhang F, Wang J, Li S, Zhang S (2009) Dispersions of carbon nanotubes in sulfonated poly-bis(benzimidazobenzisoquinolinones) and their proton conducting composite membranes. Polymer 50:3600–3608
21. Liu L, Tong C, He Y, Zhao Y, Lü C (2015) Enhanced properties of quaternized graphenes reinforced polysulfone based composite anion exchange membranes for alkaline fuel cell. J Membr Sci 487:99–108
22. Moghadassi AR, Koranian P, Hosseini SM, Askari M, Madaeni SS (2014) Surface modification of heterogeneous cation exchange membrane through simultaneous using polymerization of PAA and multi walled carbon nanotubes. J Ind Eng Chem 20:2710–2718
23. Nedelcu G, Protesescu L, Yakunin S, Bodnarchuk MI, Grotevent MJ, Kovalenko MV (2015) Nano Lett 15(8):5635
24. Perez-Aguilar MV, Munoz-Sandoval E, Diaz-Florez PE, Rangel-Mendez JR (2010) Adsorption of Cadmium and Lead onto oxidized nitrogen-doped mutiwalled carbon nanotubes: equilibrium and kinetics. J Nanopart Res 12:467–480
25. Park J, Park J, Lee J, Oh A, Baik H, Lee K (2018) ACS Nano 12(8):7996
26. Park J, Zheng H, Jun YW, Alivisatos AP (2009) J Am Chem Soc 131(39):13943
27. Pillay K et al (2011) Nanomaterials for the removal and recovery of heavy metal ions from industrial effluents. Ph.D. thesis, University of the Witwatersrand
28. Pillay K, Cukrowska EM, Coville NJ (2009) Multi-walled carbon nanotubes as adsorbents for the removal of parts per billion levels of hexavalent chromium from aqueous solution. J Hazard Mater 169:1067–1075
29. Pillay K, Cukrowska EM, Coville NJ (2013) Improved uptake of mercury by Sulphur containing carbon nanotubes. Microchem J 108:124–130
30. Powell AE, Hodges JM, Schaak RE (2016) J Am Chem Soc 138(2):471
31. Rao GP, Lu C, Su F (2007) Sorption of divalent metal ions from aqueous solution: a review. Sep Purif Technol 58:224–231
32. Robinson RD, Sadtler B, Demchenko DO, Erdonmez CK, Wang LW, Alivisatos AP (2007) Science 317(5836):355
33. Sadtler B, Demchenko DO, Zheng H, Hughes SM, Merkle MG, Dahmen U, Wang LW, Alivisatos AP (2009) J Am Chem Soc 131(14):5285
34. Saruyama M, So YG, Kimoto K, Taguchi S, Kanemitsu Y, Teranishi T (2011) J Am Chem Soc 133(44):17598
35. Sharma PP, Gahlot S, Bhil BM, Gupta H, Kulshrestha V (2015) An environmentally friendly process for the synthesis of an fGO modified anion exchange membrane for electro-membrane applications. RSC Adv 5:38712–38721
36. Tseng CY et al (2011) Sulfonated polyimide proton exchange membranes with graphene oxide show improved proton conductivity, methanol crossover impedance, and mechanical properties. Adv Energy Mater 1:1220–1224
37. Zaytseva O, Neumann G (2016) Carbon nanomaterials: production, impact on plant development, agricultural and environmental applications. Biol Technol Agric 3(17):1–26
38. Zhang J, Chernomordik BD, Crisp RW, Kroupa DM, Luther JM, Miller EM, Gao J, Beard MC (2015) ACS Nano 9(7):7151

39. Zhang R, Zhang Y, Zhang Q, Xie H, Qian W, Wei F (2013) Growth of half-meter long carbon nanotubes based on Schulz-Flory distribution. ACS Nano 7:6156–6161
40. Zuo X et al (2009) Preparation of organic–inorganic hybrid cation-exchange membranes via blending method and their electrochemical characterization. J Membr Sci 328:23–30
41. Zuo X, Yu S, Shi W (2012) Effect of some parameters on the performance of eletrodialysis using new type of PVDF–SiO_2 ion-exchange membranes with single salt solution. Desalination 290:83–88

Heavy Metals Removal Using Carbon Based Nanocomposites

Heavy Metals Adsorption by CBNs

A. Vijaya Bhaskar Reddy, V. Madhavi, Akil Ahmad, and G. Madhavi

Abstract The rapid growth of population, industrialization and urbanization tends to deteriorate the quality of available water bodies. This scenario became vulnerable to the environment, and it has alarmed the community to resolve this issue. A large variety of pollutants occur from the above activities, specifically heavy metals creating severe toxicity to human and other living organisms. A variety of heavy metals removal techniques have been established in recent decades, among which a simple adsorption using carbon nanocomposites was found effective for the eradication of heavy metal pollutants. Therefore, present chapter discussed the heavy metals removal procedures using graphene and CNTs with a special emphasis on their functionalization. This chapter summarised the toxicity of heavy metals to plants, aquatic life and human health along with some effective remediation approaches. Further, it clearly described the cutting edge methods reported for the heavy metals elimination from aqueous media using graphene and CNT nanocomposites through adsorption. At last, future recommendations that are required to upgrade the capacity of adsorbents by using environmentally safe CBNs sorbents were proposed following short conclusions.

A. Vijaya Bhaskar Reddy (✉)
Quality Control Division, Ultra International Limited, Sahibabad, Ghaziabad, Uttar Pradesh 201020, India
e-mail: vijay.dr555@gmail.com

Centre of Research in Ionic Liquids and Chemical Engineering Department, Universiti Teknologi PETRONAS, Seri Iskandar, Perak 32610, Malaysia

V. Madhavi
Department of Chemistry, BVRIT-H, JNTUH, Hyderabad 500090, India

A. Ahmad
School of Industrial Teknologi, Universiti Sains Malaysia, Penang 11800, Malaysia

G. Madhavi
Department of Chemistry, Sri Venkateswara University, Tirupati 517502, India

M. Jawaid et al. (eds.), *Environmental Remediation Through Carbon Based Nano Composites*, Green Energy and Technology,
https://doi.org/10.1007/978-981-15-6699-8_12

1 Introduction

The accessibility of clean and potable water is most vital for all living beings to sustain their life on earth. But, the rapid population growth, continued industrialization, civilization and improper disposal of waste are being caused to deteriorate the water quality across the global water bodies. Approximately, 700 million people on this planet have no access to clean water [35], whose surrounding water bodies are already contaminated with several harmful pollutants including dyes, pesticides, pharmaceuticals and heavy metals [95]. All these pollutants are harmful to the environment and living forms. Particularly, the presence of heavy metals pollutants has created severe toxicity to water bodies in recent years. According to the literature, heavy metals occur naturally and contains five times greater density than the water. Even though, many water bodies contain low quantities of heavy metals, their toxicity towards the environment and living organisms is substantial [54]. Despite the fact that some heavy metals are essential for biological processes (i.e., Cu, Fe, Zn, Ni etc.,), all heavy metals are harmful to the living organisms at high concentrations and create acute and chronic toxicity. In particular, heavy metals including cadmium (Cd), mercury (Hg), nickel (Ni), arsenic (As), lead (Pb), chromium (Cr), zinc (Zn), cobalt (Co) and selenium (Se) are highly toxic even if they exist at trace concentrations [57]. Therefore, the raising concentrations of heavy metals in water resources is currently a matter of concern, specifically a large amount of industrial heavy metal effluents joining water bodies without any pretreatment.

These heavy metals generally originate from natural and human induced activities. The natural occurrence of heavy metals includes volcanic eruptions, forest fires, sea-salt sprays, biogenic sources, rock weathering and so on. Whereas, the anthropogenic activities including agriculture, mining and metallurgical processes, industrial wastewater, metal plating, and runoffs release heavy metal pollutants into different compartments of the environment [107]. Various routes of entry of heavy metal pollutants into water bodies is presented in Fig. 1. These heavy metals occur by means of oxides, hydroxides, sulphates, phosphates, sulphides and organic compounds. When, heavy metal effluents released into drinking water sources, they can predominantly enter the human body by food and water, and to a lesser extent by inhalation of air. Absorption through the skin is another common route of exposure in adults working/living in industrial areas [33]. On other hand, ingestion is the most prevalent cause of heavy metals exposure in kids and infants. Accordingly, both natural and human induced operations polluting the water bodies at a high risk.

Considering the severe toxicity of heavy metal pollutants, a variety of heavy metal removal approaches based on adsorption, reverse osmosis, precipitation and coagulation have been established over the years [20, 26, 55, 101, 108]. Even though, some of the proposed heavy metal removal techniques are effective, their cost and drawbacks avoided their commercial use for the *in situ* applications. Among all the techniques, adsorption was found best efficient for this purpose. After the discovery of carbon nanocomposites, heavy metals adsorption by various carbon nanocomposites including graphene and CNTs has gained significant attention [29, 101].

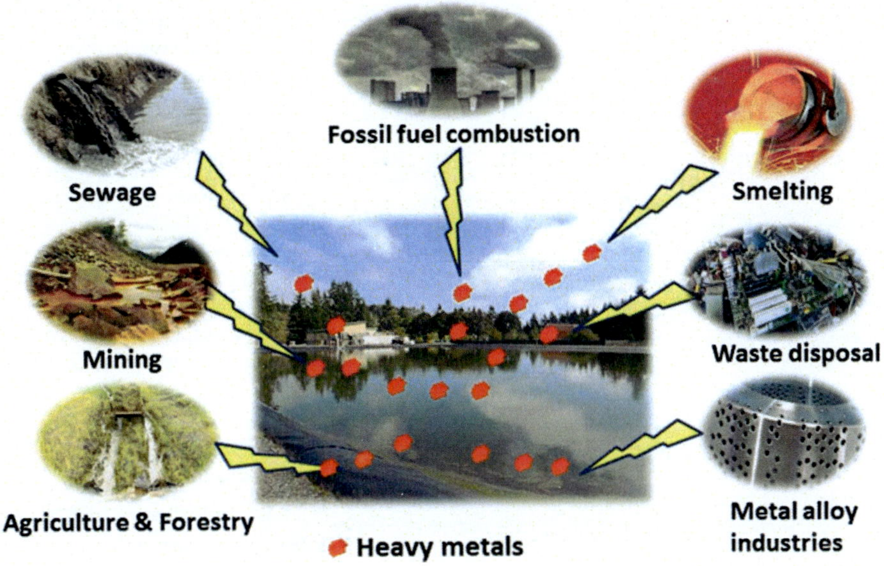

Fig. 1 Schematic representation for the entry of heavy metal pollutants into water bodies

Consequently, a large number of studies have been reported for the heavy metals' adsorption using various functionalized graphene and CNTs composites. Considering the significance and wide applicability of carbon based nanocomposites (CBNs) for heavy metals removal, this chapter is intended to describe various removal methods established using different combinations of carbon nanocomposites. A specific focus has given to summarise heavy metal removal methods using functionalized CNTs and graphene. A detailed recommendations and future perspectives that helps to improve the removal efficiency of CBNs is presented eventually.

2 Adverse Effects of Heavy Metals on Plants and Biota

The heavy metals present in aqueous systems even at low concentration are lethal to human and aquatic organisms that induce severe oxidative stress. Further, heavy metals are most persistent and remain permanently in the marine environments [94]. Also, heavy metals in aquatic systems cause significant ecotoxicology and produce devastating effects on the aquatic environment [43]. The following sections described the potential effects of heavy metals on plants, human and aquatic life. A detailed summary of the ill effects of heavy metals on plants, human and aquatic life are illustrated in Fig. 2.

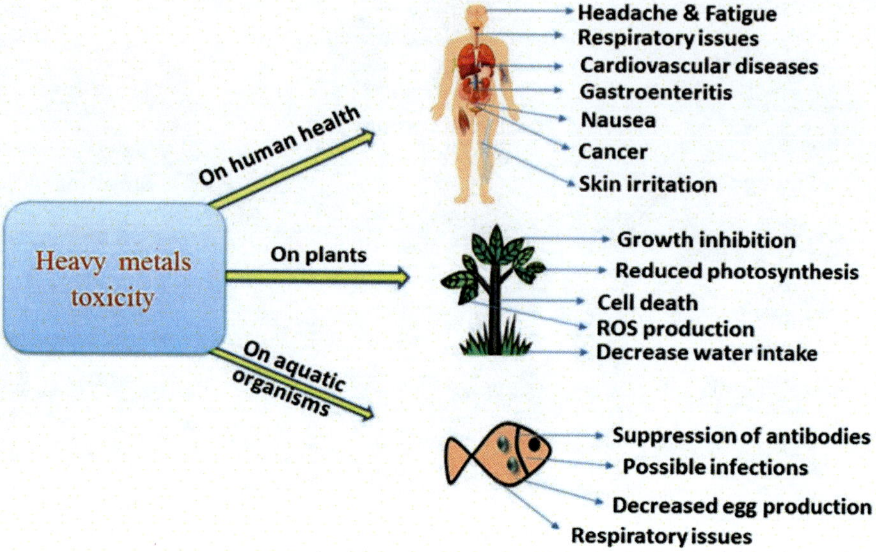

Fig. 2 The major adverse effects of heavy metals on human, plants and aquatic organisms

2.1 *Heavy Metals Impact on Plants*

Plants do not require Cd, As, Hg, Se and Pb heavy metals for any of the physiological functions. But they need some of the heavy metals such as Cu, Ni, Fe, Co, Mn, Mo and Zn at trace concentrations for the normal growth and metabolism [4]. However, these elements become poisonous to plants if their concentration exceeds the optimal values. The plant roots are main carriers that absorb heavy metals from soil and water. When plants uptake heavy metals from water at high concentrations, they may experience greater health risks and continuous accumulation of heavy metals in plants further create potential threat to human through food chains [27]. The agriculture runoff containing heavy metals enter into water bodies and affects the aquatic plants. However, the absorption of heavy metals and their accumulation in plant tissues vary with factors including pH, soil organic matter, moisture content, temperature and nutrients availability. For instance, *Beta vulgaris* (Spinach) absorb more Cd, Cr, Zn and Mn during summer, contrary more Cu, Ni, and Pb during winter [86]. This is expected that, the higher decomposition of organic matter during summer provides excessive heavy metals in soil and water for possible plants uptake. Further, heavy metals accumulation in plants vary with the nature of plant species and their efficiency in absorbing metals. Overall, heavy metals are potentially toxic to plants that reduce plant growth, affect photosynthesis, cause yield depression and chlorosis, disturbs plant metabolism and may even reduce nutrients uptake capacity and molecular nitrogen fixation ability leguminous plants [85].

2.2 Heavy Metals Impact on Aquatic Life

The discharge of heavy metal effluents into water bodies through waste disposal generally associate with particulate matter and eventually settle down with sediment parts. Therefore, surface sediments are good reservoirs for organic and inorganic pollutants in aquatic environments [76, 97]. Subsequently, sediment bound heavy metals can be absorbed by aquatic organisms and rooted aquatic microphytes [104]. The aquatic organisms adversely affect by heavy metals accumulation and transfers the heavy metals toxicity to upper classes through food chain. Hence, carnivores (human) at the top of the food chain receive most of the accumulated heavy metals from aquatic animals (particularly from fish) through their food [84]. The existence of heavy metals in water bodies induce the formation of reactive oxygen species (ROS) that can destruct all aquatic organisms including fish. Therefore, food is an important commodity for human, but it affects the human health when contaminated by a vast number of persistent organic and heavy metal pollutants. As a result, consumption of fish that is contaminated with elevated amounts of heavy metals is a serious concern, which may induce severe health impacts [17]. Among all, Hg is the most important pollutant that presents in fish and other marine organisms at high concentrations and create potential hazards to humans. Bacterial methylation of Hg produces me–Hg, which is more toxic among all forms of Hg and nearly all Hg remains in fish muscles as me–Hg. In fish, heavy metals usually bound to proteins and transport through the blood. Heavy metals accumulate to greater extent when the organs and tissues of fish comes in contact with them. Potentially, metal pollutants can enter into aquatic organisms through their food, skin, gills, non-food particles and oral consumption of water [88]. The heavy metal pollutants transport through the blood after their absorption and reaches a storage point either in bone or liver. When heavy metal pollutants processed by liver, they may be stored there or excreted with bile or moved back into blood for possible excretion by gills or kidneys or stored in fat [5].

2.3 Heavy Metals Impact on Human Health

The consumption of metal contaminated food is the major route of exposure in human, which can potentially reduce essential nutrients in the body, and lower immunological defense of the body, and develop disabilities associated with malnutrition and upper gastrointestinal cancer rates [13]. Particularly, heavy metals become toxic when body is not capable to metabolize and allow them to accumulate in soft tissues. The continuous intake of heavy metals even at trace concentrations provide undesirable effects on human, and all such impacts become perceivable only after few years of exposure [71]. For instance, Pb enters the human body through drinking water and its presence in human body create difficulties in pregnancy, increase blood pressure, damage gastrointestinal tract and urinary tract, neurological disorders and may cause severe and permanent brain damage [77]. Also, Pb restricts the growth of grey matter of the brain in children aged below 2–3 years. Next, Cr is another heavy metal and it is one of the most abundant elements in earth crust. It exists mainly in two different oxidation states i.e., Cr(III) and Cr(VI). Among these, Cr(VI) is more toxic to all living beings due to its high water solubility, strong oxidizing and corrosive nature. Next, Hg is acute toxic heavy metal that has no biological importance in human and animals. Inorganic Hg causes spontaneous abortion, congenital malformation and gastrointestinal disorders, whereas me-Hg causes acrodynia (Pink disease), stomatitis, gingivitis, neurological disorders and damage to brain and CNS and congenital malformation [19]. Similar to Hg and Cr, Ar is another lethal heavy metal that provides severe toxicity when combined with protein. Ar make complexes with co-enzymes and suppress the production of adenosine triphosphate (ATP) during respiration. Further, it is highly carcinogenic in all of its oxidation states and a high level of exposure to Ar may even result to death. The toxicity of Ar induces a disorder that is similar to Guillain-Barre syndrome, which is an anti-immune disorder that causes muscle weakness [14]. Additionally, there are several other toxic heavy metals namely Cd, Ni, Zn and Cu whose impact on human body is severe at elevated concentrations.

3 Heavy Metal Removal Technologies

A wide variety of heavy metals removal technologies including oxidation, precipitation, ion-exchange, reverse osmosis, photocatalysis and flocculation-coagulation have been developed in the past few decades. Among all, the most commonly employed techniques were briefly discussed below and presented in Fig. 3.

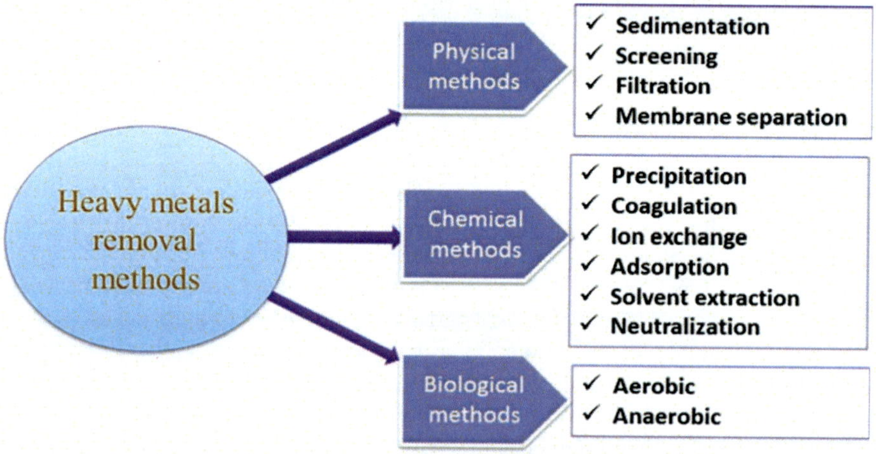

Fig. 3 The most significantly used heavy metal removal techniques

3.1 Chemical Precipitation

Chemical precipitation proceeds by the addition of chemical reagents, and subsequent elimination of precipitated solids from treated water. Usually, heavy metals precipitation can be accomplished by adding coagulants like lime, alum and iron salts. It is widely used water treatment technique across the world considering its low processing costs [62]. Based on literature studies, precipitation classified into two categories i.e., hydroxide precipitation and sulfide precipitation. Hydroxide precipitation employs sodium hydroxide, limestone, sodium carbonate and sodium decanoate as precipitants to form insoluble metal-hydroxide precipitates [8, 65]. However, a major breakthrough was made in recent years replacing hydroxide precipitants with sulfide precipitants that have provided superior metal precipitation rates. Recent studies have showed that cooperative bioleaching and sulfide precipitation methods have removed more than 99% of Zn and 75% of Fe using Na_2S [103]. Wastewater streams utilize biogenic sulfide to reduce sulfate into sulfide using the sulfate-reducing bacteria, which then binds with metal ions to form insoluble precipitates. The resulted sulfide precipitates do not readily dissolve in water and therefore they stabilize the metal ions as metal-sulfides. However, a complete removal of metal ions from wastewater is impractical using the simple precipitation processes. Consequently, different treatment processes are necessary to remove complex heavy metals from wastewaters to the acceptable levels [10].

3.2 Electrodialysis

Electrodialysis (ED) is a membrane separation process, wherein ions transport through the semi permeable membranes in presence of the applied electric potential. The transportation rate and ions direction vary with their charge and mobility. The membranes can be either cationic or anionic based, which means either negative or positive ions can flow through them [92]. Cation selective membranes possess negatively charged matter and allows only positively charged ions to pass through them by restricting negatively charged ions. The specific advantages of ED made this process as interesting alternative for wastewater treatment. In a recent study, the authors synthesized a Nafion 450 membrane to transport Cr from tannery effluent. The used Nafion 450 membrane provided good results due to its lower electric resistance [63]. Further, a few other ED methods reported for the separation of metal ions from electroplating wastewater [18, 42, 61]. Although this technique is efficient for the removal of low molecular weight metal (ionic) components, its efficiency for the removal of higher molecular weight, non-ionic and less mobile ionic species is practically not appreciable.

3.3 Coagulation/Flocculation

Coagulation and flocculation are inevitable procedures that play an essential role in drinking water and wastewater treatment. Generally, a chemical or coagulant adds to the contaminated water in coagulation process, wherein coagulant joins with colloidal material and forms small aggregates called "flocs" [36]. Also, the suspended matter present in water attracts to join with these flocs. In flocculation process, mixing of water continues to boost the flocs to form large size precipitates that can easily settle out. Several studies reported the basic mechanism occurred in the removal of metal contaminants through coagulation-flocculation. However, several parameters including initial pH and turbidity of water, alum concentration and flocculation time influence the removal efficiency [15, 95, 99]. Even though, it is an effective technique for the removal of heavy metal pollutants, the high operational cost is its major disadvantage. In some cases, a high amount of coagulant and flocculant are necessary to achieve sufficient level of flocculation. In such cases, the resulted high quantity of physico-chemical sludge needs to be processed externally [2].

3.4 Ultrafiltration

In ultrafiltration, the organic and heavy metal pollutants can be separated using membranes with pore size between 0.1 and 0.001 micron. Generally, ultrafiltration is capable to eliminate high molecular-weight substances, organic and inorganic polymeric compounds [37]. Briefly, ultrafiltration is a pressure-driven purification approach, where low molecular weight substances travel with water and pass through the membrane, but colloids, particles and macromolecules retain by the membrane. In this approach, size exclusion is the primary removal mechanism, but the surface chemistry of particles affects the membrane performance [30]. In literature, there are several ultrafiltration methods reported for the removal of heavy metal pollutants [10, 102]. However, a simple ultrafiltration effectively removes suspended matter and bacteria. The ultrafiltration membranes are sensitive to oxidative chemicals such as nitric acid, sulphuric acid, peroxides and persulfate and also the performance is pH dependent.

3.5 Reverse Osmosis

Reverse osmosis (RO) is another membrane separation method, wherein pressure applies at the concentrated side of the membrane forcing the treated water into diluted side, the rejected pollutants from the concentrated side being sent to the rejected water. This is the reversal of normal osmosis process, where solvent flows from low solute concentration to high solute concentration in presence of no external pressure

[60]. The membranes used in RO possess a dense barrier layer within the polymer matrix where the prime separation occurs. However, a high pressure is necessary to apply at the high concentration side of membrane in RO process. Several researchers have reported the use of RO process in wastewater treatment [32, 70]. Even though, RO is an effective heavy metal removal technique, it also removes dissolved essential elements from water such iron, magnesium, calcium and sodium. Moreover, chlorine presence in water can damage the RO membrane and there is no particular mechanism to know when to replace the RO membranes.

3.6 Adsorption

Despite the fact that the aforementioned methods have provided satisfactory results, they suffer from several drawbacks. For instance, chemical precipitation and flocculation-coagulation produces large amounts of hazardous waste that needs further treatment. Ion exchange is an effective process, but its poor recyclability limited the use of this technology. Similarly, the cost, material regeneration, energy requirements and the disposal of residual material are the major constraints of membrane filtration. On other hand, photocatalytic methods require extremely long durations for heavy metals oxidation [1]. However, adsorption was evolved as the alternative treatment process minimizing the abovementioned drawbacks. This approach is superior for the removal of heavy metal ions over other conventional methods considering its high efficiency at low concentration of pollutants, the availability of broad range of adsorbents, its possible regeneration and simplicity. As heavy metals are highly persistent, their removal by immobilization on to suitable adsorbent is preferred choice [56, 75].

Adsorption occurs when a gas or liquid solute concentrates on the surface of a solid or liquid adsorbent forming a molecular or atomic film. It is considered as the most suitable approach for the heavy metals removal from contaminated waters owing to its removal capacity even at trace concentrations. In the process of adsorption, there are two types of mechanisms involved. In physisorption, the adsorbate binds with absorbent by weak van der Waal forces. But in chemisorption, adsorbate molecules bind with adsorbent surface through strong chemical bonds [100]. However, the quality of adsorption relies on the adsorbent surface and its interaction with the pollutants. There are several adsorbents reported for the effective removal of heavy metals, which includes clay minerals, biomaterials, zeolites, modified chitosan, manganese oxides, peat, peanut hulls, sewage sludge ash, granular biomass, fly ash and extracellular polymeric substances are few to mention [41]. However, most of the above adsorbents provided insufficient adsorption, and therefore, further use of these adsorbents for wastewater treatment was restricted. Besides, activated carbon (AC) is the extensively studied adsorbent for the removal of different heavy metal ions. It is highly porous, and amorphous solid consisting micro crystallites with a graphite lattice [67]. However, with the strict drinking water regulations and increasing pollution of water bodies, the adsorption capacity of AC was found not sufficient for the removal of organic and metal pollutants. As a result, researchers developed several CBNs including CNTs and graphene nanocomposites along with their functionalization. All these materials have provided superior removal efficiency and improved reusability. The heavy metals removal from contaminated waters using two most studied carbon based nanocomposites i.e., CNTs and graphene materials is discussed in the following sections with detailed case studies [50].

4 Carbon Nanocomposites for Heavy Metals Removal

The revolution of nanotechnology has created opportunities to develop novel nanosorbents for the effective adsorption processes. Particularly, owing to their remarkable chemical and physical properties, CBNs were emerged as potential adsorbents for the removal of organic and heavy metal pollutants. Among many CBNs, graphene and CNTs based composites have drawn special attention [38]. Moreover, mesoporous CBNs with controlled pore size (between 2 and 50 nm) and pore structure are the desirable materials to achieve efficient adsorption of heavy metals. The basic structure of various carbon based materials is presented in Fig. 4. The detailed study about the graphene and CNTs and their functionalization with latest research reports were discussed in the following sections.

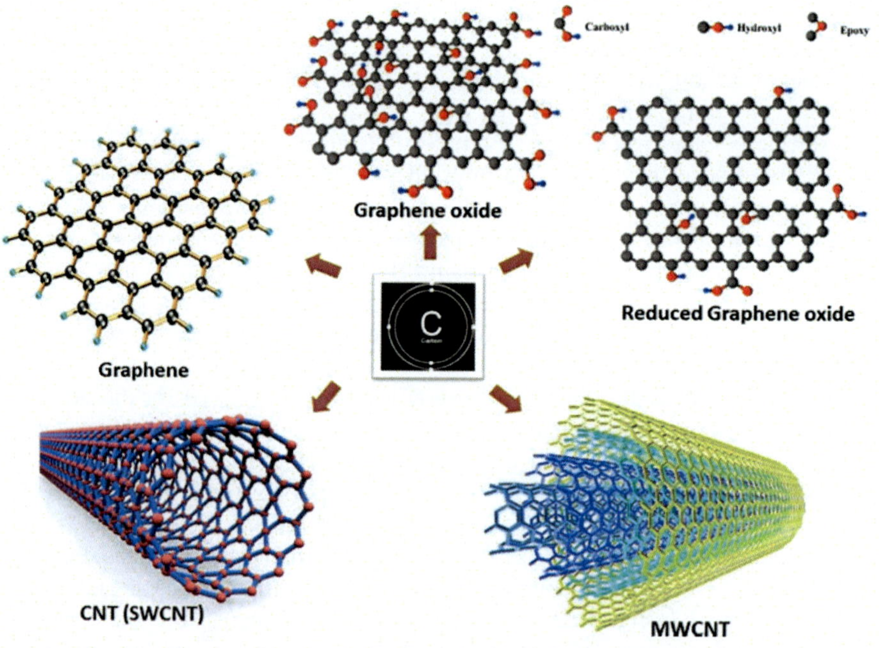

Fig. 4 The basic chemical structures of graphene, GO, rGO, SWCNT and MWCNT

4.1 CNT Based Nanocomposites

The discovery CNTs was considered as one of the finest inventions in nanotechnology, which was first presented by Iijima in 1991. In CNTs, one or more graphene sheets enfolds around themselves to form a cylindrical shape CNT with the length greater than 20 mm and radius less than 100 nm. There are two types of CNTs i.e., (i) multi walled carbon nanotubes (MWCNTs), contain more than one graphene sheet with diameter ranges from 2 to 50 nm subjecting the number of graphene tubes, and (ii) single walled carbon nanotube (SWCNTs), contain only one graphene sheet with a diameter of 1–2 nm [59]. However, CNTs are spreading rapidly in the environment by virtue of their large number of applications. Further, a great improvement in the preparation methods helping the researchers to synthesize large amounts of size-controlled CNTs for variety of applications. Both the classes of CNTs i.e., SWCNTs and MWCNTs possess extraordinary chemical and physical properties, and they were recognized as one of the strongest materials due to the bond between carbon atoms in the sp^2 direction [34]. CNTs have been examined widely over the past two decades for many applications due to their remarkable characteristics.

4.1.1 Adsorption Properties of CNTs

The debate on adsorption capacity of CNTs is a growing subject in terms of experimental and theoretical interest. The extremely porous structure of CNTs provides high specific surface area, light mass density and firm interactions with the pollutant molecules. Therefore, CNTs application for the removal of hazardous organic and heavy metal pollutants from gas and aqueous systems has been studied extensively. Various number of adsorption experiments have been established for the removal of organic compounds, small molecules and heavy metal ions using CNTs [24, 73]. Fundamentally, the adsorption capacity of CNTs mainly depends on the available adsorption sites. CNTs possess four adsorption sites namely grooves, interstitial channels, internal sites and outside the surface of CNTs as shown in Fig. 5 [81]. Considerable experiments have been conducted to examine the occupied sites by adsorbate molecules after adsorption. It is found that the adsorption rate is much faster on external sites than on internal sites under the exact conditions of pressure and temperature. Further, it was also found that the portion of opened and unblocked CNTs significantly influence the overall adsorption capacity. Further, a study reported that solution pH and heavy metals concentration in aqueous media influenced the adsorption capacity of CNTs [90]. In other study, the authors described that the sorption mechanism was controlled by the surface properties, ion exchange capacity and electrochemical potential of CNTs [31].

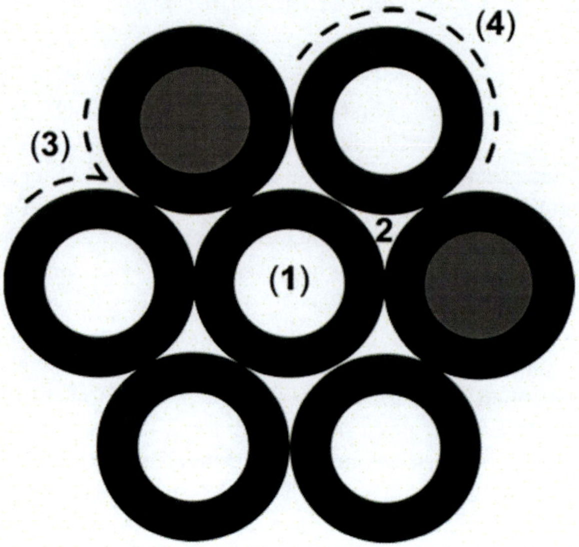

Fig. 5 The possible adsorption sites for heavy metal ions on CNTs. Reprinted with permission from [81] copyrights 2006 American Chemical Society

4.1.2 Adsorption Mechanism of CNTs with Heavy Metals

The remarkable chemical properties of CNTs including large specific surface area (150–1500 m^2/g, which is too higher than fullerenes), chemical stability and the availability of well-developed mesopores makes CNTs as effective adsorbents for the removal of heavy metals. Further, the modification/functionalization of CNTs either covalently or non-covalently using various organic molecules improves the adsorption capacity and selectivity of materials [89]. The mechanism involves in the heavy metals' adsorption by CNTs is very complicated and is attributed by electrostatic attraction, physical adsorption, precipitation and chemical interaction between heavy metals and the surface functional groups of CNTs. Nevertheless, it is commonly believed that the chemical interaction between heavy metals and surface functional groups of CNTs is the major adsorption mechanism [82]. CNTs can also form composites (e.g., Fe_2O_3, ZrO_2) through the coprecipitation method, which are effective for the removal of Cr, Ni, Hg, Pb, Cu and As [68]. But, the adsorption of heavy metal ions with these composites was found to be pH dependent, and hence desorption and recovery of absorbents can be easily accomplished by changing the solution pH.

4.1.3 Functionalized CNT Composites

Functionalization is the key step that enhances the adsorption capacity of CNTs. In spite of the fact that CNTs are significantly better absorbents over other conventional adsorbents, their capacity, selectivity and sensitivity can be greatly altered by introducing some new functional groups to their surface. In other words, the surface of CNTs can be triggered and functionalized by adding additional materials, which creates them as more effective adsorbents for the removal of pollutants [83]. In fact, CNTs may contain different functional groups such as –OH, –C =O, and –COOH depends on the synthesis and purification procedures. The functionalization of CNTs can be accomplished adding catalysts like Ni, Pd and Pt after their oxidation. The functionalized groups can also be removed by heating the CNTs at 2200 °C. Functionalization alters the properties of CNTs, meaning that CNTs before functionalization tends to be hydrophobic and it gets reversed after functionalization. The functionalized CNTs can be used in variety of applications including material science, environmental engineering, and electrical engineering because of their excellent chemical, physical and electrical properties [48]. There are basically two types of functionalization i.e., covalent and non-covalent. In covalent functionalization, the functional groups attached to the skeleton of CNTs covalently by chemical reaction. Where as in non-covalent functionalization, functional groups cover the walls of CNTs.

More recently, ionic liquids (ILs) classified as green solvents and have attracted considerable interest for the functionalization of CNTs owing to their special characteristics [64, 96]. The first report regarding the application of IL for functionalization in nanotechnology was introduced in 2001 [23]. Thereafter, several studies reported the use of ILs as substitutes for organic solvents, with strong acids as a functionalization agent. The benefit of using ILs over organic solvents is that they conduct a non-destructive reaction that retains the properties of CNTs [21]. Besides, deep eutectic solvents (DESs) were also emerged as alternative functionalization agents. In general, DESs made up of two or more compounds, and the resulted DES melting point is lower than either of the individual components. They are produced by mixing a hydrogen bond donor (HBD) with salt. DESs can be prepared from different types of salts (organic and inorganic) and HBDs [9]. There are a number of functionalized and pure CNTs reported for the successful adsorption of heavy metals ions from contaminated waters and the most important methods with the key remarks were described in Table 1.

4.2 Graphene Based Nanocomposites

Graphene is a thin two-dimensional carbon based nanomaterial, which is a basic building block for all graphitic materials including fullerenes, nanotubes or graphite. Like CNTs, graphene also possesses unique physical, chemical, structural and mechanical properties, which makes this material a very tool in many fields. Graphene exists in different forms like pristine graphene, graphene oxide (GO) and reduced GO [74]. Over the past few years, the use of graphene and other graphene based materials in the environmental remediation and water treatment has tremendously increased owing to their high specific surface area, chemical stability and the presence of active functional groups on their surface [50]. Unlike CNTs, graphene necessitates special oxidation processes to introduce hydrophilic groups on its surface. The preparation of GO nanosheets from graphite through the modified Hummer's method brings provide different oxygen-containing functional groups including –COOH, –C=O, and –OH, on GO nanosheets surface. These functional groups essentially increase the heavy metal sorption capacity of materials.

Table 1 Most recent methods reported for the removal of heavy metals from aqueous media using CNTs based composites

S. No	Adsorbent	Adsorbates	Key achievements	References
1	m-MWCNTs-rTl-Cyn	Cr, Fe, Pb, Cu	m-MWCNT-rTl-Cyn was stable for 30 days Removed ≥84% cyanates and >30% of metals Retained >94% of activity after 10 cycles	[72]
2	MWCNT-PEI/PAN	Pb, Cu	232.7 mg/g for Pb and 112.5 mg/g for Cu Chemisorption involved in metals uptake Good reusability after five cycles	[22]
3	MWCNT-PDA	Pb, Cu	Adsorption capacities 318.47 and 350.87 mg/g for Cu(II) and Pb(II) respectively Adsorption process involved typical chemical adsorption and intraparticle diffusion	[106]
4	MWCNTs-selenophosphoryl	Pb, Zn, Cd, Cu, Ni, Co	The material was reusable for many times Material was highly selective for Pb removal Removal affinity Pb>>Cd>Zn, Cu, Ni, Co Pb removal occurred through physisorption	[44]
5	pristine-MWCNTs	Zn	100% of Zn removed through complexation Synthesized membrane is highly reusable Effectively removed other metals ions	[3]

(continued)

Table 1 (continued)

S. No	Adsorbent	Adsorbates	Key achievements	References
6	MWCNTs-5,7-dinitro-8-quinolinol	Zn, Fe, Cu, Pb	Modified-MWCNTs provided relatively higher adsorption capacity Best fitted with pseudo-second order model	[78]
7	MWCNTs-8-hydroxyquinoline	Cu, Pb, Cd, Zn	Best fitted with pseudo second order model Modification of MWCNTs greatly enhanced the adsorption capacities	[79]
8	DTC-MWCNT	Cd, Cu, Zn	Best fitted with pseudo second order model DTC-MWCNT provided high adsorption capacities in order as Cd(II) > Cu(II) > Zn(II) Chemisorption and physisorption involved	[47]
9	pristine-MWCNTs	Ni	Effectively removed Ni(II) from wastewater Efficiency was highly pH dependent Promising and reusable for metals removal	[28]
10	CNT sheets	Pb, Cd, Co Cu, Zn	Efficiency followed Pb > Cd > Co > Zn > Cu Oxidation of CNTs increased efficiency Promising alternatives for metals removal	[91]
11	Chitosan-MWCNts	Cu, Zn, Cd, Ni	Chitosan-MWCNTs provided high efficiency Removal order Cu(II) > Cd(II)≈Zn(II) > Ni(II) Material is highly reusable for many cycles	[80]

(continued)

Table 1 (continued)

S. No	Adsorbent	Adsorbates	Key achievements	References
12	MWCNTs	Cd	Probe sonication well dispersed the CNTs and maximized adsorbent-adsorbate interactions Provided adsorption capacity of 181.8 mg/g	[12]
13	CNTs/NPAA	Cu, Cd	Hydrophobic pores of CNTs adsorbed metals Water pH influenced the removal efficiency	[58]
14	CNTs	Cu, Cd, Zn, Pb, Co, Mn	Removal order Cu > Pb > Co > Zn > Mn Best fitted with Freundlich adsorption model	[90]

4.2.1 Adsorption Mechanism of Graphene with Heavy Metals

The high surface area-to-mass ratio of GO makes this material as ideal candidate for the adsorption of heavy metals from aqueous systems. Recently, a large number of studies confirmed the high sorption capacity of graphene materials for wide range of heavy metals. Actually, multi-layered graphene oxide (MLGO) was observed to possess remarkable sorption capacity for the removal of many heavy metals including Cd (106.3 mg/g), Pb (842 mg/g) and U (97.5 mg/g) [51]. However, the studies concluded that heavy metals adsorption on MLGO varies with the solution pH and ionic strength. In a study, the authors reported that Pb(II) adsorption mechanism was highly pH dependent, wherein the adsorption was controlled by the combination of outer sphere electrostatic attraction and inner sphere covalent bonding. Outer sphere adsorption was dominated at lower pH, and inner sphere adsorption was dominant at higher pH values. In contrast, the adsorption mechanism for Cd and U was found to be independent of solution pH. In case of Cd, the adsorption was dominated by electrostatic outer shell attraction between Cd(II) and MLGO surface at all measured pH values. The U-MLGO systems has showed refined difference within the measured pH range, but overall adsorption was found to be dominated by an inner sphere bond. Overall, heavy metals adsorption on to MLGO varies with the type of metal [87]. In another study, the authors synthesized EDTA-GO adduct for the effective removal of Pb(II) from aqueous solutions, for which they have found the adsorption capacity of 479 ± 46 mg/g, which is significantly more than the GO adsorption capacity of 328 ± 39 mg/g. In this study, the authors concluded that the superior performance of EDTA-GO and improved removal efficiency was preceded by ion-exchange reaction process [45].

4.2.2 Functionalized Graphene Nanocomposites

Due to the tunable surface chemistry, non-corrosive property, high surface area and presence of oxygen-containing active functional groups, graphene materials have been appeared as novel and effective adsorbents with enhanced properties. Functionalization of graphene for the environmental applications is mostly carried out by chemical methods/processes (i.e., chemical oxidation and deposition, electrochemical, sol-gel, microemulsion and hydrothermal methods) [66]. These chemical reactions are more selective and straight forward than other processes, which can integrate functional groups with CBNs to achieve multiple functions [52]. The functional groups present on the surface of graphene materials plays very significant role in the adsorption process. There are a number of functionalized graphene and graphene oxide materials reported for the successful adsorption of heavy metals ions from aqueous systems and the most important methods with the key remarks were described in Table 2.

5 Future Research Perspectives

The two major CBNs namely graphene and CNTs have offered considerable advantages in the environmental remediation due to their impressive physical, chemical and electrical properties. In addition, their application involved in many other fields including electrical engineering, medical and material sciences. Certainly, much progress has been achieved in the last few years regarding the use of CBNs in heavy metals adsorption. Despite their high cost, the application of CBNs as adsorbents could be beneficial in future considering their high adsorption capacities over conventional adsorbents. In addition, several academicians and researchers are diversifying the synthesis and modification of CBNs by innovative processing techniques, which can possibly reduce the cost of CBNs in near future. However, there are still some constrains needs to be addressed before utilizing the CBNs in full-fledged manner for the adsorption of emerging pollutants. Firstly, the strong interactive forces among carbonic nanostructures provide quick aggregation and poor dispersibility of CBNs in aqueous solution and reduce the number of available sites for metal pollutants. Even though, researchers have addressed this problem with surface modification, most of the surface modification methods conducted by conventional procedures utilized large amounts of chemicals and solvent.

Furthermore, the basic graphene and CNTs may contain some degree of toxicity and therefore, the practical application of CBNs in drinking water purification depends on the development of cost-effective and environmentally benign CBNs. Further, the chemistry of real water is very complicated, and it may contain different kinds of pollutants that leads to produce serious environmental damages from their combined toxicity and relative mobility. Research on the removal of combined pollutants by CBNs is still lacking, therefore researchers need to consider the potential co-existing pollutants during the removal of a given pollutant. Also, the development of functionalized CBNs that are capable to remove both organic and metal pollutants simultaneously from water is highly desirable. Considering the effective use of zerovalent iron (ZVI) for the removal of heavy metals and organic pollutants, functionalization of CBNs with ZVI materials could produce effective materials that can remove heavy metal and halogenated organic pollutants combinedly. Although

Table 2 Recent methods reported for the removal of heavy metals from aqueous media using various graphene based composites

S. No.	Adsorbent	Adsorbates	Key achievements	References
1	Graphene-oxide	Pb, Ni, Cr	Cr and Pb were completely removed by GO The adsorption rate was effective until pH 8.0 A multilayer adsorption occurred on sorbent	[7]
2	GO–EDTA	Cu, Pb	GO-EDTA improved anti-microbial properties No cytotoxicity was found for human cells GO-EDTA adsorption capacity 454.6 mg/g (Pb) and 108.7 mg/g for Cu	[16]
3	GO foam	Zn, Fe, Pb, Cd	GOF provided very high surface area 578.4 m^2/g Superior adsorption and good reusability Data fitted with Langmuir isotherm model	[46]
4	CS/GO-SH	Cu, Pb, Cd	CS/GO-SH showed excellent adsorption ability Adsorption followed pseudo second order kinetics Possessed 85% recovery after three cycles of use	[49]

(continued)

Table 2 (continued)

S. No.	Adsorbent	Adsorbates	Key achievements	References
5	PEI-PD/GO	Cu, Cd, Pb, Hg	PEI-PD/GO exhibited improved performance PEI-PD/rGO aerogels are highly recyclable	[25]
6	mGO/SiO_2@coPPy-Th	Cu, Pb, Zn, Cd	Resulted high specific surface area and can be easily separated using magnet Promising material for metals extractio from different sample sources	[40]
7	GO-SF aerogels	Ag	Showed high adsorption capacity over GO A monolayer adsorption involved Langmuir isotherm fitted with adsorption	[98]
8	GO-SA aerogels	Cu, Pb	Found effective for heavy metals removal Pseudo-second-order model described adsorption Monolayer adsorption involved in metal removal	[39]
9	rGO/magnetite/Ag	Cd, Ni, Zn, Co, Pb, Cu	Provided high adsorption over pure GO Removal affinity Cu, Zn > Ni > Co > Pb, Cd Easy separation and high reusability	[69]
10	MnO_2/rGO	Pb, Cd, Ag, Cu, Zn	Provided high adsorption capacity of 356.37 mg/g for Pb removal High adsorption capacity after several cycles	[105]

(continued)

Table 2 (continued)

S. No.	Adsorbent	Adsorbates	Key achievements	References
11	GO-SiO_2	Pb, As	Use of ILs helped to reduce π–π stacking and Van Der waals interactions among GO particles High adsorption was found for modified GO Synergistic effects between GO and SiO_2 improved the adsorption capacity	[11]
12	GO/CMC	Ag, Cu, Pb	CMC addition prevented over stacking of GO Adsorption followed pseudo-second-order kinetics	[53]
13	GO-CA	Pb, Hg, Cd	Functionalized GO had high adsorption capacity Adsorption followed pseudo-second-order kinetics Provided good reusable capacity	[6]
14	HP-β-CD-GO	Pb, CuPb, Cu	Adsorption capacity of 50.39 and 17.91 mg/g obtained for Pb and Cu respectively More than 85% reusability after three cycles	[93]

functionalized CBNs have been proved to be potential adsorbents, their environmental implications like fate, transportation and ecotoxicity must be examined to minimize the potential adverse effects. Most importantly, functionalized CBNs need to be tested in actual wastewater rather than their use in simulated water that only contain selected pollutants. Finally, the recycling and reusability of the CBNs needs to be addressed effectively.

6 Conclusions

It is very familiar that the presence of heavy metals in water bodies create severe toxicity to all living organisms as they tend to bioaccumulate in plants, human and aquatic life. Unlike the organic pollutants, heavy metals are persistent and cannot be degraded. As a result, researchers proposed adsorption as an effective technique for the removal of heavy metals from aqueous systems. In recent years, various CBNs have emerged as superior sorbents for heavy metals elimination owing to their remarkable physiochemical properties. It is therefore, this chapter highlighted the applications of various CBNs including CNTs and graphene based materials for the removal of heavy metal pollutants including Hg, Cr, Pb, Cd, Ni, Cu and As. At first, the sources and entry of heavy metals into water bodies followed by their toxicity to plants and living beings is presented clearly. Then, recent applications of CNTs and graphene nanocomposites for heavy metals adsorption from aqueous solutions is described following a discussion on their functionalization advantages. In all reported studies, CBNs have showed excellent adsorption capacity for the removal of selected heavy metal ions, which is mainly due to large specific surface area, strong van der Waal interactions between metal ions and CBNs, and the availability of well-defined adsorption sites. Further, the sorption mechanism was appeared to proceed through the chemical interaction between metal ions and the surface functional groups of CBNs. In addition, process parameters such as solution pH, surface acidity and temperature were influenced the performance of CBN sorbents. Further, some studies have well described the possible reusability of spent CBNs for the desorption of heavy metal ions. Moreover, future research recommendations to develop a cost-effective and environmentally benign CBNs are presented.

References

1. Abdullah N, Yusof N, Lau WJ, Jaafar J, Ismail AF (2019) Recent trends of heavy metal removal from water/wastewater by membrane technologies. J Indust Eng Chem 76:17–38
2. Alexander JT, Faisal IH, Al-aboud TM (2012) Chemical coagulation-based processes for trace organic contaminant removal: Current state and future potential. J Environ Manage 111:195–207
3. Ali S, Shah IA, Ahmad A, Javed N, Haiou H (2019) Ar/O_2 plasma treatment of carbon nanotube membranes for enhanced removal of zinc from water and wastewater: a dynamic sorption filtration process. Sci Total Environ 655:1270–1278
4. Amari T, Ghnaya T, Abdelly C (2017) Nickel, cadmium and lead phytotoxicity and potential of halophytic plants in heavy metal extraction. S Afr J Bot 111:99–110
5. Amde M, Liu J, Tan ZQ, Bekana D (2017) Transformation and bioavailability of metal oxide nanoparticles in aquatic and terrestrial environments. Review. Environ Pollut 230:250–267
6. Arshad F, Selvaraj M, Zain J, Fawzi B, Abu Haija M (2019) Polyethylenimine modified graphene oxide hydrogel composite as an efficient adsorbent for heavy metal ions. Sep Purif Technol 209:870–880
7. Arthi G, Krishnan R, Thangavel S, Venugopal G, Kim SJ (2015) Removal of heavy metal ions from pharma-effluents using graphene-oxide nanosorbents and study of their adsorption kinetics. J Indust Eng Chem 30:14–19

8. Aziz HA, Adlan MN, Ariffin KS (2008) Heavy metals (Cd, Pb, Zn, Ni, Cu and Cr(III)) removal from water in Malaysia: Post treatment by high quality limestone. Bioresour Technol 99:1578–1583
9. Bagh G, Shahbaz FS, Mjalli K, Hashim MA, AlNashef IM (2015) Zinc (II) chloride-based deep eutectic solvents for application as electrolytes: preparation and characterization. J Mol Liq 204:76–83
10. Barakat MA, Schmidt E (2010) Polymer-enhanced ultrafiltration process for heavy metals removal from industrial wastewater. Desalination 256:90–93
11. Barik B, Kumar A, Nayak PS, Satish Achary LK, Lipeeka R, Priyabrat D (2020) Ionic liquid assisted mesoporous silica-graphene oxide nanocomposite synthesis and its application for removal of heavy metal ions from water. Mater Chem Phys 239:122028
12. Bhanjana G, Dilbaghi N, Kim KH, Sandeep K (2017) Carbon nanotubes as sorbent material for removal of cadmium. J Mol Liq 242:966–970
13. Bolan S, Kunhikrishnan A, Seshadri B, Choppala G, Naidu R, Bolan NS, Ok YS, Zhang M, Li CG, Li F, Noller B, Kirkham MB (2017) Sources, distribution, bioavailability, toxicity, and risk assessment of heavy metal(loid)s in complementary medicines. Environ Int 108:103–118
14. Bora BK, Ramos-Crawford AL, Sikorskii A, Michael JB, Didier M, Dieudonne MN, Abdon Mukalay WM, Daniel OL, Desire TK (2019) Concurrent exposure to heavy metals and cognition in school-age children in Congo-Kinshasa: a complex overdue research agenda. Brain Res Bull 145:81–86
15. Bu F, Gao B, Shen X, Wenyu W, Qinyan Y (2019) The combination of coagulation and ozonation as a pre-treatment of ultrafiltration in water treatment. Chemosphere 231:349–356
16. Carpio IEM, Mangadlao JD, Nguyen HN, Rigoberto CA, Debora FR (2014) Graphene oxide functionalized with ethylenediamine triacetic acid for heavy metal adsorption and anti-microbial application. Carbon 77:289–301
17. Castro-Gonzalez MI, Mendez-Armenta M (2008) Heavy metals: implications associated to fish consumption. Environ Toxicol Phar 26:263–271
18. Choi KH, Jeoung TY (2002) Removal of zinc ions in wastewater by electrodialysis. Korean J Chem Eng 19:107–113
19. Chowdhury S, Jafar Mazumder MA, Al-Attas O, Hussain T (2016) Heavy metals in drinking water: occurrences, implications, and future needs in developing countries. Sci Total Environ 569–570:476–488
20. Daabrowski A, Hubicki Z, Podkoscielny P, Robens E (2004) Selective removal of the heavy metal ions from waters and industrial wastewaters by ion-exchange method. Chemosphere 56:91–106
21. Deng MJ, Chen PY, Leong TI, Sun WI, Jeng Kuei C, Wen TT (2008) Dicyanamide anion based ionic liquids for electrodeposition of metals. Electrochem Commun 10:213–216
22. Deng S, Liu X, Liao J, Hui L, Fang L (2019) PEI modified multiwalled carbon nanotube as a novel additive in PAN nanofiber membrane for enhanced removal of heavy metal ions. Chem Eng J 375:122086
23. Deshmukh RR, Rajagopal R, Srinivasan KV (2001) Ultrasound promoted C–C bond formation: heck reaction at ambient conditions in room temperature ionic liquids. Chem Commun 17:1544–1545
24. Devi P, Rajput P, Thakur A, Kim KH, Kumar Praveen (2019) Recent advances in carbon quantum dot-based sensing of heavy metals in water. TrAC Trend Anal Chem 114:171–195
25. Dong Z, Zhang F, Wang D, Liu X, Jian J (2015) Polydopamine-mediated surface functionalization of graphene oxide for heavy metal ions removal. J Solid State Chem 224:88–93
26. Dongsheng Z, Wenqiang G, Guozhang C, Luo S, Jiao W, Liu Y (2019) Removal of heavy metal lead(II) using nanoscale zero-valent iron with different preservation methods. Adv Powder Technol 30:581–589
27. Edelstein M, Ben-Hur M (2018) Heavy metals and metalloids: Sources, risks and strategies to reduce their accumulation in horticultural crops. Sci Hortic-Amsterdam 234:431–444
28. Elsehly EM, Chechenin NG, Makunin AV, Motaweh HA, Vorobyeva EA, Bukunov KA, Leksina EG, Priselkova AB (2016) Characterization of functionalized multiwalled carbon

nanotubes and application as an effective filter for heavy metal removal from aqueous solutions. Chin J Chem Eng 24:1695–1702
29. Fiyadh SS, Al-Saadi MA, Jaafar WZ, Mohamed KO, Sabah SF, Nuruol SM, Lai SH, Ahmed ES (2019) Review on heavy metal adsorption processes by carbon nanotubes. J Clean Prod 230:783–793
30. Gao W, Liang H, Ma J, Mei H, Zhong LC, Zheng SH, Gui-bai L (2011) Membrane fouling control in ultrafiltration technology for drinking water production: a review. Desalination 272:1–8
31. Gao Z, Bandosz TJ, Zhao Z, Mei H, Qiu J (2009) Investigation of factors affecting adsorption of transition metals on oxidized carbon nanotubes. J Hazard Mater 167:357–365
32. Greenlee LF, Lawler DF, Freeman BD, Benoit M, Philippe M (2009) Reverse osmosis desalination: water sources, technology, and today's challenges. Water Res 43:2317–2348
33. Harsha Vardhan K, Senthil Kumar P, Panda RC (2019) A review on heavy metal pollution, toxicity and remedial measures: current trends and future perspectives. J Mol Liq 290:111197
34. Herrero-Latorre C, Barciela-García J, García-Martín S, Pena Crecente RM (2018) Graphene and carbon nanotubes as solid phase extraction sorbents for the speciation of chromium: a review. Anal Chim Acta 1002:1–17
35. Hossain F (2019) Natural mechanism to console global water, energy, and climate change crisis. Sustain Energy Techn 35:347–353
36. Huang X, Wan Y, Shi B, Jian S (2020) Effects of powdered activated carbon on the coagulation-flocculation process in humic acid and humic acid-kaolin water treatment. Chemosphere 238, Article 124637
37. Huang Y, Feng X (2019) Polymer-enhanced ultrafiltration: Fundamentals, applications and recent developments. J Membrane Sci 586:53–83
38. Ihsanullah AA, Al-Amer AM, Tahar L, Mohammed JA, Mustafa SN, Majeda K, Muataz AA (2016) Heavy metal removal from aqueous solution by advanced carbon nanotubes: critical review of adsorption applications. Sep Purif Technol 157:141–161
39. Jiao C, Xiong J, Tao J, Xu S, Zhang D, Lin H, Chen Y (2016) Sodium alginate/graphene oxide aerogel with enhanced strength-toughness and its heavy metal adsorption study. Int J Biol Macromol 83:133–141
40. Molaei K, Bagheri H, Asgharinezhad AA, Ebrahimzade H, Shamsipur M (2017) SiO_2-coated magnetic graphene oxide modified with polypyrrole– polythiophene: a novel and efficient nanocomposite for solid phase extraction of trace amounts of heavy metals. Talanta 167:607–616
41. Kashif Uddin M (2017) A review on the adsorption of heavy metals by clay minerals, with special focus on the past decade. Chem Eng J 308:438–462
42. Kevin LG, John FS (1988) Use of electrodialysis to remove heavy metals from water. Sep Sci Technol 23:2231–2267
43. Kim JJ, Kim YS, Kumar Vijay (2019) Heavy metal toxicity: an update of chelating therapeutic strategies. J Trace Elem Med Bio 54:226–231
44. Konczyk J, Zarska S, Ciesielski W (2019) Adsorptive removal of Pb(II) ions from aqueous solutions by multi-walled carbon nanotubes functionalized by selenophosphoryl groups: kinetic, mechanism, and thermodynamic studies. Coll Surf A 575:271–282
45. Kumar M, Chung JS, Hur SH (2019) Graphene composites for lead ions removal from aqueous solutions. Appl Sci 9:2925
46. Lei Y, Chen F, Luo Y, Zhang L (2014) Synthesis of three-dimensional graphene oxide foam for the removal of heavy metal ions. Chem Phys Lett 593:122–127
47. Li Q, Yu J, Zhou F, Jiang X (2015) Synthesis and characterization of dithiocarbamate carbon nanotubes for the removal of heavy metal ions from aqueous solutions. Coll Surf A 482:306–314
48. Li WK, Shi YP (2019) Recent advances and applications of carbon nanotubes based composites in magnetic solid-phase extraction. TrAC Trend Anal Chem 118:652–665
49. Li X, Zhou H, Wu W, Shudan W, Yan X, Yafei K (2015) Studies of heavy metal ion adsorption on chitosan/sulfydryl functionalized graphene oxide composites. J Coll Interf Sci 448:389–397

50. Lim JY, Mubarak NM, Abdullah EC, Sabzoi N, Mohammad K, Nizamuddin S (2018) Recent trends in the synthesis of graphene and graphene oxide based nanomaterials for removal of heavy metals—a review. J Indust Eng Chem 66:29–44
51. Liu X, Ma R, Wang X (2019) Graphene oxide-based materials for efficient removal of heavy metal ions from aqueous solution: a review. Environ Pollut 252:62–73
52. Lotfi RMS, Rezaei M, Babaie A (2019) Investigation of the effect of graphene oxide functionalization on the physical, mechanical and shape memory properties of polyurethane/reduced graphene oxide nanocomposites. Diam Relat Mater 95:195–205
53. Luo J, Fan C, Xiao Z, Tianshu S, Xiaodong Z (2019) Novel graphene oxide/carboxymethyl chitosan aerogels via vacuum-assisted self-assembly for heavy metal adsorption capacity. Coll Surf A 578:123584
54. Madhavi V, Prasad TNVKV, Vijaya Bhaskar Reddy A, Reddy BR, Madhavi G (2013) Application of phytogenic zerovalent iron nanoparticles in the adsorption of hexavalent chromium. Spectrochim Acta A 116:17–25
55. Madhavi V, Prasad TNVKV, Reddy BR, Vijaya Bhaskar Reddy A, Madhavi G (2014) Conjunctive effect of CMC-zero valent iron nanoparticles and FYM in the remediation of chromium-contaminated soils. Appl Nanosci 4:477–484
56. Madhavi V, Reddy AVB, Reddy KG, Madhavi G (2012) A simple method for the determination of efficiency of stabilized Fe° nanoparticles for detoxification of chromium (VI) in water. J Chem Pharma Res 4:1539–1545
57. Madhavi V, Reddy AVB, Reddy KG, Madhavi G, Prasad TNVKV (2013) An overview on research trends in remediation of chromium. Res J Recent Sci 2:71–83
58. Mahdi M, Hamed A, Vahid K (2017) Synthesis of highly ordered carbon nanotubes/nanoporous anodic alumina composite membrane and potential application in heavy metal ions removal from industrial wastewater. Mater Today 4:4906–4911
59. Mallakpour S, Khadem E (2016) Carbon nanotube–metal oxide nanocomposites: Fabrication, properties and applications. Chem Eng J 302:344–367
60. Matin A, Rahman F, Shafi HZ, Zubair SM (2019) Scaling of reverse osmosis membranes used in water desalination: phenomena, impact, and control; future directions. Desalination 455:135–157
61. Min KJ, Choi SY, Jang D, Lee J, Park KY (2019) Separation of metals from electroplating wastewater using electrodialysis. Energ Source Part A 41:2471–2480
62. Moller A, Grahn A, Welander U (2003) Precipitation of heavy metals from landfill leachates by microbially-produced sulphide. Environ Technol 25:69–77
63. Moura CA, Bertuol DA, Ferreira CA, Franco DRA (2012) Study of chromium removal by the electrodialysis of tannery and metal-finishing effluents. Int J Chem Eng Article ID 179312:1–7
64. Mustahil NA, Baharuddin SH, Abdullah AA, Reddy AVB, Mutalib AMI, Moniruzzaman M (2019) Synthesis, characterization, ecotoxicity and biodegradability evaluations of novel biocompatible surface active lauroyl sarcosinate ionic liquids. Chemosphere 229:349–357
65. Naim R, Kisay L, Park J, Qaisar M, Zulfiqar AB, Noshin M, Jamil K (2010) Precipitation chelation of cyanide complexes in electroplating industry wastewater. Int J Environ Res 4:735–740
66. Nandanapalli KR, Devika M, Lee S (2019) Functionalization of graphene layers and advancements in device applications. Carbon 152:954–985
67. Nasir AM, Goh PS, Abdullah MS, CheerNg Be, Ismail AF (2019) Adsorptive nanocomposite membranes for heavy metal remediation: recent progresses and challenges. Chemosphere 232:96–112
68. Ntim SA, Mitra S (2012) Adsorption of arsenic on multiwall carbon nanotube-zirconia nanohybrid for potential drinking water purification. J Colloid Interface Sci 375:154–159
69. Park CM, Wang D, Han J, Jiyong H, Chunming S (2019) Evaluation of the colloidal stability and adsorption performance of reduced graphene oxide–elemental silver/magnetite nanohybrids for selected toxic heavy metals in aqueous solutions. Appl Surf Sci 471:8–17
70. Perez-Gonzalez A, Urtiaga AM, Ibanez R, Ortiz I (2012) State of the art and review on the treatment technologies of water reverse osmosis concentrates. Water Res 46:267–283

71. Rai PK, Lee SS, Zhang M, Tsang YF, Kim KH (2019) Heavy metals in food crops: health risks, fate, mechanisms, and management. Environ Int 125:365–385
72. Ranjan B, Pillai S, Permaul K, Singh S (2019) Simultaneous removal of heavy metals and cyanate in a wastewater sample using immobilized cyanate hydratase on magnetic-multiwall carbon nanotubes. J Hazard Mater 363:73–80
73. Rasheed T, Adeel M, Nabeel F, Muhammad B, Hafiz MNI (2019) TiO_2/SiO_2 decorated carbon nanostructured materials as a multifunctional platform for emerging pollutants removal. Sci Total Environ 688:299–311
74. Reddy AVB, Moniruzzaman M, Reddy YVM, Madhavi G (2019) Chapter 18—Graphene-based nanomaterials for the removal of pharmaceuticals in drinking water sources. In: Graphene-based nanotechnologies for energy and environmental applications, pp 329–358
75. Reddy AVB, Yusop Z, Jaafar J, Reddy YVM, Aris AB, Zaiton AM, Juhaizah T, Madhavi G (2016) Recent progress on Fe-based nanoparticles: synthesis, properties, characterization and environmental applications. J Environ Chem Eng 4:3537–3553
76. Reddy AVB, Yusop Z, Jaafar J, Jamil NH, Majid ZA, Aris AB (2018) Development and validation of capillary electrophoresis method for simultaneous determination of six pharmaceuticals in different food samples combining on-line and off-line sample enrichment techniques. Food Anal Methods 11:533–545
77. Ross DAN, Guzman HM, Potvin C, Vincent JH (2017) A review of toxic metal contamination in marine turtle tissues and its implications for human health. Reg Studies Mar Sci 15:1–9
78. Salam ETA, Abou El-Nour KM, Awad AA, Orabi AS (2020) Carbon nanotubes modified with 5,7-dinitro-8-quinolinol as potentially applicable tool for efficient removal of industrial wastewater pollutants. Arab J Chem 13:109–119
79. Salam MA, Al-Zhrani G, Kosa SA (2014) Removal of heavy metal ions from aqueous solution by multi-walled carbon nanotubes modified with 8-hydroxyquinoline: Kinetic study. J Indust Eng Chem 20:572–580
80. Salam MA, Makki MSI, Abdelaal MYA (2011) Preparation and characterization of multi-walled carbon nanotubes/chitosan nanocomposite and its application for the removal of heavy metals from aqueous solution. J Alloy Compd 509:2582–2587
81. Sandeep A, Mota JPB, Rostam-Abadi M, Mark JR (2006) Theoretical and experimental investigation of morphology and temperature effects on adsorption of organic vapors in single-walled carbon nanotubes. J Chem Phys 110:7640–7647
82. Santhosh C, Velmurugan V, Jacob G, Soon Kwan J, Andrews Nirmala G, Amit B (2016) Role of nanomaterials in water treatment applications: a review. Chem Eng J 306:1116–1137
83. Sarkar B, Mandal S, Tsang YF, Kumar Pawan, Kim KH, Yong SO (2018) Designer carbon nanotubes for contaminant removal in water and wastewater: a critical review. Sci Total Environ 612:561–581
84. Sfakianakis DG, Renieri E, Kentouri M, Tsatsakis AM (2015) Effect of heavy metals on fish larvae deformities: a review. Environ Res 137:246–255
85. Shahid M, Dumat C, Khalid S, Eva S, Tiantian X, Nabeel KN (2017) Foliar heavy metal uptake, toxicity and detoxification in plants: a comparison of foliar and root metal uptake. J Hazard Mater 325:36–58
86. Sharma RK, Agrawal M, Marshall F (2007) Heavy metal contamination of soil and vegetables in suburban areas of Varanasi, India. Ecotox Environ Safe 66:258–266
87. Showalter AR, Duster TA, Szymanowski JES, Chongzheng N, Jeremy BF, Bruce AB (2016) Sorption mechanisms of metals to graphene oxide. J Phys: Conf Ser 712:012094
88. Shruti P, Rajasekhar B (2013) Toxicological evaluation of microcystins in aquatic fish species: current knowledge and future directions. Aquatic Toxicol 142–143:1–16
89. Sitko R, Zawisza B, Malicka E (2012) Modification of carbon nanotubes for preconcentration, separation and determination of trace-metal ions. TrAC Trend Anal Chem 37:22–31
90. Stafiej A, Krystyna P (2007) Adsorption of heavy metal ions with carbon nanotubes. Sep Purif Technol 58:49–52
91. Tofighy MA, Toraj M (2011) Adsorption of divalent heavy metal ions from water using carbon nanotube sheets. J Hazard Mater 185:140–147

92. Tongwen X (2002) Electrodialysis processes with bipolar membranes (EDBM) in environmental protection—a review. Resour Conserv Recy 37:1–22
93. Usman TM, Xintai S, Mengqi Z, LiaoY WuR, Chen D (2019) Preparation of hydroxypropyl-cyclodextrin-graphene/Fe_3O_4 and its adsorption properties for heavy metals. Surf Interf 16:43–49
94. Uwizeyimana H, Wang M, Chen W, Kifayatullah K (2017) The eco-toxic effects of pesticide and heavy metal mixtures towards earthworms in soil. Environ Toxicol Phar 55:20–29
95. Vijaya Bhaskar Reddy A, Madhavi V, Reddy KG, Madhavi G (2013) Remediation of chlorpyrifos-contaminated soils by laboratory-synthesized zero-valent nano iron particles: Effect of pH and aluminium salts. J Chem. Article ID 521045:1–7
96. Vijaya Bhaskar Reddy A, Moniruzzaman M, Goto M (2019) Ionic liquids for pretreatment of biomass. 3rd edn. Comprehensive biotechnology, vol 2, pp 190–198
97. Vijaya Bhaskar Reddy A, Yusop Z, Jaafar J, Azmi BA, Zaiton AM (2017) Simulation of a conventional water treatment plant for the minimization of new emerging pollutants in drinking water sources: process optimization using response surface methodology. RSC Adv 7:11550–11560
98. Wang S, Ning H, Hu N, Kaiyan H, Shayuan W, Xiaopeng W, Liangke W, Jie L, Alamusi SA (2019) Preparation and characterization of graphene oxide/silk fibroin hybrid aerogel for dye and heavy metal adsorption. Compos Part B-Eng 163:716–722
99. Wei H, Gao B, Ren J et al (2018) Coagulation/flocculation in dewatering of sludge: a review. Water Res 143:608–631
100. Yadav VB, Gadi R, Kalra S (2019) Clay based nanocomposites for removal of heavy metals from water: a review. J Environ Manage 232:803–817
101. Yang X, Wan Y, Zheng Y, Feng H, Zebin Y, Jun H, Hailong W, Yong SO, Yinshan J, Bin G (2019) Surface functional groups of carbon-based adsorbents and their roles in the removal of heavy metals from aqueous solutions: a critical review. Chem Eng J 366:608–621
102. Yaqub M, Lee SH (2019) Heavy metals removal from aqueous solution through micellar enhanced ultrafiltration: a review. Environ Eng Res 24:363–375
103. Ye MY, Li GJ, Yan PF, Jie R, Li Z, Dajian H, Shuiyu S, Shaosong H, Yujian Z (2017) Removal of metals from lead-zinc mine tailings using bioleaching and followed by sulfide precipitation. Chemosphere 185:1189–1196
104. Yu F, Yang C, Zhu Z, Xueting B, Jie M (2019) Adsorption behavior of organic pollutants and metals on micro/nano plastics in the aquatic environment. Sci Total Environ 694:133643
105. Zeng T, Yu Y, Li Z, Jiangtao Z, Zeyuan K, Yang J, Yehe W, Anjun W, Chang P (2019) 3D MnO_2 nanotubes@ reduced graphene oxide hydrogel as reusable adsorbent for the removal of heavy metal ions. Mater Chem Phys 231, 105-108
106. Zhan W, Gao L, Fu X, Sajid HS, Gang S, Xiaoping Y (2019) Green synthesis of amino-functionalized carbon nanotube-graphene hybrid aerogels for high performance heavy metal ions removal. Appl Surf Sci 467–468:1122–1133
107. Zhao Y, Xu M, Liu Q, Wang Z, Zhao L, Chen Y (2018) Study of heavy metal pollution, ecological risk and source apportionment in the surface water and sediments of the Jiangsu coastal region, China: a case study of the Sheyang Estuary. Mar Pollut Bull 137:601–609
108. Zhu Y, Fan W, Zhou T, Li X (2019) Removal of chelated heavy metals from aqueous solution: a review of current methods and mechanisms. Sci Total Environ 678:253–266

Removal of Air Pollutants Using Graphene Nanocomposite

Sapna Nehra, Rekha Sharma, and Dinesh Kumar

Abstract Currently, environmental pollution becomes a global issue because of rapid industrial and socioeconomic development in developing countries. The quality of air is determined by many factors like temperature, humidity, and the concentration of the pollutants. These factors affect the quality of the air and continuously contaminate the fresh air. Wastewater treatment is also an urgent need to regulate the air pollution present in the environment. In the present studies, graphene, composite, nanofibers, and adsorbents are trending for remediation of water pollutants found in water. At the same time, researchers tried to control air pollution by using the same materials. In this chapter, we mainly focus on the air pollution and their respective pollutants as, carbon dioxide (CO_2), nitrogen oxide (NO_x), sulfur dioxide SO_2, particulate matter (PM2.5 and PM10), lead, and the volatile organic compounds (VOCs). These are the main constituents of air pollution. Several filters and ion-based composites, hybrids functionalization, and synthesis methodologies are using toward keeping the indoor and outdoor air quality.

Keywords Particulate matter · Electrospinning · Coating method · Graphene · Nanocomposite

S. Nehra
Department of Chemistry, Dr. K.N. Modi University, Newai, Rajasthan 304022, India
e-mail: nehrasapna111@gmail.com

R. Sharma
Department of Chemistry, Banasthali Vidyapith, Banasthali, Rajasthan 304022, India
e-mail: sharma20rekha@gmail.com

D. Kumar (✉)
School of Chemical Sciences, Central University of Gujarat, Gandhinagar 382030, India
e-mail: dinesh.kumar@cug.ac.in

M. Jawaid et al. (eds.), *Environmental Remediation Through Carbon Based Nano Composites*, Green Energy and Technology,
https://doi.org/10.1007/978-981-15-6699-8_13

Abbreviations

ACFs	Activated carbon fiber
rGO	Reduced graphene oxide
GF-ASS	Graphite furnace atomic absorption spectroscopy
GQDs	Graphene quantum dots
MWCNT	Multiwalled carbon nanotube
PMS	Peroxymonosulfate
HM	Hydrated manganese oxide
PS	Polystyrene
PAN	Polyacrylonitrile
IMA-rGO	Ion-mediated assembled reduced graphene oxide
MSp@SiO_2NH_2	3-aminopropyltrimethoxysilane functionalized magnetic sporopollenin
HEPA	High-efficiency particulate air filters
SCR	Selective catalytic reduction
MDEA	Methyl diethanolamine
SOA	Secondary organic aerosol
SPR	Surface plasmon resonance
ACI	Activated carbon injection
WFGD	Wet flue gas desulfurization
MDEA	Methyl diethanolamine
BGCs	Bismuth oxybromide and graphene nanocomposite
NBOC/GQDs	N-doped $Bi_2O_2CO_3$/graphene quantum dots composite.

1 Introduction

Air pollution becomes a very tough problem because of the development of industry and economy in both developed and developing countries. Many air pollutants are emitted from the coal industries and the burning of fuels, which has been causing severe threats to human health and the present ecosystem. Major air pollutants, including nitric oxide (NO_x), sulfur dioxide (SO_2), lead Pb(II), particulate matters (PMs), carbon dioxide and monoxide contaminate the fresh atmospheric air shown in Fig. 1. Particularly SO_2 and NO_x are the chief sources of formation of photochemical [60, 83]. These specific gases are damaging to human health and cause dangerous diseases like lung cancers, leukemia, and so on. Rising in the level of noxious pollutants in the last few decades speedily deteriorates the quality of fresh air. Outdoor polluted air enters the house and pollutes the indoor air quality. Therefore, nowadays efficient high-efficiency particulate air filters (HEPA) and activated carbons are using

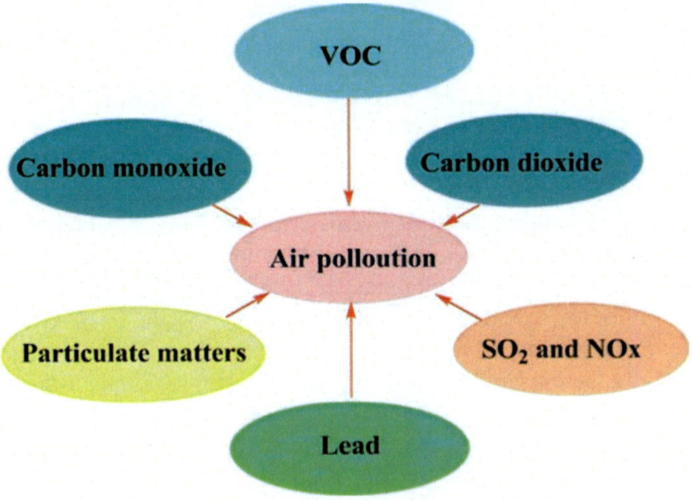

Fig. 1 Various sources to cause air pollution

as indoor air filters to clean the air and keep safe from the dangerous health-damaging pollutants [80].

There are many techniques which are involved to target the toxic air pollutant like NO_x, SO_2 like selective catalytic reduction (SCR) [31, 76] activated carbon injection (ACI) and wet flue gas desulfurization (WFGD) [54, 58, 82]. Although sorting treatment strategy could reach large removal of NO_x, and SO_2, these techniques still have the shortcomings of exorbitant cost in the building as well as high operation, complexity, occupational area, and low stability. But the conventional absorption or adsorption techniques are one of the main research directions in air pollution control.

1.1 CO_2 Mitigation and Air Treatment

Continuously, CO_2 emissions are increasing day by day because of the combustion of fossil fuels, which leads to global warming additional environmental problems. Recent reports say that today, the concentration of CO_2 in the atmosphere has jumped from 270 to 390 ppm after the industrial revolution [30]. In the past few years, CO_2 has become a primary anthropogenic greenhouse gas emitted from the burning of fossil fuels [6, 64, 65]. Other acid gases, including carbon dioxide and hydrogen sulfide, are found in natural and industrial gases [5, 9, 14, 49, 59]. Therefore, in front of researchers and scientists stayed a challenge to deal with the problem of global warming and to resolve the threat of global warming, we need to minimize the concentration of CO_2 in the atmosphere by an efficient method. In the literature, several studies have made a tremendous effort in the removal of CO_2 via removal and capturing as well.

Alghamdi et al., 2018 synthesized the nitrogen inserted GO by utilizing nitrogen that comprises polymers like polypyrrole, polyaniline, and copolymer, i.e., polypyrrole–polyaniline mixed with acids like H_2SO_4, HCl, and C_6H_5–SO_3–K, which are stimulated by adding diverse amount of KOH further carbonized at 650 °C. The resulting N-GOs were proved via passing several techniques like XRD, TEM, XPS, SEM, BET, and TGA-DSC. According to BET analysis, the porosity of the N-GOs prepared in different ratios with KOH was determined. The porosity was found to be from 1 to 3.5 nm and 50–200 nm with a ratio of 1:4 and 1:2, respectively. X-ray diffraction analysis determined the development of layered like the structure of involved graphene in synthesis. Among various-doped copolymers, the C_6H_5–SO_3–K-doped polypyrrole displayed higher surface area of 2870 m^2g^{-1}. The nitrogen graphene composite shows the outstanding CO_2 capture with varied dopant as PPy/Ar-1 and PPy/Ar-2. The 1:2 dopant ratios of polymer and KOH, in N-doped known as PPy/Ar-1, displayed the porosity in between the range from 50 to 150 nm. Alternatively, 1400 m^2g^{-1} surface area was obtained and adsorbed the 1.3 mmol g^{-1} of CO_2. Same the 1:4, a ratio of polymer and KOH renamed as PPy/Ar-2, found the 1–3.5 nm porosity, and approximately 3000 m^2g^{-1} surface area shows the 3 mmol g^{-1} adsorption of CO_2. Out of all prepared, N-GOs, the N-GOs gained from introducing the polymer PPy doped with acid C_6H_5–SO_3–K displayed the better adsorption behavior. The present study confirmed the facile synthesis of N-GOs could be possible with other polymers, which bought from the recyclable material. Therefore, it can be useful for attempting both environmental issues recycling of polymeric waste and air pollution [2].

Irani et al., 2018 employed a novel approach by amine-modified rGO/MDEA nanofluid to enhance the CO_2 adsorption. The solvothermal method was processed to synthesize the NH_2-rGO/MDEA and entirely analyzed by XRD, XPS, SEM, FTIR, and EDX to examine the structure of NH_2-rGO/MDEA. For CO_2, 16.2% adsorption was reported. Both temperature and partial pressure showed the reverse direct relationship with temperature and partial pressure. Here, the mixing of 0.1 wt% GO to the same solution enhances 9.6% absorption capacity [24].

Cao et al., 2015 synthesized the UiO-66/GO for the adsorption of CO_2, and the adsorption was estimated through the static volumetric method. To check the reversibility of the CO_2, adsorption–desorption cycle was conducted. Several techniques were used to characterize the adsorbents as follows XRD, TGA, SEM, and FTIR. The BET results confirmed that the UiO-66/GO-5 composite has a higher surface area than the bare composite UiO-66. Therefore, uptake of CO_2 UiO-66/GO-5 was 3.37 mmol g^{-1} at 298 K, and 1 bar pressure means 48% more than the UiO-66 sample and determined by the micropore volume and the BET surface area of composite. The adsorption uptake of CO_2 is also greater than the other conventional activated carbons and the zeolites. Six adsorption–desorption cycles were well performed without degradation of composite prove the stability of the UiO-66/GO-5 composite. In addition, UiO-66/GO-5 displays outstanding regeneration constancy and recyclability. So, this study reveals that the composite has a good potential to absorb the CO_2 in from the environment and reduce the surrounding global warming [8].

Sui et al., 2015 saw the adsorption capacity of CO_2 by using the hydrothermally reduced graphene oxide (HRGO) and obtained 2.4 mmol g^{-1} at 1.0 bar and 273 K temperature. Further, the sample was examined through different tools like XRD, BET, XPS, Raman, and FTIR spectroscopy. During synthesis, the hydrothermal method was applied to disperse the graphene oxide at different temperatures. The prepared HRGO possesses a 3D porous network but also shows maximum surface area with large pore volume. The surface functionalities in graphene sheets supply a higher adsorption capacity in HRGO for CO_2 mitigation. At 100 °C, synthesized HRGO exhibited higher adsorption capacity compared to two others specific temperature 80 and 120 °C synthesized adsorbent as HRGO-80, HRGO-120. Further analysis illustrates the polar groups and H-bonding among CO_2 molecule and the oxygen comprising functional group. They handled the adsorption of CO_2 [67].

1.2 Role of Nitrogen Oxides (NO_x) and Their Removal

Basically, the nitrogen oxides are the sum of NO and NO_2 which are assumed as the main precursor of the formation of SOA which enhances the mass concentration of $PM_{2.5}$. Reports suggested that in Europe and North America, the generation of NO_x emissions from fossil fuel feasting is reducing while in China, raising the biomass of NO_x by the consumption of fossil fuels particularly in urban and industrial areas [46]. The resulting fumes and gases in the atmosphere generated from the fossils and industries had an adverse effect on human health. The NO_x causes an adverse effect on the nervous system of the human body and a mixture of photochemical smog resulting from the NO_x smog [63].

To diminish the adverse effects of NO_x emissions over the biotic and abiotic ecosystem, several efficient strategies have considerable importance in eradicating NO_x from the atmosphere [81]. Generally, employed ammonia used selective catalytic reduction for NO_x control, though this is the strategy for NO_x removal with high 100 ppm concentration at elevated temperatures [42, 62, 68].

Zhu et al., 2019 prepared a ternary Bi-BiOI/graphene composite photocatalyst by using a solvothermal strategy for oxidative elimination of nitrogen oxide in the presence of visible light irradiation. During preparation, the BiOI microspheres were activated by asynchronous coupling of bismuth, and graphene was done at different range of temperature from 160 to 200 °C. Based on temperature conditions, BiOI renamed as 160BOI, 170BOI, 180BOI, 190BOI, and 200BOI correspondingly. Among all 180BOI than BiOI, and binary Bi-BiOI and BiOI/graphene composites show the higher photocatalytic activity for the oxidation of NO. With a half hour irradiation of visible light ternary photocatalyst prepared at 180 °C gained 51.8% of nitrogen oxide oxidation. The upright in the efficiency of photocatalyst depends on the significant transmission of photo-generated e^- from BiOI and Bi to graphene, resulted in significant separation of the photo-generated e^- hole pairs and the SPR effect of Bi NPs in the composite photocatalyst [88].

Ai et al., 2011 studied the BGCs displays the better activity on photocatalytic elimination of gaseous NO to pure BiOBr under visible light irradiation with the wavelength over 420 nm. They reported that the rate constant of photocatalytic removal for NO of bromide graphene nanocomposite was two times higher than that of pure bismuth oxybromide. Simple solvothermal methods including GO, bismuth nitrite, and cetyltrimethylammonium bromide were used for the formation of bromine graphene nanocomposite. At the preparation time, the simultaneous reduction of GO and the formation of BiOBr nanocrystals occurred. With the help of characterization, outcomes confirm the risen in the photocatalytic activity of the BGCs nanocomposites to more effective charge transfer and separation occurred among the BiOBr and graphene, not to their light absorption extension in the visible region and higher surface area [1].

Hu et al., 2018 successfully fabricated the 3D aerogel of CNQDs/GO-$InVO_4$ (CNQDs = graphitic C_3N_4 quantum dots, GO = graphene oxide) via the hydrothermal method. They obtained very stable and recyclable macro-material, which played a significant role in the removal of nitrogen monoxide. CNQDs reported the average size diameter of 3.0 nm through the exfoliation of g-C_3N_4 step by step, via fabricated on the superficial of GO homogeneously by electrostatic, π-π stacking and hydrogen bonding interactions. The largest adsorption capacity of 65% was reported at the 600 ppb level in the presence of visible light irradiation via synergistic heterojunction. On behalf of ESR experiments and energy bands, calculations determined the photocatalytic mechanism. Overall template-free hydrothermal method and exfoliation technology were proved to best fabrication for the novel 3D CNQDs/GO-$InVO_4$ aerogel [21].

Jia et al., 2019 designed the $(BiO)_2CO_3$-BiO_2-x-graphene photocatalyst for the elimination of NO under the exposure of solar light. With the help of analytical tools, determine the physical properties of the ternary composite and their corresponding light absorption and highly capable e^- hole separation. Optimized results declare the BOC-BiO_{2-x}(35 wt%)-GR displayed the excellent performance for NO than the bare BOC, BiO_{2-x}, and BOC-BiO_{2-x} binary composites. Further, the Z-scheme charge transfer should be dominant over the heterojunction interface. According to DFT, the Fermi scale results assured the formation of energy band structure among the BOC and BiO_{2-x} is more favored the transference of photo-generated electrons from the conduction band of BOC to the valence bond of BiO_{2-x}. This transference also can be further increased by highly conductive GR sheets [27].

Liu et al., 2017 were the first time synthesized the nitrogen mixed NBOC/GQDs beneath the ambient conditions. It shows the remarkable perfection than the pristine NBOC in visible light performing photocatalytic exclusion of indoor NO air pollutants at parts-per-billion level. XRD and TEM characterizations display the graphene quantum dots (GQDs) that have no more effect over the structure as well as morphology of N-doped $Bi_2O_2CO_3$(NBOC), while GQDs might change the surface of NBOC confirmed through XPS analysis. Altogether, results of UV-Vis, photoluminescence spectroscopy, diffuse reflectance spectroscopy (DRS), and photo-electrochemical manifest that the efficiency of both light harvesting and charge separation of NBOC/GQDs during the photocatalytic process is enhanced [40].

Chen et al., 2017 also, the first time, successfully employed the single-step hydrothermal approach for the synthesis of nitrogen mixed $(BiO)_2CO_3$(NBOC)/GO composite which obtained from the 3D ordered microspheres. During synthesis, citrate ions played a chief role in N doping. The resulted composite was used to destroy the toxic NO gaseous pollutant in the presence of visible light irradiation at parts-per-billion concentration. The experimental results confirmed they got NBOC–GO composite having 1.0 wt% graphene oxide showed the 4.3 times higher the photocatalytic NO removal compared to the pristine $(BiO)_2CO_3$. NBOC–GO composite prevents the formation of toxic NO_2 intermediate, which shows the selective conversion of NO. About regular doping of N and GO, notably increase the catalytic efficiency of NBOC–GO composite [10].

By increasing, the 74.6% removal of NO and successfully preventing the production of NO_2. The good photocatalytic activity primarily attributed to optical absorption tendency and raised the separation performance of photo-generated charge carriers in NBOC–GO composite. The ESR and theoretical results of the band structure tell that NO removal is subjected by oxidation with ·OH and $·O_2-$ radicals. Overall, the present study confirms a method to synthesize highly stable and selective Bi-containing composite to control the air pollution control and gave a novel vision about the accompanying photocatalytic mechanisms [11].

Li et al., 2018 reported the NH_2-MIL-125(Ti) through the facile microwave solvothermal route. The GO, $C_{12}H_{28}O_4Ti$, and 2-aminoterephthalic acid were used as precursors in the formation of (NH_2-MIL-125(Ti)) composite. Further, the photocatalytic oxidation activity test of NO was performed at ambient temperature. The obtained composite was thoroughly characterized via FESEM, FTIR, BET, TEM, PLS, and XPS, respectively. In the presence of microwave irradiation, the surface of graphene oxide becomes highly active and behaves as the microwave antenna to readily absorb the microwave energy than the formation of hot spots occurred over the surface of GO. Generated hot spots make possible the depth characterization of the NH_2-MIL-125(Ti) crystals, which further supported by the few techniques like Raman, UV-vis, XRD, and many more. Because of well dispersity over the surface of graphene oxide, strong interaction occurred among the NH_2-MIL-125(Ti) crystal and GO. The high electronic conductivity of GO and the strong interaction profiting such GO/NH_2-MIL-125(Ti) hybrid showed photocurrent intensity, visible light absorption, enhanced metal–organic framework crystalline, an electron carrier density and lower electron–hole recombination rate, photo-generated, compared to the pure NH_2-MIL-125(Ti). Thus, the obtained hybrid system behaves as the highly efficient composite with long-lasting robustness in the presence of visible light irradiation having over 420-nm-wavelength for the photocatalytic oxidation nitric oxide and the acetaldehyde [36].

1.3 Role of a Particulate in Air Pollution and Their Removal

Other than CO_2, SO_2 NO_2 gases PM has been a chief role in the generation of air pollution. These toxicants are the consequence of fast urbanization and industrialization in metropolitan cities in developing countries. All air pollutants influence human health [3, 15, 17, 35, 41, 47]. The particular matter (PM) is the part of airborne, which evolved from the diverse manufacturing processes, and it can be found in the form of solid and liquid aerosols. These aerosols are the combination of the air and gases including CO_2, SO_2 NO_2, and ozone [7, 19, 91]. According to the aerodynamic diameter's, particulate categorizes into the two different forms as PM2.5 and PM10, when the diameter is less than or equal to 2.5 and less than or equal to 10 μm known as $PM_{2.5}$ and PM_{10}, respectively [79]. Out of $PM_{2.5}$ and PM_{10}, $PM_{2.5}$ refers to more health hazardous which causes a severe attack on human health after the direct exposure with the $PM_{2.5}$ immediately attack over the respiratory and cardiovascular systems [13, 26, 51]. The $PM_{2.5}$ pollutants exhibited the large surface area which enhances the tendency of loading the pathogenic substance in large extent attributed the sickness and endanger global public health [34, 61, 70, 87]. So, not only should rigorous regulations be necessary, but more active air filters also form to diminish the negative effect of air pollution on public health. Investigations over to regulate the particulate pollution suggested they at large-scale low cost, durable, with higher efficiency air purification devices should be manufactured and make available to the public to deal with the severe problem. The efficient air filtration membrane is needed to overcome the existing air filters to control the air pollution, which was expensive, less efficient, and ineffective capture for ultrafine particles [16].

Jung et al. deal with toxic particulate air pollution namely $PM_{2.5}$. They synthesized the ion-mediated assembled reduced graphene oxide (IMA-rGO) filters, which removes the $PM_{2.5}$ with high-efficiency and low-pressure drop. Here the two-dimensional material of reduced graphene oxide was introduced, which has higher surface area per weight of graphene and efficiently works compare to nanofiber, nanowires. During the manufacturing of filter aggregates, the reduced graphene oxide over the foam of the copper mesh can remove the outdoor and indoor PM simultaneously. After the synthesis, the samples were characterized by XPS, SEM, EDAX, and FTIR. The higher surface area and porous structure in filters attribute the passing of air with a higher rate and minimal pressure drop. The higher surface area supplies sufficient space to accommodate the pollutants in their vacant void and easily adsorbed and attains the desired level of removal. Five repeated consecutive regeneration cycles with minimal loss in efficiency keep PM removal up to 99% proving the robustness of the filter. The ion-mediated reduced graphene oxide filter showed twice the outstanding quality factor than best reported in the literature [28].

Zhang et al., 2019 present the PAN/GO nanofiber to capture PM pollutants, which comprise PAN and GO via facile and unique electrospinning approach. It proved higher adsorption efficiency and removal efficiency for $PM_{2.5}$. The elimination efficacy of up to 99.6% was reported under 460 $\mu g\ m^{-3}$ of $PM_{2.5}$ concentration. PAN/GO nanofibers still showed the 99.1% removal efficiency after 100 h of adsorption, which means outstanding and long-lasting adsorption tendency. Therefore, the PAN/GO nanofibers could be utilized as window screens to protect from the outdoor hazards and can remove the $PM_{2.5}$ pollutants which existing indoor. The nanofibers are also used in air filtration masks for persons to eliminate $PM_{2.5}$ pollutants during the inhalation process. So, PAN/GO nanofibers attributed the large implication in the air filtration industry, to give the human being cleaner and healthier living environment [86].

Mao et al., 2019 worked for the exclusion of both $PM_{2.5}$ and PM_{10}, which causes severe damage for the biotic and abiotic environment. They designed the thermally stable PM filter through the in situ fabrication of the ZIF-8 on a 3D framework of rGA via natural drying. The main part of the novel filter is reduced graphene structure attributes the maximum surface area, and well-arranged porous web gives the space to binding sites to zeolite imidazole framework-8 (ZIF-8). The uptake efficiency for the $PM_{2.5}$ was 99.3%, and PM_{10} was 99.6%, correspondingly, at ambient circumstances. Reported higher efficiency at severe circumstances, $PM_{2.5}$ exhibited greater than 98.8%, and PM_{10} was over 99.1%, at 200 °C with a flow velocity of 30 $Lmin^{-1}$. The ZIF-8/rGA filters could be regenerated through a simple washing method. Therefore, the present study provides the new approach to develop new generation air filters shows the rapid, efficient, and sustainable treatment of air pollution in the presence of the adverse circumstance [43].

Zou et al., 2019 used graphene oxide as an air filter to capture the particulate matter present in the environment in the form of $PM_{2.5}$. Because of the specific porous structure of graphene oxide easily to form the porous connected web of composite by a simple coating method, the synthesized interconnected porous web structure captured and showed 99.46% removal efficiency with low 7 Pa pressure drop, high-quality factor 0.75 Pa^{-1} when the wind velocity found to be 0.1 ms^{-1}. The results of SEM and EDS confirmed the successfully capture the $PM_{2.5}$ by graphene oxide membrane and give a new insight to control the air pollution [90].

1.4 Role of Lead in Air Pollution and Their Elimination

Lead ions Pb(II) usually existed in all three forms, such as air, water, and soil [45]. Amidst all heavy metals, Pb(II) is the most toxic element and has a negative effect on the environment and human beings [4]. It is a minute quantity led the severe damage in the human organ system like nerve, kidney, liver, and even cancer [53]. It goes into the human body through inhalation and ingestion. Fossil fuel emission and the industrial fume are the biggest source of the presence of Pb(II) in the air.

Ravishankar et al., 2016 prepared the magnetic sorbent by using the polystyrene polymer-based graphene oxide known as $PS@Fe_3O_4@GO$ adsorbent. Various analytic techniques, for instance, SEM, TEM, AFM, Zeta, and UV-vis spectroscopy, were utilized to analyze the morphology, size, and adsorption rate, respectively. Four adsorption–desorption cycles were employed to conclude the efficacy of the adsorbent after utilization. The adsorption capacity was reduced to 40.96% in the fourth cycle; from the first cycle, adsorption efficiency was 93.78%. Reduction in the adsorption rate could be because of the incomplete desorption of the active site present in the nanoadsorbent. The obtained adsorption experimental results show the better fitting with Langmuir adsorption isotherm and gained the corresponding adsorption capacity was 73.52 mg g^{-1} with 93.78% removal at pH 6. Kinetic results data were well-fitted; the first order of kinetics and process was spontaneous. The probable mechanism of adsorption well-illustrated by the FTIR and the XPS study, which confirmed the bond formation occurred in the oxygen and Pb(II) ion [55].

Wan et al., 2016 examined the exclusion of Pb(II) from the wastewater by using hydrated manganese oxide graphene oxide composite known as HMO@GO. In this study, the fabrication of hydrated manganese oxide and enhanced adsorption efficiency overcome the low adsorption capacities of bare GO. The higher adsorption capacity more than the 500 mg g^{-1} was demonstrated for the Pb(II) in the presence of other interfering ions like Ca(II) and Mg(II) in an aqueous system established the selectivity toward the Pb(II). In desorption cycle, the 1 kg dose of the synthesized adsorbent was utilized to treat the 22 m^3 of artificial industrial drainage containing 5 ppm of lead and 40 m^3 drinking water with 0.5 ppm concentration of lead to their corresponding limits 0.1 mg L^{-1}, 10 μg L^{-1} for wastewater and drinking water. The graphene oxide in nanocomposite forms the laminated structure, which exhibited negligible pore diffusion and showed a fast-kinetic sorption rate of more than 20 min. The HMO@GO can be regenerated with 0.3 HCl solutions and easily separable due to their magnetic behavior under the external magnetic field [73].

Hassan et al., 2020 firstly demonstrated the MSp@SiO_2NH_2 as a very efficient adsorbent for the elimination of the lead. The obtained composite was fully characterized through various techniques such as FTIR, TGA, FESEM, EDX, and VSM. The optimization of the composite adsorption capacity was attained of 323.5 mg g^{-1} in the range from 50 to 200 ppm initial lead concentration. During the adsorption experiment, the initial lead concentration and residual concentration after the adsorption were estimated by using the GF-ASS. The mode of adsorption isotherm was well described by the Freundlich–Langmuir, and Temkin models. Among all three isotherm models, Langmuir shows the best fitting with the adsorption isotherm data with the high regression coefficient value model, i.e., $R^2 = 0.9994$. PSO followed the rate of kinetics with the regression value of $R^2 = 1.00$ and thermodynamic evaluation determines the endothermic and spontaneous nature of the adsorption process. Other than kinetics and thermodynamics studies, the influence of other coexisting ions was also determined on the adsorption of lead. The successive cycles of the regeneration and reusability found the reusable tendency of the adsorbent till 10 times. Therefore, the prepared adsorbent exhibited the large potential for the exclusion of lead from aqueous sample of wastewater [20].

1.5 VOC Contribution in Air Pollution and Their Mitigation

Volatile organic compounds (VOCs) are emitted as in gaseous form a solid and liquid matters. It includes many organic chemicals that include the benzene, toluene, and xylenes, and aldehydes can be a formaldehyde and acetaldehyde. VOCs origin sources the storage, transport, fossil fuel usage, and refining show the harmful effect on the ecological environment and human health [25, 52]. For the VOC production in an environment not only the air is source beside air soil and water can be the source of the origin of VOCs pollutants [56, 72]. VOC concentration found in indoor ten times higher than the outdoor air. Other than organic chemicals, paints, varnishes, wax, cosmetics, disinfection cleaner, and other variety of household products have an adverse effect on human health [18]. Various VOC eliminations have been employed by researchers such as adsorption, membrane separation, and condensation, considered as the nondestructive method and the photocatalytic incineration, ozone catalytic oxidation, biological degradation, and oxidation comes in the destruction method category, though the destruction method is not an efficient method and found the hybrid treatments which are very effective. In between all techniques photocatalytic degradation and adsorption, methods are a very promising method even at a low concentration of VOC and exhibited higher removal efficiency, required low energy [22, 23]. The adsorption method can augment VOC pollutants from the gaseous phase to a solid phase to enhance the activity of photocatalytic degradation. Alternatively, during the photocatalytic method, the VOC pollutant oxidizes to CO_2 and H_2O at ambient conditions of temperature and pressure. Several carbon-based materials were studied for the removal of VOCs listed in Table 1.

2 Conclusion

In his chapter, we have discussed all varieties of air pollution, causing pollutants and their corresponding suitable targeting method. The earlier method has the same shortcoming as expensive in nature and not much very efficient. Day by day in research, scientists were trying to generate an efficient method to face the problem of air pollution. Air filters at the domestic level are utilized to keep clean the indoor air. Air filters mask been prepared for the employees who are working in the industry. Theses filters easily capture the toxic gasses and other particulates from the air and prevent the entry of toxic gases during the inhalation process. Efforts are performed in an effective manner, but still have some drawbacks like expensive and not as efficient as per requirement. Therefore, we need to discover the hybrid material with good efficacy and should be environmentally friendly.

Table 1 Variety of carbon-based materials for the removal of VOC

Carbon-based composite	Targeted VOCs compound	Photocatalytic efficacy (%)	References
TiO_2/activated carbon fiber	Toluene at 460, 877, and 1150 ppm	81, 62, 57	[33]
TiO_2/activated carbon fiber	Formaldehyde	83.6	[39]
MnO_2/MWCNT	Formaldehyde	43	[44]
Graphene hydrogel-AgBr@rGO	Bisphenol A	91.4	[10]
g-C_3N_4/biochar	2-Mercaptobenzothiazole	90.5	[89]
Wet scrubber coupled with UV/PMS process	Ethyl acetate and toluene	98.3 and 96.5	[78]
Reduced graphene-TiO_2	Formaldehyde	88.3	[84]
TiO_2ACFs	Toluene	100	[71]
Ce-GO-TiO_2	Formaldehyde	83.6	[32]
N-doped graphene Fe_2O_3	Acetaldehyde	55	[74]
Biochar/Fe_3O_4	Carbamazepine 30 ppm	50	[57]
g-C_3N_4/biochar	p-nitrophenol	70	[50]
Graphene oxide-TiO_2	2-ethyl-1-hexanol	55.1	[12]
TiO_2 activated carbon	Propene 100 ppm	60	[48]
Graphene-based nanomaterials	Toluene and xylene at various concentrations 30, 50, and 100 ppm	92.7–98.3% for Toluene and xylene 96.7–98%	[37]
Amorphous TiO_2 and graphene	Toluene	–	[85]
Graphene oxide	Methanol	–	[69]
Al-decorated porous graphene	Carbonyl	–	[38]
Mesoporous graphene	Toluene at 120 ppm	260.0 mg g^{-1}	[75]
Metal oxides in graphene composites	For both GO-$Ni(OH)_2$ and rGO-SnO_2, GO, GO-$Co(OH)_2$	Approximately 23, 19.1, 18.8 mg g^{-1}	[29]
Ball-milled biochar	Acetone, ethanol, and chloroform	23.4–103.4 mg g^{-1}	[77]
Mesoporous carbon composites	Toluene, Ethyl benzene and O-Xylene	1820.8, 1092.5, 52.9, and 47.1 mg g^{-1}	[66]

Acknowledgements The authors are thankful to Dr. K.N. Modi University, Banasthali Vidyapith, and the Central University of Gujarat.

References

1. Ai Z, Ho W, Lee S (2011) Efficient visible-light photocatalytic removal of NO with hene naposites. J Phys Chem C 115:25330–25337
2. Alghamdi A, Alshahrani A, Khdary N, Alharthi F, Alattas H, Adil S (2018) Enhanced CO_2 adsorption by nitrogen-doped graphene oxide sheets (N-GOs) prepared by employing polymeric precursors. Materials 11:578
3. Andruse, Rosenfeld D (2008) Aerosol–cloud–precipitation interactions. Part 1. The nature and sources of cloud-active aerosols.Earth Sci Rev 89:13–41
4. Aroua MK, Leong SP, Teo LY, Yin CY, Daud WM (2008) Real-time deteion of kilyof adsorption of lead (II) onto palm shell-based activated carbon using ion selective electrode. Bioresou Technol 99:5786–5792
5. Bishnoi S, Rochelle GT (2000) Absorption of carbon dioxide into aqueous piperazine: reaction kinetics, mass transfer and solubility. Chem Eng Sci 55:5531–5543
6. Boyd AD, Hmielowski JD, David P (2017) Public perceptions of carbon capture and storage in canada results of a national survey. Int J Greenh Gas Control 67:1–9
7. Brook RD, Rajagopalan S, Pope CA III, Brook JR, Bhatnagar A, Diez-Roux AV, Holguin F, Hong Y, Luepker RV, Mittleman MA, Peters A (2010) Particulate matter air pollution and cardiovascular disease: an update to the scientific statement from the American Heart Association. Circulation 121:2331–2378
8. Cao Y, Zhao Y, Lv Z, Song F, Zhong Q (2015) Preparation and enhanced CO_2 adsorption capacity of UiO-66/graphene oxide composites. J Ind Eng Chem 27:102–107
9. Chen A, Yu Y, Zhang Y, Zang W, Yu Y, Zhang Y, Shen S, Zhang J (2014) Aqueous-phase synthesis of nitrogen-doped ordered mesoporous carbon nanospheres as an efficient adsorbent for acidic gases. Carbon 80:19–27
10. Chen F, An W, Liu L, Liang Y, Cui W (2017) Highly efficient removal of bisphenol A by a three-dimensional graphene hydrogel-AgBr@ rGO exhibiting adsorption/photocatalysis synergy. Appl Catalysis B Environ 217:65–80
11. Chen M, Huang Y, Yao J, Cao JJ, Liu Y (2018) Visible-light-driven N-$(BiO)_2CO_3$/Graphene oxide composites with improved photocatalytic activity and selectivity for NO_x removal. Appl Surf Sci 430:137–144
12. Chun HH, Jo WK (2016) Adsorption and photocatalysis of 2-ethyl-1-hexanol over graphene oxide–TiO_2 hybrids post-treated under various thermal conditions. Appl Cata B Environ 180:740–750
13. Dominici F, Peng RD, Bell ML, Pham L, McDermott A, Zeger SL, Samet JM (2006) Fine particulate air pollution and hospital admission for cardiovascular and respiratory diseases. J Am Med Assoc 295:1127–1134
14. Espinal L, Poster DL, Wong-Ng W, Allen AJ, Green ML (2013) Measurement, standards, and data needs for CO_2 capture materials: a critical review. Environ Sci Technol 47:11960–11975
15. Fang M, Chan CK, Yao X (2009) Managing air quality in a rapidly developing nation: China. Atmos Environ 43:79–86
16. Fisk WJ, Faulkner D, Palonen J, Seppanen O (2002) Performance and costs of particle air filtration technologies. Indoor Air 12:223–224
17. Gurjar BR, Butler TM, Lawrence MG, Lelieveld J (2008) Evaluation of emissions and air quality in megacities. Atmos Environ 42:1593–1606
18. Harb P, Locoge N, Thevenet F (2018) Emissions and treatment of VOCs emitted from wood-based construction materials: Impact on indoor air quality. Chem Eng J 354:641–652

19. Harrison RM, Yin J (2000) Particulate matter in the atmosphere: which particle properties are important for its effects on health? Sci Total Environ 249:85–101
20. Hassan AM, Ibrahim WAW, Bakar MB, Sanagi MM, Sutirman ZA, Nodeh HR, Mokhter MA (2020) New effective 3-aminopropyltrimethoxysilane functionalized magnetic sporopollenin-based silica coated graphene oxide adsorbent for removal of Pb (II) from aqueous environment. J Environ Manage 253:109658
21. Hu J, Chen D, Li N, Xu Q, Li H, He J, Lu J (2018) Fabrication of graphitic-C_3N_4 quantum dots/graphene-$InVO_4$ aerogel hybrids with enhanced photocatalytic NO removal under visible-light irradiation. Appl Catalysis B Environ 236:45–52
22. Huang Y, Hu H, Wang S, Balogun MS, Ji H, Tong Y (2017) Low concentration nitric acid facilitate rapid electron–hole separation in vacancy-rich bismuth oxyiodide for photo-thermo-synergistic oxidation of formaldehyde. Appl Catalysis B Environ 218:700–708
23. Huang Y, Li K, Lin Y, Tong Y, Liu H (2018) Enhanced efficiency of electron–hole separation in $Bi_2O_2CO_3$ for photocatalysis via acid treatment. Chem Cat Chem 10:1982–1987
24. Irani V, Tavasoli A, Vahidi M (2018) Preparation of amine functionalized reduced graphene oxide/methyl diethanolamine nanofluid and its application for improving the CO_2 and H_2S absorption. J Colloid Interface Sci 527:57–67
25. Iranpour R, Cox HH, Deshusses MA, Schroeder ED (2005) Literature review of air pollution control biofilters and biotrickling filters for odor and volatile organic compound removal. Environ Progress 24:254–267
26. Janssen NA, Fischer P, Marra M, Ameling C, Cassee FR (2013) Short-term effects of PM2. 5, PM10 and PM2.5–10 on daily mortality in the Netherlands. Sci Total Environ 463:20–26
27. Jia Y, Li S, Gao J, Zhu G, Zhang F, Shi X, Huang Y, Liu C (2019) Highly efficient $(BiO)_2CO_3$-BiO_{2-x}-graphene photocatalysts: Z-Scheme photocatalytic mechanism for their enhanced photocatalytic removal of NO. Appl Catalysis B Environ. 240:241–252
28. Jung W, Lee JS, Han S, Ko SH, Kim T, Kim YH (2018) An efficient reduced graphene-oxide filter for PM2.5 removal. J Mater Chem A 6:16975–16982
29. Khan A, Szulejko JE, Samaddar P, Kim KH, Eom W, Ambade SB, Han TH (2019) The effect of diverse metal oxides in graphene composites on the adsorption isotherm of gaseous benzene. Environ Res 172:367–374
30. Kintisch E (2008) The greening of synfuels. Science 320:306–308
31. Krishna K, Makkee M (2005) Coke formation over zeolite and CeO_2-zeolite and its influence on selective catalytic reduction of NO_x. Appl Catal B Environ 62:35–44
32. Li J, Zhang Q, Lai AC, Zeng L (2016) Study on photocatalytic performance of cerium–graphene oxide–titanium dioxide composite film for formaldehyde removal. phys status solidi (a) 213:3157–3164
33. Li M, Lu B, Ke QF, Guo YJ, Guo YP (2017) Synergetic effect between adsorption and photodegradation on nanostructured TiO_2/activated carbon fiber felt porous composites for toluene removal. J Hazard Mater 333:88–98
34. Li N, Xia T, Nel AE (2008) The role of oxidative stress in ambient particulate matter-induced lung diseases and its implications in the toxicity of engineered nanoparticles. Free Radical Biol Med 44:1689–1699
35. Li Q, Li X, Jiang J, Duan L, Ge S, Zhang Q, Deng J, Wang S, Hao J (2016) Semi-coke briquettes: towards reducing emissions of primary PM2.5, particulate carbon, and carbon monoxide from household coal combustion in China. Sci Rep 6:19306
36. Li X, Le Z, Chen X, Li Z, Wang W, Liu X, Wu A, Xu P, Zhang D (2018) Graphene oxide enhanced amine-functionalized titanium metal organic framework for visible-light-driven photocatalytic oxidation of gaseous pollutants. Appl Catalysis B Environ 236:501–518
37. Lim ST, Kim JH, Lee CY, Koo S, Jerng DW, Wongwises S, Ahn HS (2019) Mesoporous graphene adsorbents for the removal of toluene and xylene at various concentrations and its reusability. Sci rep 9:1–12
38. Liu B, Zhao W, Jiang Q, Ao Z, An T (2019) Enhanced adsorption mechanism of carbonyl-containing volatile organic compounds on Al-decorated porous graphene monolayer: a density functional theory calculation study. Sustain Mater Technol 21:e00103

39. Liu RF, Li WB, Peng AY (2018) A facile preparation of TiO_2/ACF with CTi bond and abundant hydroxyls and its enhanced photocatalytic activity for formaldehyde removal. Appl Surf Sci 427:608–616
40. Liu Y, Yu S, Zhao Z, Dong F, Dong XA, Zhou Y (2017) N-Doped $Bi_2O_2CO_3$/graphene quantum dot composite photocatalyst: enhanced visible-light photocatalytic no oxidation and in situ drifts studies. J Phys Chem C 121:12168–12177
41. Mahowald N (2011) Aerosol indirect effect on biogeochemical cycles and climate. Science 334:794–796
42. Mansouri M, Atashi H, Tabrizi FF, Mansouri G, Setareshenas N (2014) Fischer-Tropsch synthesis on cobalt–manganese nanocatalyst: studies on rate equations and operation conditions. Int J Ind Chem 5:1–9
43. Mao J, Tang Y, Wang Y, Huang J, Dong X, Chen Z, Lai Y (2019) Particulate matter capturing via naturally dried zif-8/graphene aerogels under harsh conditions. IScience 16:133 – 44
44. Miao JL, Li CB, Liu HH, Zhang XX (2018) MnO_2/MWCNTs Nanocomposites as highly efficient catalyst for indoor formaldehyde removal. J Nano Nanotechnol 18:3982–3990
45. Mohan S, Kumar V, Singh DK, Hasan SH (2017) Effective removal of lead ions using graphene oxide-MgO nanohybrid from aqueous solution: isotherm, kinetic and thermodynamic modeling of adsorption. J Environ Chem Eng 5:2259–2273
46. Nan D, Liu J, Ma W (2015) Electrospun phenolic resin-based carbon ultrafine fibers with abundant ultra-small micropores for CO_2 adsorption. Chem Eng J 276:44–50
47. Nel A (2005) Air pollution-related illness: effects of particles. Science 308:804–806
48. Ouzzine M, Romero-Anaya AJ, Lillo-Rodenas MA, Linares-Solano A (2014) Spherical activated carbon as an enhanced support for TiO_2/AC photocatalysts. Carbon 67:104–118
49. Pal P, Banat F (2016) Comparison of thermal degradation between fresh and industrial aqueous methyldiethanolamine with continuous injection of H_2S/CO_2 in high pressure reactor. J Nat Gas Sci Eng 29:479–487
50. Pi L, Jiang R, Zhou W, Zhu H, Xiao W, Wang D, Mao X (2015) g-C_3N_4 modified biochar as an adsorptive and photocatalytic material for decontamination of aqueous organic pollutants. Appl Surf Sci 358:231–239
51. Polichetti G, Cocco S, Spinali A, Trimarco V, Nunziata A (2009) Effects of particulate matter (PM10, PM2. 5 and PM1) on the cardiovascular system. Toxicology 261:1–8
52. Qian Q, Gong C, Zhang Z, Yuan G (2015) Removal of VOCs by activated carbon microspheres derived from polymer: a comparative study. Adsorption 21:333–341
53. Senthilkumar R, Vijayaraghavan K, Thilakavathi M, Iyer PVR, Velan M (2007) Application of seaweeds for the removal of lead from aqueous solution. Biochem Eng J 33:211–216
54. Rallo M, Lopez-Anton MA, Contreras ML, Maroto-Valer MM (2012) Hg^0 policy and regulations for coal-fired power plants. Environ Sci Pollut Res 19:1084–1096
55. Ravishankar H, Wang J, Shu L, Jegatheesan V (2016) Removal of Pb (II) ions using polymer based graphene oxide magnetic nano-sorbent. Process Saf Environ Prot 104:472–80
56. Ryu CM, Farag MA, Hu CH, Reddy MS, Wei HX, Paré PW, Kloepper JW (2003) Bacterial volatiles promote growth in Arabidopsis. Proc Natl Acad Sci 100:4927–4932
57. Shan D, Deng S, Zhao T, Wang B, Wang Y, Huang J, Yu G, Winglee J, Wiesner MR (2016) Preparation of ultrafine magnetic biochar and activated carbon for pharmaceutical adsorption and subsequent degradation by ball milling. J Hazard Mater 305:156–163
58. Sharma VK, Sohn M, Anquandah GAK, Nesnas N (2012) Kinetics of the oxidation of sucralose and related carbohydrates by ferrate(VI). Chemosphere 87:644–648
59. Shen KP, Li MH (1992) Solubility of carbon dioxide in aqueous mixtures of monoethanolamine with methyldiethanolamine. J Chem Eng Data 37:96–100
60. Sheng LY (2016) China statistical yearbook. China Statistics Press Beijing
61. Shi Y, Ji Y, Sun H, Hui F, Hu J, Wu Y, Fang J, Lin H, Wang J, Duan H, Lanza M (2015) Nanoscale characterization of PM2.5 airborne pollutants reveals high adhesiveness and aggregation capability of soot particles. Sci Rep 5:11232
62. Skalska K, Miller JS, Ledakowicz S (2010) Trends in NO_x abatement: a review. Sci Total Environ 408:3976–3989

63. Son B, Yang W, Breysse P, Chung T, Lee Y (2004) Estimation of occupational and nonoccupational nitrogen dioxide exposure for Korean taxi drivers using a microenvironmental model. Environ Res 94:291–296
64. Songolzadeh M, Soleimani M, Takht Ravanchi M, Songolzadeh R (2014) Carbon dioxide separation from flue gases: a technological review emphasizing reduction in greenhouse gas emissions. Sci World J 2014:1–34
65. Srisang W, Pouryousefi F, Osei PA, Decardi-Nelson B, Tontiwachwuthikul Akachuku A, Idem PR (2018) CO_2 capture efficiency and heat duty of solid acid catalyst-aided CO_2 desorption using blends of primary-tertiary amines. Int J Greenh Gas Control 69:52–59
66. Srivastava I, Singh PK, Gupta T, Sankararamakrishnan N (2019) Preparation of mesoporous carbon composites and its highly enhanced removal capacity of toxic pollutants from air. J. Environ Chem Eng 7:103271
67. Sui ZY, Han BH (2015) Effect of surface chemistry and textural properties on carbon dioxide uptake in hydrothermally reduced graphene oxide. Carbon 82:590–598
68. Sun Y, Zhao Z, Dong F, Zhang W (2015) Mechanism of visible light photocatalytic NO_x oxidation with plasmonic Bi cocatalyst-enhanced $(BiO)_2CO_3$ hierarchical microspheres. Phys Chem Phy 17:10383–10390
69. Tai XH, Chook SW, Lai CW, Lee KM, Yang TCK, Chong S, Juan JC (2019) Effective photoreduction of graphene oxide for photodegradation of volatile organic compounds. RSC Adv 9:18076–18086
70. Taner S, Pekey B, Pekey H (2013) Fine particulate matter in the indoor air of barbeque restaurants: elemental compositions, sources and health risks. Sci Total Environ 54:79–87
71. Tian MJ, Liao F, Ke QF, Guo YJ, Guo YP (2017) Synergetic effect of titanium dioxide ultralong nanofibers and activated carbon fibers on adsorption and photodegradation of toluene. Chem Eng J 328:962–976
72. Vandenbroucke AM, Morent R, De Geyter N, Leys C (2011) Non-thermal plasmas for non-catalytic and catalytic VOC abatement. J Hazard Mater 195:30–54
73. Wan S, He F, Wu J, Wan W, Gu Y, Gao B (2016) Rapid and highly selective removal of lead from water using graphene oxide-hydrated manganese oxide nanocomposites. J Hazard Mater 314:32–40
74. Wang H, Raziq F, Qu Y, Qin C, Wang J, Jing L (2015) Role of quaternary N in N-doped graphene–Fe_2O_3 nanocomposites as efficient photocatalysts for CO_2 reduction and acetaldehyde degradation. RSC Adv 5:85061–85064
75. Wang Y, Li Z, Tang C, Ren H, Zhang Q, Xue M, Xiong J, Wang D, Yu Q, He Z, Wei F (2019) Few-layered mesoporous graphene for high-performance toluene adsorption and regeneration. Environ Sci Nano 6:3113–3122
76. Wu ZB, Jiang BQ, Liu Y (2008) Effect of transition metals addition on the catalyst of manganese/titania for low-temperature selective catalytic reduction of nitric oxide with ammonia. Appl Catal B Environ 79:347–355
77. Xiang W, Zhang X, Chen K, Fang J, He F, Hu X, Tsang DC, Ok YS, Gao B (2020) Enhanced adsorption performance and governing mechanisms of ball-milled biochar for the removal of volatile organic compounds (VOCs). Chem Eng J 385:123842
78. Xie R, Ji J, Guo K, Lei D, Fan Q, Leung DY, Huang H (2019) Wet scrubber coupled with UV/PMS process for efficient removal of gaseous VOCs: Roles of sulfate and hydroxyl radicals. Chem Eng J 356:632–640
79. Xing YF, Xu YH, Shi MH, Lian YX (2016) The impact of PM2.5 on the human respiratory system. J Thorac Dis 1:E69–E74
80. Xiong X, Ji N, Song C, Liu Q (2015) Preparation functionalized graphene aerogels as air cleaner filter. Procedia Eng 121:957–960
81. Yang D, Qi SH, Devi NL, Tian F, Huo ZP, Zhu QY. Wang J (2012) Characterization of polycyclic aromatic hydrocarbons in PM2.5 and PM10 in Tanggu district, tianjinbinhai new area. China Front Earth Sci 6:324–330
82. Yang HQ, Xu ZH, Fan MH, Bland AE, Judkins RR (2007) Adsorbents for capturing Hg^0 in coal-fired boiler flue gas. J Hazard Mater 146:1–11

83. You CF. Xu XC (2010) Coal combustion and its pollution control in China. Energy 35:4467–4472
84. Yu L, Wang L, Sun X, Ye D (2018) Enhanced photocatalytic activity of rGO/TiO_2 for the decomposition of formaldehyde under visible light irradiation. J Environ Sci 73:138–146
85. Yue L, Cheng R, Ding W, Shao J, Li J, Lyu J (2019) Composited micropores constructed by amorphous TiO_2 and graphene for degrading volatile organic compounds. Appl Sur Sci 471:1–7
86. Zhang C, Yao L, Yang Z, Kong ES, Zhu X, Zhang Y (2019) Graphene oxide-modified polyacrylonitrile nanofibrous membranes for efficient air filtration. ACS Appl Nano Mater 2:3916–3924
87. Zhang L, Jin X, Johnson AC, Giesy JP (2016) Hazard posed by metals and As in PM2.5 in air of five megacities in the Beijing-Tianjin-Hebei region of China during APEC. Environ Sci Pollu Res 23:17603–17612
88. Zhu G, Hojamberdiev M, Zhang S, Din ST, Yang W (2019) Enhancing visible-light-induced photocatalytic activity of BiOI microspheres for NO removal by synchronous coupling with Bi metal and graphene. Appl Surf Sci 467:968–978
89. Zhu Z, Fan W, Liu Z, Yu Y, Dong H, Huo P, Yan Y (2018) Fabrication of the metal-free biochar-based graphitic carbon nitride for improved 2-Mercaptobenzothiazole degradation activity. J Photochem Photobiol A Chem 358:284–293
90. Zou W, Gu B, Sun S, Wang S, Li X, Zhao H, Yang P (2019) Preparation of a graphene oxide membrane for air purification. Mater Res Express Sep 6:105624
91. Zuo F, Zhang S, Liu H, Fong H, Yin X, Yu J, Ding B (2017) Free standing polyurethane nanofiber/nets air filters for effective PM capture. Small 46:1702139

Preparation of Carbon-Based Photo-catalyst for Degradation of Phenols

Umairah Abd Rani, Law Yong Ng, Ching Yin Ng, Chee Sien Wong, and Ebrahim Mahmoudi

Abstract Semiconductor photo-catalyst is one of the most efficient initiatives that can be used for photo-catalytic degradation of organic pollutants in wastewater. However, the recombination of photo-generated electrons and hole pairs can reduce the capability of semiconductor photo-catalysts in the degradation activity. Therefore, carbon quantum dots (CQDs) can be suggested to be one of the carbon-based photo-catalysts that can minimize the recombination of photo-generated electrons and hole pairs due to their nanosizes, high fluorescent intensity, large surface area, strong photo-luminescent, and chemical inertness. Besides, CQDs can be considered as an efficient photo-catalyst in the photo-catalytic degradation of organic pollutants because they possess large band gaps, strong tunable photo-luminescent, and electron reservoir properties. Sustainable raw materials can be used for the fabrication of CQDs because they are cost-effective, eco-friendly, and effective in minimizing waste production. CQDs can be fabricated using laser ablation, microwave irradiation, hydrothermal reaction, electrochemical oxidation, and reflux method. These methods undergo several chemical reactions such as oxidation, carbonization, pyrolysis, and polymerization processes to produce CQDs. Therefore, CQDs-based photo-catalysts are promising nanomaterials that can be used for the photo-catalytic degradation of

U. A. Rani · L. Y. Ng (✉)
Department of Chemical Engineering, Lee Kong Chian Faculty of Engineering and Science, Universiti Tunku Abdul Rahman, Jalan Sungai Long, Bandar Sungai Long, 43000 Kajang, Selangor, Malaysia
e-mail: lyng@utar.edu.my

C. Y. Ng
Department of Chemical Engineering, Faculty of Engineering, Technology and Built Environment, UCSI University (Kuala Lumpur Campus), No. 1, Jalan Menara Gading, UCSI Heights (Taman Connaught), 56000 Cheras, Kuala Lumpur, Malaysia

C. S. Wong
Chemical Engineering Section, Malaysian Institute of Chemical and Bioengineering Technology, Universiti Kuala Lumpur, 78000 Alor Gajah, Melaka, Malaysia

E. Mahmoudi
Department of Chemical and Process Engineering, Faculty of Engineering and Built Environment, Universiti Kebangsaan Malaysia, 43600 Bangi, Selangor Darul Ehsan, Malaysia

M. Jawaid et al. (eds.), *Environmental Remediation Through Carbon Based Nano Composites*, Green Energy and Technology,
https://doi.org/10.1007/978-981-15-6699-8_14

phenol molecules. This chapter will introduce the properties of CQDs as carbon-based photo-catalysts, the raw materials, and methods used in the fabrication of CQDs as well as their functions and applications in the degradation of phenols. The mechanism of phenol degradation will also be described in this chapter.

Keywords Carbon quantum dot · Carbon-based photo-catalyst · Phenol degradation · Semiconductor · Photo-luminescence

1 Introduction

Nowadays, water pollution is one of the serious environmental issues due to the expansion of industrialization and the development of the population. The problem happens when waste materials from chemical industries and domestic activities are discharged into the water resources. Aromatic pollutants are the organic pollutants that are often encountered in the environment, and the majority of the aromatic pollutants found are phenolic compounds [62, 96].

Phenolic compounds and its derivatives in the wastewater are mainly discharged from pharmaceutical industries, manufacturing processes of paper mills, fungicides production, coal industries, and polymeric resin productions. They are carcinogenic, teratogenic, and mutagenic, which can affect the growth of aquatic organisms, human health, and water resources, even at low concentrations [101]. Therefore, proper treatment is required to degrade phenols in wastewater.

Since phenols are the hazardous refractory pollutants that possess benzene rings with the characteristics of high toxicity and non-biodegradability, they can hardly be degraded into smaller molecules. Photo-catalysis, thus, is a fascinating alternative to the degradation of phenolic compounds [20, 104, 109]. Photo-catalysis is an eco-friendly process for the treatment of wastewater without producing harmful side-products, which use light energy as a source of energy.

Photo-catalytic degradation of phenols using carbon-based photo-catalysts is a favorable method since they are widely available in nature, cost-effective, and eco-friendly as they can utilize the renewable solar energy directly [105]. Hence, fluorescent semiconductor carbon quantum dots have great potential to be used as photo-catalysts for phenol degradation due to their chemical stability, high photo-luminescent characteristics, and the unique quantum confinement properties [97, 119].

2 Introduction to Carbon Quantum Dots (CQDs)

Carbon quantum dots (CQDs) are generally defined as a fascinating class of carbon nanoparticles with sizes smaller than 10 nm [15, 51, 129]. CQDs are semiconductor nanoparticles that are entirely made up of carbon-based materials. Besides, they possess fluorescent properties due to their strong quantum confinement effect, highly tunable photo-luminescent, and optoelectronic behaviors [41]. Besides, CQDs are also possessed a high number of oxygenated-functional groups in their structure, which they can easily dissolve in aqueous solutions.

Besides, CQDs have great potentials in wastewater treatment due to high stability, good biocompatibility, low toxicity, high quantum yield, low fabrication cost, and excellent photo-stability [54]. CQDs can also display moderate photo-luminescent signal, photo-induced electron transfer, excellent semiconductor properties, high fluorescent activity, chemical inertness due to their quantum confinement effect, and optical stability properties [102, 108]. Therefore, CQDs are possible to be used in the treatment of organic and inorganic pollutants through the photo-catalytic degradation process.

Electron transfers and reservoir properties of CQDs can be applied to separate photo-generated electrons [53]. It can also display excellent photo-luminescent quantum yield and possess extraordinary visible-light-sensitive photo-catalytic performance [122]. Hence, they can utilize solar energy from ultraviolet to visible range due to their nanoscale characteristics [84]. Moreover, CQDs possess the superior ability for charge transport (good performance in trapping and transferring of electrons) and can inhibit the recombination of photo-generated charges effectively [1]. Therefore, CQDs have great potential to be applied as carbon-based photo-catalyst for photo-catalytic degradation of phenols.

3 Raw Materials Used to Fabricate CQDs

According to past studies, CQDs have been fabricated from various carbon sources, such as citric acid [1, 18, 54, 119, 129, 143], red lentils [46], strawberry powders [140], cholesterol [42], gluconic acid [63], broccoli [7], biomass [40], phenylenediamine [67], Prosopis juliflora leaves [80], glucose [37], urea [14], and gelatin [77].

Nevertheless, very limited studies reported the use of organic waste products to produce CQDs. Thus, discoveries are indispensable in fabricating CQDs that are made of waste materials, which is one of the initiatives to minimize waste products from industries. Based on past studies, glucose and citric acid are the most common raw materials used in the fabrication of CQDs because they consist of oxygenated-functional groups [54, 78, 143]. However, the raw material can be replaced with green carbon sources that contain similar functional groups. For instance, organic carbon wastes can be obtained from plant wastes, fruit materials, waste of cereals, and palm oil industry wastes, which are having high carbon and oxygen contents.

CQDs are fabricated by surface functionalization of organic or inorganic molecules through the carbonization and oxidation of raw materials [42]. In the fabrication process, carbon precursors are dehydrated and then carbonized to produce CQDs. The advantages of using sustainable carbon sources or waste organic products to fabricate CQDs are cost-effectiveness, high yield, biodegradability, low toxicity, and wide availability in nature. Also, the application of organic waste products as raw materials in the CQDs fabrication can reduce soil pollution [38].

The formation of the oxygen-containing functional groups in the CQD structures can produce excitation-dependent fluorescence emission. Therefore, chemical structure and photo-luminescence properties of CQDs depend highly on the chemical structure of the raw materials and the fabrication methods used [1, 139]. Besides, the physicochemical properties of CQDs can be varied by altering the number of precursors and solvents as well as duration and temperature of reaction [42]. Therefore, CQDs have high potential as attractive components in the multifunctional applications of photo-catalysis, fluorescent probes, optoelectronic devices, medical diagnosis, and water purification technologies.

Besides, fabrication of CQDs using natural precursors without the addition of acid or other chemicals is strongly encouraged due to their high quantum effects, strong luminescent properties, and low environmental impact [19]. Chemical modification of organic and inorganic precursors can improve the physical properties, chemical structures, and fluorescent effects of CQDs [72]. In the fabrication of CQDs, the sp^2 carbon linkages of raw materials can be converted to the smallest units through the oxidation treatment [5].

Polycyclic aromatic hydrocarbon molecules and carbohydrate extraction from vegetables are reliable precursors for the fabrication of high-quality CQDs because they contain more oxygen-containing functional groups such as hydroxyl, carboxyl, and carbonyl groups [129]. They also contain vitamins and glucosinolates, which could be nitrogen and sulfur sources as natural doping for the resulted CQDs [95]. Besides, this kind of carbon precursor is easily dissolved in ultrapure water as a medium reaction. For example, previous studies have used lemon juice [68] and biowaste lignin [108] as raw materials in the fabrication of CQDs.

4 Methods Used to Fabricate CQDs

Various methods and techniques have been used to fabricate CQDs. One of the methods is the reflux method. A study has added a certain amount of wood soot in a concentrated nitric acid, which is followed by refluxing at 140 °C for 12 h [74, 142]. After the reaction ends, sodium carbonate was used to dilute the acidic medium before it was dialyzed by using a dialysis membrane for 2 days to remove the excess acidic solutions. The use of concentrated acids in the fabrication of CQDs is for functionalization and defragmentation of the precursor materials to the smallest units [5, 100]. This method can be considered as a simple reaction. However, it is time-consuming for the additional purification process due to the use of concentrated acid.

The reflux process also promotes a large amount of oxygenated-functional groups (hydroxyl, carboxyl, and carbonyl) on the CQD structures [110] due to the carbonization and oxidation reactions. The emergence of these functional groups will increase the solubility of CQDs in aqueous media, and the resultant nanosizes possessed a stable photo-luminescent and high optical properties that are suitable to be used in photo-catalytic degradation of organic pollutants.

Another method that has been used to produce CQDs is the ultrasonic process [125]. In this method, raw materials are dissolved in water, followed by the addition of solvents such as sodium hydroxide and ethanol [139]. Some researchers used hydrogen peroxide to oxidize the carbon materials [99]. Ultrasonic energy is applied to generate alternate low- and high-pressure waves in the reaction medium to produce high energy that can destroy carbon bonds, thus resulting in the formation of small vacuum bubbles [120]. Besides, ultrasonic energy waves can cut macroscopic precursors materials to the smallest particle sizes of CQDs [132].

Besides, a study has applied the electrochemical exfoliation method [111] to fabricate CQDs in which this method uses two graphite rods as electrodes. By using graphite as electrodes, a negative and positive charge can be imparted to the carbon materials, contributing to the intercalation of oppositely charged ions and assisting in exfoliation [135]. The electrochemical exfoliation process generates a high number of oxygen molecules and fosters intercalation of hydroxide ions between graphite layers, with minimal hole defects of CQDs [2]. This method can produce CQDs with high quantum yields and strong luminescent properties [31, 121].

Based on a previous study, CQDs have been synthesized from ammonium citrate by heating treatment at 180 °C for 3 h under atmospheric air [92]. This method is also known as an oxidation process of molecular precursors, but it does not require concentrated acid. The product obtained was purified by using a dialysis membrane to remove the excess molecular precursors. Another study has used lemon juice to be heated up in the air at 100 °C for 45 min in a beaker to produce CQDs [103]. Lemon juice contains ascorbic acid, maleic acid, and citric acid. Hence, through the heating and carbonization process, these raw materials can be transformed into a high fluorescent CQDs.

Laser ablation technique is among other methods to fabricate CQDs by irradiating the precursors immersed in water. The benefit of using this technique is easy to control morphology and particle sizes of CQDs. Hence, a study has reported the application of this technique by using crystalline graphite micro-particles [8]. The starting material was dissolved in water, and *ns* pulsed fiber laser was employed. The interaction between the laser beams and carbon materials can produce a high temperature and pressure plasma plume at the interface of carbon materials and the surrounding medium. Heat energy caused the fragmentation of carbon materials into smaller particles [120].

CQDs have also been prepared using a microwave irradiation method in which it undergoes polymerization and carbonization [4]. The microwave irradiation method is an effective method due to shorter fabrication time and a faster reaction rate. Interaction of electrical dipole moment with microwave irradiations resulted in homogeneous energy distribution, which generates heat energy that can be used to produce CQDs [131].

In a previous report, CQDs were prepared by adding a certain amount of glucose into the water and heated under microwave reactor (200 W, 100 °C) for only 1 min [81]. When the reaction was exposed to microwave irradiation, the color of the solution slowly changed from transparent to dark brown, indicating the formation of CQDs through the dehydration process [131]. After cooling, the product was dialyzed (using 300 Da membrane) for 5 days to obtain the purified CQDs. This method can control the particle size and morphology of particle surfaces by altering the heating time. For example, a study has reported the synthesis of CQDs via microwave heating within 15 min, which has produced nearly spherical CQDs of 3.4 nm [13], while heating for 30 min has produced CQDs of 2.7 nm with uniform morphology [114].

CQDs can also be fabricated via hydrothermal treatment by using a Teflon autoclave reactor [46, 107]. This synthesis method is recognized as the most promising method for the fabrication of CQDs with controllable particle size, well-defined morphology, and can contribute to the high crystalline CQDs [66]. Besides, this fabrication method involves heating of precursors at high temperature and pressure, which allows interactions among precursors and water molecules [117]. Based on a previous study [82], CQDs were fabricated using the hydrothermal method in which rice residue and lysine were mixed in deionized water and heated up at 200 °C for 12 h. After cooling, the product was centrifuged to remove large particles.

The formation of CQDs can be achieved by oxidizing carbon sheets to produce more oxygen-containing functional groups through the self-assembly process [6]. The sizes of CQDs can be reduced during the reaction, which produces strong fluorescent emission [44]. The hydrothermal method can produce a high quantity of quantum dots. High production yields are important as they can contribute to enhancing the photo-luminescent properties with better surface passivation of CQDs [39]. A previous study has used pulp-free lemon juice as a carbon source, which was heated at 200 °C for 6 h and attained 31% of quantum yield [21]. Lemon juice has been used in the fabrication of highly efficient CQDs with superior optical properties because it is renewable and contains mostly hydrocarbon compounds.

The hydrothermal method is an easy route to fabricate highly luminescent CQDs without the addition of any chemicals. A study has produced blue fluorescent CQDs from red lentils, and the quantum yield obtained was 13.2% [46]. Red lentils are rich in fats, protein, and carbohydrates, which are good natural sources containing nitrogen and carbon. Red lentils were dissolved in deionized water and heated up at 200 °C for 5 h.

Based on the previous study, Ginko leaves have also been used to synthesize CQDs using the hydrothermal process [38]. It was ground to powder and dissolved in water before transferred into a Teflon reactor. The reactor was placed in the oven at 200 °C for 10 h. A light brown solution was obtained as CQDs. In another work, CQDs have been synthesized via hydrothermal method using ascorbic acid and deionized water [26]. The mixture was heated at 180 °C for 3 h similar to another report in which vitamin C was used as raw material under the hydrothermal route at 180 °C for 5 h [138]. During the process, vitamin C was dehydrated due to the polymerization and carbonization of organic molecules to produce high luminescent CQDs.

5 Methods Used to Characterize CQDs

Various analyses and instruments can be used to characterize the CQDs. Properties of functional groups on the CQD surfaces can be evaluated by Fourier transform infrared spectroscopy (FTIR). Crystalline phase and interlayer spacing of the CQD structures can be evaluated by X-ray powder diffraction (XRD). Transmission electron microscopy (TEM) can be performed to observe the properties on surface morphology, nanostructure, and particle size of CQDs. Furthermore, UV-Vis absorption spectra over a range of 200–800 nm can be used to obtain the optical properties of CQDs, while photo-luminescence (PL) can be used to analyze the emission and excitation light within a detection range of 200–1000 nm.

5.1 *Fourier Transform Infrared Spectroscopy (FTIR)*

FTIR analysis has been used to confirm the functional groups of CQDs, as shown in Fig. 1a in a frequency range of 4000–1000 cm^{-1}. Based on the previous report, the FTIR spectrum for CQDs showed peaks at 3400 cm^{-1} (assigned to O–H groups stretching vibration), 1654 cm^{-1} (attributed to C=O stretching), and 1035 cm^{-1} (attributed to C–O stretching) [46]. Another report has observed two peaks at 1725 cm^{-1} and 1071 cm^{-1}, which were assigned to the stretching vibration of C=O groups and C–O stretching, respectively [88]. Results indicated the presence of hydrophilic functional groups such as hydroxyl, carboxyl, and carbonyl groups that were bonded to the aromatic ring structure of CQDs, which were derived from the precursor materials. It can be concluded that the CQD surfaces contain oxygenated-functional groups, which make them be easily dissolved in various solvents. Various

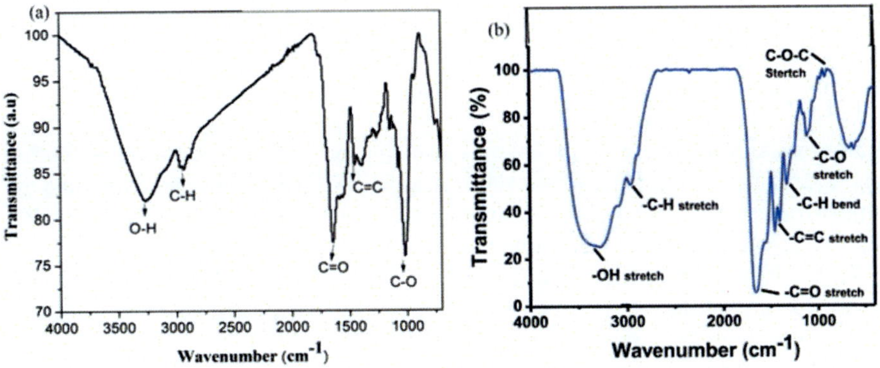

Fig. 1 FTIR spectra for **a** CQDs produced from red lentils. Reprinted with permission from Khan et al. [46]. Copyright 2019 Elsevier; **b** CQDs produced from gelatin. Reprinted with permission from Parthiban et al. [77]. Copyright 2018 Elsevier

functional groups on the CQDs surfaces can contribute to high quantum yield in which they will become more reactive, especially when exposed to the UV-light irradiation [15, 59].

Based on the FTIR spectra, a broad absorption band appeared at 3200 cm^{-1} represented the stretching vibrations of the O-H group that verified the presence of hydroxyl groups on the CQD surfaces [82]. Another two peaks appeared at 1510 cm^{-1} and 2850 cm^{-1}, which corresponded to C–C in aromatic rings and C–H bending, respectively [130]. A study has reported the presence of an sp^2 hybridized graphitic network of the aromatic C=C stretching vibrations, which were observed at 1400 cm^{-1}, as shown in Fig. 1b [77]. Based on their results, another two peaks at 2960 cm^{-1} and 1300 cm^{-1} were assigned to the stretching and bending vibration of the C–H bond. The presence of more hydrophilic functional groups attributed to enhance the proton conductivity by forming an additional proton conductivity pathway [140]. Observation of important functional groups in the FTIR spectrum revealed the successful oxidation of the precursor materials during the fabrication process. The appearance of hydrophilic groups indicated that CQDs possess high water solubility [11]. These characteristics are responsible for the efficient dispersion of CQDs in the aqueous solution.

5.2 *Transmission Electron Microscopy (TEM)*

TEM has been used to investigate the nanostructure of the synthesized CQDs and to estimate the particle sizes of CQDs. CQDs are well-dispersed with particle sizes ranging between 2 and 12 nm with an average diameter within 1–5 nm [5, 48, 73]. Based on a previous report, CQDs showed a nearly spherical shape, having good dispersibility and uniformly sized particles, which were mostly distributed in the range of 3–5 nm [134].

The nanostructures of CQDs are homogeneity in the particle size distribution, and their morphology was quasi-spherical and exhibit uniform dispersion [23, 68]. The size distribution was investigated by measuring approximately 500 particles randomly. The diameter distribution corresponding to the Gaussian fitting curve showed an average diameter of 1.7 nm, as displayed in Fig. 2 [52]. From another report [26], the particle sizes of the single dispersed CQDs are around 5 nm, and the lattice stripes with a spacing of 0.23 nm. It was found to have a good water solubility and still can be distinguished despite the low crystallinity.

Figure 3a shows TEM image and size distribution of CQDs from a previous work, which demonstrated that CQDs were well-dispersed and approximately 2–5 nm in size with the most probable diameter of 3.5 nm [67, 73]. High resolution-transmission electron microscopy (HR-TEM) image in Fig. 3b shows that the CQDs formed graphitic crystals, and the observed lattice spacing was around 0.321 nm. This result was in good agreement with the lattice planes of graphitic carbon [76]. The morphology results of CQDs were also confirmed by other studies [54, 141].

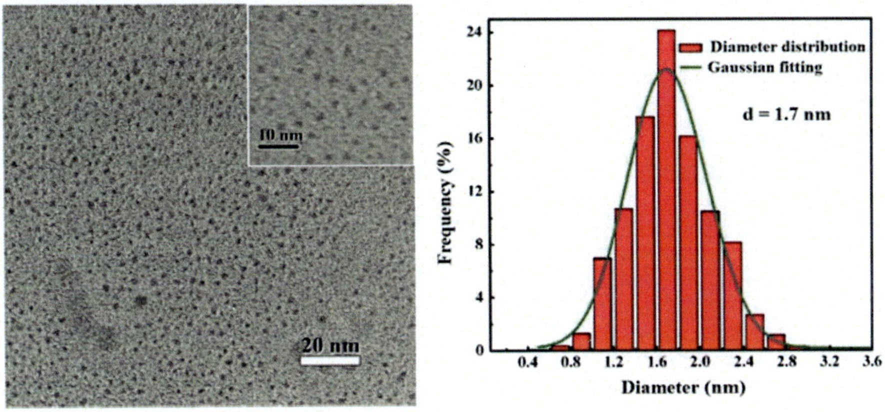

Fig. 2 TEM image of CQDs and the particle size distribution of CQDs. Reprinted with permission from Liang et al. [52]. Copyright 2013 Elsevier

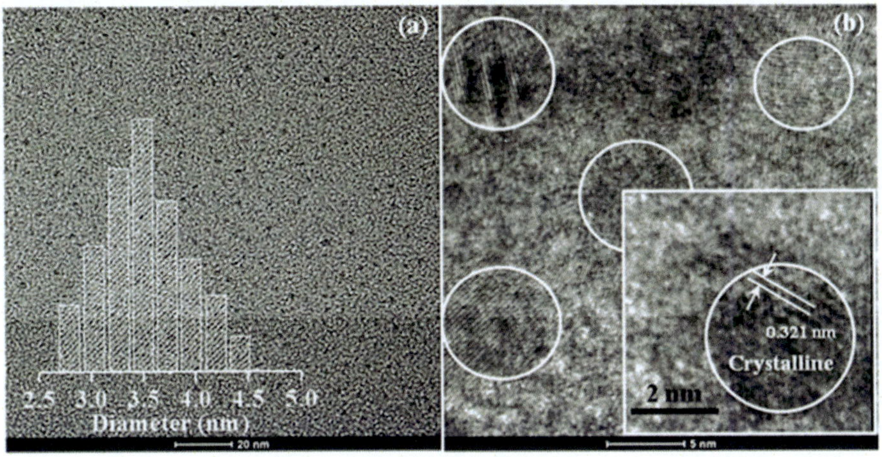

Fig. 3 **a** TEM image with size distribution and **b** HRTEM of CQDs with a crystalline structure. Reprinted with permission from Xu et al. [128]. Copyright 2016 ACS Publications

Also, the amplified HR-TEM image of CQDs revealed the coexistence of crystalline properties, which were spherical with the amorphous surface. The dispersibility of CQDs in aqueous solution was excellent, without large aggregates, indicating that CQDs were covered by several oxygenated-functional groups [99]. The lattice structure of CQDs was continuous, which has confirmed the crystallinity properties. A study has reported the lattice spacing to be 0.21 nm, which is parallel to the distance between graphite planes [21]. This result showed that the crystal structure of CQDs was mainly composed of sp^2-hybridized carbon atoms.

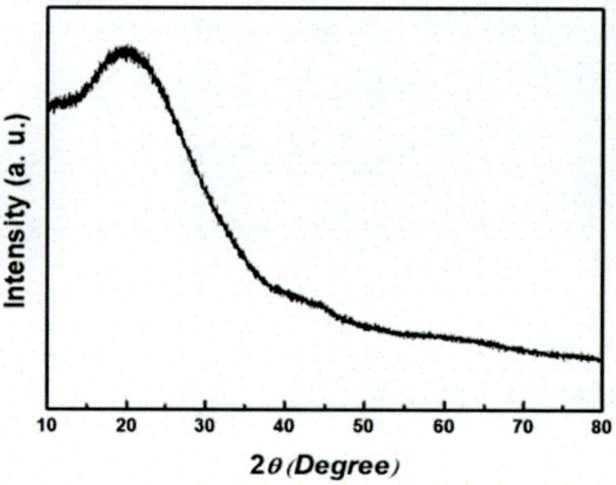

Fig. 4 XRD pattern of CQDs. Reprinted with permission from Lim et al. [54]. Copyright 2018 Elsevier

5.3 X-Ray Powder Diffraction (XRD)

Powder XRD analysis can be conducted on the CQDs to analyze the crystallinity structure of CQDs. Based on Fig. 4, the XRD spectrum of CQDs exhibited a broad diffraction peak centered at around $2\theta = 23.7°$, which corresponded to the lattice spacing of a graphitic structure [54]. The presence of graphitic structure in CQDs was due to the presence of an sp^2 carbon plane, which was in good agreement with the XRD spectrum that has appeared at $2\theta = 25°$ [70, 137]. This result suggests that CQDs can be considered as a crystalline graphitic carbon structure with the interlayer spacing of 0.32 nm [46].

XRD pattern from another study has reported a specific peak appeared at 22°, which was attributed to the highly amorphous nature of CQDs due to the large interlayer spacing of CQDs (4.0 Å) [77]. This interlayer suggested the presence of amorphous and crystalline nature in the CQDs surface due to the appearance of a broad peak centered at 20.25°. This result showed that CQDs possess high stacking compatibility in the carbon materials [37].

5.4 Photo-luminescence (PL) Spectrophotometer

PL spectrophotometer has been carried out to study the optical properties of CQDs. The absorption peak of CQDs appeared at 370 nm, while the maximal emission peak of CQDs was observed at 450 nm [12]. As displayed in Fig. 5a, when the excitation wavelength varied from 360 nm to 460 nm, the emission peak appeared at 440 nm, which decreased continuously and shifted gradually to the longer wavelength at 528 nm [85]. This phenomenon occurred due to the photo-induced electrons and

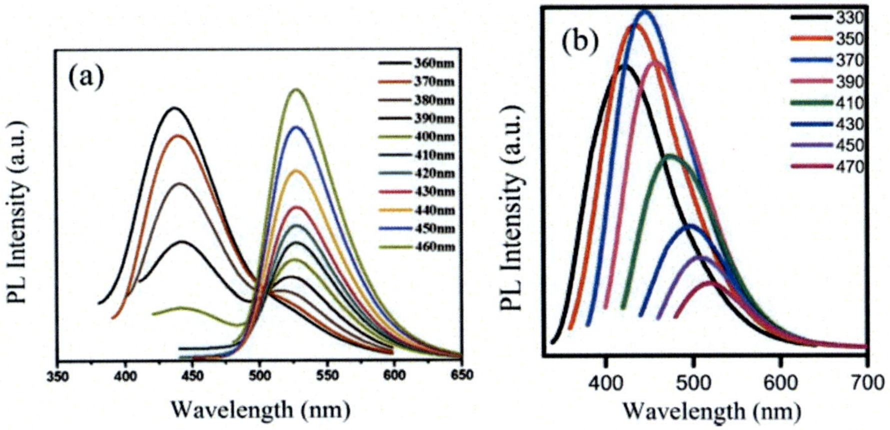

Fig. 5 PL spectra of CQDs: **a** CQDs fabricated from citric acid. Reprinted with permission from Qiao et al. [85]. Copyright 2019 Elsevier; **b** CQDs fabricated from Pomelo. Reprinted with permission from Ramar et al. [91]. Copyright 2018 Elsevier

holes that were present in the CQDs at different emissive energy traps [91]. This fluorescence emission spectra showed a slight excitation-dependent emission behavior. Different excitation wavelengths will exhibit different emission colors such as blue, red, and green [1]. Various functional groups on the CQD surfaces might be responsible for the extension of various fluorescent emission color. A report has found that CQDs can exhibit excitation-independent PL behaviors (Fig. 5b) with strong emission wavelengths observed in the range of 420–510 nm [91]. The PL emission behavior depended on the number of particles excited by the particular excitation wavelength.

5.5 *UV-Vis Spectrophotometer*

UV-Vis light absorption measurements can analyze the linear optical absorption properties of CQDs. In a previous study, the UV-Vis spectrum of CQDs has two different peaks at 233 nm and 282 nm, as shown in Fig. 6a [7]. The shoulder peak at 233 nm could be assigned to π–π^* transition of aromatic C=C and C–C bonds, which were originating from the aromatic π system. A strong peak at 282 nm could be attributed to the n–π^* transition of the C=O groups due to the successful oxidation of carbon-based raw materials.

As shown in Fig. 6b, a peak at 267 nm has been detected, which can be attributed to the π–π^* transition of conjugated C=C onto CQD surfaces [48]. Generally, the absorption band below 300 nm could be attributed to analogous features of conjugated C=C bonds corresponding to the carbon-core [94]. A peak at 360 nm could be

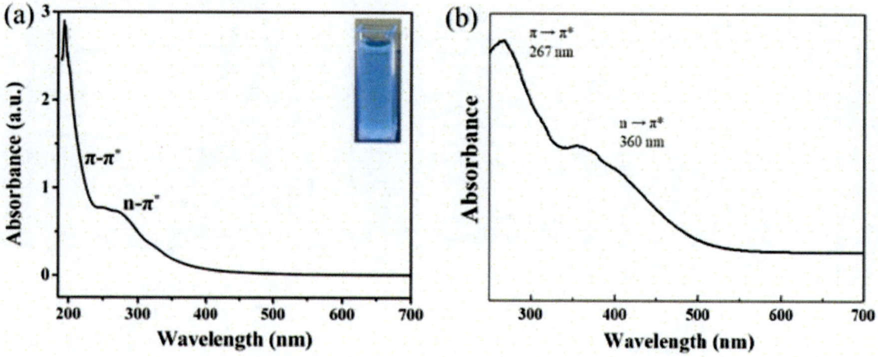

Fig. 6 UV-Vis spectra of CQDs: **a** the glowing blue color of CQDs when exposed under a UV lamp. Reprinted with permission from Arumugam and Kim [7]. Copyright 2018 Elsevier; **b** recorded at two significant absorption peaks at 267 and 360 nm. Reprinted with permission from Lei et al. [48]. Copyright 2019 Elsevier

attributed to the n-π* transition of C=O and O–H groups that were present on the CQDs surface. UV-Vis spectrum of CQDs was found to be almost similar to the raw materials obtained from carbon-based precursors.

6 CQDs-Based Photo-Catalyst for Degradation of Phenols

Degradation of phenol in the presence of carbon-based photo-catalyst under UV-light irradiation is known to be an effective photo-catalytic activity. Commonly, carbon nanoparticles are used as photo-catalyst for phenol degradation due to their chemical stability, small particle sizes, and large surface area [17]. When they are exposed to UV-light irradiation, electrons are excited to the conduction bands, which will generate positively charged holes in the valence bands [36]. This process will also produce hydroxyl radicals that can be used to attack phenol molecules during the degradation process [3, 126].

The use of carbon-based photo-catalyst materials may reduce the recombination rate of active electrons and holes [115, 118] due to their capability to separate the photo-generated electron–hole pairs [75, 124]. However, the advantages of photo-catalytic activity are cost-effective, stable photo-catalyst, chemicals free, sustainable approach, and facile process. Besides, this technique does not require non-renewable energy consumption as it can utilize sustainable solar energy [74].

CQDs are a class of carbon-based nanomaterials that can be effectively used as photo-catalyst for degradation of phenols due to their lower band gaps, pronounced quantum confinement, and edge effects [24, 127]. Owing to the smaller particle sizes of CQDs, they exhibit unique photo-luminescence optical behavior, electron reservoir properties, and photo-induced electron transfer [10]. Also, CQDs are not only able to accelerate the separation of photo-generated electrons, but they can also enlarge the absorption range of UV-light or sunlight irradiation [124]. Therefore, CQDs are the promising carbon-based photo-catalysts that can be used for the photo-catalytic degradation of organic pollutants such as phenol derivatives.

CQDs have been incorporated into another semiconductor photo-catalyst to activate the substrate and lower the activation energy to enhance the capability of UV-light absorption [43]. Besides, the combination of CQDs-based photo-catalyst with other semiconductor photo-catalyst can be an effective way to improve the performance of the photo-catalytic degradation process [50].

Based on a previous study, nitrogen-doped CQDs (NCQDs), which were supported by aluminum oxide (Al_2O_3) as a catalyst, have been synthesized for degradation of phenols due to its high stability against thermal treatment at high temperature [32]. NCQDs were fabricated via the hydrothermal method by using fumaronitrile as a raw material, which was then mixed with deionized water and heated up to 225 °C for 10 min. After cooling, the mixture solution was dialyzed by using a 0.22 μm filtration membrane to obtain NCQDs.

Based on that study [32], NCQDs supported on Al_2O_3 were prepared by the impregnation method. A certain amount of NCQDs solution was impregnated with 1.0 g of Al_2O_3. The mixture was dried at 80 °C for 5 h and underwent carbonization at a preset temperature (500, 600, or 700 °C) for 1 h under a nitrogen atmosphere to obtain the photo-catalyst (NCQDs supported by Al_2O_3).

The degradation experiment was evaluated by liquid-phase catalytic degradation of phenols in which the catalyst was dispersed into phenol solution, followed by stirring at room temperature for 1 h. The reaction solution was collected and analyzed using UV detection wavelength at 230 nm. NCQDs were highly dispersed on the surface of Al_2O_3 due to strong electrostatic attractive interactions between NCQDs and Al_2O_3. NCQDs containing nitrogen atom contributed to a large number of surface defects and could result in highly active photo-catalyst functionalization. The doping of CQDs with nitrogen can significantly affect the function of CQDs as the electron transfers happened between the CQDs and the nitrogen atom [25].

Results of that study [32] revealed that the photo-catalysts were able to degrade 81.5% of phenol derivatives (bisphenol F) within 120 min, indicating that the supported NCQDs on Al_2O_3 catalyst exhibited much higher degradation activity of phenol molecules. In contrast, in the absence of the catalyst, the bisphenol degradation was only 2.8% within 120 min, indicative of negligible degradation activities. Surface modification of CQDs can enhance their capability in the photo-catalytic degradation activity in which it can contribute to a higher degree of oxidation state, which resulted in the reduction of CQDs energy band gaps [83].

Figure 7 showed the effect of various concentrations of catalyst on the bisphenol F degradation in which a higher concentration of catalyst exhibited higher degradation of phenol. The efficiency of catalytic activity of catalyst (NCQDs supported by Al_2O_3) could be attributed to the highly exposed active sites from carbon-based photo-catalyst (NCQDs). Besides, the surface defects, graphitic structures, and nitrogen-containing CQDs could act as an active catalytic site for phenol degradation [35].

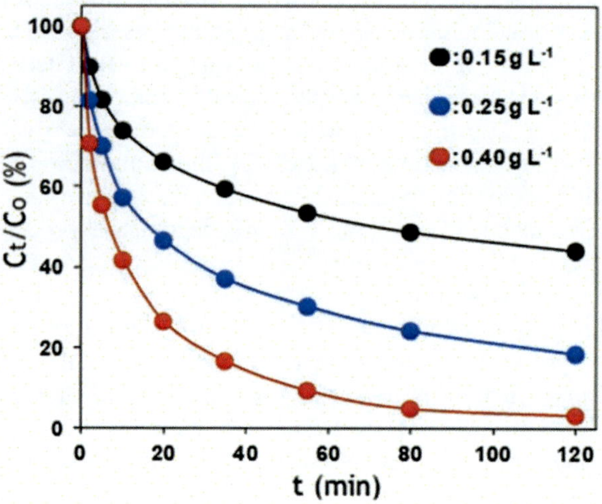

Fig. 7 Effect of catalyst (NCQDs supported by Al_2O_3) concentration on bisphenol F degradation. Reprinted with permission from Hou et al. [32]. Copyright 2018 Elsevier

Also, CQDs can facilitate the transferability of the photo-induced electron to perform as an efficient photo-catalyst. Therefore, a study has reported that CQDs can be used to produce stable hydrogen-bonding catalysts for phenol degradation [65] due to their excellent photo-luminescent properties while containing carboxylic and hydroxyl functional groups. According to that report, green synthesis of CQDs was prepared through hydrothermal carbonization. Carbohydrate-based materials (Gum Tragacanth) were used as natural precursors due to their large surface areas, thermal stability, and possessing various surface functional groups, which contribute to the production of high-quality CQDs. Hence, CQDs degraded more than 99% of phenols within 20 min (Fig. 8) due to the high photo-catalytic activity of quantum dot materials as well as the presence of hydrogen peroxide as an oxidizing agent [28].

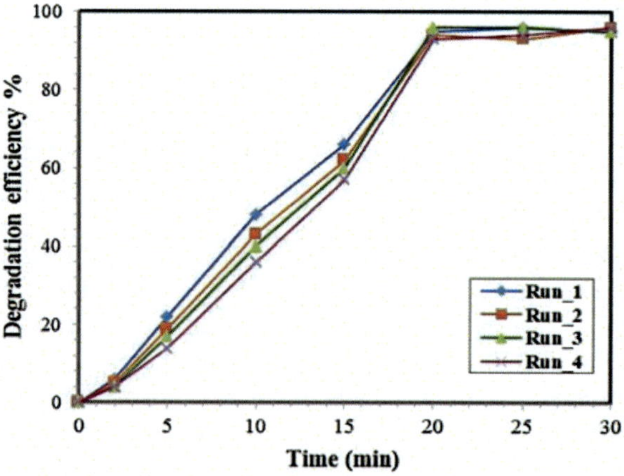

Fig. 8 Phenol degradation percentage within 20 min of reaction. Reprinted with permission from Meghdad et al. [65]. Copyright 2018 Elsevier

Certain amounts of hydrogen peroxide can be added to generate a high number of free hydroxyl radicals to accelerate the phenol degradation activity [28, 55]. Hydroxyl radical has highly oxidative potential and can effectively oxidize the organic pollutants into side-products such as water and carbon dioxide [113, 133].

A study has reported a good combination of titanium dioxide (TiO_2) and CQDs (TiO_2/CQDs) as an efficient photo-catalyst to enhance the quality of photo-catalyst semiconductor in which ~ 99% of phenol has been degraded after 6 h of irradiation by UV lamp [107]. In that work, CQDs were fabricated from citric acid, which produced a narrow size distribution and uniform spherical shape of CQDs. The combination of TiO_2 and CQDs showed higher photo-catalytic activity due to the high crystallinity of CQDs that was beneficial for the transfer of electrons from TiO_2 to CQDs as well as leading to a higher degradation of phenol derivatives. The presence of CQDs can facilitate to inhibit the recombination rate of charge carriers during the degradation process as well as can improve the ability of light-harvesting of the photo-catalysis system [107].

Besides, TiO_2/CQDs also exhibited higher photo-catalytic degradation of phenol compared to pure TiO_2 because CQDs can act as solid mediators to provide a good transfer channel for charge carriers [16]. The high percentage of phenol degradation indicated that CQDs had promoted separation efficiency of electron–hole pairs [26], which act as electron reservoirs to trap photo-generated electrons from the conduction band. Figure 9 displayed the degradation of phenol by using TiO_2/CQDs photo-catalyst.

Moreover, electron transferability of CQDs could be related to their crystalline properties and electrical conductivity [29]. Therefore, CQDs-based photo-catalyst is a good candidate as a multifunctional component in the photo-catalyst materials. Researchers have coupled nitrogen-doped CQDs (N-CQDs) and silver carbonate (Ag_2CO_3) crystals to produce more effective delocalization of the photo-generated charges during the degradation process [116], which contributed to the large enhancement in the photo-catalytic performance.

Fig. 9 Schematic illustration of the mechanism for photo-catalytic degradation of phenol over TiO_2/CQDs under UV-light irradiation. Reprinted with permission from Shen et al. [107]. Copyright 2018 Elsevier

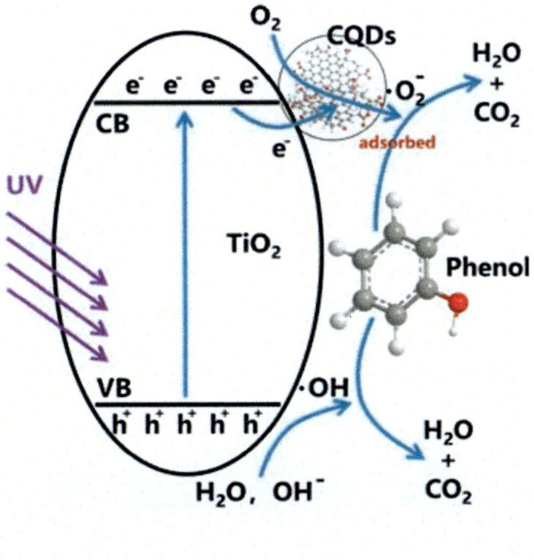

N-CQDs were produced from citric acid and urea in which the mixture was dissolved in deionized water and heated at 200 °C for 5 h. Then, the carbon-based photo-catalyst of N-CQDs and Ag_2CO_3 was produced through the precipitation method by using $AgNO_3$ and Na_2CO_3. Both of them are soluble in water, and they reacted with each other to form Ag_2CO_3 in which its solubility was very low. It was then precipitated to form a carbon-based photo-catalyst (N-CQDs/Ag_2CO_3). This CQDs-based photo-catalyst was used to degrade the phenol molecules under a 350-W Xenon lamp irradiation. It was found that over 50% of phenol was degraded by using 3 mL of photo-catalyst (N-CQDs/Ag_2CO_3).

Based on Table 1, the degradation rate using the Ag_2CO_3 in the absence of N-CQDs of phenol was very low. This result may indicate the great performance of CQDs as they contain carbon-based photo-catalyst with a strong ability to accelerate the transfer of photo-generated electrons. Moreover, the doping of nitrogen atom onto CQDs can enhance the delocalization of the photo-generated charge effectively [32, 64].

N-CQDs/Ag_2CO_3 photo-catalyst can act as an efficient photo-catalyst due to the combination of N-CQDs with the Ag_2CO_3 as the light absorption ability of the resulting combination could be greatly increased [55]. In particular, N-CQDs/Ag_2CO_3 photo-catalyst is also good for light-harvesting to generate active free radicals ($\cdot O^{2-}$ and $\cdot OH$), which would facilitate the degradation of phenol molecules [58]. Moreover, the presence of N-CQDs in photo-catalytic degradation can enhance photo-catalytic performance as they have a strong ability to accelerate the transfer of photo-generated electrons [27]. N-CQDs can also exhibit optical properties that make them suitable to be used for photo-catalytic systems.

The quantum effect of CQDs can contribute to the broadband optical absorption, which can enhance the photo-catalytic performance by acting as a light absorber [136]. The extended photo-responding range and highly efficient charge separation of CQDs can enhance the photo-catalytic activity. Besides, CQDs also possess unique electronic properties such as high stability against photo-bleaching and endow to the efficient utilization of solar and visible lights [123]. Since the carbon materials have great absorption and excellent conductivity, they are significant to be used in the fabrication of CQDs-based photo-catalyst with prominent morphology of amorphous carbon. Hence, the production of CQDs modifying sphere-flower on nitrogen-doped reduced graphene oxide was fabricated through the hydrothermal method [60]. This

Table 1 Phenol degradation rate using different amount of photo-catalyst

Type of photo-catalyst	Degradation (%)
Ag_2CO_3	10
1 mL of (N-CQDs/Ag_2CO_3)	18
3 mL of (N-CQDs/Ag_2CO_3)	50
5 mL of (N-CQDs/Ag_2CO_3)	25
10 mL of (N-CQDs/Ag_2CO_3)	15

Reprinted with permission from Tian et al. [116]. Copyright 2017 Elsevier

CQDs-based photo-catalyst can enhance the specific surface area, thus reducing the recombination of photo-carriers as well as can enlarge the visible light absorption range. Thus, they have successfully degraded 80% of phenol derivatives under visible light irradiation within 6 h [60]. This result indicated that the CQDs-based photo-catalyst could function as a strong visible light absorber with a narrow bandgap as well as providing more active sites, which contribute to enhancing the performance of photo-catalytic degradation [106].

Considering the analogous π-conjugated structure of graphitic carbon nitride and CQDs, a combination of these materials can potentially produce high photo-catalytic performance [136]. Furthermore, CQDs have been widely used as light absorbers to produce hydroxyl and oxygen radicals. For instance, the decoration of CQDs and graphitic carbon nitride (g-C_3N_4) were fabricated via a facile impregnation thermal method [34]. By using this photo-catalyst, the reaction rate of phenol degradation was 3.7 times faster than without CQDs, indicating that CQDs have boosted the photo-catalytic activity, resulting in the separation of electrons and holes [91].

Figure 10a showed the intermediate products formed during the phenol degradation. The phenols and the intermediate products were gradually decreased within the irradiation time due to the ring cleavage process. The degraded phenol was analyzed using total organic carbon (TOC) measurement. Based on Fig. 10b, about 87% of phenol by average was degraded within 200 min in the presence of CQDs-based composites. At the same duration, but in the absence of CQDs, only 43% of phenol was degraded. This observation indicated that the CQDs-based photo-catalyst possessed a greater mineralization efficiency of phenol molecules. The CQDs and g-C_3N_4 composite exhibited much higher photo-catalytic degradation of phenol as CQDs possess quantum effects with broadband optical absorption.

When CQDs were combined with semiconductor nanoparticles, electrons from the conduction band of nanoparticles would be transferred to CQD surfaces and caused the separation of electron–hole pairs. This activity was facilitated by the

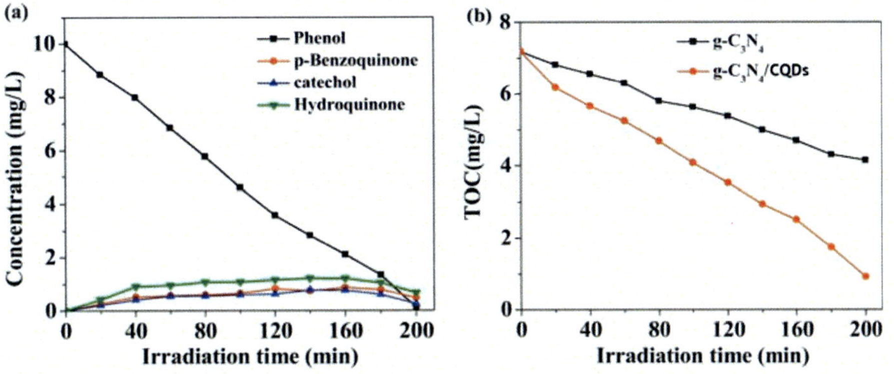

Fig. 10 **a** Degradation of intermediate phenol products, which decreased gradually within 120 min and **b** total organic carbon of degraded phenols by two different photo-catalysts. Reprinted with permission from Hui et al. [34]. Copyright 2016 Elsevier

photo-induced property of CQDs [60]. Due to the nanosizes of CQDs, the modified CQDs could have enough interface to combine with semiconductors. The charge transfer durations and channels have been prolonged by CQDs [61].

A study has reported the photo-catalytic degradation of phenol under solar light irradiation using CQDs/TiO_2 photo-catalyst composite [30]. CQDs were fabricated using citric acid, glycerol, and cow urine through the carbonization reaction. The use of cow urine in the fabrication of CQDs is to improve the fluorescence properties of CQDs. Results showed that the CQDs/TiO_2 photo-catalyst was able to degrade about 93% of phenol molecules within 6.5 h of photo-catalysis due to the synergetic effects between TiO_2 and CQDs [9]. Figure 11 showed a photo-degradation mechanism in which CQDs/TiO_2 is photo-catalyst. In this process, CQDs act as dispersing support to control the morphology of CQDs/TiO_2 nanoparticles by preventing agglomeration of TiO_2 nanohybrid.

During the photo-catalytic degradation activity under UV-light irradiation, CQDs absorb light and re-emit shorter wavelengths. The shorter wavelength will excite the CQDs/TiO_2 to generate electron–hole pairs. Thus, CQDs act as acceptors and transporters for photo-generated electrons [53]. The electron–hole pairs formed after the excitement of electrons were trapped by hydroxyl groups at the catalyst surfaces to yield OH· radicals, and it can be used for the phenol degradation [58]. Reactive oxidative species are generated due to the photo-induced electrons. Meanwhile, OH·, with strong oxidation capacity, is produced due to the photo-induced holes [112]. Therefore, the combination of two semiconductor nanoparticles can produce high photo-catalytic performance [87], improve the heterojunction constructed between CQDs and other semiconductor nanoparticles while producing strong physical–chemical interaction.

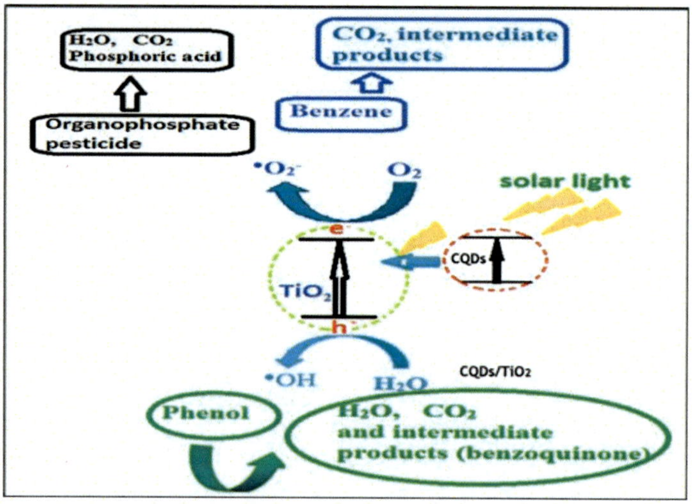

Fig. 11 Photo-catalytic mechanism of phenol and other organic pollutants by CQDs/TiO_2 photo-catalyst. Reprinted with permission from Hazarika and Karak [30]. Copyright 2016 Elsevier

7 Roles of CQDs as Photo-Catalyst for Degradation of Phenols

CQDs with sizes below 10 nm possess good fluorescent properties, chemical stability, superiority water solubility, and unique photo-induced electron transfers [58]. Applications of CQDs in photo-catalytic technologies have been explored due to their rapid photo-generation carrier transfers, excellent electron reservoir capacity as well as effective optical absorption of UV-light irradiation. Besides, CQDs have acted as a matrix to support the semiconductors due to their broadband optical absorptions and strong photo-luminescent emissions [93]. CQDs play an important role in enhancing the photo-catalytic degradation of phenols as good electrons donor and electrons acceptor [57]. CQDs might also serve as an intermedium to generate strong oxidative holes and reductive electrons. CQDs can be employed in the photo-catalysis as photo-sensitizers and electron reservoirs in which they can accelerate the photo-catalytic reaction and improve the visible light activity of wide bandgap photo-catalysts [60].

Also, CQDs can improve the visibility and UV-light response of photo-catalysts, leading to the enhancement of light absorption and photo-catalytic activity based on their unique photo-electric properties [1]. Moreover, CQDs can efficiently improve the photo-induced charges separation for the destruction of organic pollutants and reduction of contaminants.

The most significant factors that affect the photo-catalytic performance are the separation and recombination processes of electron–hole pairs on the photo-catalyst surfaces. The transfer rates of electron–hole pairs and recombination process can be affected by CQDs due to their nanosizes, large band gaps [71], and strong photo-luminescent properties [1, 93].

CQDs can perform as both electron donors and acceptors in enhancing the photo-catalytic activity [1]. Therefore, CQDs can act as an electron reservoir to trap electrons and promote the separation of electron–hole pairs. The excessive photo-generated electrons can activate the adsorbed oxygen on the surface of the photo-catalysts to generate superoxide radical anions [74]. Meanwhile, the holes could be reacted with water molecules to produce active hydroxyl radicals to reduce the recombination rate of photo-generated charges effectively. Hence, the presence of CQDs in the photo-catalytic degradation of phenol compounds could reduce the band gaps of the photo-catalyst and provide more reactive sites [34].

Besides, CQDs can also extend the lifetime of charge carriers owing to electronic interactions between semiconductors and CQDs as well as to increase the surface charge transfers [27]. In particular, CQDs have remarkable up-conversion ability to convert lower energy photons into higher energy photons, which has been applied as a spectral converter to effectively utilize the full spectrum of sunlight [58].

Light-converting properties of CQDs can be used to convert longer wavelengths into shorter wavelengths, which can excite the photo-catalyst reaction to form more electron–hole pairs. This activity has been related to the up-conversion photo-luminescent of CQDs and excellent optical absorption in the presence of UV-light

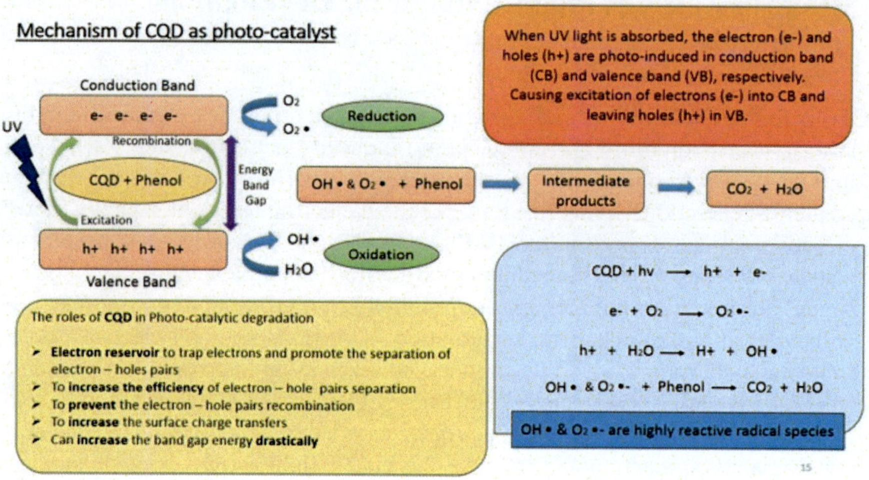

Fig. 12 Mechanism of CQDs as photo-catalysts in the phenol degradation activity

irradiation [87]. CQDs also can expand the light utilization range from UV irradiation to visible range as well as to enhance the photo-induced electron transfers [79].

Besides, CQDs also have a high absorption coefficient and fast electrons transportation that could be employed to achieve good photo-catalytic performance [45]. They can also be used to absorb light over the entire wavelength ranges of the solar spectrum, which can contribute to an excellent photo-catalytic activity. The light-converting properties of CQDs can improve the effectiveness of solar light usage by CQD-based composites, consequently enhancing their photo-catalytic activities [15, 56]. The roles of the CQDs in the photo-catalytic degradation of phenol have been summarized in Fig. 12.

8 Mechanism of Phenol Degradation

The photo-catalytic degradation in the presence of CQDs-based photo-catalyst involves free radical reaction upon initiated by UV-light irradiation [98]. Due to the up-conversion fluorescent emissions of CQDs, they can transfer two or lower energy photons to the higher energy photons by absorbing longer-wavelength multi-photon [65]. Besides, hydroxyl radicals, superoxide radicals, photo-generated holes, and electrons are reactive radical species responsible for photo-catalytic activity in the presence of UV-light irradiation [8].

The degradation of phenols generally depends on the type of reactive oxygen species involves in the process, which in turn corresponds to the type of photo-catalysts and the source of light irradiation [69]. CQDs-based photo-catalytic activity will undergo three phases. In the first phase, the absorption of light contributes to the formation of electron–hole pairs. Next, the separation and transfer of electron–hole pairs will cause the generation of reactive species. Finally, the reactive species will lead to subsequent photo-catalytic reactions.

The photo-catalytic degradation of phenols begins when CQDs are exposed under UV-light irradiation. When the irradiation energy exceeds the energy difference between the valence and conduction band of the semiconductor, electron–hole pairs will be generated [89]. Electrons and holes can effectively induce the photo-catalytic reactions, resulting in the degradation of phenol molecules [60].

Phenol molecules are easily absorbed into the surfaces of CQDs resulting from the enlarged surface areas [107]. Then, electrons are excited and transferred from the valence band to the conduction band of CQDs [33]. This process involves oxidation and reduction reactions, which will generate a high amount reactive of radical species like superoxide and hydroxyl radicals [64].

This reaction also contributes to the formation of photo-generated holes in the conduction band. The holes that are left in the valence band can oxidize phenol molecules [53]. The holes might also be captured by hydroxyl ions or water molecules to generate active hydroxyl radicals [22]. The photo-generated charges will cause redox reactions on the particle surfaces and increase the free radicals production, which in turn increases the degradation of phenol molecules [89].

Electron–hole pairs formed after the excitement of electrons can be trapped by hydroxyl radicals on the photo-catalyst surfaces. The hydroxyl radicals possess electrophilic features, which tend to attack the electrons of phenol molecules [96]. Meanwhile, the photo-electrons that combined with the dissolved oxygen molecules will produce reactive oxygen radicals [104].

Besides, the dissolved oxygen molecules will also react with the excited electrons to form superoxide radical anions, which will then oxidize phenols directly [86, 87]. The superoxide radical anions are highly reactive, and they can generate hydroxyl radicals. When hydroxyl radicals react with protons, they will produce hydro-peroxyl radicals that can be used in phenol degradation [22].

The overall photo-catalysis generates highly energetic reactive radical molecules, as shown in Fig. 13. These radical molecules will act as a strong oxidizing agent to degrade the phenol molecules in which the benzene rings of phenol structure are opened to form organic acids as intermediate products like acetic acid, formic acid, and oxalic acid [47, 49]. Next, the mineralization of the organic acids could produce the final degraded products such as water, carbon dioxide, and other inorganic compounds, which are non-toxic products [90].

9 Conclusion and Prospect of CQDs

Phenols, as the biologically carcinogenic, teratogenic, and highly toxic substances, are widely available in industrial wastewater. Currently, photo-catalytic degradation of phenols is a promising sustainable technology with low energy consumption and

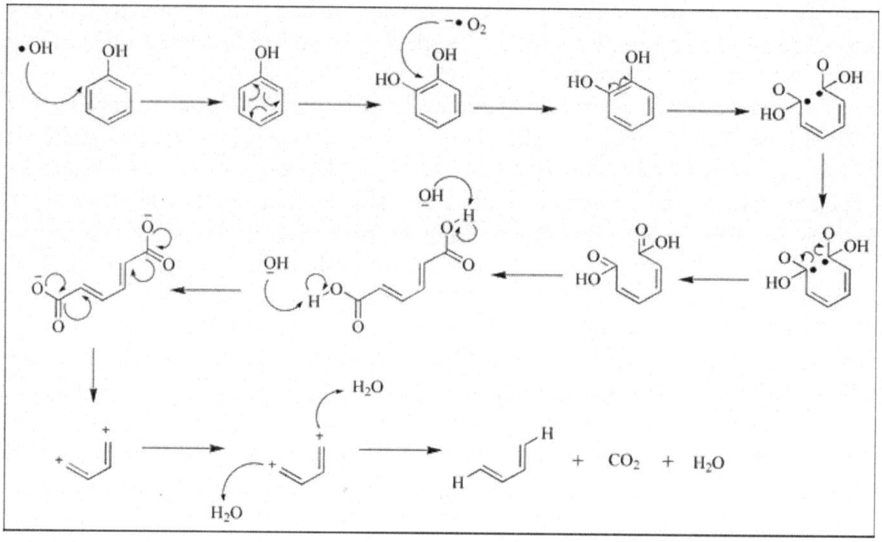

Fig. 13 Proposed mechanism of photo-catalytic degradation of phenols

high effectiveness. Photo-catalytic degradation of phenols occurs when the photo-generated electrons and holes are effectively separated during photo-reaction.

CQDs have been suggested to be an efficient photo-catalyst for the degradation of phenols in wastewater in the presence of UV, visible, or solar light irradiations. During the photo-catalytic degradation process, oxidation and reduction reactions occur and produce highly reactive radical species that will attack and degrade phenol molecules. Various operating factors such as light intensity, oxidizing agents, the concentration of photo-catalysts and phenols, electron acceptors as well as the presence of carbon-based photo-catalysts in the reaction can affect the photo-catalytic degradation rate of phenols significantly.

Roles of CQDs in the photo-catalytic degradation of phenols are to act as electron reservoirs in trapping electrons and promoting separation of electron–hole pairs, which are targeted to ensure the efficient separation of electron–hole pairs, to prevent the recombination of electron–hole pairs, to improve surface charge transfers, and to increase bandgap energy during photo-catalysis. Therefore, CQDs could efficiently induce charge delocalization and cause the electron transfer ability of CQDs-based photo-catalyst can be strongly promoted.

Raw materials that are popularly used to fabricate CQDs including citric acid, ascorbic acid, graphitic, lemon juice, fruits, strawberry powders, glycerin, vitamin C, biomass, and urea. However, the fabrication of CQDs from sustainable materials (plant-based sources and organic carbon wastes) should be encouraged due to its cost-effectiveness, wide availability in nature, low toxicity, and eco-friendliness. Besides, it can also help to reduce the consumption of chemicals and waste productions.

CQDs can be fabricated through various methods such as hydrothermal reaction, laser irradiation, reflux method, heating process, electrochemical treatment, and microwave irradiation. However, the hydrothermal reaction is the most commonly used method to fabricate CQDs. This method is working based on a water system and can be considered to be one of the most facile and cost-effective methods as it requires low energy consumption and involves simple step preparation. Although it requires a long duration of the heating (5–12 h), it does not require any additional purification process because the product obtained is high in purity and has good water solubility. It also has high quantum yields, uniform particle sizes, and exhibit strong fluorescent properties.

Currently, CQDs-based photo-catalyst has revealed their potential in wastewater treatments. Various glowing color and tunable fluorescent emissions of CQDs can be further studied to improve the performance of CQDs in photo-catalytic degradation. Besides, more studies are required on the modeling of photo-reactor to optimize its design, which can be applied in pollutant degradation.

Also, further study may include an evaluation of the physico-chemical properties of CQDs and photo-luminescent effect of CQDs based on the carbon precursors and methods used during their fabrication process. Researches that are supported with theoretical interpretations may also facilitate the future studies of CQDs for their application as nanomaterials in the environmental applications. Some structural and chemical manipulations of CQDs, such as the incorporation of suitable functional groups to their surfaces, can be considered in tuning the features of CQDs in the future for wider application. The introduction of heteroatom into CQDs can effectively disorientate the electron network, which can modulate the physicochemical properties of CQDs. For example, surface modification of CQDs can be used in the photo-catalytic degradation of other organic pollutants such as textile dyes, heavy metals, agrochemicals, pharmaceuticals, and polycyclic aromatic hydrocarbons.

Acknowledgements This work was supported by the Unit Perancangan Ekonomi Negeri Selangor, Selangor State Government (Geran Kecil Alam Sekitar Tahun 2019), and UCSI University (Pioneer Scientist Incentive Fund (PSIF) Proj-In-FETBE-047).

References

1. Aghamali A, Khosravi M, Hamishehkar H, Modirshahla N, Behnajady MA (2018) Synthesis and characterization of high efficient photoluminescent sunlight driven photocatalyst of N-carbon quantum dots. J Lumin 201:265–274
2. Ahirwar S, Mallick S, Bahadur D (2017) Electrochemical method to prepare graphene quantum dots and graphene oxide quantum dots. ACS Omega 2:8343–8353
3. Al-hamdi AM, Sillanpää M, Bora T, Dutta J (2016) Efficient photocatalytic degradation of phenol in aqueous solution by SnO_2: Sb nanoparticles. Appl Surf Sci 370:229–236
4. Alarfaj N, El-Tohamy M, Oraby H (2018) CA 19-9 pancreatic tumor marker fluorescence immunosensing detection via immobilized carbon quantum dots conjugated gold nanocomposite. Int J Mol Sci 19:1162

5. Algarra M, González-Calabuig A, Radotić K, Mutavdzic D, Ania CO, Lázaro-Martínez JM, Jiménez-Jiménez J, Rodríguez-Castellón E, del Valle M (2018) Enhanced electrochemical response of carbon quantum dot modified electrodes. Talanta 178:679–685
6. Allahbakhsh A, Bahramian AR (2018) Self-assembly of graphene quantum dots into hydrogels and cryogels: dynamic light scattering, UV–Vis spectroscopy and structural investigations. J Mol Liq 265:172–180
7. Arumugam N, Kim J (2018) Synthesis of carbon quantum dots from Broccoli and their ability to detect silver ions. Mater Lett 219:37–40
8. Arvind S, Mohapatra PK, Kalyanasundaram D, Kumar S (2019) Self-functionalized ultra-stable water suspension of luminescent carbon quantum dots. Mater Chem Phys 225:23–27
9. Atchudan R, Nesakumar T, Immanuel J, Perumal S, Vinodh R, Lee YR (2018) In-situ green synthesis of nitrogen-doped carbon dots for bioimaging and TiO_2 nanoparticles@nitrogen-doped carbon composite for photocatalytic degradation of organic pollutants. J Alloys Compd
10. Bajorowicz B, Kobyla MP, Go A, Nadolna J, Zaleska-medynska A, Malankowska A (2018) Quantum dot-decorated semiconductor micro- and nanoparticles: a review of their synthesis, characterization and application in photocatalysis. Adv Colloid Interface Sci 256:352–372
11. Basavaiah K, Tadesse A, RamaDevi D, Hagos M, Battu G (2018) Facile green synthesis of fluorescent carbon quantum dots from citrus lemon juice for live cell imaging. Asian J Nanosci Mater 1:36–46
12. Bharathi D, Krishna RH, Siddlingeshwar B, Divakar DD, Alkheraif AA (2019) Understanding the interaction of carbon quantum dots with CuO and Cu_2O by fluorescence quenching. J Hazard Mater 369:17–24
13. Che Y, Pang H, Li H, Yang L, Fu X, Liu S, Ding L, Hou J (2019) Microwave-assisted fabrication of copper-functionalized carbon quantum dots for sensitive detection of histidine. Talanta 196:442–448
14. Chen J, Gao Y, Hu X, Xu Y, Lu X (2019) Detection of hydroquinone with a novel fluorescence probe based on the enzymatic reaction of graphite phase carbon nitride quantum dots. Talanta 194:493–500
15. Cheng Y, Bai M, Su J, Fang C, Li H, Chen J, Jiao J (2019) Synthesis of fluorescent carbon quantum dots from aqua mesophase pitch and their photocatalytic degradation activity of organic dyes. J Mater Sci Technol 35:1515–1522
16. Chongyang L, Li X, Li J, Sun L, Zhou Y, Guan J, Wang H, Huo P, Yan Y (2019) Carbon dots modifying sphere-flower $CdIn_2S_4$ on N-rGO sheet multi-dimensional photocatalyst for efficient visible degradation of 2,4-dichlorophenol. 1–12
17. Cuevas A, Campos BB, Romero R, Algarra M, Vázquez MI, Benavente J (2019) Eco-friendly modification of a regenerated cellulose based film by silicon, carbon and N-doped carbon quantum dots. Carbohydr Polym 206:238–244
18. Cui P, Kuai Y, Wu Q, Zheng Y, Liu X (2019) Synthesis of a fluorescent cation surfactant derived from carbon quantum dots. Mater Lett 235:161–163
19. Devi P, Rajput P, Thakur A, Kim K, Kumar P (2019) Recent advances in carbon quantum dot-based sensing of heavy metals in water. TrAC Trends Anal Chem 114:171–195
20. Dewidar H, Nosier SA, El-Shazly AH (2018) Photocatalytic degradation of phenol solution using Zinc Oxide/UV. J Chem Heal Saf 25:2–11
21. Ding H, Zhou X, Qin B, Zhou Z, Zhao Y (2019) Highly fluorescent near-infrared emitting carbon dots derived from lemon juice and its bioimaging application. J Lumin 211:298–304
22. Ehsan MF, Bashir S, Hamid S, Zia A, Abbas Y, Umbreen K, Ashiq MN, Shah A (2018) One-pot facile synthesis of the ZnO/ZnSe heterostructures for efficient photocatalytic degradation of azo dye. Appl Surf Sci 459:194–200
23. Fan H, Zhang M, Bhandari B, Yang C (2020) Food waste as a carbon source in carbon quantum dots technology and their applications in food safety detection. Trends Food Sci Technol 95:86–96
24. Fan J, Li D, Wang X (2016) Effect of modified graphene quantum dots on photocatalytic degradation property. Diam Relat Mater 69:81–85

25. Feng L, Qin Z, Huang Y, Peng K, Wang F, Yan Y, Chen Y (2020) Boron-, sulfur-, and phosphorus-doped graphene for environmental applications. Sci Total Environ 698:134239
26. Guan Z-C, Jin P, Liu Q, Wang X, Chen L-F, Xu H, Song G-L, Du R-G (2019) Carbon quantum dots/Ag sensitized TiO2 nanotube film for applications in photocathodic protection. J Alloys Compd 797:912–921
27. Guo Y, Cao F, Li Y (2018) Solid phase synthesis of nitrogen and phosphor co-doped carbon quantum dots for sensing Fe^{3+} and the enhanced photocatalytic degradation of dyes. Sens Actuators B Chem 255:1105–1111
28. Haji S, Benstaali B, Al-Bastaki N (2011) Degradation of methyl orange by UV/H_2O_2 advanced oxidation process. Chem Eng J 168:134–139
29. Han SH, Yeom YS, Ko JG, Kang HC, Yoon HG (2015) Effect of the degree of crystallinity on the electrical properties of MWCNT filled poly (ethylene-co-ethyl acrylate)/LDPE blend composites prepared by melt mixing. Compos Sci Technol 117:351–356
30. Hazarika D, Karak N (2016) Photocatalytic degradation of organic contaminants under solar light using carbon dot/titanium dioxide nanohybrid, obtained through a facile approach. Appl Surf Sci 376:276–285
31. He M, Guo X, Huang J, Shen H, Zeng Q, Wang L (2018) Mass production of tunable multi-color graphene quantum dots from an energy resource of coke by a one-step electrochemical exfoliation. Carbon N Y 140:508–520
32. Hou J, Li H, Tang Y, Sun J, Fu H, Qu X, Xu Z, Yin D, Zheng S (2018) Supported N-doped carbon quantum dots as the highly effective peroxydisulfate catalysts for bisphenol F degradation. Appl Catal B Environ 238:225–235
33. Hu X, Sun Z, Song J, Zhang G, Zheng S (2018) Facile synthesis of nano-TiO_2/stellerite composite with efficient photocatalytic degradation of phenol. Adv Powder Technol 29:1644–1654
34. Hui Z, Zhao L, Geng F, Guo LH, Wan B, Yang Y (2016) Carbon dots decorated graphitic carbon nitride as an efficient metal-free photocatalyst for phenol degradation. Appl Catal B Environ 180:656–662
35. Jaafarzadeh N, Ghanbari F, Ahmadi M (2017) Catalytic degradation of 2,4-dichlorophenoxyacetic acid (2,4-D) by nano-Fe2O3 activated peroxymonosulfate: Influential factors and mechanism determination. Chemosphere 169:568–576
36. Jacob JM, Rajan R, Aji M, Kurup GG, Pugazhendhi A (2019) Bio-inspired ZnS quantum dots as efficient photo catalysts for the degradation of methylene blue in aqueous phase. Ceram Int 45:4857–4862
37. Javed M, Saqib ANS, Ata-ur-Rehman Ali B, Faizan M, Anang DA, Iqbal Z, Abbas SM (2019) Carbon quantum dots from glucose oxidation as a highly competent anode material for lithium and sodium-ion batteries. Electrochim Acta 297:250–257
38. Jiang X, Qin D, Mo G, Feng J, Yu C, Mo W, Deng B (2019) Ginkgo leaf-based synthesis of nitrogen-doped carbon quantum dots for highly sensitive detection of salazosulfapyridine in mouse plasma. J Pharm Biomed Anal 164:514–519
39. Jindal S, Giripunje SM (2018) An insight into electronic and optical properties of multilayer graphene quantum dots synthesized by hydrothermal approach. Synth Met 239:36–42
40. Jing S, Zhao Y, Sun RC, Zhong L, Peng X (2019) Facile and high-yield synthesis of carbon quantum dots from biomass-derived carbons at mild condition. ACS Sustain Chem Eng 7:7833–7843
41. Joshi PN, Mathias A, Mishra A, Mathias A (2018) Synthesis of ecofriendly fluorescent carbon dots and their biomedical and environmental applications. Mater Technol 33:672–680
42. Kalaiyarasan G, Joseph J (2019) Cholesterol derived carbon quantum dots as fluorescence probe for the specific detection of hemoglobin in diluted human blood samples. Mater Sci Eng C 94:580–586
43. Kandi D, Martha S, Thirumurugan A, Parida KM (2017) CdS QDs-decorated self-doped γ-Bi 2 MoO 6: a sustainable and versatile photocatalyst toward photoreduction of Cr(VI) and degradation of phenol. ACS Omega 2:9040–9056

44. Kaur M, Kaur M, Sharma VK (2018) Nitrogen-doped graphene and graphene quantum dots: a review onsynthesis and applications in energy, sensors and environment. Adv Colloid Interface Sci 259:44–64
45. Ke J, Li X, Zhao Q, Liu B, Liu S, Wang S (2017) Upconversion carbon quantum dots as visible light responsive component for efficient enhancement of photocatalytic performance. J Colloid Interface Sci 496:425–433
46. Khan ZMSH, Rahman RS, Shumaila Islam S, Zulfequar M (2019) Hydrothermal treatment of red lentils for the synthesis of fluorescent carbon quantum dots and its application for sensing Fe^{3+}. Opt Mater (Amst) 91:386–395
47. Kim SR, Ali I, Kim JO (2019) Phenol degradation using an anodized graphene-doped TiO_2 nanotube composite under visible light. Appl Surf Sci 477:71–78
48. Lei CW, Hsieh ML, Liu WR (2019) A facile approach to synthesize carbon quantum dots with pH-dependent properties. Dye Pigment 169:73–80
49. Li H, Ji J, Cheng C, Liang K (2018) Preparation of phenol-formaldehyde resin-coupled TiO_2 and study of photocatalytic activity during phenol degradation under sunlight. J Phys Chem Solids 122:25–30
50. Li J, Liu K, Xue J, Xue G, Sheng X, Wang H, Huo P, Yan Y (2019) CQDS preluded carbon-incorporated 3D burger-like hybrid ZnO enhanced visible-light-driven photocatalytic activity and mechanism implication. J Catal 369:450–461
51. Li Z, Guo S, Yuan Z, Lu C (2017) Carbon quantum dot-gold nanocluster nanosatellite for ratiometric fluorescence probe and imaging for hydrogen peroxide in living cells. Sens Actuators B Chem 241:821–827
52. Liang Q, Ma W, Shi Y, Li Z, Yang X (2013) Easy synthesis of highly fluorescent carbon quantum dots from gelatin and their luminescent properties and applications. Carbon N Y 60:421–428
53. Liang Z, Yang J, Zhou C, Mo Q, Zhang Y (2018) Carbon quantum dots modified BiOBr microspheres with enhanced visible light photocatalytic performance. Inorg Chem Commun 90:97–100
54. Lim H, Liu Y, Kim HY, Son DI (2018) Facile synthesis and characterization of carbon quantum dots and photovoltaic applications. Thin Solid Films 660:672–677
55. Lin X, Liu C, Wang J, Yang S, Shi J, Hong Y (2019) Graphitic carbon nitride quantum dots and nitrogen-doped carbon quantum dots co-decorated with $BiVO_4$ microspheres: a ternary heterostructure photocatalyst for water purification. Sep Purif Technol 226:117–127
56. Liu B, Liu X, Li L, Li J, Li C, Gong Y, Niu L, Zhao X, Sun CQ (2018) $ZnIn_2S_4$ flowerlike microspheres embedded with carbon quantum dots for efficient photocatalytic reduction of Cr(VI). Chinese J Catal 39:1901–1909
57. Liu C, Li X, Li J, Sun L, Zhou Y, Guan J, Wang H, Huo P, Ma C, Yan Y (2019) Carbon dots modifying sphere-flower $CdIn_2S_4$ on N-rGO sheet muti-dimensional photocatalyst for efficient visible degradation of 2,4-dichlorophenol. J Taiwan Inst Chem Eng 99:142–153
58. Lu K, Quan Q, Zhang N, Xu Y (2016) Multifarious roles of carbon quantum dots in heterogeneous photocatalysis. J Energy Chem 25:927–935
59. Lu M, Duan Y, Song Y, Tan J, Zhou L (2018) Green preparation of versatile nitrogen-doped carbon quantum dots from watermelon juice for cell imaging, detection of Fe^{3+} ions and cysteine, and optical thermometry. J Mol Liq 269:766–774
60. Ma D, Zhong J, Li J, Burda C, Duan R (2019) Preparation and photocatalytic performance of MWCNTs/BiOCl: Evidence for the superoxide radical participation in the degradation mechanism of phenol. Appl Surf Sci 480:395–403
61. Ma H, Ma W, Chen JF, Liu XY, Peng YY, Yang ZY, Tian H, Long YT (2018) Quantifying visible-light-induced electron transfer properties of single dye-sensitized ZnO entity for water splitting. J Am Chem Soc 140:5272–5279
62. Maallah R, Moutcine A, Laghlimi C, Smaini MA, Chtaini A (2019) Electrochemical biosensor for degradation of phenol in the environment. Sens Bio-Sens Res 24:100279
63. Madrakian T, Maleki S, Gilak S, Afkhami A (2017) Turn-off fluorescence of amino-functionalized carbon quantum dots as effective fluorescent probes for determination of isotretinoin. Sens Actuators B Chem 247:428–435

64. Martins NCT, Ângelo J, Girão AV, Trindade T, Andrade L, Mendes A (2016) N-doped carbon quantum dots/TiO_2 composite with improved photocatalytic activity. Appl Catal B Environ 193:67–74
65. Meghdad P, Moradi S, Shahlaei M, Farhadian N (2018) Application of carbon dots as efficient catalyst for the green oxidation of phenol: kinetic study of the degradation and optimization using response surface methodology. J Hazard Mater 353:444–453
66. Mo Y, Tan Z, Sun L, Lu Y, Liu X (2020) Hydrothermal method. Dict Gems Gemol 812:440
67. Mohapatra S, Bera MK, Das RK (2018) Rapid "turn-on" detection of atrazine using highly luminescent N-doped carbon quantum dot. Sens Actuators B Chem 263:459–468
68. Monte SS, Andrade SIE, Lima MB, Araujo MCU (2019) Synthesis of highly fluorescent carbon dots from lemon and onion juices for determination of riboflavin in multivitamin/mineral supplements. J Pharm Anal 9:209–216
69. Mukthar Ali M, Arya Nair JS, Sandhya KY (2019) Role of reactive oxygen species in the visible light photocatalytic mineralization of rhodamine B dye by P25–carbon dot photocatalyst. Dye Pigment 163:274–284
70. Murugan N, Prakash M, Jayakumar M, Sundaramurthy A, Sundramoorthy AK (2019) Green synthesis of fluorescent carbon quantum dots from Eleusine coracana and their application as a fluorescence 'turn-off' sensor probe for selective detection of Cu2+. Appl Surf Sci 476:468–480
71. Najafian H, Manteghi F, Beshkar F, Salavati-Niasari M (2019) Enhanced photocatalytic activity of a novel NiO/Bi_2O_3/Bi_3ClO_4 nanocomposite for the degradation of azo dye pollutants under visible light irradiation. Sep Purif Technol 209:6–17
72. Namdari P, Negahdari B, Eatemadi A (2017) Synthesis, properties and biomedical applications of carbon-based quantum dots: an updated review. Biomed Pharmacother 87:209–222
73. Omer KM, Tofiq DI, Ghafoor DD (2019) Highly photoluminescent label free probe for Chromium (II) ions using carbon quantum dots co-doped with nitrogen and phosphorous. J Lumin 206:540–546
74. Ong CB, Ng LY, Mohammad AW (2018) A review of ZnO nanoparticles as solar photocatalysts: synthesis, mechanisms and applications. Renew Sustain Energy Rev 81:536–551
75. Ovejero D, Barco B, Faraldos M (2019) An approach on the comparative behavior of chloro/nitro substituted phenols photocatalytic degradation in water
76. Pan J, Sheng Y, Zhang J, Huang P, Zhang X, Feng B (2015) Photovoltaic conversion enhancement of a carbon quantum dots/p-type $CuAlO_2$/n-type ZnO photoelectric device. ACS Appl Mater Interfaces 7:7878–7883
77. Parthiban V, Panda SK, Sahu AK (2018) Highly fluorescent carbon quantum dots-Nafion as proton selective hybrid membrane for direct methanol fuel cells. Electrochim Acta 292:855–864
78. Piri M, Sepehr E, Rengel Z (2019) Citric acid decreased and humic acid increased Zn sorption in soils. Geoderma 341:39–45
79. Pirsaheb M, Asadi A, Sillanpää M, Farhadian N (2018) Application of carbon quantum dots to increase the activity of conventional photocatalysts: a systematic review. J Mol Liq 271:857–871
80. Pourreza N, Ghomi M (2019) Green synthesized carbon quantum dots from Prosopis juliflora leaves as a dual off-on fluorescence probe for sensing mercury (II) and chemet drug. Mater Sci Eng, C 98:887–896
81. Prekodravac J, Vasiljević B, Marković Z, Jovanović D, Kleut D, Špitalský Z, Mičušik M, Danko M, Bajuk-Bogdanović D, Todorović–Marković B (2019) Green and facile microwave assisted synthesis of (metal-free) N-doped carbon quantum dots for catalytic applications. Ceram Int 45:17006–17013
82. Qi H, Teng M, Liu M, Liu S, Li J, Yu H, Teng C, Huang Z, Liu H, Shao Q, Umar A, Ding T, Gao Q, Guo Z (2019) Biomass-derived nitrogen-doped carbon quantum dots: highly selective fluorescent probe for detecting Fe^{3+} ions and tetracyclines. J Colloid Interface Sci 539:332–341

83. Qiang T, Han M, Wang X (2019) Waterborne polyurethane/carbon quantum dot nanocomposite as a surface coating material exhibiting outstanding luminescent performance. Prog Org Coat 138:105433
84. Qiankun C, Chen L, Qi J, Tong Y, Lv Y, Xu C, Ni J, Liu W (2019) Photocatalytic degradation of amoxicillin by carbon quantum dots modified $K_2Ti_6O_{13}$ nanotubes: effect of light wavelength. Chinese Chem Lett 30:1214–1218
85. Qiao G, Lu D, Tang Y, Gao J, Wang Q (2019) Smart choice of carbon dots as a dual-mode onsite nanoplatform for the trace level detection of Cr_2O_{72}. Dye Pigment 163:102–110
86. Qu A, Xie H, Xu X, Zhang Y, Wen S, Cui Y (2016) High quantum yield graphene quantum dots decorated TiO_2 nanotubes for enhancing photocatalytic activity. Appl Surf Sci 375:230–241
87. Que Q, Xing Y, He Z, Yang Y, Yin X, Que W (2017) Bi_2O_3/carbon quantum dots heterostructured photocatalysts with enhanced photocatalytic activity. Mater Lett 209:220–223
88. Rahmanian O, Dinari M, Karimi M (2018) Carbon quantum dots/ layered double hydroxide hybrid for fast and efficient decontamination of Cd (II): the adsorption kinetics and isotherms. Appl Surf Sci 428:272–279
89. Rajabi HR, Arjmand H, Kazemdehdashti H, Farsi M (2016) A comparison investigation on photocatalytic activity performance and adsorption efficiency for the removal of cationic dye: quantum dots vs. magnetic nanoparticles. J Environ Chem Eng 4:2830–2840
90. Rajender G, Kumar J, Giri PK (2018) Interfacial charge transfer in oxygen deficient TiO_2-graphene quantum dot hybrid and its influence on the enhanced visible light photocatalysis. Appl Catal B Environ 224:960–972
91. Ramar V, Moothattu S, Balasubramanian K (2018) Metal free, sunlight and white light based photocatalysis using carbon quantum dots from Citrus grandis: a green way to remove pollution. Sol Energy 169:120–127
92. Ratnayake SP, Mantilaka MMMGPG, Sandaruwan C, Dahanayake D, Murugan E, Kumar S, Amaratunga GAJ, de Silva KMN (2019) Carbon quantum dots-decorated nano-zirconia: a highly efficient photocatalyst. Appl Catal A Gen 570:23–30
93. Ren Z, Liu X, Chu H, Yu H, Xu Y, Zheng W, Lei W, Chen P, Li J, Li C (2017) Carbon quantum dots decorated $MoSe_2$ photocatalyst for Cr(VI) reduction in the UV–vis-NIR photon energy range. J Colloid Interface Sci 488:190–195
94. Rodríguez D, Algarra M, Tarelho LAC, Frade J, Franco A, de Miguel G, Jiménez J, Rodríguez-Castellón E, Luque R (2018) Catalyzed microwave-assisted preparation of carbon quantum dots from lignocellulosic residues. ACS Sustain Chem Eng 6:7200–7205
95. Romero V, Vila V, de la Calle I, Lavilla I, Bendicho C (2019) Turn–on fluorescent sensor for the detection of periodate anion following photochemical synthesis of nitrogen and sulphur co–doped carbon dots from vegetables. Sens Actuators B Chem 280:290–297
96. Rosman N, Salleh WNW, Ismail AF, Jaafar J, Harun Z, Aziz F, Mohamed MA, Ohtani B, Takashima M (2018) Photocatalytic degradation of phenol over visible light active $ZnO/Ag_2CO_3/Ag_2O$ nanocomposites heterojunction. J Photochem Photobiol A Chem 364:602–612
97. Sabet M, Mahdavi K (2019) Green synthesis of high photoluminescence nitrogen-doped carbon quantum dots from grass via a simple hydrothermal method for removing organic and inorganic water pollutions. Appl Surf Sci 463:283–291
98. Safardoust H, Salavati M (2017) Degradation of methylene blue as a pollutant with N-doped graphene quantum dot/ titanium dioxide nanocomposite. J Clean Prod 148:31–36
99. Saikia M, Hower JC, Das T, Dutta T, Saikia BK (2019) Feasibility study of preparation of carbon quantum dots from Pennsylvania anthracite and Kentucky bituminous coals. Fuel 243:433–440
100. Sakho Ehadji M, Oluwafemi OS, Perumbilavil S, Philip R, Kala MS, Thomas S, Kalarikkal N (2016) Rapid and facile synthesis of graphene oxide quantum dots with good linear and nonlinear optical properties. J Mater Sci Mater Electron 27:10926–10933
101. Samsudin MFR, Bacho N, Sufian S, Ng YH (2019) Photocatalytic degradation of phenol wastewater over Z-scheme $g\text{-}C_3N_4/CNT/BiVO_4$ heterostructure photocatalyst under solar light irradiation. J Mol Liq 277:977–988

102. Saud PS, Pant B, Alam A-M, Ghouri ZK, Park M, Kim H-Y (2015) Carbon quantum dots anchored TiO_2 nanofibers: effective photocatalyst for waste water treatment. Ceram Int 41:11953–11959
103. Schneider EM, Bärtsch A, Stark WJ, Grass RN (2019) Safe one-pot synthesis of fluorescent carbon quantum dots from lemon juice for a hands-on experience of nanotechnology. J Chem Educ 96:540–545
104. Scott T, Zhao H, Deng W, Feng X, Li Y (2019) Photocatalytic degradation of phenol in water under simulated sunlight by an ultrathin MgO coated Ag/TiO_2 nanocomposite. Chemosphere 216:1–8
105. Shahbazi R, Payan A, Fattahi M (2018) Preparation, evaluations and operating conditions optimization of nano TiO_2 over graphene based materials as the photocatalyst for degradation of phenol. J Photochem Photobiol A Chem 364:564–576
106. Sharma S, Dutta V, Singh P, Raizada P, Rahmani-Sani A, Hosseini-Bandegharaei A, Thakur VK (2019) Carbon quantum dot supported semiconductor photocatalysts for efficient degradation of organic pollutants in water: a review. J Clean Prod 228:755–769
107. Shen T, Wang Q, Guo Z, Kuang J, Cao W (2018) Hydrothermal synthesis of carbon quantum dots using different precursors and their combination with TiO_2 for enhanced photocatalytic activity. Ceram Int 44:11828–11834
108. Shi Y, Liu X, Wang M, Huang J, Jiang X, Pang J, Xu F, Zhang X (2019) Synthesis of N-doped carbon quantum dots from bio-waste lignin for selective irons detection and cellular imaging. Int J Biol Macromol 128:537–545
109. Sin JC, Lim CA, Lam SM, Mohamed AR, Zeng H (2019) Facile synthesis of novel ZnO/Nd-doped BiOBr composites with boosted visible light photocatalytic degradation of phenol. Mater Lett 248:20–23
110. Souza DR, Caminhas LD, de Mesquita JP, Pereira FV (2018) Luminescent carbon dots obtained from cellulose. Mater Chem Phys 203:148–155
111. Yulei S, Ling W, Shengkui Z, Qingxia L (2019) Carbon quantum dots/TiO nanosheets with dominant (001) facets for enhanced photocatalytic hydrogen evolution. Appl Surf Sci 480:810–816
112. Sun S, Zhao R, Xie Y, Liu Y (2019) Photocatalytic degradation of aflatoxin B1 by activated carbon supported TiO_2 catalyst. Food Control 100:183–188
113. Tammina SK, Mandal BK, Kadiyala NK (2018) Photocatalytic degradation of methylene blue dye by nonconventional synthesized SnO_2 nanoparticles. Environ Nanotechnol Monit Manag 10:339–350
114. Tammina SK, Yang D, Koppala S, Cheng C, Yang Y (2019) Highly photoluminescent N, P doped carbon quantum dots as a fluorescent sensor for the detection of dopamine and temperature. J Photochem Photobiol B Biol 194:61–70
115. Teymourinia H, Salavati-Niasari M, Amiri O, Safardoust-Hojaghan H (2017) Synthesis of graphene quantum dots from corn powder and their application in reduce charge recombination and increase free charge carriers. J Mol Liq 242:447–455
116. Tian J, Liu R, Liu Z, Yu C, Liu M (2017) Boosting the photocatalytic performance of Ag_2CO_3 crystals in phenol degradation via coupling with trace N-CQDs. Cuihua Xuebao/Chinese J Catal 38:1999–2008
117. Tian P, Tang L, Teng KS, Lau SP (2018) Graphene quantum dots from chemistry to applications. Mater Today Chem 10:221–258
118. Tsuji M, Matsuda K, Tanaka M, Kuboyama S, Uto K, Wada N, Kawazumi H, Tsuji T, Ago H, Hayashi J (2018) Enhanced photocatalytic degradation of methyl orange by Au/TiO_2 nanoparticles under neutral and acidic solutions. ChemistrySelect 3:1432–1438
119. Wang L, Wang Y, Hu Y, Wang G, Dong S, Hao J (2019) Magnetic networks of carbon quantum dots and Ag particles. J Colloid Interface Sci 539:203–213
120. Wang R, Lu K, Tang Z, Xu Y (2017) Recent progress in carbon quantum dots: synthesis, properties and applications in photocatalysis. J Mater Chem A 5:3717–3734
121. Wang Y, Hu A (2014) Carbon quantum dots: synthesis, properties and applications. J Mater Chem C 2:6921

122. Wang Z, Zhao X, Guo Z, Miao P, Gong X (2018) Carbon dots based nanocomposite thin film for highly efficient luminescent solar concentrators. Org Electron 62:284–289
123. Xie R, Zhang L, Xu H, Zhong Y, Sui X, Mao Z (2016) Construction of up-converting fluorescent carbon quantum dots/$Bi_{20}TiO_{32}$ composites with enhanced photocatalytic properties under visible light. Chem Eng J
124. Xie R, Zhang L, Xu H, Zhong Y, Sui X, Mao Z (2017) Construction of up-converting fluorescent carbon quantum dots/$Bi_{20}TiO_{32}$ composites with enhanced photocatalytic properties under visible light. Chem Eng J 310:79–90
125. Xie X, Yang Y, Xiao Y-H, Huang X, Shi Q, Zhang W-D (2018) Enhancement of photoelectrochemical activity of Fe2O3 nanowires decorated with carbon quantum dots. Int J Hydrogen Energy 43:6954–6962
126. Xinyue W, Sun Y, Yang L, Shang Q, Wang D, Guo T, Guo Y (2019) Novel photocatalytic system Fe-complex/TiO_2 for efficient degradation of phenol and norfloxacin in water. Sci Total Environ 656:1010–1020
127. Xiyang M, Xiang Q, Liao Y, Wen T, Zhang H (2018) Visible-light-driven CdSe quantum dots/graphene/TiO_2 nanosheets composite with excellent photocatalytic activity for *E. coli* disinfection and organic pollutant degradation. Appl Surf Sci 457:846–855
128. Xu X, Bao Z, Zhou G, Zeng H, Hu J (2016) Enriching photoelectrons via three transition channels in amino-conjugated carbon quantum dots to boost photocatalytic hydrogen generation. ACS Appl Mater Interfaces 8:14118–14124
129. Xue B, Yang Y, Sun Y, Fan J, Li X, Zhang Z (2019) Photoluminescent lignin hybridized carbon quantum dots composites for bioimaging applications. Int J Biol Macromol 122:954–961
130. Yan Y, Kuang W, Shi L, Ye X, Yang Y, Xie X, Shi Q, Tan S (2019) Carbon quantum dot-decorated TiO_2 for fast and sustainable antibacterial properties under visible-light. J Alloys Compd 777:234–243
131. Yang P, Zhu Z, Chen M, Chen W, Zhou X (2018) Microwave-assisted synthesis of xylan-derived carbon quantum dots for tetracycline sensing. Opt Mater (Amst) 85:329–336
132. Yang W, Yang H, Ding W, Zhang B, Zhang L, Wang L (2016) High quantum yield ZnO quantum dots synthesizing via an ultrasonication microreactor method. Ultrason Sonochem 33:106–117
133. Ye Q, Huang Y, Grier S, Miller SA (2019) Oxidative degradation pathways of beclabuvir hydrochloride mediated by hydrogen peroxide and UV/Vis light. J Pharm Biomed Anal 172:388–394
134. Yu J, Han L, Liu SG, Ju YJ, Gao X, Li NB, Luo HQ (2019) Green fluorescent carbon quantum dots as a label-free probe for rapid and sensitive detection of hematin. Spectrochim Acta Part A Mol Biomol Spectrosc 212:167–172
135. Yu P, Lowe SE, Simon GP, Zhong YL (2015) Electrochemical exfoliation of graphite and production of functional graphene. Curr Opin Colloid Interface Sci 20:329–338
136. Zhang H, Zhao L, Geng F, Guo L, Wan B, Yang Y (2015) Carbon dots decorated graphitic carbon nitride as an efficient metal-free photocatalyst for phenol degradation. Elsevier BV
137. Zhang Y, Wang L, Yang M, Wang J, Shi J (2019) Carbon quantum dots sensitized $ZnSn(OH)_6$ for visible light-driven photocatalytic water purification. Appl Surf Sci 466:515–524
138. Zhang Z, Wu L, Wang P, Zhang Y, Wan S, Guo X, Jin W, Zhang J (2018) Carbon quantum dots modified $La_2Ti_2O_7$ nanosheets for visible light photocatalysis. Mater Lett 230:72–75
139. Zhao L, Hou H, Fang Z, Gao F, Wang L, Chen D, Yang W (2019) Hydrogenated TiO_2 Nanorod arrays decorated with carbon quantum dots toward efficient photoelectrochemical water splitting. ACS Appl Mater Interfaces 11:19167–19175
140. Zhao Y, Duan J, He B, Jiao Z, Tang Q (2018) Improved charge extraction with N-doped carbon quantum dots in dye-sensitized solar cells. Electrochim Acta 282:255–262
141. Zhao Y, Zhang Y, Liu X, Kong H, Wang Y, Qin G, Cao P, Song X, Yan X, Wang Q, Qu H (2017) Novel carbon quantum dots from egg yolk oil and their haemostatic effects. Sci Rep 7:4452

142. Zhong Q, Chen Y, Su A, Wang Y (2018) Synthesis of catalytically active carbon quantum dots and its application for colorimetric detection of glutathione. Sens Actuators B Chem 273:1098–1102
143. Zhou L, Qiao M, Zhang L, Sun L, Zhang Y, Liu W (2019) Green and efficient synthesis of carbon quantum dots and their luminescent properties. J Lumin 206:158–163

Synthesis of Carbon Nanofibers and Its Application in Environmental Remediation

Ritu Painuli, Praveen Kumar Yadav, Sapna Raghav, and Dinesh Kumar

Abstract Owing to the inimitable properties of the carbon nanofibers (CNFs), for instance, the enhanced surface-to-volume ratio, nanoscale diameter, physical, mechanical, and chemical properties, they have excellent capabilities in science, biomedicine, energy storage, and environmental science. Carbon fibers prepared from various synthetic techniques have different carbon morphologies and structures. The carbon fibers prepared from electrospinning, chemical vapor deposition with the consequent chemical treatment have flat, mesoporous, and porous surfaces. Along with this, the carbon fibers can be altered with the several materials to expand their application in various fields. Thus, in this chapter, we concentrate on the synthesis and design along with the application of the carbon nanofibers. The synthesis routes of CNFs like chemical vapor deposition (CVD), substrate method, phase separation, electrospinning, etc., have been introduced. In addition, the synthesis of carbon nanocomposites has also been discussed. In addition, the application of the prepared carbon fibers in the various environmental fields has also been explored.

Keywords Carbon fibers · Synthesis · Nanocomposites · Applications

R. Painuli · S. Raghav
Department of Chemistry, Banasthali Vidyapith, Banasthali, Tonk 304022, India
e-mail: ritsjune8.h@gmail.com

S. Raghav
e-mail: sapnaraghav04@gmail.com

P. K. Yadav
Academy of Science and Innovative Research (AcSIR), Chemical and Food BND Group, IRM (BND), CSIR–National Physical Laboratory, New Delhi 110012, India
e-mail: kpraveen.yadav414@gmail.com

D. Kumar (✉)
School of Chemical Sciences, Central University of Gujarat, Gandhinagar, India
e-mail: dinesh.kumar@cug.ac.in

M. Jawaid et al. (eds.), *Environmental Remediation Through Carbon Based Nano Composites*, Green Energy and Technology,
https://doi.org/10.1007/978-981-15-6699-8_15

1 Introduction

In recent years, progress in nanoscience has led to the creation of many nanomaterials (NMs) for sensing applications [76]. Among the various nanomaterial (NM), one-dimensional (1D) materials have gained noteworthy potential [23]. The 1D materials enable short paths for the electrons transfer and encourage electrolyte penetration along the axis of nanofiber [86]. This enhances the sensing application of the nanofibers. Owing to the inimitable optical, electrical, and mechanical properties of CNTs, they have been extensively used for preparing the sensors and biosensors [53]. Besides CNTs, carbon nanofibers (CNFs) have also been widely used or examined because of their unique physical, chemical properties [11, 56]. CNFs possess a high potential for the modification or alteration of surface to form functional hybrid CNF-based NMs which have been utilized in the areas of medicines [71], nanodevices, tissue engineering [1], sensors [38, 50], energy storage [10], and environmental science [52, 66].

CNFs are the filaments present in the nanometer range, organized in graphene layers with a specific alignment parallel to the fiber axis. According to the angle between the growth axis and graphene layers, they are usually classified into three categories, i.e., fishbone, parallel, and platelet. Their arrangement can be found via transmission electron microscopy (TEM). In the CNFs, the regular arrangement among the sheets of a graphene is ~3.4 Å, which is very near to that of graphite diameter, i.e., 0.335 nm. This is the reason that the CNFs are mentioned as graphite nanofibers. The properties of CNFs can be differentiated by seeing the structure derived from the powdered material, the structure of the distinct nanofibers, and the agglomeration of filaments [63]. The difference in the structure of CNFs and CNTs cannot be easily distinguished from the TEM. Under the theoretical definition, nanotubes are synthesized either by the single graphene wrapped in the cylindrical tube, i.e., single-walled CNT or many sheets wrapped together, i.e., multi-walled CNT. In contrast, in CNFs, the graphene layers may not be continuous. In terms of properties, CNTs possess excellent thermal, electrical conductivities, better mechanical resistance, and enhanced structural features. However, the main drawbacks associated with the CNFs are their complex scalability and their excessive cost. CNFs can be divided on the basis of their purpose by concerning the mechanical property necessities and tensile strength and Young's modulus [54]. CNFs are simultaneously categorized as ultrahigh strength and ultrahigh modulus. The CNFs are also classified as super-high strength owing to their high tensile strength. The mechanical properties of carbon fiber can differ even on having an undistinguishable origin and equivalent thickness. Therefore, the main dissimilarity is determined by the arrangement of the fiber. The excellent electrical conductivity of CNFs is of the significant consideration for many applications ranging from electronics to composites. In this chapter, we focus on preparing CNFs by thermal chemical vapor deposition, gas-phase flow catalytic method, spray method, plasma-enhanced chemical vapor deposition, substrate method, electrospinning, phase separation, and templating. The second part explored the preparation of CNFs nanocomposites. In the third part, various applications of CNFs towards gas sensors, sensors for small molecules, air filtrations, sensors for small molecules, etc., are deliberated. Last, the conclusions and outlooks of the CNFs preparations and its applications are given.

2 Synthesis of Carbon Nanofibers

Owing to the many advantages of CNFs, for instance, enhanced surface area, less density, high specific modulus, excellent strength, good thermal and electrical conductivity, etc, the CNFs have their applications in areas of sensing, adsorbent, electrochemistry, adsorbent, storage, etc. [17, 72]. The following represents the methods that have been used for preparing CNFs. Figure 1 depicts the various methods for preparing CNFs.

2.1 Thermal Chemical Vapor Deposition

For the fabrication of CNFs by the chemical vapor (CVD) deposition method, thermal decomposition of the cost-effective hydrocarbon is carried out over a metal catalyst at a constant temperature of 500–100 °C [82]. According to the fashion by which the catalyst added or present, the CVD method can be categorized into the main types: substrate method and spray method.

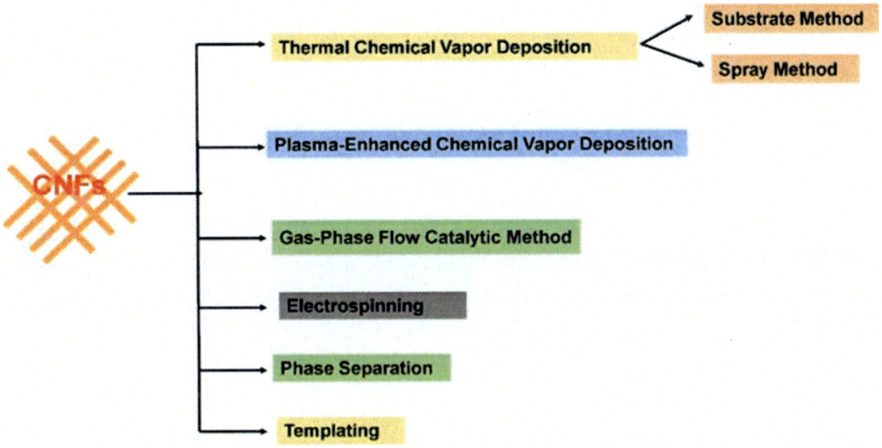

Fig. 1 Various methods for preparing CNFs

2.1.1 Substrate Method

In the substrate method, the SiO_2 fibers or ceramic are utilized as the substrate for the uniform dispersion of the catalyst particles (in their nanosized form) over its surface. At the surface of the catalyst, the H_2 gas is pyrolyzed, then the deposition of the carbon occurs, and further, it is grown to obtain the carbon fibers in the nanoform. Enrique et al. developed high-purity CNFs by using nickel as the catalyst at 599 °C, and for the carbon source, they used $CH_4/C_2H_6/H_2$. Along with the synthesis of CNFs, the effect of various conditions, for instance, temperature, and carbon sources over the layer thickness, porosity, and uniformity of CNFs were also explored [18]. However, by this method, the CNFs are prepared for excellent purity. Since the fabrication of catalyst at the nanoscale is tedious and the product and catalyst cannot be separated in time, therefore it is difficult to obtain large scale production of carbon fibers with this method.

2.1.2 The Spray Method

In this method, the catalyst is mixed with the organic solvent like benzene, and this mixture is sprayed into a reaction chamber with high temperature, to obtain the CNFs. The growth of the carbon fiber, by the spray method, depends on the continuous injection of the catalyst helpful for industrial or large-scale production also [19]. The drawbacks related with this method are the irregular dispersal of the catalyst particles and the difficulty in controlling the ratio of hydrocarbon gases. These issues ultimately lead to the lower production of the CNFs, with a certain amount of carbon black.

2.2 *Plasma-Enhanced Chemical Vapor Deposition (PECVD)*

Deposition with the help of the plasma is also very important because the plasma possesses high-energy electrons, which offers an activation energy that is required in the CVD process. The electron collision with the gaseous molecules starts the excitation, decomposition, ionization, and compounding of gaseous molecules which produce chemical groups with excellent activity [22, 67]. This method can fabricate the aligned carbon fibers and demands a high cost of production with low production efficacy.

2.3 *Gas-Phase Flow Catalytic Method*

In this method, the catalyst precursor is heated directly, followed by the introduction into the reaction compartment along with the hydrocarbon gas. The hydrocarbon gas and the catalyst are decompositions at the two different temperature zones. Then, the catalyst that is decomposed is aggregated into the nanosized particles. Finally, the carbon fibers are synthesized at the nanoscale catalyst particles [8]. Subsequently, the catalyst particles that are decomposed from the organic compound can be disseminated in a 3D space. By using the method, the volatilization quantity can be simply managed. Hence, the amount of fabrication of carbon fibers in less amount of time is enormous, while the uninterrupted fabrication of carbon fibers can be obtained.

2.4 *Electrospinning*

In the 1930s, a revolutionary technology, i.e., electrospinning technology, was first introduced. It has received widespread attention over the years and has been using for preparing carbon fibers [25, 77]. In this process, a high voltage of static electricity is utilized for charging the polymer solution or melt (Fig. 2). In the presence of an electric field, a Taylor cone is formed by the charged polymers at the spinning port. The former Taylor cone then gets drafted or accelerated. The moving jet is progressively drafted and dispersed. The fibers deposited on the collecting plate are of nanosize because of the fast motion. This results in the formation of the fibrous mat, the same as that of woven fabric. The former fiber matrix is the air oxidized and carbonized in the N_2 environment to attain the carbon fibers.

In comparison to that of other methods available for the manufacturing of carbon fibers, the electrospinning methods possess the following advantages: (I) This method uses high voltages, but the consumption of current is less so that the energy utilization is very less (II) a nanofiber nonwoven fabric can be openly manufactured. The nanofibers formed by this process can be easily made into a nonwoven fabric in the 2D expanded form, as of which, no additional processing is needed after the spinning process. Specifically, the generation of numerous spinning amplified the manufacturing of nanofibers and also upgraded fabrication efficacy, (III) the electrospinning

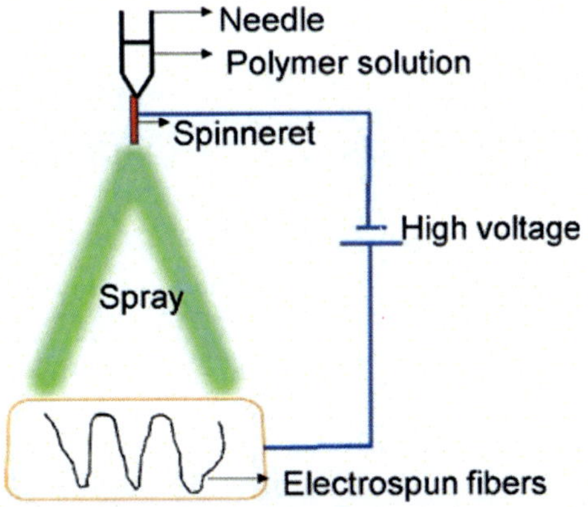

Fig. 2 Schematic representation of electrospinning technique for the synthesis of CNFs

method permits the spinning at room temperature (RT). As a result, a solution to having a low thermal stability compound can also be spun. The raw materials of diverse types have used synthetic polymers, i.e., polyamide, polyester, along with a natural high molecular mass like silk, DNA, collagen for the fabrication of carbon fibers by the electrospinning process.

2.5 *Phase Separation*

The phase separation is a new technique that comprises gelation, dissolutions, and extraction with the help of various drying, solvent, freezing processes that will lead to the formation of nanoporous foam. Converting the solid polymer into nanoporous foam takes a comparatively long time. The procedure of self-arrangement of randomly dispersed components results in the formation of the systematized assembly or configuration. Local interactions among the constituents themselves cause such an organization. Like that of phase separation, this technique is time–taking in the manufacturing of the continuous polymer nanofibers. Therefore, the electrospinning method is the most proper process for the manufacturing of continuous nanofibers from different polymers [87].

2.6 *Templating*

The different techniques that have extensively used for the fabrication of nanofibers include drawing, template synthesis, phase separation, and self-assembly [16, 24, 49]. CNFs fabricated by using this template technique are used to make solid or hollow nanofibers of the broad range of raw materials, comprising metals, semiconductors, carbons, electronically conducting polymers, etc. However, the manufacturing of one-by-one continuous nanofiber is not achievable by using this technique of nanofiber synthesis.

3 Preparation of CNF Composites

The complete efficiency of the CNF composites is usually administrated via the dispersion of carbon fibers into a matrix of a polymer. Hence, the role of dispersion is very important in the fabrication of the CNF composites. There are only two methods that govern CNF dispersion in polymer: the sonication process in less viscous solutions and the mixing process. Owing to features like cost-efficacy, straightforwardness, and obtainability, melt mixing method is the most efficiently used method for preparing CNF composites. The methods like a mini–max molder, Haake torque rheometer, and extrusion or roll mill [48, 60] all belong to the method melt mixing, where a trimmed mixing state is essential for gaining the proper dispersion conditions in the polymer matrix. The high shear mixing will cause a comparatively better CNT dispersion. The aspect ratio governs most of performing the CNF polymer composites. It has been observed that the decreases in the aspect ratio lead to a decrease in the properties of the CNT polymer composites [3]. Thus, an examination of the comparatively less shear mixing technique without the change of the dispersion is still a hurdle in their fabrication via the melt mixing method [47].

The promising method, i.e., the chemical surface treatment, helps the dispersion in the polymer where the compatibility among the polymer matrix and grafting functional group are the chief features that allow the dispersion of CNF and the overall performance of the CNF polymer composites. Usually, the surface of the CNF is treated by soaking it in H_2SO_4/HNO_3 at different temperatures, i.e., followed by the acylation. Then, the functional group adhered to the surface of the carbon fibers via the reaction between the functional groups and the oxidized CNF. By using triamines or diamines as the linker molecules, Li et al. synthesized and characterized the surface-treated CNF [40]. For forming the CNF–C(O)–NH– structure, the amine groups (a bridging compound) links the–NH_2 and CNF. The CNF/ethylene/propylene copolymer composite was synthesized by Kelarakis et al. [32]. The surface of as-prepared CNFs was oxidized by HNO_3/H_2SO_4 and then reduced by sodium borohydride for the formation of the structure of CNF–OH, which was then dispersed in absolute ethanol for forming the CNF–O– structure. In this method, before being mixed in the hardener, the CNFs are dispersed in the liquid epoxy form via sonication. Acetone or different solutions are used to help with the sonication effect. External cooling devices are used to minimize the increasing temperature through the sonication process in most cases. The nanocomposites preparation by the CNF and SC–15 epoxy was demonstrated by Pervin and coworkers [58]. The ultrasonication of SC–15 epoxy and carbon fibers with high intensity was done for performing the mixing process. After the completion of the sonication process, the mixture was filled with hardener, and then the mechanical stirring of high speed was done, followed by the preservation at RT. Choi and coworkers showed the preparation of CNF/nanocomposite [14]. The dispersion of CNF into acetone was carried out via the stirring and sonication process at the RT, followed by the addition of epoxy resin into the CNF acetone solution with continuous stirring and sonication. Then, the mixture is heated to remove the acetone, followed by the addition of the hardener. Finally, it is preserved at RT.

4 Applications

Carbon nanofibers (CNFs) are widely used in various industries such as biomedicine, analytical science, and environmental science because they exhibit exceptional chemical and physical properties. Besides this, these CNFs have a high surface-to-volume ratio, low defects, high electrical and thermal conductivity, good electron transferability, and easily modifiable surfaces. These properties of CNFs extend its application as sensors for detecting gas, biomolecule, strain, and pressure. CNF–based NMs are in great demand because of their novel characteristics, which make CNF-based NMs potential candidate for various sensing processes [30, 2, 79, 13]. Based on the target materials, CNFs-based NMs applications are as follows.

4.1 Gas Sensors

Li et al. [41] used a solid-phase graphitization method assisted with electrospinning and prepared a one-dimensional CNFs composed of graphitic nanorolls, which act as excellent RT sensors for explosive gases. These CNFs are sensitive to carbon monoxide, methane, hydrogen, and ethanol at RT. They detected carbon monoxide gas at low ppm concentrations [41]. Similarly, Zhang et al. [85] reported ZnO–CNFs composite-based H_2S sensor. The H_2S sensor showed high stability, selectivity, and linear response for H_2S in 50–102 ppm range [85]. In addition, some other workers, Claramunt et al. [15], did similar work for the detection of NH_3. They deposited metal NPs–decorated CNFs on Kapton for the detection of NH_3 [15]. The results showed that by controlling the percentage of Pd and Au, the sensitivity of CNFs to NH_3 could be improved. Moreover, the sensor showed a response time of up to 5 min within a temperature range of 110–120 °C. Moreover, on comparing with the spectroscopic sensors such as quartz-enhanced photoacoustic and mid-infrared sensors [33, 74] which possess the capability of quick detection at RT with no reagent, the operation temperature of Au and Pd NPs decorated CNFs was much higher. To overcome the limitation of the detection temperature, Lee et al. [37] developed a NO_2 gas sensor with a detection limit of 1 ppm. It comprises Wo_3 nanomodule-decorated hybrid carbon nanofibers. This sensor offers a higher sensing surface area. At the material surface, WO^{2+} is associated with the oxygen of NO_2, which helps in the exposure of NO_2 gas at RT [37].

4.2 Strain/Pressure Sensors

A pressure sensor is a device that is used to convert the pressure into an electric signal. This sensor is applied to gases and liquids. Silicon piezoresistive pressure sensor and silicon capacitive pressure sensor that come under the conventional microelectromechanical system (MEMS) have a high potential as sensors because of their several advantages such as they are accurate, power consumption is less, and they are cost-effective. Despite several advantages, they have some limitations and also, for example, they perform poorly in high-intensity piezoresistive measurements. CNFs are also utilized in health monitoring because of their high electrical conductivity, toughness, strain capacity, and low cost [75, 78]. Zhu et al. developed an electrically conductive polymer nanocomposite using the solvent-assisted casting method that can be utilized as strain sensors with large mechanical deformation. Two elastomers (VM_1, VM_2) with somewhat different compositions have been utilized as the hosting polymer matrix. It is used to manufacture the conductive PNCs strengthened with CNFs. The dielectric performance of the PNCs has been compared. Zhu et al. developed an electrically conductive polymer nanocomposite using the solvent-assisted casting method that can be utilized as strain sensors with large mechanical deformation. Two elastomers (VM1, VM2) with somewhat different compositions have been utilized as the hosting polymer matrix. It is used to manufacture the conductive PNCs strengthened with CNFs. The dielectric performance of the PNCs has been compared. Unique negative permittivity was observed in the composites with the CNF concentration. Additionally, when an extremely large strain is applied, they showed appreciable resistivity, which makes it useful for sensing applications [88]. Similar work was also done by Azhari and coworkers. They also developed a piezoresistive sensor by mixing 1% carbon nanotubes and 15% CNFs. The sensor overcomes the limitation of traditional cement-based sensors and offers more accuracy and reproducibility. The load amplitudes provided by the sensor are up to 30 k, and the gauge factor is 445 [6]. A CNF cement-based composite was developed by Bazea et al. They also found that by adding 2 wt% CNFs to cement, a gauge factor of 190 can be obtained [7]. Hu et al. developed a highly sensitive strain sensor. The sensor is made up of metal (Ag)-coated CNFs and epoxy composites. When they compared the two sensors with and without Ag coating, they found that sensor with Ag coating shows higher strain sensitivity and better conductivity [26]. Tallman et al. by electrical impedance tomography (EIT) studied CNF/polyurethane (PU) nanocomposites for distributed strain sensing and tactile imaging, and for exploring the effect of CNFs filling volume fraction on piezoresistive response. They also revealed that the change in strain was because of a 12.5–15% filling volume fraction [69]. Yan and coworkers developed a flexible strain sensor with the help of carbon/graphene composites nanofiber yarn/thermoplastic polyurethane, with high stability and average gauge factor of >1700 under an applied strain of 2% [78].

4.3 *Small Molecules Sensors*

CNF-based NMs are widely used in many industries. There use is not limited to strain sensing and for the detection of gas molecules only. They can also be utilized for the detection of small molecules. Huang et al. developed a CNF loaded with palladium nanoparticle (Pd/CNFs) by the combination of two processes. One is electrospinning, and the other is thermal treatment processes. Scanning electron microscopy (SEM) and transmission electron microscopy (TEM) studies were done to characterize the nanoparticles. The electrochemical study (CV and EIS) showed that Pd/CNFs have high electron transfer ability and high electric conductivity. The Pd/CNF-modified carbon paste electrode (Pd/CNF–CPE) showed a direct and mediator fewer responses to H_2O_2 and NADH at low potentials. The Pd/CNF–CPE exhibits high sensitivity, wider linear range response, it is highly reproducible, and these properties make it a suitable and promising candidate for amperometric H_2O_2 or NADH sensor. The sensor was used for the detection of ascorbic acid (AA), uric acid (UA), and dopamine (DA) [29]. The detection limit of Pd/CNFs-based electrodes for DA was 0.2 μM, UA was 0.7 μM, and AA was 15 μM. The linear range was 0.5–160 μM, 2–200 mM, and 0.05–4 mM, respectively. There are many groups of researchers who have worked on Pd NP-loaded CNFs modified carbon paste electrode for sensing different molecules. For example, Liu et al. [43] used a similar electrode for oxalic acid detection with a linear range from 0.2 to 45 nM and a very low detection limit of 0.2 mM. Similarly, Liu et al. by the electrospinning process developed Ni/CNFs composite electrode for the detection of glucose [44]. The electrode is overly sensitive, stable, and catalytically active. The detection limit of the sensor for glucose was 1 μM. Li et al. by one-pot polymerization process synthesized a magnetic composite of Ni NP-loaded CNFs, the neurotransmitter dopamine, laccase. The magnetic composite is high selectivity towards catechol and showed a detection limit of 0.69 μM for catechol and linear range from 1 to 9100 μM [39]. Table 1 represents nanomaterial-assisted CNFs for the detection of small molecules [46].

4.4 Biomacromolecules Sensors

The CNFs have many active sites and high surface area. These properties of CNFs help in protein and enzyme adsorption. The high surface area and numerous active sites of CNFs helps not only in the protein and enzyme adsorption, but CNFs can also provide direct electron transfer and stabilize the enzyme activity [84]. Therefore owing to their wide range of potential, CNFs are the most suitable substrate for the sensor development [59]. Periyaruppan et al. developed a carbon nanofiber-based nanoelectrode arrays for the label-free detection of cardiac troponin–I. The sensor helps in the early detection of the detection of myocardial infarction, a heart disease [57]. The sensor is highly sensitive, which shows the linear response ranges and the detection limit of 0.2 ng/mL. Vamvakaki et al. [73] developed a highly stable electrochemical sensor to protect the protein from the protease attack. They synthesized silica (biomimetically) and encapsulate the CNF-immobilized enzyme acetylcholine esterase to protect it from degradation by thermal denaturation and protease attack. Hence, increase the shelf life of the protein over 3.5 months under continuous polarization. [73]. Arumugam et al. [5] advanced an electrochemical biosensor for the detection of *E. coli* O157:H7. Similarly, Gupta et al. [22] developed a label-free nanoelectrode array based on vertically aligned CNFs for the detection of C–reactive protein with a detection limit of 90 pM. Their study revealed that the concentration of the C–reactive protein causes the increase in charge of transfer resistance as well as a decrease in redox current [22]. Later, Swisher et al. [68] developed an electrochemical biosensor to measure the activity of the protease. This sensor is based on enhanced AC voltammetry using carbon nanofiber nanoelectrode arrays.

Table 1 Nanomaterial-assisted CNFs for the detection of small molecules [36]

Sensor	Detected molecule	Limit of detection	References
CNFs	Trp, Tyr,Cys	0.1 μM	[70]
CNFs	HQ, CC	0.4 μM(HQ), 0.2 μM (CC)	[20]
CNFs	DA	0.08 μM	[51]
Pd–HCNFs	Glucose, H_2O_2	0.03 mM (glucose), 3 μM(H_2O_2)	[31]
PtNP–CNFs	H_2O_2	11 μM	[35]
Pt/CNFs	H_2O_2	0.6 μM	[45]
Pd/CNFs	H_2O_2 and NADH	0.2 μM(H_2O_2)	[28]
Ag–Pt/pCNFs	Dopamine	0.11 μM	[27]
CNF–PtNP	H_2O_2	1.9 μM	[38]
ZNF–CNFs	H_2S	1–10 ppm	[85]
CuCo–CNFs	Glucose	1 μM	[42]
Co_3O_4/CNFs	H_2O_2	0.5 μM	[55]
Wo_3–CNFs	NO_2	1 ppm	[37]
CuO/rGO/CNFs	Glucose	0.1 μM	[80]
Pd–Ni/CNFs	Sugar	7–20 nM	[21]
$Ni(OH)_2$/ECF	glucose	0.1 μM	[12]

CNFs: Carbon nanofibers, Pd–HCNFs: Palladium–helical carbon nanofibers, PtNP–CNFs: Platinum NP-decorated carbon nanofibers, Pt/CNFs: Platinum NP-loaded carbon nanofibers, Pd/CNFs: Palladium NP-loaded carbon nanofibers, Ag–Pt/pCNFs: nanoporous carbon nanofibers decorated with Ag–Pt bimetallic NPs, CNF–PtNP: nanoporous carbon nanofibers decorated with platinum nanoparticles, ZNF–CNFs: Nanoporous carbon nanofibers decorated with platinum nanoparticles, CuCo–CNFs: bimetallic CuCo NPs anchored and embedded in CNFs, Co_3O_4/CNFs: Co_3O_4 nanoparticles on mesoporous carbon nanofibers, Wo_3–CNF: Wo_3 nanomodule-decorated hybrid carbon nanofibers, CuO/rGO/CNFs: CuO nanoneedle/reduced graphene oxide/carbon nanofibers, Pd–Ni/CNFs: Pd–Ni alloy NP/carbon nanofibers composites, $Ni(OH)_2$/ECF: $Ni(OH)_2$ nanoplatelet/electrospun carbon nanofiber hybrids

The enhanced AC voltammetry properties help in measuring the proteolytic cleavage by proteases of the surface-attached tetrapeptides.

4.4.1 Fuel Cell Systems

The fuel cell is an electrochemical cell that acts like a battery that converts the chemical energy say (hydrogen) of the fuel into electricity through a redox reaction [4]. There is an urgent need for technologies which can replace fossil fuel-based systems. There are different fuel cells based on the electrolyte, for example, polymer, alcohol, and alkaline electrolyte-based fuel cells. The alcohol electrolyte-based fuel cell is also known as direct alcohol fuel cells, and if a cell is fed with carbon, then it is called as direct carbon fuel cells [4]. Despite these, some other fuel cell is also available like phosphoric acid, solid oxide [81], and molten carbonate electrolyte-based fuel cells. The fuel cell can be chosen based on its application, such as durability, temperature, specific energy required, response time, power density, and others. During catalytic reactions in fuel cells, the mesoporous property of CNFs reduces the resistance of inner pore diffusion of products or reactants [9], electrical conductivity, and the metal–support interaction as well. In fuel cell, electrocatalysts are used to increase the rate of reaction. There are many electrocatalysts that are used in a fuel cell. For example, platinum-based electrocatalysts, which are grown on a carbon, shows the ability for energy conversion in the electrochemical process. The support material such as carbon has some property that determines the durability and activity of catalysts. Important criteria of fuel cell electrodes designing are to utilize a high concentration of metal in the catalyst for a certain power ty so that the ohmic drop can be minimized in the catalytic layer. The low surface area (<200 m^2/g) of CNF, which supports for fuel cell catalyst, is a major disadvantage. And because of the low surface area, the proper dispersion of the high number of noble-metal nanoparticles is difficult [34]. The metal deposition method on CNFs looks critical to attaining a good dispersion. Hence, the microemulsion and colloidal methods are more competence to synthesize the Pt catalysts with a smaller size [64]. The carbon fiber support with low surface area is also encouraged to ease the corrosion in fuel cell applications due to the carbon support. Their mesoporous structure also decreases mass transport constraints. Another exciting application of the CNF–supported catalyst is the Pt–Ru catalyst, which is utilized for alcohol oxidation in direct alcohol fuel cells. In comparison with Pt catalysts, this catalyst oxidizes carbon monoxide (CO) at a more negative potential owing to the effect of Ru, which oxidizes CO to CO_2 by the adsorption of oxygen [4]. Sebastián et al. offered different CNFs as the support for Pt–Ru catalysts for the anodic electrochemical reaction of a direct alcohol fuel cell. For example, highly graphitic CNFs as support in Pt–Ru catalyst are utilized, which is suitable for methanol oxidation, while these CNFs exhibit low activity toward the ethanol oxidation. Hence, the importance of pore volume is very high in CNFs because highly porous CNFs can oxidize the ethanol also [65]. Using CNFs in electrodes can expand the performance of the direct alcohol fuel cell owing to many advantages like no parasitic load, operation at RT, operation at a low concentration of methanol, and low catalyst loading in the cathode and anode [83]. The CNF oxidation also signifies a substantial rise in the electro-oxidation of methanol [62]. The maximum support can be attained by balancing three parameters, i.e., an improved metal–support interaction, a sufficient electrochemical surface area, and good methanol diffusion through the catalyst pores [61, 62].

4.5 Air Filtration Applications

The nanofiber membranes have been used in environmental monitoring for air filtration from the old times. As we know the fact that industrialization and globalization are causing a harmful effect on the environment because of which the quality of air has deteriorated in many places, there is an urgent need requiring regeneration of air through filtration and other processes for better quality filtration media. Air filtration has a wide range of applications; they remove particulate materials from work environments and supply protection from toxic agents. Today, nanomaterials are used as nanofiber mats for air filtration applications. There are several companies that pioneered the use of nanofibers in air filtration. There are several advantages of nanofibers over conventional filtration media. The nanofibers have tiny dimensions and thus offer better efficiency than conventional filtration fibers. In addition, for nanometer-sized fibers, the pressure drop is reduced because of a decrease in drag force on the fiber. The occurrence of slip flow also results in more contaminants passing. The nanofibers have a high surface-to-volume ratio that makes them beneficial to adsorb contaminants from the air and made nanofiber membranes an increasingly popular choice in air filtration applications.

5 Conclusion and Future Perspective

The fabrication routes, along with the environmental application of the CNFs, have been discussed in this chapter. It has been observed that because of the excellent physical, chemical, and optical properties of carbon fibers, they can be utilized in various areas. Owing to the enhanced chemical inertness and mechanical strength, the carbon fiber-based sensors have excellent stability and selectivity to the target molecules. Normally, the carbon fiber structures depend on the shape of the catalytic nanoscale particles that have been used for preparing the CNFs. Usually, CVD and the electrospinning methods have been used for preparing the carbon fibers. As the carbon fibers fabricated from the electrospinning method possess great environmental applications. Other methods, for instance, self-assembly, chemical, hydrothermal methods, and template–based synthesis, could also be considered for preparing the carbon fibers. It is possible to produce carbon fibers based on two-dimensional and three-dimensional scaffolds. By introducing the functional nanosized building blocks in the carbon fibers assembly, more consideration is given in the good design performance energy storage materials, for instance, solar cells, batteries, fuel cell, etc.

Acknowledgements Dinesh Kumar is thankful DST, New Delhi, for financial support to this work sanctioned vide project Sanction Order F. No. DST/TM/WTI/WIC/2K17/124(C). One author, Praveen Kumar Yadav, is thankful to CSIR–National Physical Laboratory, New Delhi.

References

1. Abd El-Aziz AM, ElBackly RM, Taha NA, El-Maghraby A, Kandil SH (2017) Preparation and characterization of carbon nano fibrous/hydroxyapatite sheets for bone tissue engineering. Mater Sci Eng C 76:1188–1195

2. Adabi M, Saber R, Faridi-Majidi R, Faridbod F (2015) Performance of electrodes synthesized with polyacrylonitrile-based carbon nanofibers for application in electrochemical sensors and biosensors. Mater Sci Eng C 48:673–678
3. Al-Saleh MH, Sundararaj U (2009) A review of vapor grown carbon nanofiber/polymer conductive composites. Carbon 47:2–22
4. Arico AS, Bruce P, Scrosati B, Tarascon J-M, Van Schalkwijk W (2005) Nanostructured materials for advanced energy conversion and storage devices. Nat Mater 4:366–377
5. Arumugam PU, Chen H, Siddiqui S, Weinrich JAP, Jejelowo A, Li J, Meyyappan M (2009) Wafer–scale fabrication of patterned carbon nanofiber nanoelectrode arrays: a route for development of multiplexed, ultrasensitive disposable biosensors. Biosens Bioelectron 24:2818–2824
6. Azhari F, Banthia N (2012) Cement–based sensors with carbon fibers and carbon nanotubes for piezoresistive sensing. Cem Concr Compos 34:866–873
7. Baeza FJ, Galao O, Zornoza E, Garces P (2013) Multifunctional cement composites strain and damage sensors applied on reinforced concrete (RC) structural elements. Materials 6:841–855
8. Bauman YI, Mishakov IV, Rudneva YV, Plyusnin PE, Shubin YV, Korneev DV, Vedyagin AA (2019) Formation of active sites of carbon nanofibers growth in self–organizing Ni–Pd catalyst during hydrogen–assisted decomposition of 1,2–dichloroethane. Ind Eng Chem Res 58:685–694
9. Bessel CA, Laubernds K, Rodriguez NM, Baker RTK (2001) Graphite nanofibers as an electrode for fuel cell applications. J Phys Chem B 105:1121–1122
10. Chen LF, Feng Y, Liang HW, Wu ZY, Yu SH (2017) Macroscopic–scale three–dimensional carbon nanofiber architectures for electrochemical energy storage devices. Adv Energy Mater 7:1–31
11. Chen L, Liu L, Guo Q, Wang Z, Liu G, Chen S, Hou H (2017) Preparation of $Ni(OH)_2$ nanoplatelet/electrospun carbon nanofiber hybrids for highly sensitive nonenzymatic glucose sensors. RSC Adv 7:19345–19352
12. Chen LF, Lu Y, Yu L, Lou XW (2017) Designed formation of hollow particle–based nitrogen–doped carbon nanofibers for high–performance supercapacitors. Energy Environ Sci 10:1777–1783
13. Cho E, Perebikovsky A, Benice O, Holmberg S, Madou M, Ghazinejad M (2018) Rapid iodine sensing on mechanically treated carbon nanofibers. Sensors 18:1–17
14. Choi YK, Gotoh Y, Sugimoto KI, Song SM, Yanagisawa T, Endo M (2005) Processing and characterization of epoxy nanocomposites reinforced by cup–stacked carbon nanotubes. Polymer 46:11489–11498
15. Claramunt S, Monereo O, Boix M, Leghrib R, Prades JD, Cornet A, Merino P, Merino C, Cirera A (2013) Flexible gas sensor ray with an embedded heater based on metal decorated carbon nanofibres. Sens Actuators B Chem 187:401–406
16. Feng L, Li S, Li H, Zhai J, Song Y, Jiang L (2002) Super–hydrophobic surface of aligned polyacrylonitrile nanofibers. Angewan Chem Int Ed 41:1221–1223
17. Fu Y, Yu HY, Jiang C, Zhang TH, Zhan R, Li XW, Li JF, Tian JH, Yang RZ (2018) Ni Co alloy nanoparticles decorated on N–doped carbon nanofibers as highly active and durable oxygen electro catalyst. Adv Funct Mater 28:1–10
18. Garcia-Bordeje E, Kvande I, Chen D, Ronning M (2007) Synthesis of composite materials of carbon nanofibres and ceramic monoliths with uniform and tuneable nanofibre layer thickness. Carbon 45:1828–1838
19. Guan HJ, Zhang J, Liu Y, Zhao YF, Zhang B (2019) Rapid quantitative determination of hydrogen peroxide using an electrochemical sensor based on Pt Ni alloy/CeO_2 plates embedded in n–doped carbon nanofibers. Electrochim Acta 295:997–1005
20. Guo QH, Huang JS, Chen PQ, Liu Y, Hou HQ, You TY (2012) Simultaneous determination of catechol and hydroquinone using electrospun carbon nanofibers modified electrode. Sens Actuators B Chem 163:179–185
21. Guo QH, Liu D, Zhang XP, Li LB, Hou HQ, Niwa O, You TY (2014) Pd–Ni alloy nanoparticle/carbon nanofiber composites: preparation, structure, and superior electrocatalytic properties for sugar analysis. Anal Chem 86:5898–5905

22. Gupta R, Sharma SC (2019) Modelling the effects of nitrogen doping on the carbon nano fiber growth via catalytic plasma–enhanced chemical vapour deposition process. Contrib Plasma Phys 59:72–85
23. Hahm JI (2016) Fundamental properties of one–dimensional zinc oxide nanomaterials and implementations in various detection modes of enhanced biosensing. Annu Rev Phys Chem 67:691–717
24. Hartgerink JD, Beniash E, Stupp SI (2001) Self–assembly and mineralization of peptide–amphiphile nanofibers. Science 294:1684–1688
25. He ZX, Li MM, Li YH, Zhu J, Jiang YQ, Meng W, Zhou HZ, Wang L, Dai L (2018) Flexible electrospun carbon nanofiber embedded with TiO_2 as excellent negative electrode for vanadium redox flow battery. Electrochim Acta 281:601–610
26. Hu N, Itoi T, Akagi T, Kojima T, Xue J, Yan C, Atobe S, Fukunaga H, Yuan W, Ning H, Surina, Liu Y, Alamusi (2013) Ultrasensitive strain sensors made from metal-coated carbon nanofiller/epoxy composites. Carbon 51:202–212
27. Huang Y, Miao YE, Ji S, Tjiu WW, Liu T (2014) Electrospun carbon nanofibers decorated with Ag–Pt bimetallic nanoparticles for selective detection of dopamine. ACS Appl Mater Interfaces 6:12449–12456
28. Huang JS, Wang DW, Hou HQ, You TY (2008) Electrospun palladium nanoparticle–loaded carbon nanofibers and their electrocatalytic activities towards hydrogen peroxide and nadh. Adv Funct Mater 18:441–448
29. Huang JS, Wang DW, Hou HQ, You TY (2008) Electrospun palladium nanoparticle-loaded carbon nanofibers and their electrocatalytic activities towards hydrogen peroxide and nadh. Adv Funct Mater 18:441–448
30. Huang JS, Liu Y, You TY (2010) Carbon nanofiber based electrochemical biosensors: A review. Anal Methods 2:202–211
31. Jia XE, Hu GZ, Nitze F, Barzegar HR, Sharifi T, Tai CW, Wagberg T (2012) Synthesis of palladium/helical carbon nanofiber hybrid nanostructures and their application for hydrogen peroxide and glucose detection. ACS Appl Mater Interfaces 5:12017–12022
32. Kelarakis A, Yoon K, Somani RH, Chen X, Hsiao BS, Chu B (2005) Rheological study of carbon nanofiber induced physical gelation in polyolefin nanocomposite melt. Polymer 46:11591–11599
33. Klocke JL, Mangold M, Allmendinger P, Hugi A, Geiser M, Jouy P, Faist J, Kottke T (2018) Single–shot sub–microsecond mid–infrared spectroscopy on protein reactions with quantum cascade laser frequency combs. Anal Chem 90:10494–10500
34. Kvande I, Briskeby ST, Tsypkin M, Rønning M, Sunde S, Tunold R, Chen D (2007) On the preparation methods for carbon nanofiber supported Pt catalysts. Top Catal 45:81–85
35. Lamas-Ardisana PJ, Loaiza OA, Anorga L, Jubete E, Borghei M, Ruiz V, Ochoteco E, Cabanero G, Grande HJ (2014) Disposable amperometric biosensor based on lactate oxidase immobilised on platinum nanoparticle–decorated carbon nanofiber and poly (diallyldimethylammonium chloride) films. Biosens Bioelectron 56:345–351
36. Lao J, Huang J, Wang D, Ren Z (2004) Adv Mater 16:65–69
37. Lee JS, Kwon OS, Shin DH, Jang J (2013) Wo_3 nanonodule–decorated hybrid carbon nanofibers for NO_2 gas sensor application. J Mater Chem A 1:9099–9106
38. Li M, Liu LB, Xiong YP, Liu XT, Nsabimana A, Bo XJ, Guo LP (2015) Bimetallic MCo (M = Cu, Fe, Ni, and Mn) nanoparticles doped–carbon nanofibers synthetized by electrospinning for nonenzymatic glucose detection. Sens Actuators B Chem 207:614–622
39. Li DW, Luo L, Pang ZY, Ding L, Wang QQ, Ke HZ, Huang FL, Wei QF (2014) Novel phenolic biosensor based on a magnetic poly dopamine–laccase–nickel nanoparticle loaded carbon nanofiber composite. ACS Appl Mater Interfaces 6:5144–5151
40. Li J, Vergne MJ, Mowles ED, Zhong WH, Hercules DM, Lukehart CM (2005) Surface functionalization and characterization of graphitic carbon nanofibers (GCNFs). Carbon 43:2883–2893
41. Li W, Zhang LS, Wang Q, Yu Y, Chen Z, Cao CY, Song WG (2012) Low–cost synthesis of graphitic carbon nanofibers as excellent room temperature sensors for explosive gases. J Mater Chem 22:15342–15347

42. Li Y, Zhang M, Zhang X, Xie G, Su Z, Wei G (2015) Nanoporous carbon nanofibers decorated with platinum nanoparticles for non–enzymatic electrochemical sensing of H_2O_2. Nanomaterials 5:1891–1905
43. Liu Y, Huang JS, Wang DW, Hou HQ, You TY (2010) Electrochemical determination of oxalic acid using palladium nanoparticle–loaded carbon nanofiber modified electrode. Anal Methods 2:855–859
44. Liu Y, Teng H, Hou HQ, You TY (2009) Nonenzymatic glucose sensor based on renewable electrospun Ni nanoparticle–loaded carbon nanofiber paste electrode. Biosens Bioelectron 24:3329–3334
45. Liu Y, Wang DW, Xu L, Hou HQ, You TY (2011) A novel and simple route to prepare aptnanoparticle–loaded carbon nanofiber electrode for hydrogen peroxide sensing. Biosens Bioelectron 26:4585–4590
46. Liu LJ, Wang ZH, Yang JH, Liu GL, Li JJ, Guo L, Chen SL, Guo QH (2018) $NiCO_2O_4$ nanoneedle–decorated electrospun carbon nanofiber nanohybrids for sensitive non–enzymatic glucose sensors. Sens Actuators B Chem 258:920–928
47. Llobet E (2013) Gas sensors using carbon nanomaterials: a review. Sens. Actuators B Chem 179:32–45
48. Lozano K, Bonilla-Rios J, Barrera EV (2001) A study on nanofiber–reinforced thermoplastic composites (II): investigation of the mixing rheology and conduction properties. J Appl Polym Sci 80:1162–1172
49. Ma PX, Zhang R (1999) Synthetic nano–scale fibrous extracellular matrix. J Biomed Mater Res 46:60–72
50. Magana JR, Kolen'ko YV, Deepak FL, Solans C, Shrestha RG, Hill JP, Ariga K, Shrestha LK, Rodriguez–Abreu C (2016) From chromonic self–assembly to hollow carbon nanofibers: efficient materials in super capacitor and vapor–sensing applications. ACS Appl Mater Interfaces 8:31231–31238
51. Mao XW, Yang XQ, Rutledge GC, Hatton TA (2014) Ultra–wide–range electrochemical sensing using continuous electrospun carbon nanofibers with high densities of states. ACS Appl Mater Interfaces 6:3394–3405
52. Merdrignac-Conanec O, Moseley PT (2002) J Mater Chem 12:1779–1781
53. Meyyappa M (2016) Carbon nanotube–based chemical sensors. Small 12:2118–2129
54. Mordkovich VZ (2003) Carbon nanofibers: a new ultrahigh–strength material for chemical technology. Theor Found Chem Eng 37:429–438
55. Ni Y, Liao Y, Zheng MB, Shao SJ (2017) In–situ growth of CO_3O_4 nanoparticles on mesoporous carbon nanofibers: a new nanocomposite for nonenzymatic amperometric sensing of H_2O_2. Microchim Acta 184:3689–3695
56. Ning PG, Duan XC, Ju XK, Lin XP, Tong XB, Pan X, Wang TH, Li QH (2016) Facile synthesis of carbon nanofibers/MnO_2 nanosheets as high–performance electrodes for asymmetric supercapacitors. Electrochim Acta 210:754–761
57. Periyakaruppan A, Gandhiraman RP, Meyyappan M, Koehne JE (2013) Label–free detection of cardiac troponin–I using carbon nanofiber based nanoelectrode arrays. Anal Chem 85:3858–3863
58. Pervin F, Zhou Y, Rangari VK, Jeelani S (2005) Testing and evaluation on the thermal and mechanical properties of carbon nano fiber reinforced SC–15 epoxy. Mater Sci Eng, A 405:246–253
59. Rizwan M, Koh D, Booth MA, Ahmed MU (2018) Combining a gold nanoparticle–polyethylene glycol nanocomposite and carbon nanofiber electrodes to develop a highly sensitive salivary secretory immunoglobulin a immunosensor. Sens Actuators B Chem 255:557–563
60. Sánchez M, Rams J, Campo M, Jiménez-Suárez A, Urena A (2011) Characterization of carbon nanofiber/epoxy nanocomposites by the nano indentation technique. Compos B Eng 42:638–644
61. Sawicka K, Gouma P, Simon S (2004) Proc Electrochem Soc 8:354–358
62. Sebastián D, Lazaro MJJ, Moliner R, Suelves I, Aricò ASS, Baglio V (2014) Oxidized carbon nanofibers supporting Pt-Ru nanoparticles for direct methanol fuel cells. Int J Hydrog Energ 39:5414–5423

63. Sebastian D, Suelves I, Moliner R, Lázaro MJ (2012) Carbon nanofibers. In: Nanofibers: synthesis, properties, and applications. Nova Science Publishers Inc, UK, pp 1–40
64. Sebastián D, Suelves I, Moliner R, Lázaro MJMJ, Stassi A, Baglio V, Aricò AS (2013) Optimizing the synthesis of carbon nanofiber based electrocatalysts for fuel cells. Appl Catal B Environ 132–133:22–27
65. Sebastián D, Suelves I, Pastor E, Moliner R, Lazaro MJMJ (2013) The effect of carbon nanofiber properties as support for PtRu nanoparticles on the electrooxidation of alcohols. Appl Catal B Environ 132–133:13–21
66. Shen Y, Li L, Xiao KJ, Xi JY (2016) Constructing three–dimensional hierarchical architectures by integrating carbon nanofibers into graphite felts for water purification. ACS Sustain Chem Eng 4:2351–2358
67. Shoukat R, Khan MI (2018) Synthesis of vertically aligned carbon nanofibers using inductively coupled plasma–enhanced chemical vapor deposition. Electr Eng 100:997–1002
68. Swisher LZ, Syed LU, Prior AM, Madiyar FR, Carlson KR, Nguyen TA, Hua DH, Li J (2013) Electro chemical protease biosensor based on enhanced AC voltammetry using carbon nanofiber nano electrode arrays. J Phys Chem C 117:4268–4277
69. Tallman TN, Gungor S, Wang KW, Bakis CE (2015) Tactile imaging and distributed strain sensing in highly flexible carbon nanofiber/polyurethane nanocomposites. Carbon 95:485–493
70. Tang XF, Liu Y, Hou HQ, You TY (2010) Electrochemical determination of L–tryptophan, L–tyrosine and L–cysteine using electrospun carbon nanofibers modified electrode. Talanta 80:2182–2186
71. Teradal NL, Jelinek R (2017) Carbon nanomaterials in biological studies and biomedicine. Adv Healthc Mater 6:1–36
72. Tiwari JN, Vij V, Kemp KC, Kim KS (2016) Engineered carbon–nanomaterial–based electrochemical sensors for biomolecules. ACS Nano 10:46–80
73. Vamvakaki V, Hatzimarinaki M, Chaniotakis N (2008) Bionumetically synthesized silica–carbon nanofiber architectures for the development of highly stable electrochemical biosensor systems. Anal Chem 80:5970–5975
74. Viciani S, Cumis MS, Borri S, Patimisco P, Sampaolo A, Scamarcio G, Natale P, D'Amato F, Spagnolo V (2015) A quartz–enhanced photoacoustic sensor for H_2S trace–gas detection at 2.6 μm. Appl Phys B 119:21–27
75. Wang YL, Wang YS, Wan BL, Han BG, Cai GC, Li ZZ (2018) Properties and mechanisms of self–sensing carbon nanofibers/epoxy composites for structural health monitoring. Compos Struct 200:669–678
76. Wang L, Wu AG, Wei G (2018) Graphene–based aptasensors: from molecule–interface interactions to sensor design and biomedical diagnostics. Analyst 143:1526–1543
77. Wang SX, Yap CC, He JT, Chen C, Wong SY, Li X (2016) Electrospinning: a facile technique for fabricating functional nanofibers for environmental applications. Nanotechnol Rev 5:51–73
78. Yan T, Wang Z, Wang YQ, Pan ZJ (2018) Carbon/graphene composite nanofiber yarns for highly sensitive strain sensors. Mater Design 143:214–223
79. Yang C, Denno ME, Pyakurel P, Venton BJ (2015) Recent trends in carbon nanomaterial-based electrochemical sensors for biomolecules: A review. Anal Chim Acta 887:17–37
80. Ye DX, Liang GH, Li HX, Luo J, Zhang S, Chen H, Kong JL (2013) A novel nonenzymatic sensor based on CuO nanoneedle/graphene/carbon nanofiber modified electrode for probing glucose in saliva. Talanta 116:223–230
81. Yu F, Zhang Y, Yu L, Cai W, Yuan L, Liu J, Liu M (2016) All–solid–state direct carbon fuel cells with thin yttrium–stabilized–zirconia electrolyte supported on nickel and iron bimetal–based anodes. Int J Hydrog Energ 41:9048–9058
82. Zahid MU, Pervaiz E, Hussain A, Shahzad MI, Niazi MBK (2018) Synthesis of carbon nanomaterials from different pyrolysis techniques: a review. Mater Res Express 5:1–19
83. Zainoodin AM, Kamarudin SK, Masdar MS, Daud WRW, Mohamad AB, Sahari J (2014) High power direct methanol fuel cell with a porous carbon nanofiber anode layer. Appl Energ 113:946–954

84. Zhang TT, Xu HX, Xu ZQ, Gu Y, Yan XY, Liu H, Lu NN, Zhang SY, Zhang ZQ, Yang M (2019) A bioinspired antifouling zwitterionic interface based on reduced graphene oxide carbon nanofibers: electrochemical aptasensing of adenosine triphosphate. Microchim Acta 186:3343–3347
85. Zhang JT, Zhu ZJ, Chen CM, Chen Z, Cai MQ, Qu BH, Wang TH, Zhang M (2018) ZnO–carbon nanofibers for stable, high response, and selective H_2S sensors. Nanotechnology 29:1–28
86. Zhao QX, Zhao MM, Qiu JQ, Lai WY, Pang H, Huang W (2017) One dimensional silver-based nanomaterials: preparations and electrochemical applications. Small 13:1–18
87. Zhou F-L, Gong R-H, Porat I (2009) Mass production of nanofibre assemblies by electrostatic spinning. Polym Int 58:331–342
88. Zhu JH, Wei SY, Ryu J, Guo ZH (2011) Strain–sensing elastomer/carbon nanofiber "metacomposites". J Phys Chem C 115:13215–13222

Lead and Cadmium Toxic Metals Removal by Carbon Nanocomposites

Rekha Sharma, Kritika S. Sharma, and Dinesh Kumar

Abstract Water pollution because of the discharge of heavy metal ions in water is a severe environmental issue in the current situation. So, drinking water sanitization before human consumption is necessary for good human health. Out of diverse technologies used for heavy metal removal, adsorption on nanomaterial substrates and membrane filtration are potential techniques owing to their competence, easy functioning, cost-effectiveness, and constraint equipped area. To date, several materials have been employed in engineering these nanoadsorbents and filtration membranes. In the morphology of these developed nanoadsorbents and membranes, it can be outcome that these have good site density, charge, and surface areas and to adsorb the metal ions on their surface efficiently. In this chapter, we mainly emphasis on the submission of carbon nanocomposites to remove lead and cadmium engineered at the nanoscale, testing both advantages and limitations of these adsorbents. Finally, scopes and future scenarios of these adsorbents have been discussed.

Keywords Nanoadsorbents · Adsorption · Nanocomposites · Heavy metals · Pollutants

Abbreviations

MHT	Mechanohydrothermal
LDH	Layered double hydroxide
GO	Graphene oxide

R. Sharma
Department of Chemistry, Banasthali Vidyapith, Banasthali, Rajasthan 304022, India
e-mail: sharma20rekha@gmail.com

K. S. Sharma · D. Kumar (✉)
School of Chemical Sciences, Central University of Gujarat, Gandhinagar 382030, India
e-mail: dinesh.kumar@cug.ac.in

K. S. Sharma
e-mail: kritsharma98@gmail.com

M. Jawaid et al. (eds.), *Environmental Remediation Through Carbon Based Nano Composites*, Green Energy and Technology,
https://doi.org/10.1007/978-981-15-6699-8_16

MGL	Magnetite-graphene oxide-layered double hydroxide
Γ_m	Maximum sorption amount
2,4-D	2,4-Dichlorophenoxyacetate
C_s effect	Sorbent concentration effect
SCA	Surface component activity
A_s	Sorption capacities in mg m^{-2}
SBE	Spent bleaching earth
CNPs	Carbon nanoparticles
HATU	Coupling agent
PSO	Pseudo-second order
SWASV	Square wave anodic stripping voltammograms
DEAMTPP	Water-dispersible diethyl-4-(4-amino-5-mercapto-4H-1,2,4-triazol-3-yl) phenyl phosphonate
WHO	World Health Organization
US EPA	United States Environmental Protection Agency
PSO	Pseudo-second order
PVK-GO	Poly(N-vinylcarbazole)–graphene oxide
$GO-MnFe_2O_4$	Graphene oxide$-MnFe_2O_4$ magnetic
LDHs	Layered double hydroxides
GO	Graphene oxide
ZrRP	Zirconium resorcinol phosphate nanocomposite

1 Introduction

Recently, pollution through heavy metal is the key environmental problem that hazards the human health worldwide. Various heavy metal ions, for example, Hg(I)/Hg(II), Pb(II), As(III)/AS(V), Cr(III)/Cr(VI), Ni(I), Cu(II), Cd(II), Zn(II), and Co(II), etc., are amenably or obliquely released hooked on the rivers, lakes, streams or oceans because of rapid industrialization, for instance, batteries, tanneries, metal plating, painting, printing, mining and photographic industries, fertilizer and pesticides industries, etc., particularly in emerging nations [2, 6, 7, 35, 36, 40, 72, 84, 106]. These pollutants are probable to accumulate in living beings through food chains or drinking water because of their non-biodegradable properties [9, 37, 69, 74]. Owing to their role in emerging approximately cofactors or vitamins, the trace amount consumption of many heavy metals for human beings is very indispensable, but their extreme intake can cause hazardous effects. These effects comprise many mental and physical obstruction like diarrhea, pneumonia, kidney, nausea, weight loss, vomiting, skin degeneration, liver break down, asthma, congenital deformities, and many cancers [9, 35, 37, 62, 69, 74, 77, 84, 88]. The toxicities and their lethal effects are brought to light here, consistent with the WHO and the US EPA. To eliminate these pollutants from aqueous bodies, mainly adsorption technique is utilized using various carbon-based nanoadsorbents which is schematically shown in Fig. 1.

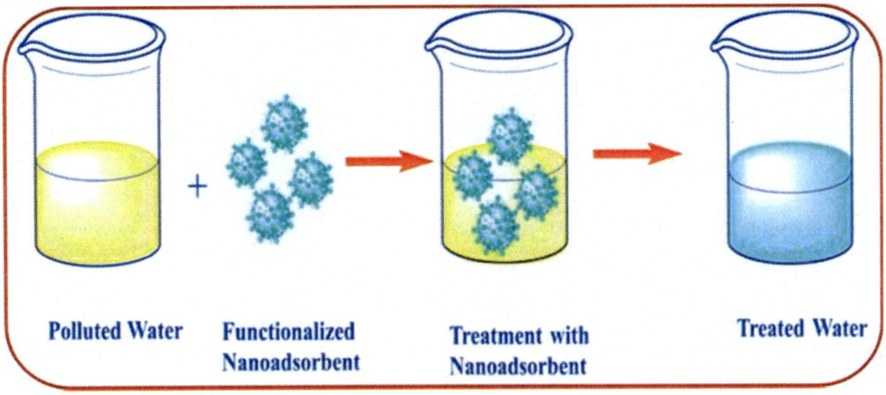

Fig. 1 Water treatment of by functionalized Nanoadsorbent

The novelty of this chapter is to describe the advanced nanoadsorbents which are viable toward the elimination of toxic metals especially, Pb(II) and Cd(II) which are dangerous to human being and aquatic species. This chapter mainly focusses on carbon-based nanomaterials having high adsorption capacity, cost-effectiveness, easy separable, and easy synthesis methods. In continuation, the toxicities of elevated concentration of lead and cadmium and their future perspectives are comprised in this chapter in subsequent sections.

1.1 Lead (Pb)

Inorganic lead (Pb) usually contaminates surface and groundwater systems because it is progressively discharged from different sources like mining, leaded gasoline, and industrial fuel [11, 38]. Acute Pb contamination causes various kinds of lead poisoning, for instance, reproductive and liver system dysfunction, severe kidney dysfunction [5, 11, 23, 38, 67, 70]. It may also cause intellectual disabilities, mainly in offspring [23, 70]. Lead is listed as second most toxic contaminant among other hazardous substances because of its other additional lethal symptoms like irritability, insomnia, muscle weakness, anemia, renal damages, and hallucination [5, 11, 23, 38, 67, 70]. In drinking water, the maximum contaminant level (MCL) of Pb ions set permissible limits of 0.015 mg L^{-1} while WHO set as 0.05 mg L^{-1} [26, 103].

1.2 Cadmium (Cd)

Cd (Cadmium) releases into environments through Ni and Cd batteries, electroplating, atomic fission plants, welding, fertilizers, and plastics and paints, etc. [29,

39, 65]. Because of chronic cadmium toxicity, the Itai-Itai disease has been found in Japan [29]. Various diseases like kidney damage, testicular tissue destruction, red blood cells destruction, and osteoporosis, and high blood pressure occurred due to Cd poisoning from environmental exposure [25, 34, 41, 42, 46, 68]. In enzyme structure because of zinc replacement by Cd in these enzymes, its catalytic activity is damaging because of changes in the stereo-structure of the enzyme [34]. The safe drinking water limits for cadmium are 0.005 mg L^{-1} set by USEPA has and Cd ions mandate as a human carcinogen [26, 103].

2 Carbon Nanocomposites Applications in Removal of Heavy Metal

2.1 Nanoadsorbents Applications in Pb Removal: Scopes/Limitations

For the removal of lead, the date tree leaves were used as an adsorbent material in aqueous solutions. The date tree leaves in their low concentration in solution have been used to remove an advanced proportion of Pb(II) aqueous ion systems. Onto the removal of Pb(II) ions, the effect of various parameters has been studied, for example, pH of the solution, ionic strength, temperature, initial metal concentration, agitation speed, adsorbent dose, and contact time. The adsorption capacity increases on increasing the Pb(II) ion initial concentration, and pH 5.8 is appropriate for the maximum adsorption at any temperature. The maximum Pb(II) ions adsorption achieved at 1 g L^{-1} adsorption dose on a powdered date tree leaf. The ionic strength, agitation speed, and time-consistent to the maximum adsorption are 0.005 M, 200 rpm, and 50 min, respectively. Temkin models were best fitted to the adsorption equilibrium. The maximum adsorption value was found at 60 °C up to 57 mg g^{-1}. The PSO model of the kinetic experiment was followed for the Pb(II) ions adsorption on the leaves of a date tree. The change in enthalpy designates very strong interaction forces among date tree leaves and Pb(II) ions for the adsorption method, which was attained at 21.59 kJ mol^{-1}. The kinetic process of adsorption is of endothermic in nature. The positive value of ΔS was found up to 95.45 J mol^{-1} K^{-1} at the solid/solution interface shows the augmented randomness, and the negative value of ΔG shows the spontaneity of the adsorption method [13].

Bushra et al., 2012 [14] synthesized poly-o-toluidine Zr(IV) tungstate composite to remove Pb(II) ions from water. The synthesized composite materials show improved reproducibility, thermal stability together with chemical stability, exchange capacity, granulometric properties, and also own improved selectivity likened to pure inorganic and organic materials. The selective adsorption of heavy metal ions was occurred on an amorphous nanocomposite cation exchanger and may withstand equitably elevated temperature. At 200 °C, the composite shows an important ion-exchange capacity and thermally stable. The synthesized composite is of analytical importance and might remove metals from the solution. Determination of Pb(II) metal ions by FAAS does not cause any previous digestion from electroplating wastewater and tap water samples. From industrial effluents to the further recovery and elimination of significant metal ions, the adsorbent material can be explored. Poly-o-toluidine Zr(IV) tungstate composite cation exchanger displays the features of a conducting material along with auspicious ion-exchanger [14].

Musico et al., 2013 [66] developed PVK–GO nanocomposites for the adsorption of Pb(II) ions from water solutions. The graphene-based polymer nanocomposites are the utmost current technological advances and auspicious composite materials that comprise inimitable properties of polymer materials and graphene-based materials in one nanohybrid material [75]. These nanohybrid materials cannot usually be attained using pure polymers or conventional composites and display substantial upgrading in possessions [48]. The PVK–GO nanocomposite among the nanohybrid materials is vital because of having dissimilar ways of fabrication, dispersion, and polymerization [73, 85, 110]. PVK–GO has imperative antimicrobial goods [16, 86]. Though, for the adsorption of toxicants, PVK–GO nanomaterials have been employed. In aqueous solution, the outcomes display that PVK–GO shows an excellent Pb(II) adsorption competence. The adsorption competence is explained by the role of surface hydroxyl and carboxylic acid groups of adsorbents. Additionally, as the concentration of GO in PVK–GO nanocomposite rises, the adsorption capacity of the adsorbent upsurge for Pb(II) ions. On the surface of Pb(II) ions and the PVK–GO, the interaction among the oxygenated functional groups influenced by the alteration in pH. As the pH of the solution upsurges, the adsorption of Pb(II) ions increases. Though, high adsorption of Pb(II) was detected at high pH from aqueous solution. The high adsorption capacity depends on precipitation of heavy metals along with the adsorbents adsorption competence. The adsorption of Pb(II) was best fitted to the Langmuir model, and shows the 887.98 mg g^{-1} of adsorption capacity within adsorption time of 90 min at pH 7.5 [66].

For effective adsorption of As(III) and Pb(II) from water, Kumar et al., 2014 [49] synthesized GO−$MnFe_2O_4$ nanohybrids materials. The reference $MnFe_2O_4$ nanoparticles and the synthesized GO−$MnFe_2O_4$ nanohybrids compared for As(III)/As(V), and Pb(II) heavy metals so far with the adsorption capacities of different adsorbents. From the results, it was demonstrated that the synthesized GO−$MnFe_2O_4$ nanohybrids adsorbent material is higher compared to other reported adsorbents for the adsorption of Pb(II), and As(III)/As(V) up to now. The virtuous adsorption capacities of GO along with NP and the amalgamation of the inimitable-layered nature of the hybrid system, including maximum surface area, are the main cause for the exceptional adsorption property. In water treatment processes, this material is very attractive because of the rapid adsorption rates and easy magnetic separation of the GONH hybrids. In the handling of water, the utilization of adsorbent could be completed thru coating it on sand particles or ceramic beads or by making a membrane. The adsorbent can be magnetically separated from the solution. To remove other contaminants, the adsorbent materials based on graphene oxide-nanoparticle hybrids are viable to discover [49].

According to Xiao et al., 2015 [105], a one-step microwave-assisted solvothermal method was utilized to develop aluminum-magnesium oxide nanocomposites to adsorb As(V) and Pb(II) ions from a solution which is a cost-effective, template-free, and simple and route. The adsorbent material could adsorb cations and anions owing to its enormous surface area. The adsorbent showed an adsorption capacity of 423 mg g^{-1} and 133 mg g^{-1} toward Pb(II) and As(V), correspondingly [105].

To minimize waste disposal, and reduce energy usage, preserve resources via a suitable, effective, repetitive, and selective recovery of Pb(II) ions Sharma et al., 2014 [87] developed exceedingly steady functionalized magnetic nanoparticles. To attain an effective and novel adsorbent, the stepwise modification of iron oxide nanoparticles was performed, which showed exceptional recital for Pb(II) ion adsorption. Because of the collective involvement of numerous factors, i.e., substrate material, anchored functional groups, physicochemical properties, and experimental conditions of the analyte, the adsorbent was only selective toward the target analyte. The rapid recovery of Pb(II) ions attained by the help of sonication without destructive the structure of the functionalized nanoparticles in the elution procedure. The synthesized AAA-NH_2-Si@MNPs have some valuable properties such as high adsorption capacity, good material stability, the cost-effectiveness of embedded MNPs, superior reusability up to five adsorption/desorption cycles, and ease of fabrication. Therefore, for the systematic recycling and recovery of Pb(II) ions, this ending covers a maintainable path by the productive incorporation of competence, reusability, and selectivity. Besides, the unique and novel applicability to lower the leachable metal content can be intended as an authoritative source in mycorrhizal-treated by ash samples so much that the environmental stress is stable [87].

Recently, LDHs and GO have been extensively utilized as probable adsorbent materials. LDHs are a class of lamellar inorganic materials, which are likewise recognized as anionic clays or hydrotalcite-like compounds (HTlc). The general formula $[M^{II}_{1-x}M^{III}_{x}(OH)_2]^{x+}[(A^{n-})_{x/n}]^{x-}\cdot mH_2O$ used for the representation of these materials. Here, di- and trivalent metal cations represented by M^{II} and M^{III}, respectively, m is the molar amount of intercalated water,x is the molar ratio of $M^{III}/(M^{II} + M^{III})$, and the interlayer anion isA^{n-} of charge$n$. For the synthesis of magnetic composites, an MHT route was utilized, containing GO, Fe_3O_4, $Mg_3Al - OH$, and LDH in a composite material. The components of the MGL composites, i.e., Fe_3O_4, LDH, and GO bonded via chemical bonding. The MGL composites displayed a strong magnetic response, good water-dispersity, and highE_R and Γ_m for both 2,4-D pollutants and Pb(II) ions. The increased RGO content can improve the higher uptake of pollutants and could upsurge theA_s of the MGL. The Pb(II) ions adsorbed more strongly compared to 2,4-D pollutants on MGLs. TheE_R increased for Pb(II) deceased for a while 2,4-D, with increasing pH from 4 to 10. The Freundlich-SCA isotherms, and Langmuir-SCA describes the adsorption of 2,4-D and Pb(II) on the MGLs by an obviousC_s-effect. For the adsorption of 2,4-D and Pb(II) on MGLs, diverse mechanisms were recognized. Because of the contribution of both the GO and LDH components, the adsorption of 2,4-D primarily attained, whereas the LDH component was responsible for the adsorption of Pb(II) via surface-induced precipitation of $Pb_3(CO_3)_2(OH)_2$. The intercalation of 2,4-D anions into the LDH gallery is the key mechanism for adsorption of

2,4-D on the LDH component, while hydrophobic and $\pi-\pi$ interactions attributed to their GO component. To treat wastewater, the MGL composites are considered as a potential adsorbent. For developing GO–LDH composite materials, the MHT method delivers an environmentally friendly and simple route [111].

From industrial water, the adsorption of Pb(II) ions, Datta and Uslu [20] used an ordinary clay material, i.e., montmorillonite, and its improved form. Trin-octylamine was utilized to modify the montmorillonite, which displayed virtuous efficacy of Pb(II) ions removal. The adsorbent dose of 2.4 g L^{-1} of Mt-TOA and 12 g L^{-1} of Mt showed an improved Pb(II) ion adsorption. The pH value of 7 was appropriate for the supreme adsorption of Pb(II) ions. The 33.10 and 3.37 mg g^{-1} are the maximum adsorption efficacy of Mt-TOA and Mt, respectively, for the Pb(II) ion adsorption from the water phase by the use of the Langmuir isotherm model. The kinetic data was best fitted to the PSO model to explain the kinetic behavior of an adsorption process. For Pb(II) ion adsorption from manufacturing effluents, the trin-octylamine modified montmorillonite adsorbent is a viable material [20].

Pourbeyram [76], developed GO–Zr-P nanocomposite to adsorb Zn(II), Cd(II), Pb(II), and Cu(II) divalent heavy metal ions. In preparing the GO–Zr-P adsorbent, graphene oxide (GO) act as a template material, Zr binder ion, and phosphate behaves as an adsorbent by a facile method. Various instrumental techniques were used to check the surface morphology and structure of the adsorbent, i.e., SEM, TEM, DLS analysis, XPS, and XRD. The appropriate pH range was 3–6 for the batch experiments. For Cd(II) ions, utmost adsorption capacity of the adsorbent was found up to 232.36 mg g^{-1}, Zn(II) 251.58 mg g^{-1}, Cu(II) 328.56 mg g^{-1} and for Pb(II) ions, it was found 363.42 mg g^{-1}at pH 6. The adsorption controlled by chemical adsorption predicted from proposed kinetic, and adsorption isotherms on adsorbent material concerning the strong surface complexation surface phosphate groups of GO–Zr-P with heavy metals and the adsorption were of monolayer coverage. After the complexation of heavy metals, the adsorbent showed good dispersibility of adsorption experiments in water. The tendency to precipitate and agglomerate was detected after adsorption. In analytical chemistry, the adsorbent applied as a solid adsorbent to adsorb different heavy metal ions from its maximum adsorption capacity [76].

Nonkumwong et al., 2016 [71] synthesized mesoporous amine-functionalized magnesium ferrite nanoparticles ($MgFe_2O_4$-NH_2NPs). Various calcination steps, for example coprecipitation [30, 56, 93] sol-gel [54] and reverse microemulsion [18], methods were employed to synthesis methods of $MgFe_2O_4$ nanoadsorbents. A pressure vessel able to comprising a dispersing solvent at increased pressure and temperature is essential, although, the hydrothermal method is not required the calcination step [21]. Intended for the formation of superparamagnetic (Ni, Mn, Co) Fe_2O_4 with high saturation magnetization values, Mohapatra et al., 2011 [63] had freshly established the simple heating method to evade such complicate instruments/steps under refluxing condition. Therefore, concurrent surface functionalization modifiers were employed, which are made of several functional groups, i.e., carboxylic, amine, and a hydroxyl group (lactic and ascorbic acid, L-lysine, and ethanolamine) by the facile refluxing condition for the synthesis of $MgFe_2O_4$ NPs with. The morphology, magnetic properties, phase formation, chemical composition, surface functional groups, pore size distribution, and specific surface area were prudently determined. The hopeful progression of which the functionalization and synthesis occur concurrently offers virtuous class magnetic nanoadsorbent as $MgFe_2O_4$ nanoparticles functionalized by amine groups. The kinetic and thermodynamic studies have also been optimized for the investigation of Pb(II) removal on amine-functionalized $MgFe_2O_4$ nanoparticles, which display rapid and outstanding adsorption [71].

The adsorption of Pb(II) ions, Liang et al., 2016 [53] synthesized the leaf powder of phoenix tree-based granular adsorbent with bentonite as the binder. Various instrumental techniques, i.e., TGA, BET, and SEM studies, were utilized to illustrate the granular adsorbent. The thermogravimetric analysis shows a mass loss in three stages. At calcination at 500 °C of the adsorbent, the obtained pore volume and maximum specific surface area were 0.276 cm^3g^{-1} and 166.3 m^2g^{-1}, respectively. Various optimizing parameters have been studied to check the adsorption behaviors for Pb(II) by 500 °C-calcined granular adsorbents such as initial metal ion concentration, adsorption time, and effects of pH. For Pb(II) ions, 71 mg g^{-1} was the maximum adsorption capacity [53].

2.2 *Recent Advances in Nanoadsorbents Applications in Pb Removal*

Currently, for the adsorption of metals, various reports have revealed the prospect of this method, e.g., goethite (α-FeOOH) and nanoscale zero-valent iron displayed high rapid removal rate and removal capability for several toxicants concurrently, [99] and nanomaterials based on titanium compounds, for example, titanate nanobelts [102] titanate nanoflowers, [31] and mesoporous titania beads, [104] likewise divulged potential submissions intended for the adsorption process. Though, in the applied drinking water treatments, their large-scale application limited because of some drawbacks of adsorbent materials, i.e., complicated preparation processes, insufficient adsorption capacity, low yields, and instability of iron-based nanomaterials. The adsorption of heavy metal-containing anions along with adsorption of heavy metal cations reported on a new titanium glycolate nanomaterial by Han et al. [28] in which the 3-D V-type strip structure is responsible for the removal and adsorption simultaneously. Furthermore, because of its cost-effective and ecologically approachable features, this material predictable to have applied submissions in water remediation and equal to be accomplished a precise simplistic one-pot hydrothermal technique. From a three-dimensional V-type stripe structure, a novel titanium-based nanomaterial was manufactured with a large surface area via a green and convenient method. The surface area of biodegradable material is found as high as 246.5 m^2g^{-1} and its own mesopore structure, several spectroscopic methods well characterized [28].

For the fast and effective elimination of Pb(II) ions, Moradi et al., 2017 [64] synthesized a novel adsorbent as Fe_3O_4@GMA–AAm magnetic nanocomposite. On adsorbents surface, the progress of the polymer chains showed in the AFM and

SEM images, and the core-shell structure could also be seen in TEM image. The Pb(II) ion removal attained within 2 min equilibrium time. The PSO model is fitted to the kinetic data, and the chemical adsorption is the rate-controlling step. The adsorption of Pb(II) ions was well fitted to the Langmuir modal. The adsorption process is endothermic and spontaneous. To remove Pb(II) ions, our synthesized Fe_3O_4@GMA–AAm nanocomposite can be a virtuous magnetic adsorbent with easy separation method, suitable core-shell structure, rapid adsorption rate within 2 min and high adsorption capacity up to 158.73 mg g^{-1} [64].

Rusmin et al., 2017. [81], established Palygorskite-iron oxide nanocomposite to remove aqueous Pb(II), which was synthesized through a coprecipitation technique. Pal-IO exhibited low isoelectric point, increased precise surface area, and significant magnetic susceptibility at 3.5, 99.8 m^2g^{-1}, and 20.2 emu g^{-1}, correspondingly. To treat the contaminated water covering 200 mg L^{-1} of Pb(II) ions, the 26.6 mg g^{-1} is the utmost adsorption capability for Pb(II) at pH 5. The recyclability of magnetic adsorbent, the best desorbing agent was EDTA-Na_2, with over 90% desorption capability. At the end of three consecutive adsorption-desorption cycles, 64% Pb(II) removal was attained. The adsorbent was found to be recyclable and magnetic stable, with the fewest leaching properties because of the robust binding of magnetite NPs on palygorskite. For purifying, specifically heavy metal cations contaminated wastewaters, the nanocomposite could develop as an auspicious material [81].

Recently, various organic and inorganic adsorbents have been extensively utilized in treatments of water and wastewater, i.e., metal oxides, carbonaceous nanomaterials, biomass, natural clay, etc., [78, 94, 101, 113, 120]. Though, because of their low efficiency and adsorption capacity, they were still incapable of having large-scale practical applications in water treatment processes. For organic dye sequestration and heavy metal ion, it is essential to perform additional research and progress innovative adsorbents having rapid removal rate and a high removal capacity [8, 52, 55, 98, 101, 115].

2.3 *Nanoadsorbents Applications in Cd Removal: Scopes/Limitations*

Owing to their influences on ecological water pollution, the detection of ultra-trace quantities of toxic metals is of significance. Because of its wide distribution and a high level of toxicity, the study of cadmium is exceedingly required as heavy metal [89]. In various industries, for example, metal plating, batteries, and alloys, cadmium metal is extensively utilized as a pigment and stabilizer in plastics [17, 83]. Due to industrial activities, the entry of cadmium into surface water has augmented the possibility of human contact to this toxic metal. The WHO has been announced the acceptable level of cadmium in water intake up to 3 mg L^{-1}, because of high toxicity even at low concentrations [82]. Various damages occur to organs thru the intake of Cd(II) in the body such as the liver, lungs, and kidney because of its high biological half-life

[27]. So, by commissioning a reliable, sensitive, and selective analytical technique, the detection of minute quantities of Cd(II) in ecological aqueous samples is of great importance [61].

For the first time via the use of SWASV to manufacture an electrochemical platform, the concurrent examination of Cd(II) and Pb(II) in solution stated by Gao et al., 2012 [24] to combine the maximum adsorption capacity of γ-AlOOH with the conductivity of graphene. The AlOOH-RGO nanocomposites were synthesized hydrothermally. GO is concurrently reduced to graphene under the hydrothermal condition, accompanied by the homogeneous precipitation of AlOOH nanoplates. In this, the RGO offered to conduct pathways, and the agglomeration of graphene averted by the use of the AlOOH nanoplates. Moreover, the RGO-dispersed AlOOH nanoplates helps to collect the target metals on the electrode surface. By this research, Gao et al., 2012 [24] show the new bridge among electrochemical behavior and adsorption. The AlOOH-RGO adsorbent is a promising material that owns outstanding practical applicability, high stability, durability, and better sensitivity in electrochemical sensing of heavy metals [24].

Though the advantages of MnO_2 coating has been highlighted in earlier research for adsorption of metal ions, they predicted the adsorption percentage deprived of a depth examination hooked on the experimental parameters and mechanism. In wastewater treatment, no research has emphasized the magnetic particles coated with MnO_2 NPs. The 3D flowerlike MnO_2-coated magnetic nanocomposite was hydrothermally synthesized without utilizing any organic surfactant and template. Fe_3O_4/MnO_2 showed an efficient and rapid removal of diverse metals, for example, Cu(II), Zn(II), Cd(II), and Pb(II) from water in a few seconds by an exterior magnetic field. The in height ionic strength, low pH, and attendance of Ca ions affect the adsorption property of the adsorbent.

The regeneration experiment has been carried out, which shows that without an essential reduction in its adsorption capacity of the composite up to five cycles can be reused. The outcomes displayed that Fe_3O_4/MnO_2 has extensive submissions used for detection and adsorption of metals because of environmental friendliness, good regeneration performance, simple fabrication procedure, and excellent removal capacity [45].

The magnetic separation amid these submissions motivate us to manufacture a magnetic discriminating collection approach. In the synthesis of recyclable water treatment agents, core-shell Ni@$Mg(OH)_2$ composite has been developed for the adsorption of metals. The authors finally attain the water treatment agent reuse by separated and collected the Ni@$Mg(OH)_2$ nanocomposites by the usage of the allure of Ni core under an external magnetic field. To adsorb metals from enchanting wastewater advantage of the ferromagnetic properties and high surface area of Ni@$Mg(OH)_2$ nanocomposites, Zhang et al., 2015 [114] intended a magnetically decomposable approach. The results demonstrated that Ni@$Mg(OH)_2$ nanocomposites had low capacity fading and high removal efficiency [111].

From aqueous solutions, to adsorb Cu(II), Pb(II), and Cd(II), the SBE based Attapulgite/carbon nanocomposites (APT/C) adsorbent developed through a one-step calcination. The adsorption competence for Cd(II) up to 46.72 mg g^{-1}, for Cu(II)

32.32 mg g^{-1}, and for Pb(II) 105.61 mg g^{-1} likewise exhibited faster adsorption equilibrium. The synthesized nanocomposite showed an outstanding adsorption-desorption process up to the fifth cycle, especially for Cu(II) ions. The XPS analysis in which the peaks corresponding to Cu(II), Cd(II), Pb(II) appear after adsorption confirms the adsorption due to surface complexation, ion exchange, and electrostatic attraction among metal ion and the adsorbents [92].

Venkateswarlu and Yoon, 2015 [97], studied the removal of Cd(II) ion on biogenic Fe_3O_4 magnetic nanocomposites capped with DEAMTPP, {(DEAMTPP@Fe_3O_4 MNP)} utilizing ananas comosus peel pulp extract. The manufactured DEAMTPP@Fe_3O_4 MNPs investigated the magnetic properties, size, porosity, and structure. The adsorption procedure rests on the concentration and pH of the solution of DEAMTPP@Fe_3O_4 MNPs. The 60 mg L^{-1} concentration of Cd(II) ion solution, and pH 6 was appropriate for the maximum adsorption capacity up to 96.1% of adsorption. The maximum adsorption capacity was 49.1 mg g^{-1} which was best fitted to the Langmuir isotherm modal. Without a remarkable loss of removal efficiency, the composite can be recycled because of a ferromagnetic nature. To adsorb Cd(II) ions from industrial and environmental wastes, various properties make it an appropriate adsorbent, for example, biodegradable composition, simple separation, and reusability [97].

2.4 *Recent Advances in Nanoadsorbents Applications in Cd Removal*

Currently, for the separation of adsorbents from wastewater, the deliberated effective technique id magnetic separation, so, these adsorbents have been examined extensively [3, 4, 19, 91, 95, 107]. Furthermore, because of their synergistic effect among multiple distinct components and improved properties, magnetic core-shell adsorbents have fascinated abundant attention. Precisely, the heavy metal ions can be adsorbed by functional shell rapidly, and the magnetic core could enable fast separation. In conclusion, a monodisperse amino-functionalized magnetic $CoFe_2O_4@SiO_2$ nanosphere was fruitfully synthesized. The adsorption capacity has been improved due to the grafting of many more amino groups into a monodisperse nanosphere. For the adsorption method, various optimizing parameters have been studied, for example, initial concentration, time, reaction temperature, and pH effect by $CoFe_2O_4@SiO_2$–NH_2. The PSO and Langmuir models best fitted to the kinetic and adsorption data, respectively. The adsorbent having high chemical stability can regenerate after acid treatment up to five cycles. The synthesized adsorbent can be separable because of superparamagnetism nature [79].

Bashir et al., 2016 [10] developed novel ZrRP adsorbent for the removal of Cd(II) ions having useful adsorptive and ion-exchange properties. The results exposed that nanocomposite own novel features as a new material were synthesized by water in the oil (W/O) microemulsion technique. The adsorbent selectively adsorbs Cd(II), employed for some binary separations of Cd(II) mixtures [10].

To adsorb metals, calcium carbonate ($CaCO_3$) has been discovered as the utmost profuse materials in nature [43]. Mainly, the Cd(II) in paddy soils specifically adsorbed on $CaCO_3$ adsorbent, and at maximum flooding periods for Cd(II) polluted paddy soil, the maximum Cd(II) was in Cd–$CaCO_3$ form [44, 47, 121]. In this series, several nanostructured $CaCO_3$, together with amorphous $CaCO_3$ and their hybrids, vaterite, and calcite have been fabricated [15, 33, 51, 57, 58, 112] because minute amount of Cd(II) adsorb on natural calcite [1, 60]. For the first time, Cd(II) ions removal occurs on the nanostructured calcite changed with biochar obtained from sewage sludge. The synthesized adsorbent was efficient, low cost, and biologically approachable. For Cd(II) ions, the adsorption recital of the biochars suggestively recovered by the encumbered calcite nanoparticles. The adsorption experiment is considerably higher efficiency and lower cost, which is the main advantage of this adsorbent [122].

To adsorb toxicants on CNPs as adsorbents is even fewer unusual in the literature. Carbon nanoparticles (CNPs) have the capacity to eliminate destructive heavy metals, for example, soot from water [80]. Though, the technique intricated in making the stated CNPs was a time-taking and monotonous. Thus, studies on functionalization of CNPs and different synthesis routes show good yields, and low cost will be believable to broaden the usage of this relatively new nanocarbon. In this, a microwave-assisted carbonization process was used to synthesize CNPs from the dehydration of glucose. For the first time, adsorption of Cd(II) ions from water carried on the ethylenediamine (EDA) functionalized CNPs. The lone pairs present on nitrogen groups of EDA was the motivation for its selection for functionalization material because it can chelate with most metal ions. CNPs modified with ethylenediamine were fruitfully synthesized from glucose, using HATU as a coupling agent. In adsorption, the synthesized CNPs show promising properties such as cost-effective, easy to functionalize, and reusable. Thus, this adsorbent is the viable material for producing potential in the adsorption of several organic and inorganic pollutants [96].

The synthesis of hierarchical hollow α-Fe_2O_3 nanomaterials have still some deficiencies, despite these synthesis approaches and methods have various beneficial assets. First, in this formulation, approaches characteristically comprise the usage of lethal templates. Second, for the decomposition of the precursor, a high temperature (> 200°C) is required via calcination. Over a long reaction time, the mainstream of hollow α-Fe_2O_3 adsorbent has been solvo/hydrothermally synthesized at 180°C, which requires high energy consumption and large-scale industrialization. Therefore, it is imperatively needed to be proficient in manufacturing α-Fe_2O_3 nanostructures in a cost-effective, green, and a facile pathway. This work predicts chestnut nests and buds type of α-Fe_2O_3 hierarchical hollow spheres. At comparatively low reaction temperature, the adsorbent was solvothermal synthesized. The water/glycerol and water/2-propanol solvent systems permitted morphological tuning. In the water/alcohol soft template, the occurrence of microheterogeneity is the primary mechanism for adsorption. The hollow porous α-Fe_2O_3 architectures displayed a narrow pore size distribution and a large specific surface area.

The large number of OH groups can act as effective active adsorption sites on the surface of the nanostructured α-Fe_2O_3 particles for chelating with metals. The adsorbent was carried out for batch experiments for maximum adsorption capacity and reusability studies [109].

The adsorption of heavy metal ions especially lead and cadmium is summarized in Table 1 which includes the effect of various parameters which are appropriate for the maximum adsorption capacity.

3 Conclusions and Future Perspectives

This chapter comprises the role of various carbon-based nanocomposite materials in the effective adsorption of heavy metals under diverse situations. An examination of associated literature divulges that useful polymer-based composites have been used, such as a combination of magnetic nanoparticles, clay minerals, metal nanoparticles, carbon materials, and polymers used for the removal of metal ions from wastewater solutions. Afterward, combined inside a polymer matrix, the mechanical strength, adsorption capacity, separation from solution, and recycling performance have been suggestively enhanced for such substituents themselves are also extensively utilized for wastewater treatment. The intricate polymers are used as stabilizers along with as a support material for the other substituents, chelating constituents, and rigid frames. They incorporated various substituents also offering reducibility, operative chelating sites, and magnetism, etc. For the synthesis of the adsorbents, separate methods have been classified and deliberated. The macro/micro analysis and the batch experiments for the prediction of the adsorption method have been precised widely and deliberated in detail. The adsorption capacities have been related with synthesized adsorbents with other conventional materials. It was predicted from the outcomes that the synthesized adsorbents offer fast adsorption kinetics toward heavy metal ions, good regeneration ability, and strong chelating capabilities with the synergistic effect of the polymers. However, for practical utilization of these nanoadsorbent materials at the industrial scale, there is little research going on assessing the possibility and the manufacturing price. Shortly, the following issues must be elucidated.

- To evade subordinate contamination, the toxicity of the counterparts and the polymer in the designation of composite material must be pondered.
- To assess the adsorption capacity of precise, heavy metals individually, there must be approximately consistent conditions, for example, adsorbent/metal ion concentration, pH, temperature, etc.
- Having improved recyclability, additional robust polymer-based composites ought to be explored. The long-term performance of adsorbents should be tested for permanent usage, not only for a few cycling reuses. The long-term performance is crucial for real applications.
- For the regeneration of adsorbent, the desorption methods must be examined, and additional consistent approaches are essential to be established.

Table 1 Adsorption of lead and cadmium utilizing various carbon-based nanomaterials in aqueous media

S. No.	Adsorbate	Adsorbent	Adsorption capacity (mg g^{-1})	Conc. (mg L^{-1})	Contact time (min)	Temperature (K)	pH	References
1.	Pb(II)	Graphene	22.42	40	15	303	4.0	[100]
2.	Pb(II)	Graphene (heat treated at 973 K)	35.46	40	15	303	4.0	[32]
3.	Pb(II)	Functionalized graphene (GNSPF6)	406.4	–	4	–	5.1	[22]
4.	Pb(II)	Functionalized graphene (GNSC8P)	74.18	–	4	–	5.1	[22]
5.	Pb(II)	GO@EDTA	367	5–300	24	298 ± 2	6.8	[59]
6.	Pb(II)	GO@TiO_2	35.6 ± 1.3	–	–	–	5.6	[50]
7.	Pb(II)	GO	887.98	5–300	90	298 ± 5	7.0 ± 0.5	[66]
8.	Pb(II)	PVK–GO	1119	–	2	298	5.0	[66]
9.	Pb(II)	Poly(amidoamine) modified GO	0.0513 mmol g^{-1}	0.0193 mmol L^{-1}	24	Room temp.	–	[108]
10.	Pb(II)	FGO	842 1150 1850	–	24	293 313 333	6.0	[116–119]
11.	Cd(II)	FGO	106.3 153.6 167.5	–	–	303 313 333	6.0 ± 0.1	[116–119]
12.	Pb(II)	Graphene/c-MWCNT	104.9	50	120	Room temp.	–	[90]
13.	Cd(II)	GO–TiO_2	72.8 ± 1.6	–	–	–	5.6	[50]
14.	Cd(II)	RGO–Fe(0)/Fe_3O_4	1.91	2–6	1	298	7.0	[12]

- After its application in the adsorption of metals from aqueous solutions, the effective removal of polymer-based compounds should be confirmed because if they remain in solution, they can become the contaminants.
- The utilized adsorbents should be established for the possibility of large-scale preparation and at a low price with the ecological acquaintance and high stability. For submission on a large scale, the cost is a significant constraint.
- Toward the selective adsorption of heavy metals, additional exertions must be bestowed onto adsorbents in the occurrence with multi-metal ions.
- To adsorb of target metal ions at the low concentration from aqueous systems, the high selectivity of the adsorbent is a vital parameter.
- Various instrumental techniques should be utilized to characterize and predict the adsorption mechanism of the adsorption process, for example, EXAFS, XAS, the in situ, and theoretical simulation techniques to accomplish insight information about the interaction mechanism and the removal process at a molecular level.

Acknowledgements The authors gratefully acknowledge the support from the Ministry of Human Resource Development Department of Higher Education, Government of India under the scheme of Establishment of Center of Excellence for Training and Research in Frontier Areas of Science and Technology (FAST), for providing the necessary financial support to perform this study vide letter No, F. No. 5–5/201 4–TS.Vll. Dinesh Kumar is also thankful DST, New Delhi, for financial support to this work (sanctioned vide project Sanction Order F. No. DST/TM/WTI/WIC/2K17/124(C).

References

1. Ahmed IA, Crout NM, Young SD (2008) Kinetics of Cd sorption, desorption, and fixation by calcite: a long-term radiotracer study. Geochim Cosmochim Acta 72:1498–1512
2. Aman T, Kazi AA, Sabri MU, Bano Q (2008) Potato peels as solid waste for the removal of heavy metal copper (II) from waste water/industrial effluent. Colloids Surf B Biointerfaces 63:116–121
3. Ambashta RD, Sillanpää M (2010) Water purification using magnetic assistance: a review. J Hazard Mater 180:38–49
4. Anbia M, Kargosha K, Khoshbooei S (2015) Heavy metal ions removal from aqueous media by modified magnetic mesoporous silica MCM-48. Chem Eng Res Des 93:779–788
5. Awual MR, Hasan MM (2014) Novel conjugate adsorbent for visual detection and removal of toxic lead (II) ions from water. Micropor Mesopor Mat 196:261–269
6. Awual MR, Urata S, Jyo A, Tamada M, Katakai A (2008) Arsenate removal from water by a weak-base anion exchange fibrous adsorbent. Water Res 42:689–696
7. Babel S, Kurniawan TA (2004) Cr (VI) removal from synthetic wastewater using coconut shell charcoal and commercial activated carbon modified with oxidizing agents and/or chitosan. Chemosphere 54:951–967
8. Badawi MA, Negm NA, Kana MA, Hefni HH, Moneem MA (2017) Adsorption of aluminum and lead from wastewater by chitosan-tannic acid modified biopolymers: isotherms, kinetics, thermodynamics and process mechanism. Int J Biol Macromol 99:465–476
9. Bailey SE, Olin TJ, Bricka RM, Adrian DD (1999) A review of potentially low-cost sorbents for heavy metals. Water Res 33:2469–2479

10. Bashir A, Ahad S, Pandith AH (2016) Soft template assisted synthesis of zirconium resorcinol phosphate nanocomposite material for the uptake of heavy-metal ions. Ind Eng Chem Res 55:4820–4829
11. Bhowmick S, Chakraborty S, Mondal P, Van Renterghem W, Van den Berghe S, Roman-Ross G, Iglesias M (2014) Montmorillonite-supported nanoscale zero-valent iron for removal of arsenic from aqueous solution: kinetics and mechanism. Chem Eng J 243:14–23
12. Bhunia P, Kim G, Baik C, Lee H (2012) A strategically designed porous iron–iron oxide matrix on graphene for heavy metal adsorption. Chem Commun 48:9888–9890
13. Boudrahem F, Aissani-Benissad F, Soualah A (2011) Adsorption of lead (II) from aqueous solution by using leaves of date trees as an adsorbent. J Chem Eng Data 56:1804–1812
14. Bushra R, Shahadat M, Raeisssi AS, Nabi SA (2012) Development of nano-composite adsorbent for removal of heavy metals from industrial effluent and synthetic mixtures; Its conducting behavior. Desalination 289:1–11
15. Cai GB, Zhao GX, Wang XK, Yu SH (2010) Synthesis of polyacrylic acid stabilized amorphous calcium carbonate nanoparticles and their application for removal of toxic heavy metal ions in water. J Phys Chem C 114:12948–12954
16. Carpio IEM, Santos CM, Wei X, Rodrigues DF (2012) Toxicity of a polymer–graphene oxide composite against bacterial planktonic cells, biofilms, and mammalian cells. Nanoscale 4:4746–4756
17. Chamsaz M, Atarodi A, Eftekhari M, Asadpour S, Adibi M (2013) Vortex-assisted ionic liquid microextraction coupled to flame atomic absorption spectrometry for determination of trace levels of cadmium in real samples. J Adv Res 4:35–41
18. Chandradass J, Kim KH (2011) Solvent effects in the synthesis of $MgFe_2O_4$ nanopowders by reverse micelle processing. J Alloy Compd 509:L59–L62
19. Chen X, Lam KF, Yeung KL (2011) Selective removal of chromium from different aqueous systems using magnetic MCM-41 nanosorbents. Chem Eng J 172:728–734
20. Datta D, Uslu H (2016) Adsorptive separation of lead (Pb^{2+}) from aqueous solution using tri-n-octylamine supported montmorillonite. J Chem Eng Data 62:370–375
21. Deng H, Chen H, Li H (2007) Synthesis of crystal MFe_2O_4 (M = Mg, Cu, Ni) microspheres. Mater Chem Phys 101:509–513
22. Deng X, Lü L, Li H, Luo F (2010) The adsorption properties of Pb (II) and Cd (II) on functionalized graphene prepared by electrolysis method. J Hazard Mater 183:923–930
23. Ding G, Bao Y (2014) Revisiting pesticide exposure and children's health: focus on China. Sci Total Environ 472:289–295
24. Gao C, Yu XY, Xu RX, Liu JH, Huang XJ (2012) AlOOH-reduced graphene oxide nanocomposites: one-pot hydrothermal synthesis and their enhanced electrochemical activity for heavy metal ions. ACS Appl Mater Interfaces 4:4672–4682
25. Ghorbani M, Eisazadeh H (2013) Removal of COD, color, anions and heavy metals from cotton textile wastewater by using polyaniline and polypyrrole nanocomposites coated on rice husk ash. Compos Part B Eng 45:1–7
26. Griffiths C, Klemick H, Massey M, Moore C, Newbold S, Simpson D, Wheeler W (2012) US Environmental Protection Agency valuation of surface water quality improvements. Rev Environ Econ Policy 6:130–146
27. Gunduz S, Akman S, Kahraman M (2011) Slurry analysis of cadmium and copper collected on 11-mercaptoundecanoic acid modified TiO_2 core-Au shell nanoparticles by flame atomic absorption spectrometry. J Hazard Mater 186:212–217
28. Han W, Yang X, Zhao F, Shi X, Wang T, Zhang X, Wang C (2017) A mesoporous titanium glycolate with exceptional adsorption capacity to remove multiple heavy metal ions in water. RSC Adv 7:30199–30204
29. Horiguchi H, Teranishi H, Niiya K, Aoshima K, Katoh T, Sakuragawa N, Kasuya M (1994) Hypoproduction of erythropoietin contributes to anemia in chronic cadmium intoxication: clinical study on Itai-itai disease in Japan. Arch Toxicol 68:632–636
30. Hu J, Lo IM, Chen G (2007) Comparative study of various magnetic nanoparticles for Cr (VI) removal. Sep Purif Technol 56:249–256

31. Huang J, Cao Y, Liu Z, Deng Z, Tang F, Wang W (2012) Efficient removal of heavy metal ions from water system by titanate nanoflowers. Chem Eng J 180:75–80
32. Huang X, Hernick M (2011) A fluorescence-based assay for measuring N-acetyl-1-d-myo-inosityl-2-amino-2-deoxy-α-d-glucopyranoside deacetylase activity. Anal Biochem 414:278–281
33. Islam MS, San Choi W, Nam B, Yoon C, Lee HJ (2017) Needle-like iron oxide@ $CaCO_3$ adsorbents for ultrafast removal of anionic and cationic heavy metal ions. Chem Eng J 307:208–219
34. Järup L, Hellström L, Alfvén T, Carlsson MD, Grubb A, Persson B, Elinder CG (2000) Low level exposure to cadmium and early kidney damage: the OSCAR study. J Occup Environ Med 57:668–672
35. Javadian H (2014) Application of kinetic, isotherm and thermodynamic models for the adsorption of Co (II) ions on polyaniline/polypyrrole copolymer nanofibers from aqueous solution. Ind Eng Chem Res 20:4233–4241
36. Jiang Y, Pang H, Liao B (2009) Removal of copper (II) ions from aqueous solution by modified bagasse. J Hazard Mater 164:1–9
37. Joseph T, Dubey B, McBean EA (2015) A critical review of arsenic exposures for Bangladeshi adults. Sci Total Environ 527:540–551
38. Jouad EM, Jourjon F, Le Guillanton G, Elothmani D (2005) Removal of metal ions in aqueous solutions by organic polymers: use of a polydiphenylamine resin. Desalination 180:271–276
39. Kadirvelu K, Namasivayam C (2003). Activated carbon from coconut coirpith as metal adsorbent: adsorption of Cd (II) from aqueous solution. Adv Environ Res 7:471–478
40. Kadirvelu K, Thamaraiselvi K, Namasivayam C (2001) Removal of heavy metals from industrial wastewaters by adsorption onto activated carbon prepared from an agricultural solid waste. Bioresour Technol 76:63–65
41. Karthik R, Meenakshi S (2015) Chemical modification of chitin with polypyrrole for the uptake of Pb (II) and Cd (II) ions. Int J Biol Macromol 78:157–164
42. Kazantzis G (2004) Cadmium, osteoporosis and calcium metabolism. Biometals 17:493–498
43. Kazi TG, Tuzen M, Shah F, Afridi HI, Citak D (2014) Development of a new green non-dispersive ionic liquid microextraction method in a narrow glass column for determination of cadmium prior to couple with graphite furnace atomic absorption spectrometry. Anal Chim Acta 812:59–64
44. Khaokaew S, Chaney RL, Landrot G, Ginder-Vogel M, Sparks DL (2011) Speciation and release kinetics of cadmium in an alkaline paddy soil under various flooding periods and draining conditions. Environ Sci Technol 45:4249–4255
45. Kim EJ, Lee CS, Chang YY, Chang YS (2013) Hierarchically structured manganese oxide-coated magnetic nanocomposites for the efficient removal of heavy metal ions from aqueous systems. ACS Appl Mater Interfaces 5:9628–9634
46. Kobya M, Demirbas E, Senturk E, Ince M (2005) Adsorption of heavy metal ions from aqueous solutions by activated carbon prepared from apricot stone. Bioresour Technol 96:1518–1521
47. Kosolsaksakul P, Farmer JG, Oliver IW, Graham MC (2014) Geochemical associations and availability of cadmium (Cd) in a paddy field system, northwestern Thailand. Environ Pollut 187:153–161
48. Kuilla T, Bhadra S, Yao D, Kim NH, Bose S, Lee JH (2010) Recent advances in graphene-based polymer composites. Prog Polym Sci 35:1350–1375
49. Kumar S, Nair RR, Pillai PB, Gupta SN, Iyengar MAR, Sood AK (2014) Graphene oxide–$MnFe_2O_4$ magnetic nanohybrids for efficient removal of lead and arsenic from water. ACS Appl Mater Interfaces 6:17426–17436
50. Lee YC, Yang JW (2012) Self-assembled flower-like TiO_2 on exfoliated graphite oxide for heavy metal removal. J Ind Eng Chem 18:1178–1185
51. Li C, Qian ZJ, Zhou C, Su W, Hong P, Liu S, Ji H (2014) Mussel-inspired synthesis of polydopamine-functionalized calcium carbonate as reusable adsorbents for heavy metal ions. RSC Adv 4:47848–47852

52. Li C, Zhou L, Yang H, Lv R, Tian P, Li X, Lin F (2017) Self-assembled exopolysaccharide nanoparticles for bioremediation and green synthesis of noble metal nanoparticles. ACS Appl Mater Interfaces 9:22808–22818
53. Liang S, Ye N, Hu Y, Shi Y, Zhang W, Yu W, Yang J (2016) Lead adsorption from aqueous solutions by a granular adsorbent prepared from phoenix tree leaves. RSC Adv 6:25393–25400
54. Liu CP, Li MW, Cui Z, Huang JR, Tian YL, Lin T, Mi WB (2007) Comparative study of magnesium ferrite nanocrystallites prepared by sol–gel and coprecipitation methods. J Mater Sci 42:6133–6138
55. Liu L, Guo X, Tallon R, Huang X, Chen J (2017) Highly porous N-doped graphene nanosheets for rapid removal of heavy metals from water by capacitive deionization. Chem Commun 53:881–884
56. Liu X, An S, Zhou X, Zhang L, Zhang Y, Shi W, Yang J (2014) Comparative studies of removal of methyl green and basic fuchsin from wastewater by a novel magnetic nanoparticles Mg-ferrites. J Disper Sci Technol 35:1727–1736
57. Liu Z, Shen Q, Zhang Q, Bian L, Liu Y, Yuan B, Jiang F (2014) The removal of lead ions of the aqueous solution by calcite with cotton morphology. J Mater Sci 49:5334–5344
58. López Marzo AM, Pons J, Merkoçi A (2014) Extremely fast and high Pb^{2+} removal capacity using a nanostructured hybrid material. J Mater Chem A 2:8766–8772
59. Madadrang CJ, Kim HY, Gao G, Wang N, Zhu J, Feng H, Hou S (2012) Adsorption behavior of EDTA-graphene oxide for Pb (II) removal. ACS Appl Mater Interfaces 4:1186–1193
60. Martin-Garin A, Van Cappellen DP, Charlet L (2003) Aqueous cadmium uptake by calcite: a stirred flow-through reactor study. Geochim Cosmochim Acta 67:2763–2774
61. Mohajer S, Chamsaz M, Entezari MH (2014) Ultratrace determination of cadmium (II) ions in water samples using graphite furnace atomic absorption spectrometry after separation and preconcentration using magnetic activated carbon nanocomposites. Anal Methods 6:9490–9496
62. Mohan D, Pittman CU Jr (2007) Arsenic removal from water/wastewater using adsorbents—A critical review. J Hazard Mater 142:1–53
63. Mohapatra S, Rout SR, Panda AB (2011) One-pot synthesis of uniform and spherically assembled functionalized MFe_2O_4 (M = Co, Mn, Ni) nanoparticles. Colloids Surf A Physico Chem Eng Asp 384:453–460
64. Moradi A, Moghadam PN, Hasanzadeh R, Sillanpää M (2017) Chelating magnetic nanocomposite for the rapid removal of Pb (II) ions from aqueous solutions: characterization, kinetic, isotherm and thermodynamic studies. RSC Adv 7:433–448
65. Moreno-Castilla C, Alvarez-Merino MA, López-Ramón MV, Rivera-Utrilla J (2004) Cadmium ion adsorption on different carbon adsorbents from aqueous solutions. Effect of surface chemistry, pore texture, ionic strength, and dissolved natural organic matter. Langmuir 20:8142–8148
66. Musico YLF, Santos CM, Dalida MLP, Rodrigues DF (2013) Improved removal of lead (II) from water using a polymer-based graphene oxide nanocomposite. J Mater Chem A 1:3789–3796
67. Naseem R, Tahir SS (2001) Removal of Pb (II) from aqueous/acidic solutions by using bentonite as an adsorbent. Water Res 35:3982–3986
68. Nawrot TS, Staessen JA, Roels HA, Munters E, Cuypers A, Richart T, Vangronsveld J (2010) Cadmium exposure in the population: from health risks to strategies of prevention. Biometals 23:769–782
69. Netzer A, Hughes DE (1984) Adsorption of copper, lead and cobalt by activated carbon. Water Res 18:927–933
70. Nevin R (2009) Trends in preschool lead exposure, mental retardation, and scholastic achievement: association or causation? Environ Res 109:301–310
71. Nonkumwong J, Ananta S, Srisombat L (2016) Effective removal of lead (II) from wastewater by amine-functionalized magnesium ferrite nanoparticles. RSC Adv 6:47382–47393
72. Nriagu JO, Pacyna JM (1988) Quantitative assessment of worldwide contamination of air, water and soils by trace metals. Nature 333:134

73. Pernites R, Vergara A, Yago A, Cui K, Advincula R (2011) Facile approach to graphene oxide and poly (N-vinylcarbazole) electro-patterned films. Chem Commun 47:9810–9812
74. Peters RW (1999) Chelant extraction of heavy metals from contaminated soils. J Hazard Mater 66:151–210
75. Potts JR, Dreyer DR, Bielawski CW, Ruoff RS (2011) Graphene-based polymer nanocomposites. Polymer 52:5–25
76. Pourbeyram S (2016) Effective removal of heavy metals from aqueous solutions by graphene oxide–zirconium phosphate (GO–Zr-P) nanocomposite. Ind Eng Chem Res 55:5608–5617
77. Ray PZ, Shipley HJ (2015) Inorganic nano-adsorbents for the removal of heavy metals and arsenic: a review. RSC Adv 5:29885–29907
78. Rekha P, Muhammad R, Sharma V, Ramteke M, Mohanty P (2016) Unprecedented adsorptive removal of $Cr_2O_7{}^{2-}$ and methyl orange by using a low surface area organosilica. J Mater Chem A 4:17866–17874
79. Ren C, Ding X, Fu H, Li W, Wu H, Yang H (2017) Core–shell superparamagnetic monodisperse nanospheres based on amino-functionalized $CoFe_2O_4@SiO_2$ for removal of heavy metals from aqueous solutions. RSC Adv 7:6911–6921
80. Ruparelia JP, Duttagupta SP, Chatterjee AK, Mukherji SOUMYA (2008) Potential of carbon nanomaterials for removal of heavy metals from water. Desalination 232:145–156
81. Rusmin R, Sarkar B, Tsuzuki T, Kawashima N, Naidu R (2017) Removal of lead from aqueous solution using superparamagnetic palygorskite nanocomposite: material characterization and regeneration studies. Chemosphere 186:1006–1015
82. Şahan S, Şahin U (2012) An automated on-line minicolumn preconcentration cold vapour atomic absorption spectrometer: application to determination of cadmium in water samples. Talanta 88:701–706
83. Salah TA, Mohammad AM, Hassan MA, El-Anadouli BE (2014) Development of nano-hydroxyapatite/chitosan composite for cadmium ions removal in wastewater treatment. J Taiwan Inst Chem Eng 45:1571–1577
84. Samiey B, Cheng CH, Wu J (2014) Organic-inorganic hybrid polymers as adsorbents for removal of heavy metal ions from solutions: a review. Materials 7:673–726
85. Santos CM, Tria MCR, Vergara RAMV, Ahmed F, Advincula RC, Rodrigues DF (2011) Antimicrobial graphene polymer (PVK-GO) nanocomposite films. Chem Commun 47:8892–8894
86. Santos CM, Tria MCR, Vergara RAMV, Cui KM, Pernites R, Advincula RC (2011) Films of Highly Disperse Electrodeposited Poly (N-vinylcarbazole)–Graphene Oxide Nanocomposites. Macromol Chem Phys 212:2371–2377
87. Sharma RK, Puri A, Monga Y, Adholeya A (2014) Acetoacetanilide-functionalized Fe_3O_4 nanoparticles for selective and cyclic removal of Pb^{2+} ions from different charged wastewaters. J Mater Chem A 2:12888–12898
88. Singh R, Singh S, Parihar P, Singh VP, Prasad SM (2015) Arsenic contamination, consequences and remediation techniques: a review. Ecotoxicol Environ Safety 112:247–270
89. Stoeppler M (ed) (1992). Hazardous metals in the environment, vol 12. Elsevier
90. Sui Z, Meng Q, Zhang X, Ma R, Cao B (2012) Green synthesis of carbon nanotube–graphene hybrid aerogels and their use as versatile agents for water purification. J Mater Chem 22:8767–8771
91. Tan Y, Chen M, Hao Y (2012) High efficient removal of Pb (II) by amino-functionalized Fe_3O_4 magnetic nano-particles. Chem Eng J 191:104–111
92. Tang J, Mu B, Zheng M, Wang A (2015) One-step calcination of the spent bleaching earth for the efficient removal of heavy metal ions. ACS Sustain Chem Eng 3:1125–1135
93. Tang W, Su Y, Li Q, Gao S, Shang JK (2013) Superparamagnetic magnesium ferrite nanoadsorbent for effective arsenic (III, V) removal and easy magnetic separation. Water Res 47:3624–3634
94. Tian P, Han XY, Ning GL, Fang HX, Ye JW, Gong WT, Lin Y (2013) Synthesis of porous hierarchical MgO and its superb adsorption properties. ACS Appl Mater Interfaces 5:12411–12418

95. Tran HV, Dai Tran L, Nguyen TN (2010) Preparation of chitosan/magnetite composite beads and their application for removal of Pb (II) and Ni (II) from aqueous solution. Mater Sci Eng, C 30:304–310
96. Tshwenya L, Arotiba OA (2017) Ethylenediamine functionalized carbon nanoparticles: synthesis, characterization, and evaluation for cadmium removal from water. RSC Adv 7:34226–34235
97. Venkateswarlu S, Yoon M (2015) Rapid removal of cadmium ions using green-synthesized Fe_3O_4 nanoparticles capped with diethyl-4-(4 amino-5-mercapto-4 H-1, 2, 4-triazol-3-yl) phenyl phosphonate. RSC Adv 5:65444–65453
98. Vijayalakshmi K, Devi BM, Latha S, Gomathi T, Sudha PN, Venkatesan J, Anil S (2017) Batch adsorption and desorption studies on the removal of lead (II) from aqueous solution using nanochitosan/sodium alginate/microcrystalline cellulose beads. Int J Biol Macromol 104:1483–1494
99. Wang B, Wu H, Yu L, Xu R, Lim TT, Lou XW (2012) Template-free formation of uniform urchin-like α-FeOOH hollow spheres with superior capability for water treatment. Adv Mater 24:1111–1116
100. Wang Sh, Peng Y (2010) Natural zeolites as effective adsorbents in water and waste-water treatment. Chem Eng J 156:11–24
101. Wang X, Cai J, Zhang Y, Li L, Jiang L, Wang C (2015) Heavy metal sorption properties of magnesium titanate mesoporous nanorods. J Mater Chem A 3:11796–11800
102. Wen T, Zhao Z, Shen C, Li J, Tan X, Zeb A, Xu AW (2016) Multifunctional flexible free-standing titanate nanobelt membranes as efficient sorbents for the removal of radioactive $_{90}Sr^{2+}$ and $_{137}Cs^{+}$ ions and oils. Sci Rep 6:20920
103. WH Organization (2004) Guidelines for drinking-water quality: recommendations
104. Wu N, Wei H, Zhang L (2011) Efficient removal of heavy metal ions with biopolymer template synthesized mesoporous titania beads of hundreds of micrometers size. Environ Sci Technol 46:419–425
105. Xiao F, Fang L, Li W, Wang D (2015) One-step synthesis of aluminum magnesium oxide nanocomposites for simultaneous removal of arsenic and lead ions in water. RSC Adv 5:8190–8193
106. Yao T, Cui T, Wu J, Chen Q, Lu S, Sun K (2011) Preparation of hierarchical porous polypyrrole nanoclusters and their application for removal of Cr (VI) ions in aqueous solution. Polym Chem 2:2893–2899
107. Yavuz CT, Prakash A, Mayo JT, Colvin VL (2009) Magnetic separations: from steel plants to biotechnology. Chem Eng Sci 64:2510–2521
108. Yuan Y, Zhang G, Li Y, Zhang G, Zhang F, Fan X (2013) Poly (amidoamine) modified graphene oxide as an efficient adsorbent for heavy metal ions. Poly Chem 4:2164–2167
109. Zha R, Shi T, Zhang Z, Xu D, Jiang T, Zhang M (2017) Quasi-reverse-emulsion-templated approach for a facile and sustainable environmental remediation for cadmium. RSC Adv 7:6345–6357
110. Zhang B, Chen Y, Xu L, Zeng L, He Y, Kang ET, Zhang J (2011) Growing poly (N-vinylcarbazole) from the surface of graphene oxide via RAFT polymerization. J Polym Sci Pol Chem 49:2043–2050
111. Zhang F, Song Y, Song S, Zhang R, Hou W (2015) Synthesis of magnetite–graphene oxide-layered double hydroxide composites and applications for the removal of Pb (II) and 2, 4-dichlorophenoxyacetic acid from aqueous solutions. ACS Appl Mater Interfaces 7:7251–7263
112. Zhang J, Yao B, Ping H, Fu Z, Li Y, Wang W, Zhang F (2016) Template-free synthesis of hierarchical porous calcium carbonate microspheres for efficient water treatment. RSC Adv 6:472–480
113. Zhang M, Song W, Chen Q, Miao B, He W (2015) One-pot synthesis of magnetic Ni@ $Mg(OH)_2$ core–shell nanocomposites as a recyclable removal agent for heavy metals. ACS Appl Mater Interfaces 7:1533–1540

114. Zhang R, Chen C, Li J, Wang X (2015) Preparation of montmorillonite@carbon composite and its application for U (VI) removal from aqueous solution. Appl Surf Sci 349:129–137
115. Zhang S, Yang H, Huang H, Gao H, Wang X, Cao R, Wang X (2017) Unexpected ultrafast and high adsorption capacity of oxygen vacancy-rich WO_x/C nanowire networks for aqueous Pb^{2+} and methylene blue removal. J Mater Chem A 5:15913–15922
116. Zhao G, Jiang L, He Y, Li J, Dong H, Wang X, Hu W (2011) Sulfonated graphene for persistent aromatic pollutant management. Adv Mater 23:3959–3963
117. Zhao G, Li J, Ren X, Chen C, Wang X (2011) Few-layered graphene oxide nanosheets as superior sorbents for heavy metal ion pollution management. Environ Sci Technol 45:10454–10462
118. Zhao G, Li J, Wang X (2011) Kinetic and thermodynamic study of 1-naphthol adsorption from aqueous solution to sulfonated graphene nanosheets. Chem Eng J 173:185–190
119. Zhao G, Ren X, Gao X, Tan X, Li J, Chen C, Wang X (2011) Removal of Pb (II) ions from aqueous solutions on few-layered graphene oxide nanosheets. Dalton Trans 40:10945–10952
120. Zhao R, Li X, Sun B, Li Y, Li Y, Yang R, Wang C (2017) Branched polyethylenimine grafted electrospun polyacrylonitrile fiber membrane: a novel and effective adsorbent for Cr (VI) remediation in wastewater. J Mater Chem A 5:1133–1144
121. Zhao X, Jiang T, Du B (2014) Effect of organic matter and calcium carbonate on behaviors of cadmium adsorption–desorption on/from purple paddy soils. Chemosphere 99:41–48
122. Zuo WQ, Chen C, Cui HJ, Fu ML (2017) Enhanced removal of Cd (II) from aqueous solution using $CaCO_3$ nanoparticle modified sewage sludge biochar. RSC adv 7:16238–16243

Removal of Pesticides Using Carbon-Based Nanocomposite Materials

Shahnawaz Uddin

Abstract To keep pace with the demand for more crops for increasing global human population and to reduce human efforts at the same time, the technology and pesticides in the agriculture sector are being used without caring for their impact on the environment. Therefore, it is our first priority to remove the hazardous and toxic pesticide residues completely from the environment and food chain which is very challenging too. In order to remove or convert the pesticide residues into non-toxic form effectively, efficiently, and easily, the carbon-based nanomaterials and their composites using the adsorption process is discussed here in detail. After a brief introduction of pesticides, different techniques used for environmental remediation are discussed. Furthermore, different properties/characteristics of carbon-based nanomaterials/nanocomposites from the point of view of environmental remediation along with adsorption mechanisms have been described in an elucidate manner.

Keywords Pesticide · Environmental remediation · Adsorption · Carbon nanomaterials · Carbon nanocomposites

1 Introduction to Pesticides

Before going into the details of removal of pesticides using carbon-based nanocomposites, it will be of paramount importance to understand pesticides briefly and how do pesticide residues are created in the environment? The term "pesticide" is a composite of two words: pest (means "annoyance or nuisance") and Latin word cide (means "to kill"). Broadly, a pesticide may be defined as any material or mixture of materials used intentionally for preventing/controlling/destroying the pests or unwanted species of animals/plants that destroy during the crop production/processing/transport/marketing/storage. The pesticides are also used against the

S. Uddin (✉)
School of Physics, Universiti Sains Malaysia, Gelugor, Malaysia
e-mail: shahnawazuddin@gmail.com

University Women's Polytechnic, Aligarh Muslim University, Aligarh, India

M. Jawaid et al. (eds.), *Environmental Remediation Through Carbon Based Nano Composites*, Green Energy and Technology,
https://doi.org/10.1007/978-981-15-6699-8_17

vectors of human/animal disease, unwanted or nuisance species of plants/animals, entities causing harm during the production/processing/storage/transport/marketing of the agricultural-commodities/wood-products/animal-feedstuffs. The pesticides are also prescribed to control the insects/agents responsible for thinning or the premature fall of fruits [1]. In other words, pesticides are the chemical substances used to kill or control the pests/insects/fungi/unwanted-rodents/unwanted-plants. Pesticides are also employed in health care to kill vectors of disease (e.g., mosquitoes, bedbugs, etc.) and in the agriculture sector to kill pests (e.g., rats, locusts, etc.). By their nature, pesticides are potentially toxic to other organisms, including humans, and need to be used safely and disposed of properly [2].

Pesticides may be differentiated by a target organism (e.g., insecticides, rodenticides, fungicides, herbicides, and pediculicides), chemical structure (e.g., inorganic, organic, biological, or synthetic) and physical state (e.g., solid, liquid, or gaseous) [3, 4]. The pesticides which are microbial or biochemical in nature are known as biopesticides. There is another category of pesticides derived from plants (botanicals) include rotenoids, pyrethroids, nicotinoids, scilliroside, and strychnine [5, 6].

As we know that prevention is better than cure. It is also possible, under certain circumstances, to produce food without the use of pesticides. Therefore, we should adopt the alternative techniques and methods to save the crops from the pets without using pesticides which include cultivation practices (e.g., polyculture, crop rotation, planting crops in areas where the pests do not live, time of planting when pests will be least problematic, and use of trap crops), use of biological pest controls, genetic engineering, methods of interfering with insect breeding, and composted yard waste to control pests. These methods are becoming increasingly popular and often are safer than traditional chemical pesticides. But the aforementioned approaches for controlling the pets maybe sometimes costly, time-consuming, difficult to apply, and work on some specific types of insects [7, 8]. That's why the people adopt the pesticides for controlling or killing the pests.

1.1 Effects of Pesticides

The main goals of using pesticides are to increase agriculture production and to fight against disease-causing vectors. However, pesticide residues pollute the environment (the air, water, soil, and biomass) and harm to the living things and environment which is a very serious problem worldwide. As shown in Fig. 1, the living organisms may expose to pesticides by ingesting pesticide-contaminated water/food, or by inhaling pesticide-contaminated air, or by skin contact with pesticide-contaminated soil/agricultural-products/air/water (during bathing/washing/swimming/raining). Exposure to pesticides or pesticide residues may cause various harmful effects ranging from a simple eye/skin irritation to more serious effects on the nervous system/reproductive system or causing cancer [9, 10]. The most commonly used pesticides for gardening/agriculture are organophosphate (OP) and CA (Carbonate) pesticides. The organophosphate pesticides (OPPs) include

Fig. 1 **a** Precautions for spraying pesticides **b** use of pesticides at large-scale farming

the following chemicals: malathion, parathion, chlorpyrifos, dichlorvos, methyl parathion, diazinon, triazophos, oxydemeton, phosmet, tetrachlorvinphos, azinphos methyl, etc. OPPs are very toxic because they inhibit cholinesterase which causes neurotoxicity and affects the endocrine system in long term. The carbonate pesticides carbamylate the essential enzyme, acetylcholinesterase (AChE), and inhibit the enzyme activity. The chlorinated pesticides too have very much toxic effects on living organisms (humans/animals/plants), particularly, chlordane which resists to chemical and biological degradation very much is known as POP (persistent organic pollutant) [11–14]. Due to insufficient regulation and safety precautions, more than 90% of people die because of pesticides in the developing countries which use pesticides less than 30% of the total global production [15]. Practically, more than 90% of insecticides and herbicides sprayed are mixed into soil/water/air and creating environmental pollution at the faraway places too because winds carry the suspended pesticides-particles in air to other distant regions. Therefore, in addition to environmental pollution, pesticides affect biodiversity by reducing pollination and habitat for birds and endangered species [8, 16–18].

1.2 Maximum Permissible Limit of Pesticide Residue

To bring down the negative effect of pesticides on living organisms, pesticides should be used minimally, quick bio-degradable, or deactivated. The left out pesticides in the environment are known as "pesticide residues" [19]. The pesticide regulatory body in every country puts the restriction on the maximum permissible level of pesticide residues in the food and environment. WHO (World Health Organization) in collaboration with FAO (Food and Agriculture Organization) assesses the harmful effects of pesticides on living organisms. JMPR (Joint Meeting on Pesticide Residues), a joint WHO/FAO independent international expert scientific group, conducts and evaluates the harmful impact of pesticide residues in food. The group has developed an

International Code of Conduct on Management of Pesticides. There are two objectives of WHO relating to pesticides [2, 20]: (i) to put ban on most toxic pesticides to humans and long-lasting pesticides in the environment (ii) to protect the living organism health by setting MPL (maximum permissible limit) for pesticide residues in the environment as given below.

1. For Water: $\leq$0.1 μg/L (for an individual pesticide) and $\leq$0.5 μg/L (for all pesticides combined together) [21]
2. For Soil: $\leq$0.05 mg/kg [22]
3. For Air: Since, a few jurisdictions in the world had regulated pesticide residues in the air, one of them is the USA [23, 24]. Therefore, MPL for air could not be listed here because of lack of information available globally.

2 Introduction to Pesticide Removal from Environment

As we know that the pesticides help in boosting the crop output throughout agriculture but they also pollute the environment. Therefore, it is obligatory for humanity to protect the environment from any kind of pollution by identifying and treating the contaminants/pollutants. It is the need of the hour to employ the more advance and efficient methods/materials to remove the pesticide residues from the environment. In the recent past, nanotechnology has been more popular in sensing environmental contamination and its remediation using nanomaterials/nanocomposites having unique properties in more practical ways [11]. All over the world, scientists are trying to find out or synthesize more effective and efficient nanomaterials/nanocomposites for environmental remediation. Emerging carbon-based nanomaterials/nanocomposites (e.g., carbon nanotubes, graphene, graphene oxide, reduced graphene oxide, etc.) have flexible or tunable physical/chemical/electrical properties which are best suited for challenging environmental remediation. The environmental remediation applications of carbon-based nanomaterials/nanocomposites discussed here are both proactive (preventing environmental degradation, improving organism health, optimizing energy and resource consumption) and retroactive (remediation, transformation of pollutants, reuse of wastewater) [25]. These nanomaterials/nanocomposites have enormous potential for adsorption of pesticides from the environment because of their large surface area. The carbon nanotubes (CNTs) are indeed very useful for environmental remediation because of their shape, large surface area, large length-to-radius ratio, hydrophobic-wall and flexible-surface [25–28]. Similarly, graphene and graphene-based nanocomposites are very useful and attractive choice towards environmental remediation (such as water purification) due to their unique properties such as large surface area, stronger π-π interactions with target pollutants having aromatic rings, tunable chemical-properties, antibacterial nature and lesser cytotoxicity [29–35]. Furthermore, graphene and its derivatives in the modified form are also used to identify organic pesticides/compounds [36].

This chapter begins with the definition of pesticides, necessity of pesticides, effects of pesticide residues on the environment, brief outline of carbon-based nanomaterials/nanocomposites, and their corresponding properties befitting the adsorption process for environmental remediation. More feasible and practical applications of carbon-based nanomaterials/nanocomposites have been incorporated for the removal of pesticide residues/pollutants from the environment. This chapter also highlights the comparative benefits of functionalization of carbon-based nanomaterials to facilitate and enhance the environmental remediation effectively and efficiently. Here, the readers will be benefitted from the working principle, application, and comparative benefits of the most economical and modern practices adopted for pesticide removal using carbon-based nanocomposites and in the existing world.

3 Why Carbon-Based Nanomaterials/Nanocomposites?

The unique hybridization property of carbon and carbon's sensitivity to its structure with variations in synthesis conditions permit for tuned manipulation in the carbon-based nanostructures not yet provided by inorganic nanostructures [25]. The unique physiochemical properties of carbon-based nanomaterials/nanocomposites (e.g., shape, size, and surface area; molecular interactions, sensitivity, sorption properties, etc.) make these materials effective, specific-candidate and reactive-media for environmental remediation for very levels of pesticide residues. Although the metal oxides nanostructures are effectively used for absorption of destructive and toxic organophosphorus pesticides, they are relatively difficult and expensive to produce in fine powder form with high quality [37, 38]. Recently, the results of many studies show, many other nanostructures (like gold-based nanorods/spheres) have been employed to remove various types of pesticides (e.g., a very common organophosphorus pesticide in agriculture: dimethoate) from an aqueous solution. But these nanomaterials on a large-scale basis are very expensive for the environmental remediation [39].

4 Techniques to Remove Pesticides from Environment

Due to the higher stability or strong opposition to biological degradation, it has been observed that a large number of the pollutants/pesticide residues can't be removed completely from the environment through a biological/conventional treatment techniques. And environmental contamination (especially water pollution) has surpassed the threshold of the natural/ground filtration process because of fast industrialization [40]. Therefore, in parallel with adequate regulatory control on the use of pesticides, there is an urgent need for the identification and removal of pesticide residues from the environment effectively, efficiently, and economically. Thus, an overview of recent developments/advances has been provided for the environmental remediation using three main methods: filtration, degradation, and adsorption [21, 41–47].

4.1 Filtration Process

The most common, promising, and recent process of membrane filtration is nanofiltration (NF) in which the pressure-driven membrane has the properties between UF (ultrafiltration) and RO (reverse osmosis) membranes. In this process, the hazardous/toxic organic micropollutants (such as pesticides, dyes, and other various synthesized chemical compounds) are filtered out completely. Depending on the application/requirement, NF membranes from some manufacturers target only the specific molecules based on their molecular size or weight. The adsorption of organic chemicals on NF membrane surface is affected by the physiochemical characteristics of the membrane, characteristics of pesticides, composition of the fluid/water, and operating parameters of filtration system. During filtration process, the contaminants/compounds present in the fluid/water such as pollutant-particles, salts, colloids, soluble microbes and natural organic matter get accumulated on the surface of membrane and results in a major problem of "membrane fouling". This fouling of membrane causes slow filtration process due to a decrease in flux to permeate and an increase in cost of operation due to short life span of the membrane [48–51]. The membrane fouling also changes the characteristics of membrane surface such as zeta potential, contact-angle, surface morphology, and functionality which decreases the transport of pollutants as compared to fresh membranes [21].

4.2 Degradation Process

Through the degradation process, a pesticide is converted into simpler chemical byproducts like water, carbon dioxide, and ammonia due to chemical reactions [52]. ZVI (Zero-Valent Iron) is one of the most widely used degradation processes for treating the contaminations because of its accessibility, effectiveness, and very little generation of waste-products/secondary-pollutants [53]. Another degradation process, photocatalytic oxidation, is a very environment-friendly method for removing a wide range of organic pollutants. It is suited for pre-treatment of hazardous/toxic and non-biodegradable pollutants to increase their biodegradability [54]. In a photocatalysis degradation process, the UV or direct sun-light photo-excites the surfaces of solid semiconductor (e.g., ZnO, TiO_2, Fe_2O_3, CdS, and WO_3) and results into free electrons and positive semiconductor surfaces. The ions in an excited state and free electrons accelerate oxidation/reduction reaction and as a result fast degradation of pollutants/contaminants. Further, to increase the rate of photodegradation, a higher power UV lamp can be used. Through nanotechnology advancements, semiconductor-based photocatalysts have been improved in terms of selectivity and reactivity and as a result, a number of pesticides are being treated by photocatalysis degradation process. Due to its chemical stability, low toxicity, low cost because of abundance on earth, titanium dioxide (TiO_2) is used most widely for the photodecomposition of pesticides. Organochlorine class of pesticides has been photodecomposed by the UV irradiation on nano-TiO_2 coated films in air. In another study, the complete photodegradation of dicofol is done by irradiating UV light in the presence of TiO_2 nanoparticles and results in less toxic compounds [55–62].

4.3 *Adsorption Process*

In the adsorption process, the adhesion of atoms/ions/molecules from an adsorbate fluid (gas/liquid/dissolved solid) takes place to the surface of an adsorbent, i.e., a film of the adsorbate is formed on the surface of an adsorbent (as shown in Fig. 2). The adsorption process is different from absorption, in which a fluid (adsorbate) is dissolved or permeated by a liquid/solid (adsorbent). Furthermore, the adsorption process is a surface phenomenon while in absorption process, the whole volumes of the materials are involved [63]. Adsorption process is considered as an effective and equilibrium separation method for cleaning water in the present scenario. It is superior to other methods for water cleaning and recycling because of simplicity of design, low initial cost, flexible, easy to operate, insensitive to toxic contaminants/pollutants, and no formation of harmful byproducts. As we know that the adsorption process is a surface phenomenon, thus, the adsorption of pollutants depends on specific surface area of the adsorbent, available number of sites, porosity, and various kinds of interactions as well [21].

Therefore, after considering the pros and cons of the materials and process for environmental remediation, the adsorption process will be discussed throughout this chapter to remove the pesticides from the environment using carbon-based nanomaterials/nanocomposites.

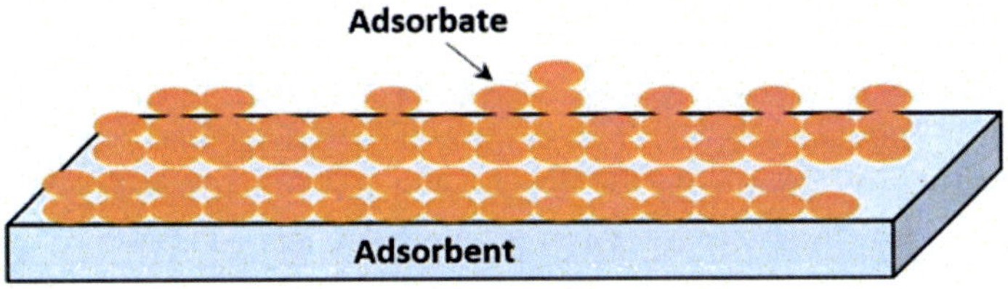

Fig. 2 Adsorbate and adsorbent in adsorption process [63]

5 Properties of Carbon-Based Nanomaterials/Nanocomposites and Adsorption Mechanisms Used in Environmental Remediation

Figure 3 shows various hybridization states of carbon-based nanomaterials. The mutable hybridization-state of carbon is responsible for the diversity of carbon-based compounds [64].

Carbon-based nanomaterials/nanocomposites combine the unique properties of sp^2-hybridized carbon-bonds with the unusual physiochemical properties of carbon at nanoscale to have distinctive properties (e.g., shape, size, chirality, surface area, molecular interactions, etc.) for environmental remediation applications [25]. The molecular manipulation of a carbon-based nanomaterial/nanocomposite means holding control over its structure and conformation (such as shape, size, chirality, number of graphitic-layers, etc.). The diameter of fullerene/tubular-nanostructure is a very critical parameter in determining their properties and applications, for example, carbon nanotubes (CNTs) with a narrow inner-diameter are useful in separation, novel molding, size exclusion processes and membrane filtration [65–71]. There are some other parameters of the environment such as ionic strength, pH value, and dispersion state of carbon nanotubes that also affect the adsorption. The surfaces of unmodified carbon nanotubes are hydrophobic and preferring the adsorption of hydrocarbons (e.g., benzene, hexane, cyclohexane, etc.) over alcohols (e.g., 2-propanol, ethanol, etc.) [72, 73].

A higher number of surface-atom sites of CNTs provide high adsorption capability and a good platform for the adsorbate-species [74–77]. The liquids having high surface-tension are not able to wet CNTs properly but water and most of the organic-solvents having substantial intermolecular-interactions with surface of CNTs induce capillary-action [78]. Physisorption or physical adsorption because of Van der Waals forces is one of the predominant mechanisms of sorption for non-functionalized carbon-based nanomaterials [79].

The organic pollutants/contaminants are sorbed directly to the carbon nanomaterials/nanocomposites surface through the same fundamental mechanisms such as hydrophobic interaction, weak dipolar forces, and dispersion that were applicable in conventional sorption systems [80, 81]. The equilibrium rates of carbon-based nanocomposite sorbents are higher as compared to conventional activated carbon sorbents due to π electron-polarizability or π-π EDA (electron–donor–acceptor)

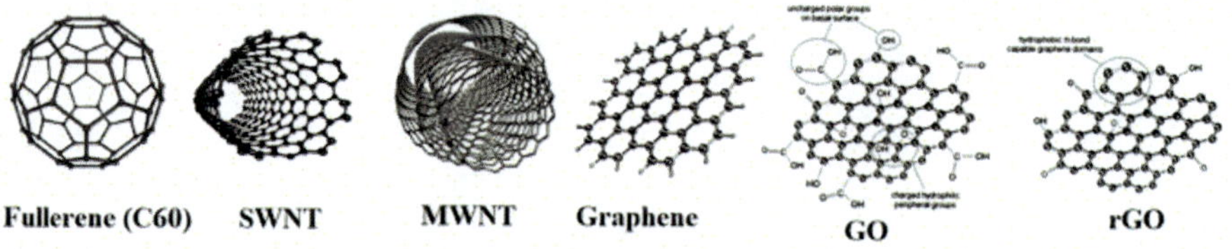

Fig. 3 Hybridization-states of carbon-based nanomaterials [25]

interactions with aromatic-sorbates [82–84], less heterogeneity of adsorption-energies [85] and no pore-diffusion as an intermediate adsorption-mechanism [86]. SWNTs (single wall nanotubes) with curved geometries enhance Van der Waals interaction which in turn amplifies the adsorption.

Prediction and optimization of pore-diameter of SWNTs for gaseous adsorption is done with the help of Monte-Carlo simulations. For example, an optimum diameter of 1.05 nm for CNT in case of adsorption of a potential greenhouse gas (tetrafluoromethane) and effectively balance the enthalpy of adsorption [87]. Similarly, the optimized carbon-based nanostructures may be predicted to remove the specific pollutants/contaminants. As we know that the potential attributes of carbon-based nanosorbents are higher capacity of sorption and fast equilibrium rates but tailoring the surface chemistry of carbon nanostructures has revolutionized their use for target-based adsorption. The optimized carbon-based nanosorbents are useful to target specific micro-contaminants, to remove pollutants with very low concentration, and to increase subsurface-mobility [88, 89]. Furthermore, the sorption rate through the aligned CNT-membranes is higher (4–5 orders of magnitude) than predicted by the conventional fluid flow theory. Theoretically, these higher flow rates through the carbon nanotubes are assumed due to hydrophobicity and less frictional forces because of smooth internal-walls of the tubes upon passage of water-molecules [53, 90–92].

The potential physiochemical properties (such as large surface area, low cytotoxicity large number of delocalized π-electrons and tunable-chemistry) of graphene and its derivatives (such as graphene oxide, reduced graphene, etc.) support graphene and its derivatives for environmental remediation by adsorbing various commonly used pesticides and other pollutants [35, 93–99]. However, the aggregation of adsorbents reduces the available surface area of graphene and as a result, decrease its adsorption capacity. Therefore, to avoid aggregation and increase the selectivity of adsorbents, the functionalization of graphene is done by selecting the molecules which are water-soluble and have an affinity towards target-adsorbents [31, 100–102].

6 Removal of Pesticides Using Carbon-Based Nanomaterials/Nanocomposites

In this section, various carbon-based nanomaterials/nanocomposites have been discussed for the removal of pesticides/pollutants from the environment in an effective and practical manner.

6.1 Use of Carbon Nanotubes (CNTs) as Adsorbent

Carbon-based nanomaterials, CNTs, consist of graphitic-carbons having single/multiple concentric tubules and are known as SWNTs (single-walled nanotubes)/MWNTs (multi-walled nanotubes), respectively. They are unique one-dimensional structures that are thermally stable with some unique physiochemical properties [103, 104]. These nanomaterials are potentially very effective in removing a variety of pesticides because porous structure of CNTs and a wide range of surface functional groups enhance the adsorption capacity of contaminants [105]. One or more adsorption mechanisms, such as covalent-bonding, hydrophobic-effect, π-π interactions, hydrogen-bonding, and electrostatic interactions may be responsible for the adsorption of organic compound on CNTs [106–108]. For example, π-π interaction is responsible for the adsorption of organic compounds having C=C bond or benzene-ring (such as polycyclic aromatic hydrocarbons (PAHs) and polar aromatic chemicals) [82, 109], hydrogen-bonding is responsible for the adsorption of the compounds having any of the functional groups –OH, $–NH_2$, –COOH and organic molecules [110] and adsorption of some organic compounds such as dyes with appropriate pH and antibiotics are due to electrostatic attraction with the functionalized CNTs [111, 112].

The herbicides, diuron, and dichlobenil have been removed by adsorption on MWNTs and as expected the adsorption increases with an increase in total surface area and pore-volume of MWNTs. But the amount of adsorption and surface-coverage of diuron are high as compared to dichlobenil, although the water-solubility, the surface area and the total molecular-volume in case of dichlobenil are low which may be assumed because of lower Van der Waals interactions of dichlobenil as compared to diuron [113, 114].

The atrazine herbicide was adsorbed with the help of surfactant treated SWNTs and MWNTs and it was observed that surfactant-treatment modified the CNTs to become more hydrophilic which in turn suppressed the adsorption of atrazine [115]. But the oxidation-treatment of MWNTs increases the adsorption of diuron due to the increase in surface area and total pore-volume of MWNTs [116]. Higher adsorption of a phenoxy-acid herbicide (4-chloro-2-methylphenoxyacetic acid or MCPA) has been shown by SWCNTs as compared to MWNTs and some other nanostructured oxide adsorbents (such as Al_2O_3, TiO_2 and ZnO, etc.) because of a mechanism known as "pseudo-second-order kinetics" (chemisorption-the adsorption is due to physicochemical-interactions between two-phases) [117, 118].

6.2 Use of Graphene and Its Derivatives (GO, rGO, etc.) as Adsorbent

Recently, a lot of attention has been paid to graphene for environmental remediation, especially for the removal of pesticides from water due to its unique physiochemical properties. There is a tremendous adsorption capacity of organophosphorus pesticides such as endosulfan, chlorpyrifos, and malathion (approximately in the range of 600–2000 mg/g) on graphene surface due to effective-interactions between graphene having large surface area and pesticides and polar-structure of water also plays an important role in adsorbent-adsorbate interaction [119]. To remove the persistent halocarbon pesticides from the water, the dehalogenation of pesticides has been done by graphene [120, 121]. Graphene and other carbon-based nanomaterials/nanocomposites adsorb contaminants/pollutants having aromatic-ring because of π-π interactions [82, 107, 109, 122]. A combination of graphene with other materials is used to improve adsorption capacity for pesticides, e.g., GCS (graphene-coated silica) is highly efficient/effective sorbent for removal of organophosphorus pesticide residues from water because of a mechanism that uses electron–donor abilities of P, N and S atoms and strong π-bonding of benzene rings [101, 102, 123–125].

Graphene oxide (GO) is the most popular derivative of graphene having a 2D-layered structure. Along with having a very large surface area, it has various prominent oxygen-containing functional groups such as carboxylic, hydroxyl, epoxide, carbonyl, etc. And because of these functional groups, GO is easily dispersed in aqueous solution and makes a homogeneous colloidal-suspension by unfolding its 2D-structure. Studies show that GO is being used more vigorously for the absorption of various pesticides/pollutants due to its interaction with adsorbates in molecular/ionic forms through any of the mechanisms such as electrostatic, π-π interaction, hydrophobic, etc. As shown in Fig. 4, for the quick and potential adsorption, the adsorbents should provide the active sites at their surfaces, and adsorbate molecules migrate as fast as possible. In another study, it is found that the adsorption of commonly used OPs such as DMT (dimethoate) and CPF (chlorpyrifos) on graphene and its derivatives to be very sensitive to the adsorbent-structure, i.e.,

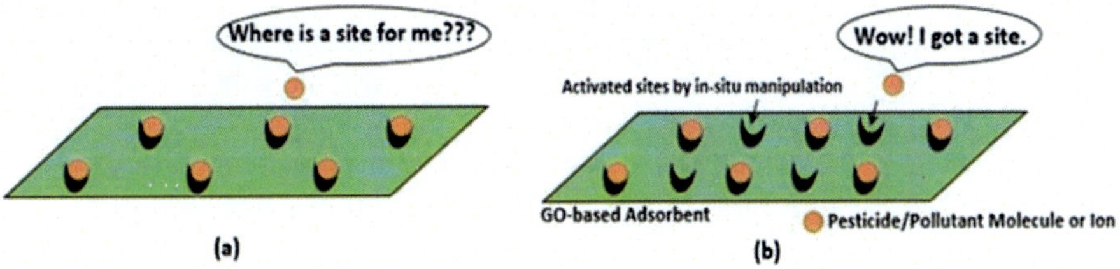

Fig. 4 Schematic diagrams: **a** adsorption process using conventional 1-Step **b** adsorption process using enhanced 2-Steps (generation of new-sites from inactive-structures simultaneously) [126]

aliphatic-DMT prefers surfaces of hydrophilic oxidized-graphene while CPF pesticide having aromatic-moiety prefers graphene's highly-ordered π-system [97, 126, 127].

Magnetically modified-graphene is employed successfully for the removal of five commonly used pesticides to support rapeseed-crops in agriculture (metazachlor, tebuconazole, λ-cyhalothrin, chlorpyrifos, and deltamethrin) from the edible rapeseed oil. As already discussed, a large system of delocalized π-electrons of graphene (adsorbent) effectively helps in the formation of potential π-π stacking interactions between aromatic rings of pesticides (adsorbates) and large surface area, noncovalent-interactions and hydrophobic-effect of graphene too support during the adsorption process. To make the whole adsorption process easy and error-free, an aqueous suspension was prepared. Additionally, to avoid the aggregation of nanomaterial, enhance the dispersion of adsorbent in solvents, and render better adsorption, magnetically modified-graphene is employed here. It also helps in separating graphene easily from the isolated-analytes in the supernatant with the help of an external-magnet [36, 128].

6.3 Use of Carbon-Based Nanocomposite as Adsorbent

Reduced graphene oxide supporting silver (rGO-Ag) nanocomposite removes chlordane (a persistent organic pesticide) from water using a 2-steps mechanism: first degradation by Ag nanoparticles and followed by adsorption of the degraded-byproducts (such as 1, 10-dichlorodecane, ether bis (2-chloroallyl) and octadecanoid acid) on rGO-surface. The whole degradation-adsorption process is very fast and takes place at room temperature [129].

7 Optimization of Physiochemical Parameters During Adsorption Process

Being a surface phenomenon, the adsorption process involves the physicochemical interaction of sorbate to sorbent. And to obtain the maximum efficiency of adsorption, different physiochemical parameters such as pH value, contact-time, and operating temperature are varied to get their optimum values which are summarized in Table 1 [130] for a reader to have a quick look. During the optimization process of different parameters, only one parameter is varied by keeping all other parameters at some appropriate constant values.

Table 1 Optimized physiochemical parameters of different pesticides for adsorption

Name of pesticide	pH value	Operating temperature (°C)	Initial concentration level (mg/L)	Average time of contact (min)	Efficiency of removal (%)	Reference
Bifenthrin	–	270	0.01	90	–	[93]
Cyhalothrin	–	270	0.01	90	–	
Permethrin	–	270	0.01	90	–	
Cypermethrin	–	270	0.01	90	–	
Phenvalerate	–	270	0.01	90	–	
Deltamethrin	–	270	0.01	90	–	
Carbofuran	2–7	30 ± 2	2×10^{-7}	15	–	[131]
Metolcarb	2–9	30 ± 2	2×10^{-7}	15	–	
Pirimicarb	2–9	30 ± 2	2×10^{-7}	15	–	
Isoprocarb	2–9	30 ± 2	2×10^{-7}	15	–	
Diethofencarb	2–9	30 ± 2	2×10^{-7}	15	–	
Chlorpyrifos	–	30 ± 2	1	720	–	[132]
Thiame-thoxam	6	30 ± 2	5×10^{-7}	10	55	Wang et al. (2012)
Imidacloprid	6	30 ± 2	5×10^{-7}	10	78	
Acetamiprid	6	30 ± 2	5×10^{-7}	10	72	
Atrazine	6–7	25 ± 2	0.01	30	84–96.4	Wu et al. (2012)
Prometon	6–7	25 ± 2	0.01	30	84–96.4	
Ametryn	6–7	25 ± 2	0.01	30	84–96.4	
Prometryn	6–7	25 ± 2	0.01	30	84–96.4	
Phonamiphos	3–11	260	10	0	<85	[101, 102]
Dimethoate	3–11	260	10	0	<85	
Phorate	3–11	260	10	0	>85	
Parathion-methyl	3–11	260	10	0	>85	
Pirimiphos-methyl	3–11	260	10	0	>85	
Malathion	3–11	260	10	0	>85	
Fenthion	3–11	260	10	0	>85	
Isocarbophos	3–11	260	10	0	>85	
Chlorfenvinphos	3–11	260	10	0	>85	
Profenofos	3–11	260	10	0	>85	
Methidathion	3–11	260	10	0	>85	
Thiacloprid	6	30 ± 2	5×10^{-7}	10	70	
Chlorpyrifos	3–9	30 ± 2	2	30	100	[119]
Endosulfan	3–9	30 ± 2	1	45	100	
Malathion	3–9	30 ± 2	2	60	100	

(continued)

Table 1 (continued)

Name of pesticide	pH value	Operating temperature (°C)	Initial concentration level (mg/L)	Average time of contact (min)	Efficiency of removal (%)	Reference
2, 4-dichlorophenoxyacetic acid	7	30 ± 2	20	140	100	[133]
Pirimicarb	6.8–10	30 ± 2	5×10^{-7}	–	–	[134]
Diethofencarb	6.8–10	30 ± 2	2.5×10^{-6}	–	–	
Carbaryl	6.8–8.2	30 ± 2	2.5×10^{-6}	–	–	
Isoprocarb	6.8	30 ± 2	2.5×10^{-6}	–	–	
Baycarb	6.8	30 ± 2	3×10^{-6}	–	–	
Baygon	6.8	30 ± 2	2.5×10^{-6}	–	–	
Chlorpyrifos	5.8	30 ± 2	5	60	–	[135]
Simeton	9	25–45	1	0	85	[124, 125]
Simazine	9	25–45	1	0	82	
Atrazine	9	25–45	1	0	98	
Ametryn	9	25–45	1	0	95	
Prometryn	9	25–45	1	0	72	
Cyprazine	11	25–45	1	0	90	

8 Conclusions and Future Perspectives

It is near to impossible to avoid the use of pesticides for the benefits of living organisms. In parallel with the regulated use of pesticides, it is of urgent importance to employ the latest technology and materials for environmental remediation. Out of various technologies and materials, nanotechnology coupled with carbon-based nanomaterials/nanocomposites (such as CNTs, graphene, and its derivatives) are considered as a very effective technique for environmental remediation due to many advantages such as (i) effective and efficient use of large surface area of carbon-based nanomaterials/nanocomposites, (ii) multiple adsorption mechanisms operate simultaneously during removal of pesticides (iii) more economical to use, (iv) functionalization of nanomaterials to avoid aggregation, (v) detecting and removing very low levels of pesticide residues, and (vi) possibility of effectively catalyzing the adsorption process. From the future point of view, we have to look into the challenges/problems and bring some nanotechnologies using carbon-based nanomaterials/nanocomposites still at the laboratory level to commercial level through collaboration among scientific facilities, research groups, and funding agencies. In the near future, through rigorous research, the carbon-based novel nanomaterials/nanocomposites will eventually solve the environmental remediation problems and make the environment safe for every living organism.

References

1. World Health Organization (2010) International code of conduct on the distribution and use of pesticides: guidelines for the registration of pesticides. World Health Organization, Technical report, Geneva
2. World health organization (2014) World Health Crises
3. Gilden RC, Huffling K, Sattler B (2010) Pesticides and health risks. J Obstet Gynecol Neonatal Nurs 39(1):103–110. https://doi.org/10.1111/j.1552-6909.2009.01092.x.PMID20409108
4. Lyznicki JM, Kennedy WR, Young DC, Skelton WD, Howe JP, Davis RM, Rinaldi RC (1997) Educational and informational strategies to reduce pesticide risks. Prev Med 26(2):191–200. https://doi.org/10.1006/pmed.1996.0122
5. Kamrin MA (1997) Pesticide Profiles: toxicity, environmental impact and fate, 1st ed. CRC, Boca Raton. ISBN 978-1566701907. OCLC 35262311
6. US EPA (2017) Types of pesticide ingredients
7. McSorley R, Gallaher RN (1996) Effect of yard waste compost on nematode densities and maize yield. J Nematology 28 (4S): 655–60. PMC 2619736. PMID 19277191
8. Miller GT, Spoolman SE (2004) Biodiversity of sustaining the earth, 6th edn. Thompson Learning Inc., Pacific Grove, CA, pp 211–216., ISBN 9780495556879, OCLC 52134759
9. US EPA (2006) Human health issues. Pesticides: health and safety
10. US EPA (2007) National assessment of the worker protection workshop #3. Pesticides: health and safety
11. Ion I, Culetu A, Gherase D, Sirbu F, Ion AC (2014) Environmental applications of carbon-based nanomaterials. acetylcholinesterase biosensors for organophosphate pesticide analysis. New Appl Nanomaterials Bucharest Academiei Române 33–51
12. Li J, Ma N, Giesy JP, Wang Z (2008) Toxicol Lett 183:65–71
13. Saito T, Miura N, Namera A, Miyazaki S, Ohta S, Oikawa H, Inokuchi S, Chromatographia (2013) 76:781 https://doi.org/10.1007/s10337-013-2482-y
14. Sinclair CJ, Boxall AB (2003) Environ Sci Technol 37:4617–4625
15. Goldmann L (2004) childhood pesticide poisoning: information for advocacy and action (Report), WHO
16. Palmer WE, Bromley PT, Brandenburg RL (2007) Wildlife and pesticides-peanuts. N Carol Coop Extension Serv
17. Tosi S, Costa C, Vesco U, Quaglia G, Guido G (2018) A survey of honey bee-collected pollen reveals widespread contamination by agricultural pesticides. Sci Total Environ 615:208–218. https://doi.org/10.1016/j.scitotenv.2017.09.226. PMID 28968582
18. Wells M (2007) Vanishing bees threaten U.S. crops. www.bbc.co.uk. BBC News, London. Retrieved 2007-09-19. Italic or bold markup not allowed in: publisher = (help)
19. McNaught AD, Wilkinson A (1987) Pesticide Residue. Compendium of chemical terminology, 2nd ed. Blackwell, Oxford. https://doi.org/10.1351/goldbook.p04520, ISBN 978-0-9678550-9-7, OCLC 901451465
20. US EPA (2011) Pesticide residues in food
21. Firozjaee TT, Mehrdadi N, Baghdadi M, Bidhendi GRN (2018) Application of nanotechnology in pesticides removal from aqueous solutions—a review. Int J Nanosci Nanotechnol 14(1):43–56
22. Silva V, Mol HGJ, Zomer P, Tienstra M, Ritsema CJ, Geissen V (2019) Pesticide residues in European agricultural soils—a hidden reality unfolded. Sci Total Environ 653:1532–1545. https://doi.org/10.1016/j.scitotenv.2018.10.441
23. Li Z, Jennings A (2017) Worldwide regulations of standard values of pesticides for human health risk control: a review. Int J Environ Res Public Health 14:826. https://doi.org/10.3390/ijerph14070826
24. US EPA (2016) regional screening levels (RSLs)-equations
25. Mauter MS, Elimelech M (2008) Environmental applications of carbon-based nanomaterials. Environ Sci Technol 42(16)

26. De Volder MFL, Tawfick SH, Baughman RH, Hart AJ (2013) Carbon nanotubes: present and future commercial applications. Science 339:535–539
27. Goh K, Chen Y (2017) Controlling water transport in carbon nanotubes. Nano Today 14:13–15
28. Qu X, Alvarez PJJ, Li Q (2013) Applications of nanotechnology in water and wastewater treatment. Water Res 47:3931–3946
29. Chang Y, Yang S-T, Liu J-H, Dong E, Wang Y, Cao A, Liu Y, Wang H (2011) In vitro toxicity evaluation of graphene oxide on A549 cells. Toxicol Lett 200:201–210
30. Liu S, Zeng TH, Hofmann M, Burcombe E, Wei J, Jiang R, Kong J, Chen Y (2011) Antibacterial activity of graphite, graphite oxide, graphene oxide, and reduced graphene oxide: membrane and oxidative stress. ACS Nano 5:6971–6980
31. Sreeprasad TS, Maliyekkal MS, Deepti K, Chaudhari K, Xavier PL, Pradeep T (2011) Transparent, luminescent, antibacterial and patternable film forming composites of graphene oxide/reduced graphene oxide. ACS Appl Mater Interfaces 3:2643–2654
32. Sreeprasad TS, Gupta SS, Maliyekkal SM, Pradeep T (2013) Immobilized graphene-based composite from asphalt: facile synthesis and application in water purification. J Hazard Mater 246–247:213–220
33. Sreeprasad TS, Maliyekkal SM, Lisha KP, Pradeep T (2011) Reduced graphene oxide–metal/metal oxide composites: facile synthesis and application in water purification. J Hazard Mater 186:921–931
34. Tuček J, Kemp KC, Kim KS, Zbořil R (2014) Iron-oxide-supported nanocarbon in lithium-ion batteries, medical, catalytic, and environmental applications. ACS Nano 8:7571–7612
35. Zhu Y, Murali S, Cai W, Li X, Suk JW, Potts JR, Ruoff RS (2010) Graphene and graphene oxide: synthesis, properties, and applications. Adv Mater 22:3906–3924
36. Sitko R, Zawisza B, Malicka E (2013) Graphene as a new sorbent in analytical chemistry. TrAC Trends Anal Chem 51:33–43
37. Carnes CL, Stipp J, Klabunde KJ, Bonevich J (2002) Synthesis, characterization, and adsorption studies of nanocrystalline copper oxide and nickel oxide. Langmuir 18:1352–1359
38. Hinklin T, Toury B, Gervais C, Babonneau F, Gislason J, Morton R, Laine R (2004) Liquid-feed flame spray pyrolysis of metalloorganic and inorganic alumina sources in the production of nanoalumina powders. Chem Mater 16:21–30
39. Momic T, Pasti TL, Bogdanovic U, Vodnik V, Mrakovic A, RakoIevic Z, Pavlovic VB, Vasic V (2016) Adsorption of organophosphate pesticide dimethoate on gold nanospheres and nanorods. hindawi publishing corporation. J Nanomaterials 2016, Article ID 8910271. http://dx.doi.org/10.1155/2016/8910271
40. Hoffmann MR, Martin ST, Choi W, Bahnemann DW (1995) Chem Rev 95:69–96. https://doi.org/10.1021/cr00033a004
41. Andrades MS, Rodriguez-Cruz MS, Sanchez-Martin MJ, Sanchez-Camazano M (2004) Effect of the modification of natural clay minerals with hexadecylpyridinium cation on the adsorption-desorption of fungicides. Int J Environ Anal Chem 84:133–141
42. Badawy MI, Ghaly MY, Gad-Allah TA (2006) Advanced oxidation processes for the removal of organophosphorus pesticides from wastewater. Desalination 194:166–175
43. Bhattacharya A, Ray P, Brahmbhatt H, Vyas K, Joshi S, Devmurari C, Trivedi J (2006) Pesticides removal performance by low-pressure reverse osmosis membranes. J Appl Polym Sci 102:3575–3579
44. Ghosh S, Das SK, Guha AK, Sanyal AK (2009) Adsorption behavior of lindane on rhizopus oryzae biomass: physico-chemical studies. J Hazard Mater 172:485–490
45. Memon GZ, Bhanger M, Akhtar M, Talpur FN, Memon JR (2008) Adsorption of methyl parathion pesticide from water using watermelon peels as a low cost adsorbent. Chem Eng J 138:616–621
46. Shawaqfeh AT (2010) Removal of pesticides from water using anaerobic-aerobic biological treatment. Chin J Chem Eng 18:672–680
47. Zhang Y, Hou Y, Chen F, Xiao Z, Zhang J, Hu X (2011) The degradation of chlorpyrifos and diazinon in aqueous solution by ultrasonic irradiation: effect of parameters and degradation pathway. Chemosphere 82:1109–1115

48. Bellona C, Marts M, Drewes JE (2010) The effect of organic membrane fouling on the properties and rejection characteristics of nanofiltration membranes. Sep Purif Technol 74:44–54. https://doi.org/10.1016/j.seppur.2010.05.006
49. Mo Y, Xiao K, Liang P, Huang X (2015) Effect of nanofiltration membrane surface fouling on organic micro-pollutants rejection: the roles of aqueous transport and solid transport. Desalination 367:103–111
50. Plakas KV, Karabelas AJ (2011) A systematic study on triazine retention by fouled with humic substances NF/ULPRO membranes. Sep Purif Technol 80:246–261
51. Plakas K, Karabelas A, Wintgens T, Melin T (2006) A study of selected herbicides retention by nanofiltration membranes—the role of organic fouling. J Membr Sci 284:291–300
52. Ochoa V, Maestroni B (2018) Pesticides in water, soil, and sediments, integrated analytical approaches for pesticide management (Chap 9). Academic Press. https://doi.org/10.1016/B978-0-12-816155-5.00009-9
53. Dujardin E, Ebbesen TW, Hiura H, Tanigaki K (1994) Capillarity and wetting of carbon nanotubes. Science 265:1850–1852
54. De Lasa H, Serrano-Rosales B (2009) Advances in chemical engineering: photocatalytic technologies. Academic Press
55. Abdennouri M, Baâlala M, Galadi A, El Makhfouk M, Bensitel M, Nohair K, Sadiq M, Boussaoud A, Barka N (2016) Photocatalytic degradation of pesticides by titanium dioxide and titanium pillared purified clays. Arabian J Chem 9:S313–S318
56. Coronado JM, Fresno F, Hernandez-Alonso MD, Portela R (2013) Design of advanced photocatalytic materials for energy and environmental applications. Springer
57. Rajeswari R, Kanmani S (2009) A study on degradation of pesticide wastewater by TIO_2 photocatalysis. J Sci Ind Res 68:1063–1067
58. Reddy PVL, Kim K-H, Song H (2013) Emerging green chemical technologies for the conversion of CH_4 to value added products. Renew Sust Energ Rev 24:578–585
59. Senthilnathan J, Philip L (2009) Removal of mixed pesticides from drinking water system by photodegradation using suspended and immobilized TiO_2. J Environ Sci Health Part B 44:262–270
60. Yang Y, Guo Y, Hu C, Wang Y, Wang E (2004) Preparation of surface modifications of mesoporous titania with monosubstituted Keggin units and their catalytic performance for organochlorine pesticide and dyes under UV irradiation. Appl Catal A General 273:201–210
61. Yu B, Zeng J, Gong L, Yang X, Zhang L, Chen X (2008) Photocatalytic degradation investigation of dicofol. Chin Sci Bull 53:27–32
62. Yu B, Zeng J, Gong L, Zhang M, Zhang L, Chen X (2007) Investigation of the photocatalytic degradation of organochlorine pesticides on a nano-TiO_2 coated film. Talanta 72:1667–1674
63. Kyzas GZ, Deliyanni EA, Matis KA (2014) Graphene oxide and its application as an adsorbent for wastewater treatment. J Chem Technol Biotechnol 89:196–205
64. Falcao EHL, Wudl F (2007) Carbon allotropes: beyond graphite and diamond. J Chem Technol Biotechnol 82:524–531
65. Bandow S, Asaka S, Saito Y, Rao A, Grigorian L, Richter E, Eklund P (1998) Effect of the growth temperature on the diameter distribution and chirality of single-wall carbon nanotubes. Phys Rev Lett 80:3779–3782
66. Cassell AM, Raymakers JA, Kong J, Dai HJ (1999) Large scale CVD synthesis of single-walled carbon nanotubes. J Phys Chem B 103:6484–6492
67. Han W, Fan S, Li Q, Hu Y (1997) Synthesis of gallium nitride nanorods through a carbon nanotube-confined reaction. Science 277:1287–1289
68. Holt J, Park H, Wang Y, Stadermann M, Artyukhin A, Grigoropoulos C, Noy A, Bakajin O (2006) Fast mass transport through sub-2-nanometer carbon nanotubes. Science 312:1034–1037
69. Jost O, Gorbunov A, Liu XJ, Pompe W, Fink J (2004) Single walled carbon nanotube diameter. J Nanosci Nanotechnol 4:433–440
70. Dresselhaus MS, Dresselhaus G, Avouris P, Smalley RE (2001) Carbon nanotubes: synthesis, structure, properties and applications vol 80. Springer, Berlin, NY

71. Popov VN (2004) Carbon nanotubes: properties and application. Mater Sci Eng Res 43:61–102
72. Liu Y, Zhang Y, Wang X, Wang Z, Lai W, Zhang X, Liu X (2018) Excellent microwave absorbing property of multiwalled carbon nanotubes with skin–core heterostructure formed by outer dominated fluorination. J Phys Chem C 122:6357–6367
73. Pan B, Xing B (2008) Adsorption mechanisms of organic chemicals on carbon nanotubes. Environ Sci Technol 42:9005–9013
74. Anson A, Lafuente E, Urriolabeitia E, Navarro R, Benito AM, Maser WK, Martinez MT (2006) Hydrogen capacity of palladium-loaded carbon materials. J Phys Chem B 110:6643–6648
75. Kongkanand A, Kuwabata S, Girishkumar G, Kamat P (2006) Single-wall carbon nanotubes supported platinum nanoparticles with improved electrocatalytic activity for oxygen reduction reaction. Langmuir 22:2392–2396
76. Krishna V, Pumprueg S, Lee SH, Zhao J, Sigmund Krishna V, Pumprueg S, Lee SH, Zhao J, Sigmund W, Koopman B, Moudgil BM (2005) Photocatalytic disinfection with titanium dioxide coated multi-wall carbon nanotubes. Process Saf Environ Prot 83:393–397
77. Lin YH, Cui XL (2006) Bontha, J. Electrically controlled anion exchange based on polypyrrole and carbon nanotubes nanocomposite for perchlorate removal. Environ Sci Technol 40:4004–4009
78. Dujardin E, Ebbesen TW, Krishnan A, Treacy MMJ (1998) Wetting of single shell carbon nanotubes. Adv Mater 10:1472–1475
79. Be´guin F, Flahaut E, Linares-Solano A, Pinson J (2006) Understanding carbon nanotubes 495
80. Allen-King RM, Grathwohl P, Ball WP (2002) New modeling paradigms for the sorption of hydrophobic organic chemicals to heterogeneous carbonaceous matter in soils, sediments, and rocks. Adv Water Resour 25:985–1016
81. Pignatello JJ, Xing BS (1996) Mechanisms of slow sorption of organic chemicals to natural particles. Environ Sci Technol 30:1–11
82. Chen W, Duan L, Zhu D (2007) Adsorption of polar and nonpolar organic chemicals to carbon nanotubes. Environ Sci Technol 41:8295–8300
83. Gotovac S, Hattori Y, Noguchi D, Miyamoto J, Kanamaru M, Utsumi S, Kanoh H, Kaneko K (2006) Phenanthrene adsorption from solution on single wall carbon nanotubes. J Phys Chem B 110:16219–16224
84. Long RQ, Yang RT (2001) Carbon nanotubes as superior sorbent for dioxin removal. J Am Chem Soc 123:2058–2059
85. Papirer E, Brendle E, Ozil F, Balard H (1999) Comparison of the surface properties of graphite, carbon black and fullerene samples, measured by inverse gas chromatography. Carbon 37:1265–1274
86. Peng XJ, Li YH, Luan ZK, Di ZC, Wang HY, Tian BH, Jia ZP (2003) Adsorption of 1,2-dichlorobenzene from water to carbon nanotubes. Chem Phys Lett 376:154–158
87. Kowalczyk P, Holyst R (2008) Efficient adsorption of super greenhouse gas tetrafluoromethane) in carbon nanotubes. Environ Sci Technol 42:2931–2936
88. Savage N, Diallo MS (2005) Nanomaterials and water purification: opportunities and challenges. J Nanopart Res 7:331–342
89. Yan H, Gong AJ, He HS, Zhou J, Wei YX, Lv L (2006) Adsorption of microcystins by carbon nanotubes. Chemosphere 62:142–148
90. Aleksandrov PS (1961) Elementary Concepts of Topology. Dover Publications, New York
91. Lecoanet HF, Bottero JY, Wiesner MR (2004) Laboratory assessment of the mobility of nanomaterials in porous media. Environ Sci Technol 38:5164–5169
92. Majumder M, Chopra N, Andrews R, Hinds BJ (2005) Nanoscale hydrodynamics: enhanced flow in carbon nanotubes. Nature 438:930
93. Chen J, Zou J, Zeng J, Song X et al (2010) Preparation and evaluation of graphene-coated solid-phase micro-extraction fiber. Anal Chim Acta 678:44–49. https://doi.org/10.1016/j.aca.2010.08.008
94. Dobrota AS, Pasti IA, Mentus SV, Skorodumova NV (2016) A general view on the reactivity of the oxygen-functionalized graphene basal plane. Phys Chem Chem Phys 18:6580–6586

95. Dobrota AS, Pasti IA, Mentus SV, Skorodumova NV (2017) A DFT study of the interplay between dopants and oxygen functional groups over the graphene basal plane-implications in energy-related applications. Phys Chem Chem Phys 19:8530–8540
96. Hu W, Peng C, Luo W, Lv M, Li X, Li D, Huang Q, Fan C (2010) Graphene-Based antibacterial paper. ACS Nano 4:4317–4323
97. Lazarević-Pašti T, Anićijević V, Baljozović M, Anićijević DV, Gutić S, Vasić V, Skorodumova NV, Pašti IA (2018) The impact of the structure of graphene-based materials on the removal of organophosphorus pesticides from water. Environ Sci Nano 6
98. Lu M, Li J, Yang X, Zhang C, Yang J, Hu H, Wang X (2013) Applications of graphene based materials in environmental protection and detection. Chin Sci Bull 58:2698–2710
99. Whitby M, Quirke N (2007) Fluid flow in carbon nanotubes and nanopipes. Nat Nanotechnol 2:87–94
100. Dervin S, Dionysiou DD, Pillai SC (2016) 2D nanostructures for water purification: graphene and beyond. Nanoscale 8:15115–15131
101. Liu X, Zhang H, Ma Y, Wu X et al (2013) Graphene-coated silica as a highly efficient sorbent for residual organophosphorus pesticides in water. J Mater Chem A 1:1875–1884. https://doi.org/10.1039/c2ta00173j
102. Liu X, Zhang H, Ma Y, Wu X, Meng L, Guo Y, Yu G, Liu Y (2013) Graphene-coated silica as a highly efficient sorbent for residual organophosphorus pesticides in water. J Mater Chem A 1:1875–1884
103. Firozjaee TT, Mehrdadi N, Baghdadi M, Bidhendi GN (2017) The removal of diazinon from aqueous solution by chitosan/carbon nanotube adsorbent. Desalin Water Treat 79:291–300
104. Ren X, Chen C, Nagatsu M, Wang X (2011) Carbon nanotubes as adsorbents in environmental pollution management: a review. Chem Eng J 170:395–410
105. Yunus IS, Harwin Kurniawan A, Adityawarman D, Indarto A (2012) Nanotechnologies in water and air pollution treatment. Environ Tec Rev 1:136–148
106. Pyrzynska K (2011) Carbon nanotubes as sorbents in the analysis of pesticides. Chemosphere 83:1407–1413
107. Yang K, Xing B (2010) Adsorption of organic compounds by carbon nanomaterials in aqueous phase: polanyi theory and its application. Chem Rev 110:5989–6008
108. Yu JG, Zhao XH, Yang H, Chen XH, Yang Q, Yu LY, Jiang JH, Chen XQ (2014) Aqueous adsorption and removal of organic contaminants by carbon nanotubes. Sci Total Environ 482:241–251
109. Smith SC, Rodrigues DF (2015) Carbon-based nanomaterials for removal of chemical and biological contaminants from water: a review of mechanisms and applications. Carbon 91:122–143
110. Yang K, Wu W, Jing Q, Zhu L (2008) Aqueous adsorption of aniline, phenol, and their substitutes by multi-walled carbon nanotubes. Environ Sci Technol 42:7931–7936
111. Ghaedi M, Kokhdan SN (2012) Oxidized multiwalled carbon nanotubes for the removal of methyl red (MR): kinetics and equilibrium study. Desalin Water Treat 49:317–325
112. Ji L, Chen W, Duan L, Zhu D (2009) Mechanisms for strong adsorption of tetracycline to carbon nanotubes: a comparative study using activated carbon and graphite as adsorbents. Environ Sci Technol 43:2322–2327
113. Chen GC, Shan X-Q, Pei Z-G, Wang H, Zheng L-R, Zhang J, Xie Y-N (2011) Adsorption of diuron and dichlobenil on multiwalled carbon nanotubes as affected by lead. J Hazard Mater 188:156–163
114. Lopez-Ramon M, Fontecha-Camara M, Alvarez-Merino M, Moreno-Castilla C (2007) Removal of diuron and amitrole from water under static and dynamic conditions using activated carbons in form of fibers, cloth, and grains. Water Res 41:2865–2870
115. Shi B, Zhuang X, Yan X, Lu J, Tang H (2010) Adsorption of atrazine by natural organic matter and surfactant dispersed carbon nanotubes. J Environ Sci 22:1195–1202
116. Deng J, Shao Y, Gao N, Deng Y, Tan C, Zhou S, Hu X (2012) Multiwalled carbon nanotubes as adsorbents for removal of herbicide diuron from aqueous solution. Chem Eng J 193:339–347

117. De Martino A, Iorio M, Xing B, Capasso R (2012) Removal of 4-chloro-2-methylphenoxyacetic acid from water by sorption on carbon nanotubes and metal oxide nanoparticles. RSC Adv 2:5693–5700
118. Wang H, Zhou A, Peng F, Yu H, Yang J (2007) Mechanism study on adsorption of acidified multi-walled carbon nanotubes to Pb(II). J Colloid Interface Sci 316:277–283. https://doi.org/10.1016/j.jcis.2007.07.075
119. Maliyekkal SM, Sreeprasad TS, Krishnan D, Kouser S, Mishra AK, Waghmare UV, Pradeep T (2013) Graphene: a reusable substrate for unprecedented adsorption of pesticides. Small 9:273–283. https://doi.org/10.1002/smll.201201125
120. Pei Z, Li L, Sun L, Zhang S, Shan XQ, Yang S, Wen B (2013) Adsorption characteristics of 1, 2, 4-trichlorobenzene, 2, 4, 6-trichlorophenol, 2-naphthol and naphthalene on graphene and graphene oxide. Carbon 51:156–163
121. Sen Gupta S, Chakraborty I, Maliyekkal SM, Mark TA, Pandey DK, Das SK, Pradeep T (2015) Simultaneous dehalogenation and removal of persistent halocarbon pesticides from water using graphene nanocomposites: a case study of lindane. Sustain Chem Eng 3:1155–1163
122. Bjork J, Hanke F, Palma CA, Samori P, Cecchini M, Persson M (2010) Adsorption of aromatic and anti-aromatic systems on graphene through π-π stacking. J Phys Chem Lett 1:3407–3412
123. Mahpishanian S, Sereshti H, Baghdadi M (2015) Superparamagnetic core-shells anchored onto graphene oxide grafted with phenylethyl amine as a nano-adsorbent for extraction and enrichment of organophosphorus pesticides from fruit, vegetable and water samples. J Chromatogr A 1406:48–58
124. Zhang C, Zhang RZ, Ma YQ, Guan WB, Wu Xiao L, Liu X, Li H, Du YL, Pan CP (2015) Preparation of cellulose/graphene composite and its applications for triazine pesticides adsorption from water. ACS Sustain Chem Eng 3:396–405. https://doi.org/10.1021/sc500738k
125. Zhang C, Zhang RZ, Ma YQ, Guan WB, Wu XL, Liu X, Li H, Du YL, Pan CP (2015) Preparation of cellulose/graphene composite and its applications for triazine pesticides adsorption from water. Sustain Chem Eng 3:396–405
126. Wang Y, Pan C, Chu W, Vipin AK, Sun L (2019) Environmental remediation applications of carbon nanotubes and graphene oxide: adsorption and catalysis. Nanomaterials 9:439, MDPI. https://doi.org/10.3390/nano9030439
127. Wang S, Li X, Liu Y, Zhang C, Tan X, Zeng G, Song B, Jiang L (2018) Nitrogen-containing amino compounds functionalized graphene oxide: Synthesis, characterization and application for the removal of pollutants from wastewater: a review. J Hazard Mater 342:177–191
128. Madej K, Janiga K, Piekoszewski W (2018) The potential of graphene as an adsorbent for five pesticides from different classes in rape oil samples using dispersive solid-phase extraction, Hindawi. J Anal Methods Chem 8. Article ID 3587860. https://doi.org/10.1155/2018/3587860
129. Sarno M, Casa M, Cirillo C, Ciambelli P (2017) Complete removal of persistent pesticide using reduced graphene oxide-silver nanocomposite, Chem Eng Trans 60. ISBN978-88-95608-50-1; ISSN 2283-9216. https://doi.org/10.3303/cet1760026
130. Paramasivan T, Sivarajasekar N, Muthusaravanan S, Subashini R, Prakashmaran J, Sivamani S, Koya PA (2019) Graphene family materials for the removal of pesticides from water. A new generation material graphene: applications in water technology. Springer International Publishing AG. https://doi.org/10.1007/978-3-319-75484-0_13
131. Wu Q, Zhao G, Feng C, Wang C, Wang Z (2011) Preparation of a graphene-based magnetic nanocomposite for the extraction of carbamate pesticides from environmental water samples. J Chromatogr A 1218:7936–7942. https://doi.org/10.1016/j.chroma.2011.09.027
132. Gupta SS, Sreeprasad TS, Maliyekkal SM, Das SK, Pradeep T (2012) Graphene from sugar and its application in water purification. ACS Appl Mater Interfaces 4:4156–4163. https://doi.org/10.1021/am300889u
133. Tang Y, Zhang G, Liu C, Luo S, Xu X (2013) Magnetic TiO_2-graphene composite as a high-performance and recyclable platform for efficient photocatalytic removal of herbicides from water. J Hazard Mater 252–253:115–122. https://doi.org/10.1016/j.jhazmat.2013.02.053

134. Shi Z, Hu J, Li Q, Zhang S, Liang Y, Zhang H (2014) Graphene based solid phase extraction combined with ultra high performance liquid chromatography-tandem mass spectrometry for carbamate pesticides analysis in environmental water samples. J Chromatogr A 1355:219–227. https://doi.org/10.1016/j.chroma.2014.05.085
135. Gupta VK, Eren T, Atar N, Yola ML, Parlak C, Karimi-Maleh H (2015) $CoFe_2O_4@TiO_2$ decorated reduced graphene oxide nanocomposite for photocatalytic degradation of chlorpyrifos. J Mol Liq 208:122–129. https://doi.org/10.1016/j.molliq.2015.04.032

Manufacturing and Characterization of Carbon-Based Nanocomposite Membrane for Water Cleaning

Gunjan Bhalla, Anupamdeep Sharma, Vaneet Kumar, Barjinder Bhalla, Saruchi, and Harsh Kumar

Abstract Nanotechnology-based wastewater treatment assures to conquer various drawbacks of the present adopted wastewater treatment methods. Wastewater treatment is an important task for environmental conservation. Among different wastewater treatment methods, the use of carbon nanomaterials (CNMs) to eliminate wide range of pollutants has appealed the most. Different types of CNMs with distinct roles like nanostructures membranes, nano-adsorbents, and nanocatalysts have been taken as cost-effective, eco-friendly, and efficient replacements to the current existing wastewater treatment methods. Fabrication, growth, and characterization of carbon nanotubes (CNTs) and its wastewater treatment applications are discussed.

Keywords Nanotechnology · Wastewater treatment · Carbonaceous nanomaterials · Nano-adsorbents · Nanocatalysts · Carbon nanotubes

---

G. Bhalla
Department of Civil Engineering, CT Group of Institutions, Shahpur Campus, Jalandhar 144020, Punjab, India

A. Sharma
CT Group of Institutions, Shahpur Campus, Jalandhar 144020, Punjab, India

V. Kumar (✉)
Department of Applied Sciences, CT Group of Institutions, Shahpur Campus, Jalandhar 144020, Punjab, India
e-mail: vaneet2106@gmail.com

Saruchi
Pushpa Gujral Science City, Kapurthala 144601, Punjab, India

H. Kumar
Department of Biotechnology, CT Group of Institutions, Shahpur Campus, Jalandhar 144020, Punjab, India

B. Bhalla
Department of Chemistry, Dr B R Ambedkar NIT, Jalandhar 144011, Punjab, India

M. Jawaid et al. (eds.), *Environmental Remediation Through Carbon Based Nano Composites*, Green Energy and Technology,
https://doi.org/10.1007/978-981-15-6699-8_18

1 Introduction

Recent advancement in nanoscience and technology has proposed different methods for wastewater decontamination. Different assortments of CNMs with extraordinary characteristics like nanostructures membranes, nano-adsorbents, and nanocatalysts have been believed to be a cost-effective, efficient, and eco-friendly alternative to the present-day wastewater treatment methods. Division of various CNMs used for the treatment of water and wastewater is revealed in Fig. 1.

The outdated techniques of wastewater treatment are not effective to fully eliminate pollutants and follow the water quality standards [1]. Furthermore, different prevalent wastewater treatment methods have abundant disadvantages such as imperfect contaminant exclusion, toxic sludge generation, and high-energy necessity [2]. A diversity range of eco-friendly, efficient, and cost-effective CNMs have been generated with exceptional properties for prospective decontamination of wastewater [3–5]. Different methods for remediation of ground water and wastewater are depicted in Fig. 2.

CNTs comprised of multi-walled carbon nanotubes (MWCNTs) and single-walled carbon nanotubes (SWCNTs) are shown in Fig. 3. They have high mechanical quality, surface region, inertness with chemicals, and water-carrying characteristic.

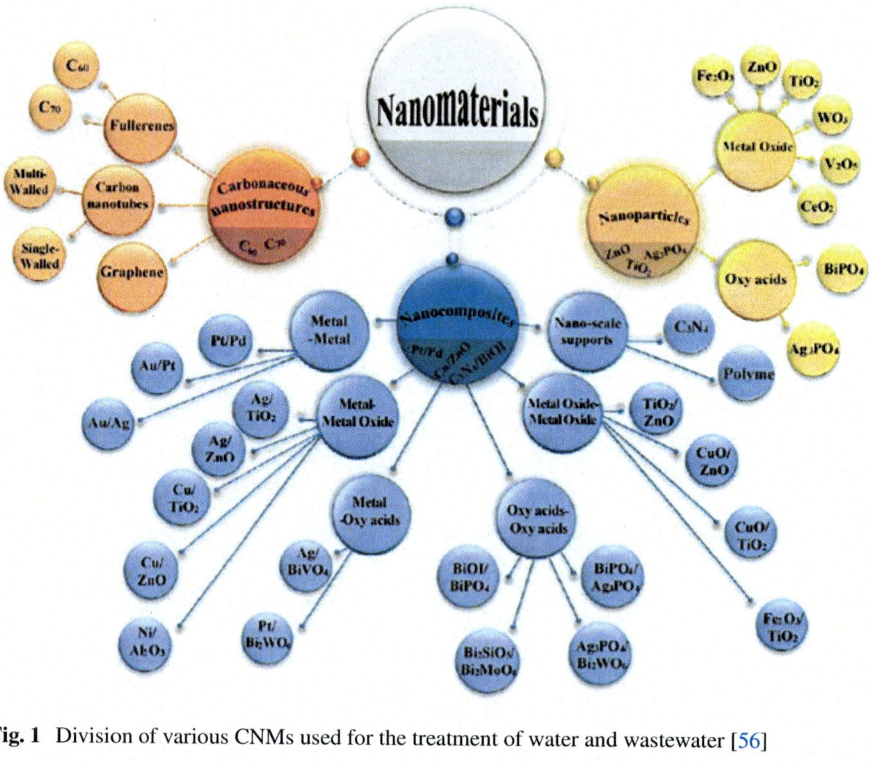

Fig. 1 Division of various CNMs used for the treatment of water and wastewater [56]

Fig. 2 Different methods for remediation of groundwater and wastewater [57]

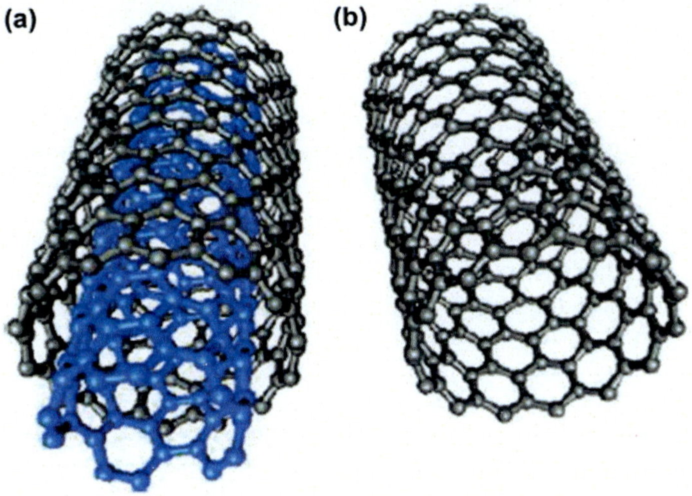

Fig. 3 Arrangement of **a** MWCNT and **b** SWCNT [58]

Hence, broad interest in development of new composite layers for wastewater treatment has been shown by various researchers [6–11]. Furthermore, CNTs exhibit promising catalytic, electrochemical and adsorption properties, therefore increasing water treatment capabilities of CNTs membranes. It can be categorized into three main divisions: nano-adsorbents, nanocatalysts and nanomembranes depending on nature of CNMs. Nano-adsorbent can be created by utilizing atoms of specific elements which have extraordinary adsorption capacity and chemically active (Kyzas and Matis 2015).

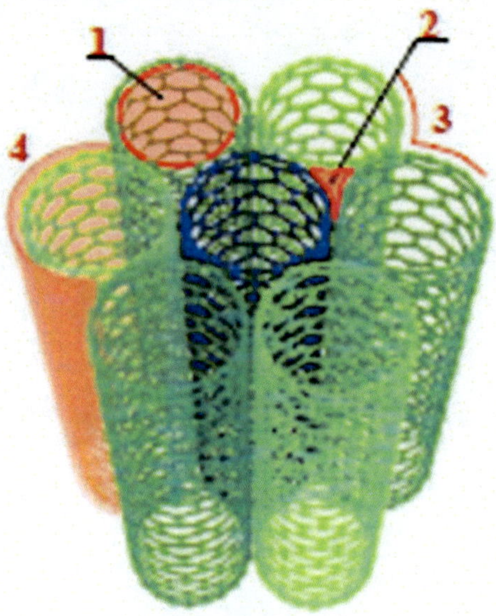

Fig. 4 Diagrammatic outlook of a partly open-ended SWNTs bunch and various adsorption locations (1) Hollow internal space (2) Interstitial network (3) Grooves on the exterior surface of CNTs, and (4) Curved exterior surface of a bunch [59]

The materials used for the formation of nano-adsorbents comprise silica, activated carbon, clay materials, modified composites compounds, and metal oxides [12]. The presence of van der Waals force causes the development of SWCNTs. Potential locations for the adsorption of various pollutants are shown in Fig. 4.

CNMs have added a significant attention as nanocatalysts for the degradation of wastewater pollutants, for example, Fenton catalysts [13], antimicrobial catalysts [14], electrocatalysts [15] for augmenting chemical-based organic contaminants oxidation [16]. Another significant sort of CNMs utilized in treatment of wastewater is by utilizing nanomembranes. Nanofiltration (NF) is widely used among different varieties of membrane-based filtration, [17–19]. NF has high efficiency, low cost, small pore sizes, and user-friendly because of that extensively used in pollutants removal from effluent ([20]; Petrinic et al. 2007; Babursah et al. 2006). Among different wastewater treatment technologies, membrane separation is broadly acknowledged as advancing technique not exclusively to reuse wastewater, yet additionally to desalinate seawater and salty water [21]. It has various advantages like energy advantages, simplicity in operation, less pollution, small operational cost, and high stability and efficiency. Existing progresses in nanotechnology in amalgamation with membrane separation technique have been acknowledged as successful method for effluent treatment [22]. It provides sun precedented degree of mechanization, needed fewer chemical and area and the arrangement permits design flexibility [23]. A key preliminary of the film innovation is the basic change among porousness and layer selectivity. More energy utilization is a major obstacle in the extensive application in the pressure-based membrane processes. Membrane fouling enhances energy intake and the process design and operation complexity. Likewise, it decreases the lifespan of membranes. The adding of nanofillers, for example, CNTs into membranes decreases the accumulation of pollutants and increases hydrophobicity, mechanical properties, and thermal stabilities [24].

A wide range of nanocomposites in the form of polymeric are used for biotherapeutic applications such as engineering of tissue, drug release, and cellular therapy. Because of unique interactions between polymer and nanoparticles, a wide range of property combinations can be engineering to imitate inhabitant tissue structure and properties. Natural and synthetic polymers are utilized to intend nanocomposites in polymeric form for biomedical applications together with cellulose, alginate, starch, collagen, chitosan, gelatin, and fibrin, poly(vinyl alcohol), poly(ethylene glycol), poly(caprolactone), poly(lactic-co-glycolic acid), and poly(glycerol sebacate).

2 CNTs Fabrication, Growth, and Characterization

2.1 CNTs Fabrication Techniques

Various methods available for CNTs fabrication are chemical vapor deposition (CVD), electric-arc discharge, and laser ablation [25]. CNTs structure and morphology can be accustomed by modifying characteristics of the existing methods of CNT fabrication. Among different procedures, CVD is the mostly used on account of its straight forwardness adaptability and affordable undertaking. Furthermore, CVD generated high-purity CNTs with structure controllability [25]. CVD is a warm dehydrogenation system. A hydrocarbon fume permits over a reactor containing metal catalyst such as Fe, Co, or Ni at high temperature (600–1200 °C) to decay the hydrocarbon [26–28]. Atrial setup for CVD-based CNTs generation is depicted in Fig. 5.

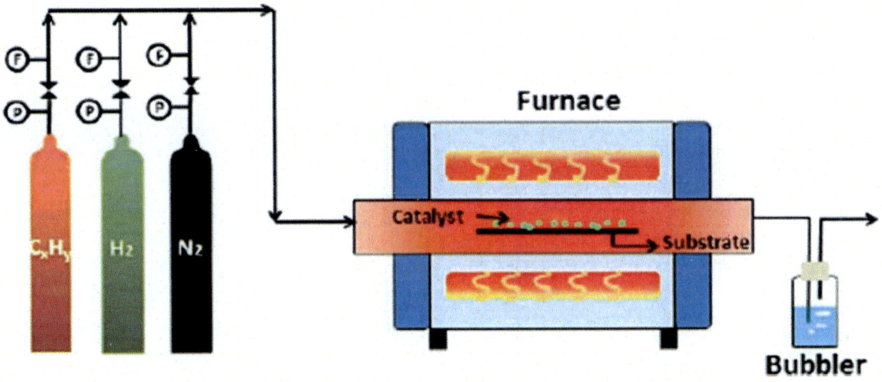

Fig. 5 A trial setup for CVD-based CNTs generation [26–28, 60]

2.2 CNTs Growth Mechanisms

Two diverse growth mechanisms: vapor–liquid–solid (VLS) and vapor–solid–solid (VSS) are shown in Fig. 6. These depend on condition of metal catalyst and dispersion system of carbon at high temperature. At higher temperature, a carbon forerunner is adsorbed and afterward decayed by nanoparticles catalyst to form carbon atoms. Carbon atoms further react to form liquid metal carbide with the help of catalyst. Afterward, at interface stage between catalyst and substrate, it decays into catalyst and carbon particles and the carbon atoms are precipitated out to form CNTs. With time, carbon atoms are released to grow CNTs at catalyst–substrate interface CNTs [29]. This process is recognized as VLS mechanism. Other development mechanism identified as VSS consisted of three successive steps. In the first stage, carbon precursor decays to give carbon particles. In secondary stage, carbon particles multiply on catalyst CNMs surface and shift toward the boundary among catalyst and substrate. In concluding stage, carbon particles precipitate formed by precipitation–nucleation–crystallization process.

Alternatively, on the basis of interaction strength among catalyst and substrates, two types of CNTs growth models, tip growth and base growth models, are involved as depicted in Fig. 7. When the interface strength is less among catalyst and substrate, carbon forerunner disintegrates into carbon atoms on the peak surface of the catalyst; afterward, carbon atoms disseminate down by the catalyst, and CNTs rise up from the base of catalyst, therefore forcing down the complete catalyst at a distance from the substrate. CNTs produce constantly as long as catalyst surface is available for more carbon precursor decomposition. Development of CNTs stops due to the deactivation of catalyst if catalyst is wholly covered by carbon atoms. This is called as tip growth.

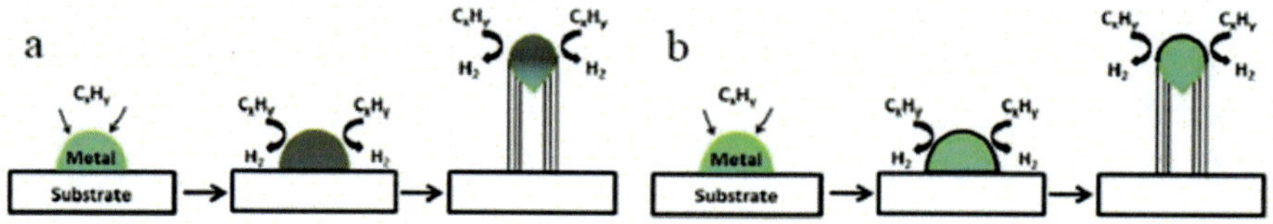

Fig. 6 CNTs growth mechanisms depended upon metal catalyst and diffusion **a** VLS mechanism and **b** VSS mechanism [30, 61–64]

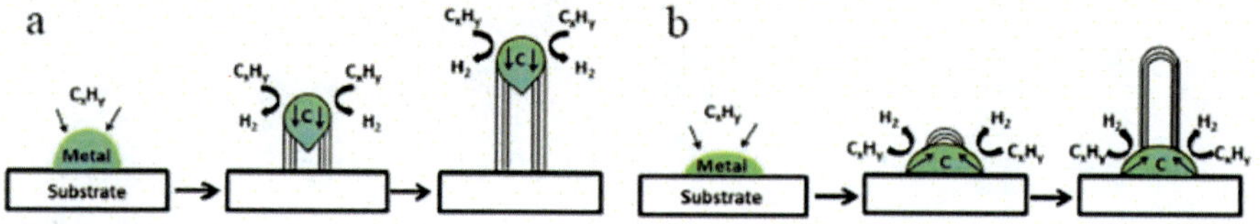

Fig. 7 CNTs growth mechanisms dependent upon interaction strength between catalyst and substrates **a** tip growth model **b** base growth model [30, 61, 63]

In another situation, when the cooperation between catalyst and substrate is large, the disintegration of carbon forerunner and the carbon atoms dispersions are alike to tip growth. A crescent dome is formed in the begining, which develops CNTs at the end. Subsequently, a catalyst which consistently fixes on the base to support CNTs development is called as base growth.

2.3 CNTs Characterization

So as to assess its characteristics and morphologies of CNTs, a great deal of endeavors utilizing different complex strategies has been created [30]. A sample of CNT membrane is revealed in Fig. 8. In short, transmission electron microscopy (TEM) and

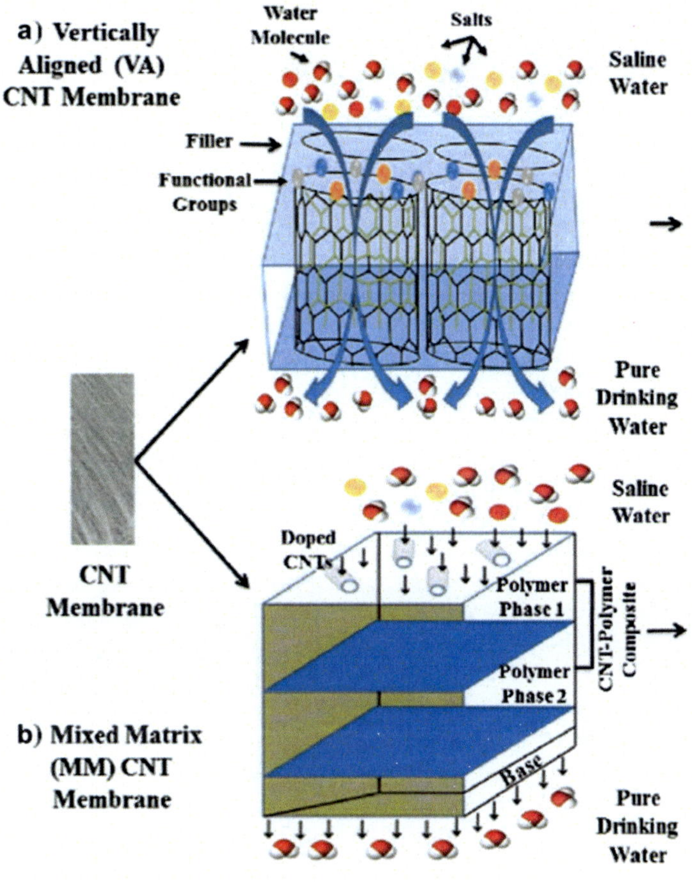

Fig. 8 Trapping of salts and movement of water molecules from salinated water through SWCNT **a** vertically aligned CNT membrane **b** mixed matrix CNT membrane [6, 7]

scanning electron microscopy (SEM) are the recognized techniques that are generally used to decide the location of tip and side wall, in addition to CNTs morphology [31–33].

Raman spectroscopy (RS) is a commanding depiction method for CNTs [34–36]. It is generally used to assess the purity and quality of CNTs. Thermogravimetric analysis (TG), X-ray photoelectron spectroscopy (XPS), and infrared spectroscopy are mainly utilized to specifically confirm the happening of CNTs fictionalizations reactions for concluding quality assessment [37, 38]. Consequently, by changing reactants and CVD readiness parameters, for example, catalyst, carbon forerunner, temperature, substrate, time, pressure, and gas flow rate helped with different characterization and fictionalizations method, enhanced CNTs for diverse uses. Figure 9 depicts structures of various CNT membranes.

3 Water Treatment Applications

All the four categories of CNMs like fullerenes, graphene-based CNMs, CNTs, and nanoporous activated carbon (NAC) have abundant prospective in numerous phases of wastewater treatment like membrane process, photocatalysis, disinfection, and adsorption. The prospective applications of CNMs in wastewater treatment are shown in Fig. 10.

Various researches on membrane nanotechnology have concentrated on generating multifunction by introducing CNMs into membranes. CNMs utilized for such uses comprise photocatalytic nanomaterials (e.g., TiO_2, bi-metallic nanoparticles), antimicrobial nanoparticles (e.g., CNTs and nano-Ag), and metal oxide nanoparticles which are hydrophilic in nature (e.g., zeolite, TiO_2, and Al_2O_3). The key objective of adding hydrophilic metal oxide CNMs is to decrease fouling by enhancing membrane hydrophilicity. The input of metal oxide nanoparticles like TiO_2 [39], zeolite [40], silica [41], alumina [42], and polymeric ultrafiltration membranes increases membrane surface hydrophobicity. They additionally help to improve the thermal and mechanical dependability of films along these lines diminishing the destructive impact of heat and compaction on penetrability of membrane [40, 43]. CNTs composite membranes combine the exceptional attributes of conventional layer materials with those of CNTs for water treatment. Evolving water treatment applications like oil–water detachment, expulsion of substantial metal particles, water desalination, and emerging pollutants in wastewater utilizing CNTs composite membranes have been progressively studied by various researchers [30]. CNTs composite membranes applications as emerging technology for water and wastewater treatment are discussed as under.

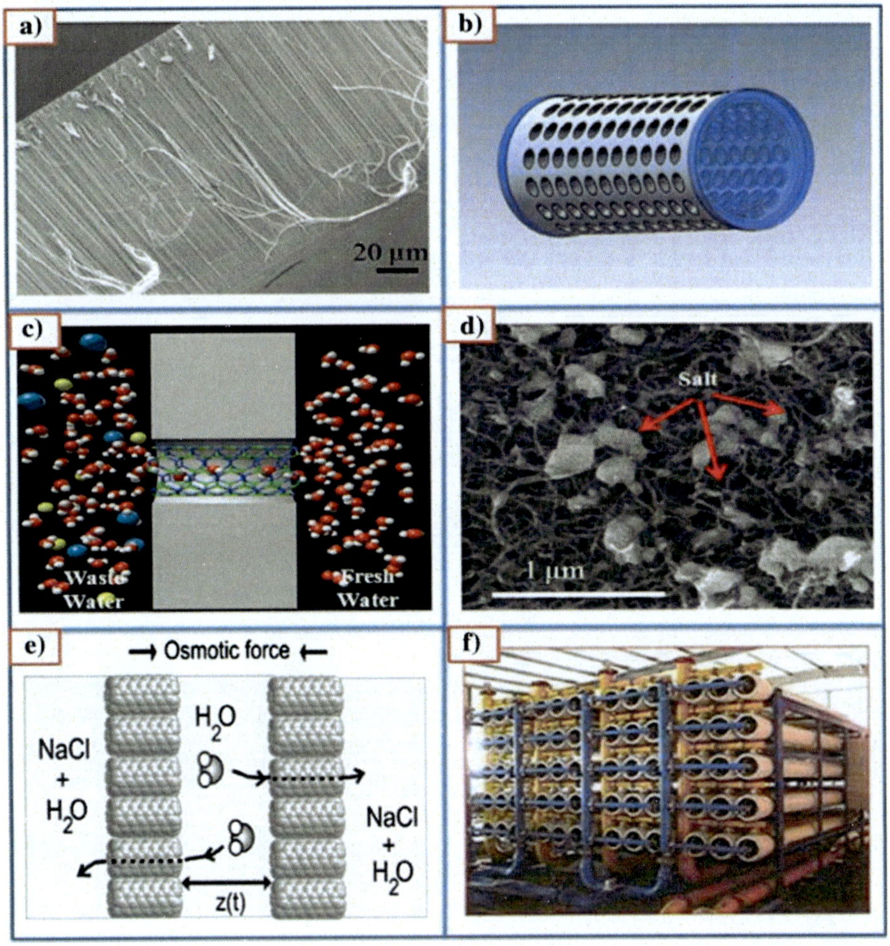

Fig. 9 CNT membranes structures **a** pristine CNT membrane cross-sectional scanning electron microscope (SEM) image **b** cylindrical CNT-based water filter **c** movement of water molecules movement through CNT channel **d** SEM image of CNT membrane surface having scattered NaCl nanocrystals **e** pure water molecules movement through CNT membrane in osmotically imbalanced compartments, and **f** engineered CNT membranes used in industries [6, 7]

3.1 Water Desalination

CNTs utilization is significantly rising because of their characteristics to increase the effectiveness of presently accessible membrane techniques, for example, membrane distillation (MD), nanofiltration (NF), and reverse osmosis (RO), ([30]; Bhadra et al. 2016; Baeket al. 2016; Son et al. 2015; [44]; Roy et al. 2014; Madaeni et al. 2013). Gethard et al. 2011 explained that CNTs can function as a sorbent to give an added

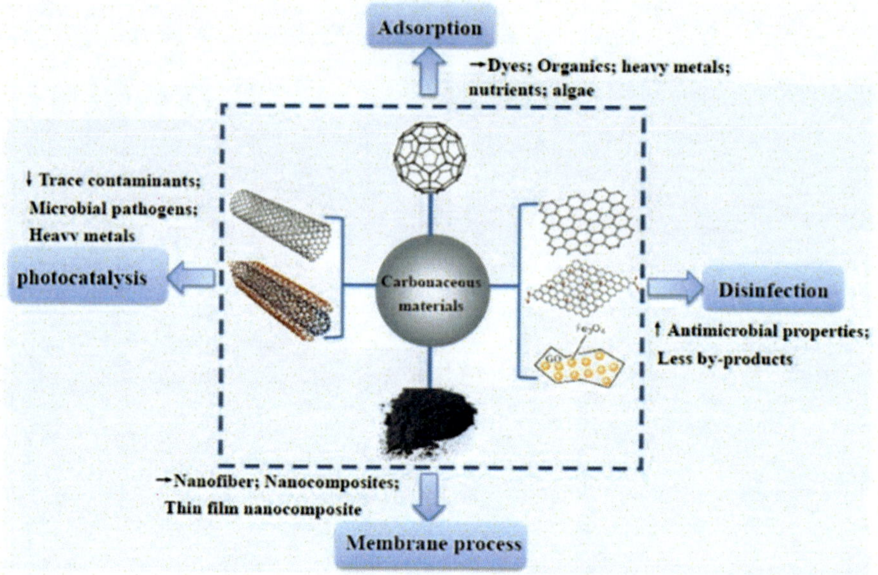

Fig. 10 Significant applications of CNMs in wastewater treatment (Shan et al. 2013)

passage way for the transportation of solute in CNTs-PVDF (polyvinylidene fluoride) membranes. CNTs-PVDF can enhance MD method at a comparatively lesser temperature and greater permeate vapor flux for extensive variety of salts as compared to conventionally used membranes.

3.2 *Oil–Water Separation*

Membrane filtration includes nanofiltration (NF), ultrafiltration (UF), and microfiltration (MF) techniques. It is mostly applied for oil–water treatment with specific benefits [45, 46]. Gu et al. [47] developed Dans Ingenious methodology for emulsified oil–water separation by developing superhydrophobic CNTs-polystyrene composite membranes, as shown in Fig. 11. The consequent membranes efficiently isolated an extensive variety of water–oil emulsions with an extraordinary refusal efficiency (>99.9%).

3.3 *Removal of Heavy Metal Ions from Wastewater*

Effluent discharge from various industries like battery, chemical, and metal plating contains huge number of toxic metal ion discharge [48]. CNTs have been used as

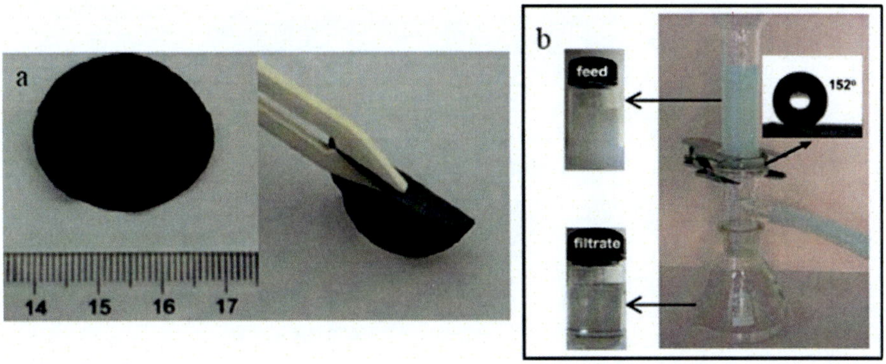

Fig. 11 Superhydrophobic CNTs-polystyrene composite membranes for oil–water separation **a** membranes and **b** oil–water separation [47]

favorable metal ion adsorbents [26–28, 49]. The main challenge is to alter CNT adsorption capability into an economical and appreciated filtration membrane for particular applications. Parham et al. explained CNT-based composite filter with 100% removal efficiency of heavy metal ions [50].

3.4 Emerging Pollutants Removal from Wastewater

Emerging pollutants like persistent organic pollutants (POPs), pharmaceutical and personal care products (PPCPs) and environmental endocrine disruptors (EEDs) persist in low concentration in environment, yet will in general reason causes extreme harm to environment and human well-being [51, 52]. A CNTs-PVDF composite layer utilized in the filtration of ibuprofen (IBU), triclosan (TCS), and acetaminophen (AAP) makes able illustration of elimination of emerging pollutants [44].

3.5 Membrane Separation Combined with Assistant Techniques

Fan et al. [53] studied removal of phenol under electrochemical assistance by CNTs-Al_2O_3 composite membranes. The results showed that the CNTs-Al_2O_3 composite membrane separation performance can be considerably enhanced under electrochemical assistance, such as electrostatic adsorption, degradation, and repulsion as shown in Fig. 12. CNTs composite membranes gave good results for water treatment by integration of ozonation catalysis and photocatalysis with membrane filtration [54, 55].

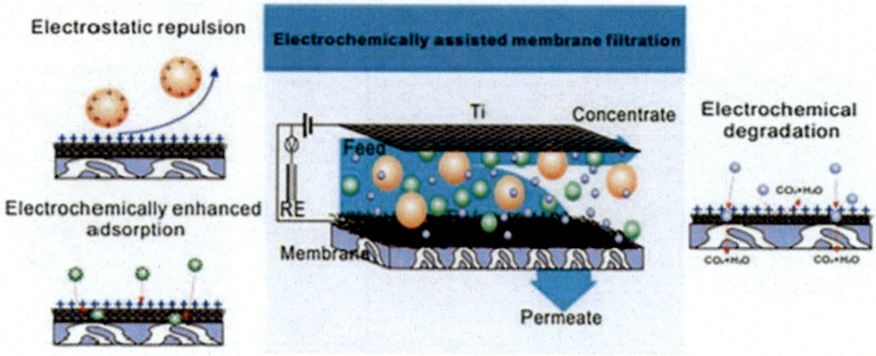

Fig. 12 CNTs-Al_2O_3 composite membrane separation mechanism under electrochemical assistance [53]

4 Conclusion and Future Perspective

Fabrication, characterization, and fictionalization of CNTs composite membranes water treatment uses are discussed. The desired characteristics of CNTs can be obtained by fictionalization and characterization. CNMs possess greater surface area, outstanding chemical, thermal, electrical, and optical activities. It has been taken as an ideal candidate to eliminate wastewater organic and chemical pollutants from wastewater. Previous studies have proved the prospective applications of CNMs for wastewater treatment at laboratory scale. Though the large-scale utilization of these CNMs still faces extensive challenges, further detailed studies are required.

- Practically all the investigations were done under laboratory conditions. Advance studies should be done under more convincing situations to estimate the applicability and efficiency of various CNMs in wastewater treatment.
- The cytotoxicity of CNMs on environment and public health has not been studied intensively. The risk assessment studies of CNMs are urgently required.
- CNTs and graphene-based CNMs accumulation in liquid phase is a major shortcoming for water decontamination. Accumulated CNMs would decrease surface area and active sites consequently affect the accessibility, resulting in declined proficiency for the removal of pollutants. Furthermore, researches should stress more on the targeted alterations and upturn the proficiency, selectivity and liking for particular contaminant in wastewater.
- Furthermore, membrane separation together with various assisted methods like adsorption, catalysis, and electrochemistry are accessible, and a profound comprehension of the blended components is inadequate. Likewise, extra logical and specialized endeavors are required.

- The existing production techniques for CNMs are complex and low-efficient. Simple, robust, and efficient fabrication methods are required.
- The potential harmful impacts of CNTs exposure to the surrounding environment are needed to be studied in detail.

References

1. Qu X, Brame J, Li Q, Alvarez PJ (2012) Nanotechnology for a safe and sustainable water supply: enabling integrated water treatment and reuse. Acc Chem Res 46(3):834–843
2. Ferroudj N, Nzimoto J, Davidson A, Talbot D, Briot E, Dupuis V, Abramson S (2013) Maghemite nanoparticles and maghemite/silica nanocomposite microspheres as magnetic Fenton catalysts for the removal of water pollutants. App Catal B Env 136:9–18
3. Brumfiel G (2003) Nanotechnology: a little knowledge. Nature 424:246–248
4. Gupta VK, Tyagi I, Sadegh H, Shahryari-Ghoshekand R, Makhlouf ASH, Maazinejad B (2015) Nanoparticles as adsorbent; a positive approach for removal of noxious metal ions: a review. Sci Technol Dev 34:195
5. Theron J, Walker JA, Cloete TE (2008) Nanotechnology and water treatment: applications and emerging opportunities. Crit Rev Microbiol 34:43–69
6. Das R, Ali ME, Hamid SBA, Ramakrishna S, Chowdhury ZZ (2014) Carbon nanotube membranes for water purification: a bright future in water desalination. Desalination 336:97–109
7. Das R, Ali ME, Hamid SBA, Ramakrishna S, Chowdhury ZZ (2014) Carbon nanotube membranes for water purification: a bright future in water desalination. Desalination 336(3):97–109
8. Goh K, Karahan HE, Wei L, Bae TH, Fane AG, Wang R, Chen Y (2016) Carbon nanomaterials for advancing separation membranes: a strategic perspective. Carbon 109:694–710
9. Goh PS, Ismail AF, Ng BC (2013) Carbon nanotubes for desalination: performance evaluation and current hurdles. Desalination 308:2–14
10. Kim J, Bruggen BV (2010) The use of nanoparticles in polymeric and ceramic membrane structures: review of manufacturing procedures and performance improvement for water treatment. Environ Pollut 158:2335–2349
11. Volder MFLD, Tawfick SH, Baughman RH, Hart AJ (2013) Carbon nanotubes: present and future commercial applications. Science 339:535–539
12. El Saliby IJ, Shon H, Kandasamy J, Vigneswaran S (2008) Nanotechnology for wastewater treatment: in brief. Encycl Life Support Syst (EOLSS)
13. Kurian M, Nair DS (2015) Heterogeneous Fenton behavior of nano nickel zinc ferrite catalysts in the degradation of 4-chlorophenol from water under neutral conditions. J Water Process Eng 8:37–49
14. Chaturvedi S, Dave PN, Shah NK (2012) Applications of nanocatalyst in new era. J Saudi Chem Soc 16:307–325
15. Dutta AK, Maji SK, Adhikary B (2014) C-Fe_2O_3 nanoparticles: an easily recoverable effective photo-catalyst for the degradation of rose benga and methylene blue dyes in the waste-water treatment plant. Mater Res Bull 49:28–34
16. Ma H, Wang H, Na C (2015) Microwave-assisted optimization of platinum-nickel nanoalloys for catalytic water treatment. Appl Catal B Environ 163:198–204
17. Blanco J, Torrades F, De la Varga M, Garcı´a-Montan˜o J (2012) Fenton and biological-Fenton coupled processes for textile wastewater treatment and reuse. Desalination 286:394–399
18. Lau WJ, Ismail A (2009) Polymeric nanofiltration membranes for textile dye wastewater treatment: preparation, performance evaluation, transport modelling and fouling control—a review. Desalination 245:321–348

19. Ouyang X, Li W, Xie S, Zhai T, Yu M, Gan J, Lu X (2013) Hierarchical CeO_2 nanospheres as highly-efficient adsorbents for dye removal. New J Chem 37(3):585–588
20. Rashidi HR, Sulaiman NMN, Hashim NA, Hassan CRC, Ramli MR (2015) Synthetic reactive dye wastewater treatment by using nanomembrane filtration. Desalin Water Treat 55(1):86–95
21. Wang K, Abdalla AA, Khaleel MA, Hilal N, Khraisheh MK (2017) Mechanical properties of water desalination and wastewater treatment membranes. Desalination 401:190–205
22. Pendergast MTM, Hoek EMV (2011) A review of water treatment membrane nanotechnologies. Energy Environ Sci 4:1946–1971
23. Qu X, Alvarez PJJ, Li Q (2013) Applications of nanotechnology in water and wastewater treatment. Water Res 47(12):3931–3946
24. Ursino C, Castro-Munor R, Drioli E, Gzara L, Albeirutty MH, Figoli A (2018) Progress of nanocomposit membranes for water treatment. Membranes 8(2):18
25. Kumar M, Ando Y (2010) Chemical vapor deposition of carbon nanotubes: a review on growth mechanism and mass production. J Nanosci Nanotechnol 10:3739–3758
26. Mubarak NM, Abdullah EC, Jayakumar NS, Sahu JN (2014) An overview on methods for the production ofcarbon nanotubes. J Ind Eng Chem 20:1186–1197
27. Mubarak NM, Sahu JN, Abdullah EC, Jayakumar NS (2014) Removal of heavy metals from wastewater using carbon nanotubes. Sep Purif Rev 43:311–338
28. Mubarak NM, Abdullah EC, Jayakumar NS, Sahu JN (2014) An overview on methods for the production of carbon nanotubes. J Ind Eng Chem 20:1186–1197
29. Xiong J, Dong X, Song Y, Dong Y (2013) A high performance Ru-ZrO_2/carbon nanotubes-Ni foam composite catalyst for selective CO methanation. J Power Sources 242:132–136
30. Ma L, Dong X, Chen M, Zhu L, Wang C, Yang F, Dong Y (2017) Fabrication and water treatment application of carbon nanotubes (CNTs)-based composite membranes: a review. Membranes 7(16):1–21
31. Belin T, Epron F (2015) Characterization methods of carbon nanotubes: a review. Mater Sci Eng 119:105–118
32. Herrero-Latorre C, Alvarez-Méndez J, Barciela-Garcia J, Garcia-Martin S, Pena-Crecente RM (2015) Characterization of carbon nanotubes and analytical methods for their determination in environmental and biological samples: a review. Anal Chim Acta 853:77–94
33. Ping D, Wang C, Dong X, Dong Y (2016) Co-production of hydrogen and carbon nanotubes on nickel foam via methane catalytic decomposition. Appl Surf Sci 369:299–307
34. Dresselhaus MS, Dresselhaus G, Jorio A (2007) Raman spectroscopy of carbon nanotubes in 1997 and 2007. J Phys Chem 111:17887–17893
35. Dresselhaus MS, Jorio A, Saito R (2010) Characterizing grapheme graphite and carbon nanotubes by Raman spectroscopy. Annu Rev Cond Matter Phys 1:89–108
36. Saito R, Hofmann M, Dresselhaus G, Jorio A, Dresselhaus MS (2011) Raman spectroscopy of graphene and carbon nanotubes. Adv Phys 60:413–550
37. Lehman JH, Terrones M, Mansfield E, Hurst KE, Meunier V (2011) Evaluating the characteristics of multiwall carbon nanotubes. Carbon 49:2581–2602
38. Shulga YM, Tien TC, Huang CC, Lo SC, Muradyan VE, Polyakova NV, Ling YC, Loutfy RO, Moravsky AP (2007) XPS study of fluorinated carbon multi-walled nanotubes. J Electron Spectrosc Rel Phen 160:22–28
39. Bae TH, Tak TM (2005) Effect of TiO_2 nanoparticles on fouling mitigation of ultrafiltration membranes for activated sludge filtration. J Membr Sci 249(1–2):1–8
40. Pendergast MTM, Nygaard JM, Ghosh AK, Hoek EMV (2010) Using nanocomposite materials technology to understand and control reverse osmosis membrane compaction. Desalination 261(3):255–263
41. Bottino A, Capannelli G, D'Asti V, Piaggio P (2001) Preparation and properties of novel organic-inorganic porous membranes. Sep Purif Technol 22–23(1–3):269–275
42. Maximous N, Nakhla G, Wong K, Wan W (2010) Optimization of Al_2O_3/PES membranes for wastewater filtration. Sep Purif Technol 73(2):294–301
43. Ebert K, Fritsch D, Koll J, Tjahjawiguna C (2004) Influence of inorganic fillers on the compaction behaviour of porous polymer based membranes. J Membr Sci 233(1–2):71–78

44. Wang Y, Zhu J, Huang H, Cho HH (2015) Carbon nanotube composite membranes for microfiltration of pharmaceuticals and personal care products: capabilities and potential mechanisms. J Membr Sci 479:165–174
45. Chen M, Zhu L, Dong Y, Li L, Liu J (2016) Waste-to-resource strategy to fabricate highly porous whisker-structured mullite ceramic membrane for simulate oil-in-water emulsion wastewater treatment. ACS Sustain Chem Eng 4:2098–2106
46. Zhu L, Chen M, Dong Y, Tang CY, Huang A, Li L (2016) A low-cost mullite-titania composite ceramic hollow fiber microfiltration membrane for highly efficient separation of oil-in-water emulsion. Water Res 90:277–285
47. Gu JC, Xiao P, Chen J, Liu F, Huang Y, Li G, Zhang J, Chen T (2014) Robust preparation of superhydrophobic polymer/carbon nanotube hybrid membranes for highly effective removal of oils and separation of water-in-oil emulsions. J Mater Chem A 2:15268–15272
48. Fu FL, Wang Q (2011) Removal of heavy metal ions from wastewaters: a review. J Environ Manag 92:407–418
49. Rao GP, Lu C, Su F (2017) Sorption of divalent metal ions from aqueous solution by carbon nanotubes: a review. Sep Purif Technol 58:224–231
50. Parham H, Bates S, Xia Y, Zhu Y (2013) A highly efficient and versatile carbon nanotube/ceramic composite filter. Carbon 54:215–223
51. Liu Y, Liu H, Zhou Z, Wang T, Ong CN, Vecitis CD (2015) Degradation of the common aqueous antibiotic tetracycline using a carbon nanotube electrochemical filter. Environ Sci Technol 49:7974–7980
52. Musteret CP, Teodosiu C (2007) Removal of persistent organic pollutants from textile wastewater by membrane processes. Environ Eng Manage J 6:175–187
53. Fan XF, Zhao HM, Liu YM, Quan X, Yu H, Chen S (2015) Enhanced permeability, selectivity, and antifouling ability of CNTs/Al_2O_3 membrane under electrochemical assistance. Environ Sci Technol 49:2293–2300
54. Oulton R, Haase JP, Kaalberg S, Redmond CT, Nalbandian MJ, Cwiertny DM (2015) Hydroxyl radical formation during ozonation of multiwalled carbon nanotubes: performance optimization and demonstration of a reactive CNT filter. Environ Sci Technol 49:3687–3697
55. Zhao H, Li H, Yu H, Chang H, Quan X, Chen S (2013) CNTs–TiO_2/Al_2O_3 composite membrane with a photocatalytic function: fabrication and energetic performance in water treatment. Sep Purif Technol 116:360–365
56. Zhao QQ, Boxman A, Chowdhry UJ (2003) Nanopart Res 5(5–6):567–572
57. Zheng K, Setyawati MI, Leong DT, Xie J (2018) Antimicrobial silver nanomaterials. Coordin Chem Rev 357:1–7
58. Zhao YL, Stoddart JF (2009) Noncovalent functionalization of single-walled carbon nanotubes. Chem Res 42(8):1161–1171
59. Nowack B, Bucheli TD (2007) Environ Pollut 150(1):5–22
60. Zhao N, He C, Jiang Z, Li J, Li Y (2006) Fabrication and growth mechanism of carbon nanotubes by catalytic chemical vapor deposition. Mater Lett 60:159–163
61. Kim SM, Jeong S, Kim HC (2013) Investigation of carbon nanotube growth termination mechanism by insitu transmission electron microscopy approaches. Carbon Lett 14:228–233
62. Kukovitsky EF, L'Vov SG, Sainov NA (2000) VLS-growth of carbon nanotubes from the vapor. Chem Phys Lett 317:65–70
63. Lo AY, Liu SB, Kuo CT (2010) Effect of temperature gradient direction in the catalyst nanoparticle on CNTs growth mode. Nanoscale Res Lett 5:1393–1402
64. Tessonnier JP, Su DS (2011) Recent progress on the growth mechanism of carbon nanotubes: a review. Chemsuschem 4:824–847
65. Kim HJ, Baek Y, Choi K, Kim DG, Kang H, Choi YS, Yoon J, Lee JC (2014) The improvement of anti-biofouling properties of a reverse osmosis membrane by oxidized CNTs. RSC Adv 4:32802–32810
66. Kim HJ, Choi K, Baek Y, Kim DG, Shim J, Yoon J, Lee JC (2014) High-performance reverse osmosis CNT/polyamide nanocomposite membrane by controlled interfacial interactions. ACS Appl Mater Interfaces 6:2819–2829

67. Kim S, Fornasiero F, Park HG, In JB, Meshot E, Giraldo G, Stadermann M, Fireman M, Shan J, Grigoropoulos CP (2014) Fabrication of flexible aligned carbon nanotube composite membranes by insitu polymerization. J Membr Sci 460:91–98
68. Shan SJ, Zhao Y, Tang H, Cui FY (2017) A mini-review of carbonaceous nanomaterials for removal of contaminants from waste water. In: IOP conferences series: earth and environmental science, vol 68, pp 1–7

Recent Advances in Preparation and Characterization of Graphene-Based Nanocomposite Membranes for Water Purification

Arun Kumar Shukla, Mohammad Azam Ansari, Javed Alam, Ali Aldalbahi, and Mansour Alhoshan

Abstract Increasing water scarcity and diminishing freshwater resources have generated the need for new water filtration technologies to supply safe potable water for domestic and industrial uses. Filtration with polymeric membranes shows good capability for water treatment; however, it has some limitations, such as selectivity and permeability, and fouling, thereby, limits the continuity of safe water. Long-term filtration stability can be improved using a new generation of nanocomposite membranes that are formed by incorporating nanomaterials into polymeric membrane matrices. A number of different nanomaterials and methods have been adopted at an accelerated pace to develop a thin-film nanocomposite (TFN) membrane. Among these nanomaterials, graphene and its functionalized derivatives have received much research attention, as this approach for developing a new TFN membrane has shown excellent performance, leading to water treatment. Graphene has superior properties that are highly advantageous for solving major problems, such as biofouling, scaling, low flux rate, selectivity, and degradation. This chapter provides an overview of the development of graphene-based nanocomposite membrane methods and physicochemical properties after the incorporation of graphene nanomaterials and their derivatives. A specific focus has been employed on improving understanding of how graphene nanomaterials can be used in a number of different ways, such as ultrafiltration, nanofiltration, and reverse osmosis for water filtration.

A. K. Shukla · J. Alam · A. Aldalbahi (✉) · M. Alhoshan
King Abdullah Institute for Nanotechnology, King Saud University, P.O. Box 2455, Riyadh 11451, Saudi Arabia
e-mail: aaldalbahi@ksu.edu.sa

M. A. Ansari (✉)
Department of Epidemic Disease Research, Institute For Research and Medical Consultation, Imam Abdulrahman Bin Faisal University, 1982, Dammam 31441, Saudi Arabia
e-mail: maansari@iau.edu.sa

M. Alhoshan
Department of Chemical Engineering, College of Engineering, King Saud University, P.O. Box 2455, Riyadh 11451, Saudi Arabia

M. Jawaid et al. (eds.), *Environmental Remediation Through Carbon Based Nano Composites*, Green Energy and Technology,
https://doi.org/10.1007/978-981-15-6699-8_19

Keywords Graphene · Nanocomposite membrane · Thin-film nanocomposite membrane · Water purification · Physicochemical properties

1 Introduction

The current intensification of the synthesis and processing of carbon-based nanomaterials and their derivatives suggest an exciting prospect for developing a new class of advanced nanocomposite membranes. Among carbon-based nanomaterials, graphene is regarded as one of the most auspicious materials due to its tightly packed, two-dimensional honeycomb lattice with excellent optical properties, well-organized carbon network building blocks, high surface area, superior thermal and mechanical properties, excellent intrinsic carrier mobility and barrier properties, and reliable processing strategies [4, 15, 17, 40]. Therefore, graphene-associated nanomaterials, such as single-layer graphene, multilayer graphene, chemically converted graphene, and graphene oxide (GO), are much more frequently considered because of their availability [31]. The oxidized functional groups of GO have increased the dispersion and minimized the aggregation of graphene into matrix phases. Hence, the presence of such functional groups (e.g., epoxide, hydroxyl, carbonyl, and carboxylic) enhances the electrical properties of GO attributable to the extensive presence of mixed sp^2/sp^3 hybridized carbon domains on the basal planes and edges [16, 22]. As an excellent candidate, nanofiller offers a perfect implementation with different polymer matrices. The presence of integrated functional groups in GO allows for exfoliation and dispersion, with strong interfacial bonding between the polymer matrix and nanofiller, which is embedded to form a nanocomposite polymeric membrane. The graphene-based nanocomposite membrane has gained huge improvement in their selective permeation and/or antifouling properties and maximized the permeate flux of the pristine membranes [26, 28]. Therefore, graphene-based nanocomposite membranes can be used to change the microstructures and transport pathways, enabling their use in various extraordinary membrane separation applications, for instance ultrafiltration, nanofiltration, and reverse osmosis [2, 3, 26]. Moreover, as mentioned above, graphene-based nanocomposite membrane fabrication was employed in different preparation methods, dependent on the selection of polymer and the membrane structure of interest, whose advantages and the pathways for practical applications are described in the next section. This chapter provides a comprehensive and systematic summary of the current state of the art of graphene-based nanocomposite membrane and assesses its applicability to water purification. First, the graphene properties and the potential of graphene-based membrane for water treatment are introduced. Then, its fabrication is systematically summarized. An in-depth description of the physical and chemical properties of membranes, including the surface properties, permeability, removal efficacy, and fouling propensity, is provided. Furthermore, their applicability and performance are assessed in altered applications, for instance ultrafiltration, nanofiltration, and reverse osmosis [7].

2 Preparation of Nanocomposite Membrane

The methods engaged to develop nanocomposite membranes are dependent on the selection of polymer, and the properties and structure of the membrane of interest. The most popular methods of preparing polymeric nanocomposite membranes are phase inversion and interfacial polymerization.

2.1 Phase Inversion Method

The graphene-based nanocomposite membranes were fabricated by phase inversion induced by the immersion precipitation method, using casting solutions containing polymer and a proper amount of GO nanoparticles in solvent [5]. This method is a demixing approach that takes a polymer solution that is initially a homogenous liquid and uses control techniques to transform it into a solid that becomes thermodynamically unstable [13, 26]. The schematic shown in Fig. 1 describes the phase inversion method for nanocomposite membrane formation.

In this method, the specific percentage of GO was added into the solvent and dispersed by sonication to improve homogeneity and reduce aggregation. After dispersing the nanoparticles in the solvent, polymer was dissolved in the dope solution by continuous stirring to obtain a homogeneous solution. After removing air bubbles, the dopant (casting solution) was cast onto an appropriate clean glass plate surface, using a doctor blade to a thickness of about 100 ± 3 μm. Subsequently, the glass plate was horizontally immersed into a coagulation bath that is occupied with water

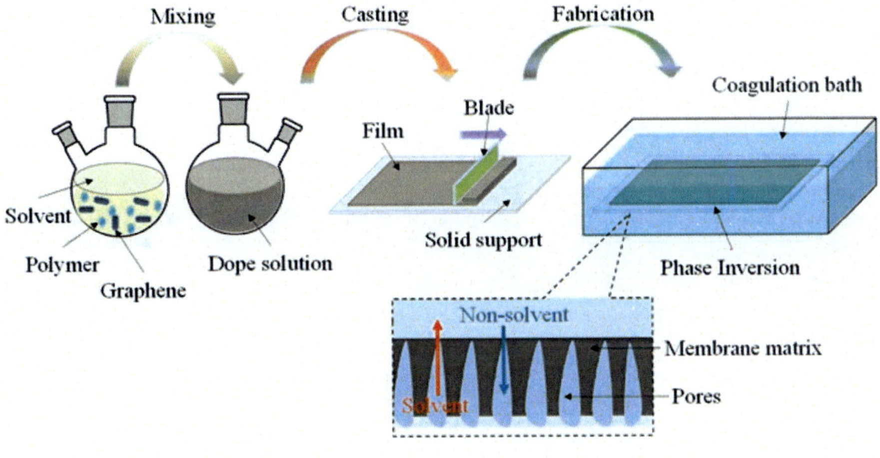

Fig. 1 Representation of graphene-based nanocomposite membrane fabrication by phase inversion process

(non-solvent) for the immersion precipitation process. For the duration of immersion precipitation, any one of the demixing processes can take place. The demixing process can be described via the exchange rate between the non-solvent and solvent during precipitation [25]. There are two types of demixing that can be applied: one type, direct demixing, arises instantly after the immersion of the casted solution in a non-solvent and leads to a faster solvent–non-solvent exchange rate to form a porous top layer with a finger-like sublayer membrane. The other, late demixing, is a slow process in which the demixing takes time, and the solvent–non-solvent exchange takes place at a slow rate afterward the immersion. Consequently, a dense top layer with fewer macrovoids and sponge-like structure forms [25]. Furthermore, the incorporation of GO nanomaterial enhances the phase separation and shows deterrent effects. From the thermodynamic perspective, the nanomaterial-based casting solution reductions the solvent power of the solution and works as a non-solvent agent. Hence, less non-solvent is needed for the phase separation of the polymeric casting solution. This phenomenon leads to direct demixing and the effects in a membrane with an asymmetric finger-like structure. The membrane formation, exhibiting structures such as those that are sponge-like and finger-like are shown in Fig. 2. After the primary phase separation and membrane solidification, the resultant nanocomposite membrane may then go through treatment, for example washing, strengthen, or drying.

2.2 *Interfacial Polymerization*

Interfacial polymerization (IP) is the best-established route for the preparation of a new type of thin-film nanocomposite (TFN) membrane, and now, this process is the most significant method for commercial fabrication in industry. For the interfacially polymerized TFN membrane, nanomaterials are contained within the polyamide dense layer or inside the support layer of the membrane to increase the membrane separation properties. IP is a type of condensation polymerization in which polymerization follows at the interface among an aqueous solution having one monomer m-phenylenediamine (MPD) and an organic solution having a second monomer trimesoyl chloride (TMC) [7]. In this method, the preparation of TFN membranes has been accomplished by few stages (see Fig. 3).

First, a diamine monomer aqueous solution was prepared via dissolving MPD and trimethylamine in the presence or absence of a surfactant in deionized water. Subsequently, MPD/GO aqueous solution was prepared with the incorporation of a specific percentage of GO in prepared MPD and dispersed by sonication to reduce aggregation; flat microporous polymeric support was soaked with the diamine monomer aqueous solution. Second, the amine-impregnated membrane was immersed in a TMC solution in hexane. Finally, after the removal of the TMC solution, the membrane was left under the ambient conditions for 2 min, which brings about

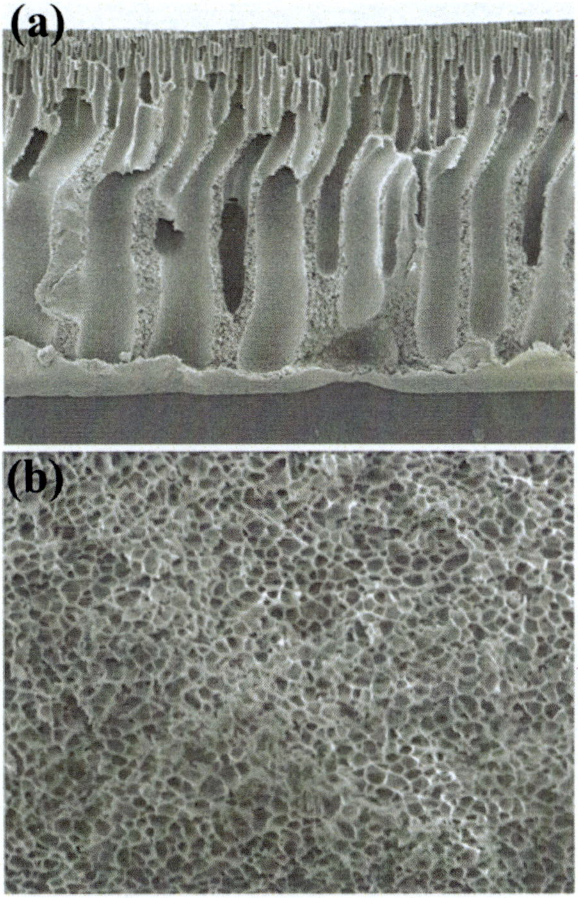

Fig. 2 Nanocomposite membrane formation **a** finger-like and **b** sponge-like structure

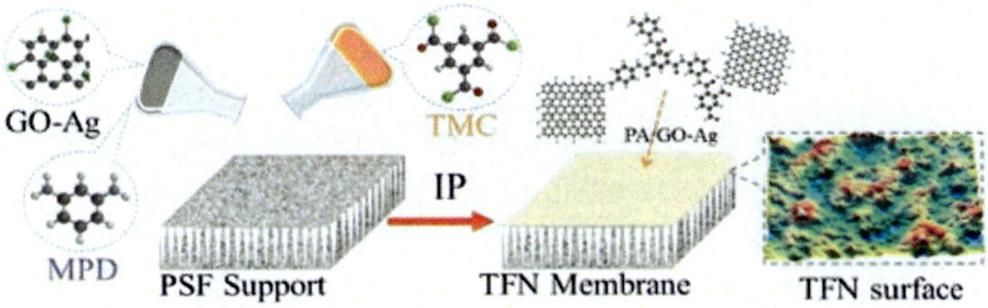

Fig. 3 Schematic illustration of interactions of GO-Ag with TMC and MPD via interfacial polymerization to development of thin-film nanocomposite membrane [2, 3]

in the formation of a polyamide nanocomposite thin film on surface of the polymeric membrane support (50–200 nm thickness) [11]. Owing to the polarity differences between the two phases, the monomers react with one another at the surface. The cross-linking of the membrane was achieved by treating it with heat at 80 °C for 5 min to ensure that the interfacial reaction has been completed. The prepared TFN membrane was rinsed thoroughly and stored in water for subsequent characterization and experiments [2, 3]. As the IP method was prominently superior in terms of the independent optimization of the microporous substrate and skin layer properties, several types of TFN membrane have been created [6]. Membrane structural morphology and the barrier layer composition are influenced by a number of elements, e.g., post-treatment conditions, reaction time, solvent type, monomer concentration, and subsequent treatment. The prepared TFN membranes show a smoother surface or better hydrophilicity because the monomers contain more functional or polar groups [7]. The enhanced membranes hydrophilicity is advantageous to the improvement of the antifouling property, water permeability, and solute rejection performance.

3 Physicochemical Characterization of Nanocomposite Membranes

Graphene-based nanocomposite membranes were characterized using various analytical tools.

3.1 Surface and Cross-Sectional Morphology

The most commonly used technique for structural characterization (e.g., membrane pre-layer and skin layer) of nanocomposite membranes is SEM. An SEM image is formed by scanning a focused electron beam across the sample and recording the intensity of the scattered or secondary electrons. Shukla et al. [27] developed a GO-based nanocomposite membrane with polyphenylsulfone polymer from the phase inversion method. The cross-sectional morphology significantly varied in many ways with pristine polyphenylsulfone membranes after the addition of GO. Primarily, the pores in the sub-porous layer of the membrane were quite thinner and seemed in straight finger structures with open ends. Furthermore, the thickness of the skin layer was decreased, and very fine oval-shaped pores were developed just under this thin-skin layer. Finally, the thickness of the spongy support layer was also compressed but appeared much denser with interconnected pores (Fig. 4).

Karkooti et al. [12] observed that nanocomposite membranes have an asymmetric structure, with a dense skin top layer supported by a porous finger-type structure, and the average skin layer thickness increased significantly. The containing of hydrophilic

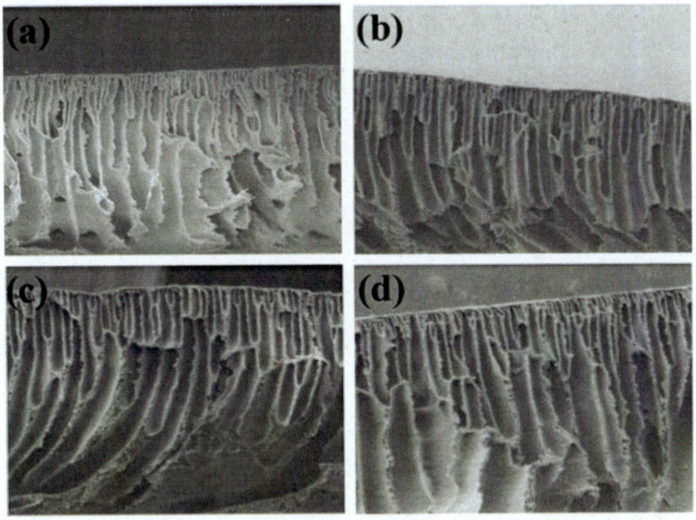

Fig. 4 SEM images of membrane cross-sectional morphologies of pristine membrane and different concentration of GO-based nanocomposite membranes [27]

additive GO tends to improve the thermodynamic instability of the membrane casting solution, which accelerates the demixing of solvent and non-solvent exchange rate in the coagulation bath. Moreover, the swelling of the polymeric membrane by hydrophilic GO prior to its solidification allows for greater passage of non-solvent to the casting nanocomposite membrane throughout the phase inversion process. Furthermore, with the differing concentrations of GO in the polymer matrix, a significant increase in the viscosity of the casting solution has countered these effects and increased the skin layer thickness owing to a reduction in the mutual diffusivities between the solvent and non-solvent [26].

The cross-sectional morphology of hollow fiber nanocomposite membranes confirmed the significant changes in the morphology of the membrane when the incorporation of GO. Figure 5 clearly shows that the cross-sectional morphology for the graphene-based nanocomposite membrane is asymmetric, consisting of a thin top layer and a porous sublayer with larger and uniform finger-like pores and macrovoids. Furthermore, the top layer of the nanocomposite membranes consists of closely packed nodules with micron-size void spaces. The hollow fiber nanocomposite outer layer turns out to be porous with a sponge-like substructure spanning about half of the cross section, and the inner side seems to be much less structurally asymmetrical [1].

As shown in Fig. 6, the top surface morphology of TFN membrane is observed, exhibiting morphologies such as "nodular" and "leaf-like." However, with polymerization time, the surface morphology changed to "hill and valley." The incorporated GO nanomaterials reacted with MPD and TMC during the polymerization reaction, and its oxygen-containing functional groups reacted only with MPD. The acyl

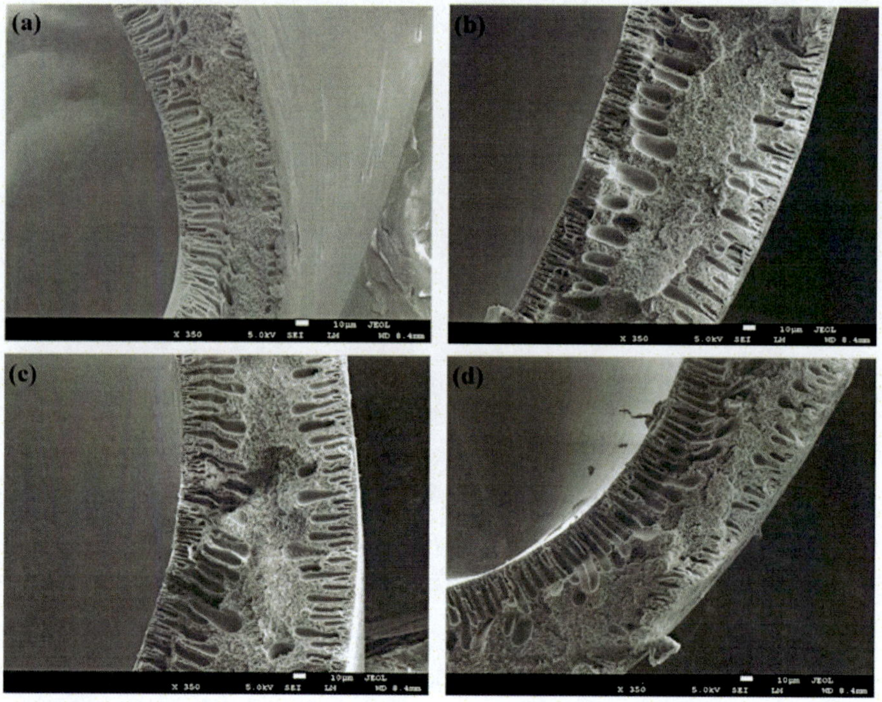

Fig. 5 SEM cross-sectional images of hollow fiber nanocomposite membranes with different concentration of GO

Fig. 6 SEM images of surface of polyamide graphene-based thin-film nanocomposite membrane [2, 3]

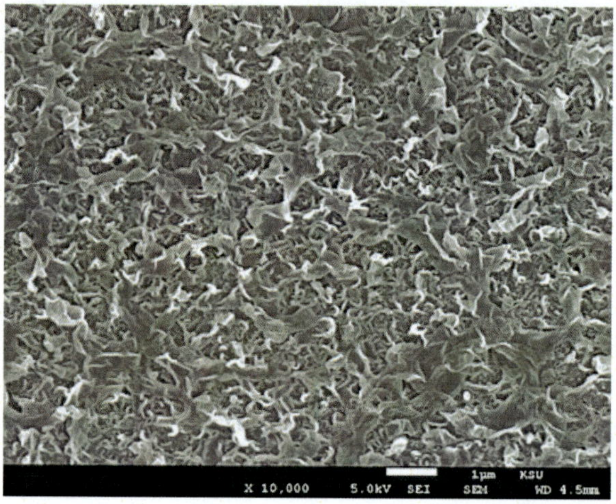

chloride groups of TMC interacted with the hydroxyl and carboxyl groups of GO nanomaterials. As suggested by Ali et al. [2, 3], the presence of hydrophilic GO nanomaterials in the organic phase enhances the miscibility of the aqueous and organic phases during interfacial polymerization. Contact between the hydrophilic nanomaterials in the organic phase and hydrated MPD from the aqueous phase results in particle hydration and the release of heat, which accelerate polymerization. With respect to the top of the thin polyamide layer, the thickness of polyamide was clearly reduced with different concentrations of GO, which is probably due to aggregation of GO at high GO loading. Aggregation can result in poor dispersion of nanomaterials in the polyamide matrix, decreasing the rate of polymerization, which leads to the formation of dense thin polyamide layers.

Then, the AFM method was used to obtain quantitative investigation of surface morphological structures, with 3-D topographic AFM images of the top surface of TFC membranes shown in Fig. 7. With the addition of GO nanomaterial into the polymer matrix, the enormous "peaks-and-valleys" are significantly exchanged by slighter ones, exhibiting a smoother surface. The decrease in the surface roughness of the nanocomposite membrane could be attributed to the functional groups, which increase the diffusion rate of the non-solvent during the phase inversion precipitation process and form a smoother surface (Fig. 7b) [27].

As Fig. 7d demonstrates, the formation of the polyamide layer with nanomaterial by the IP process increased the surface roughness of the polymeric support layer. The

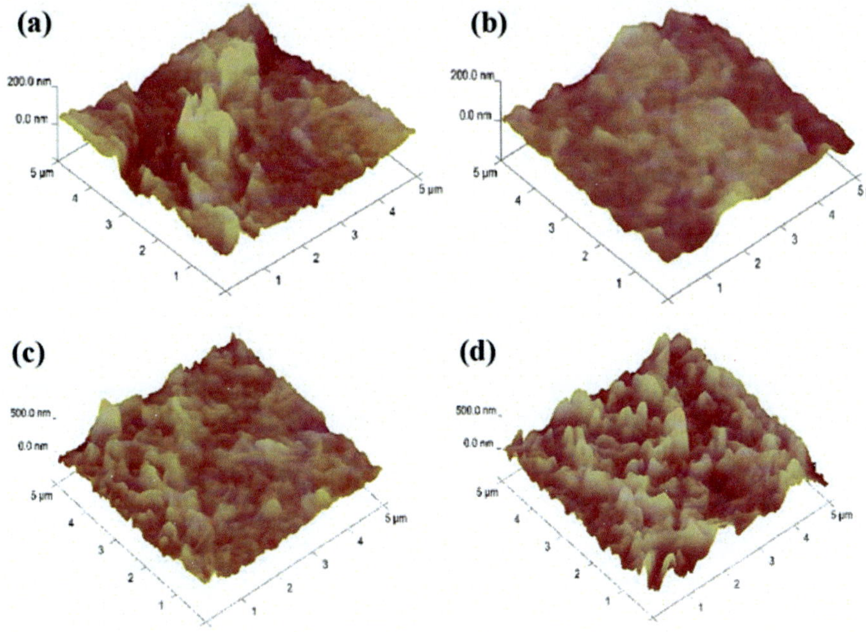

Fig. 7 AFM 3-D surface images of pristine membrane and nanocomposite [2, 3, 27]

graphene-based TFN membranes exhibited a "peaks-and-valleys" polyamide structure. The addition of graphene nanomaterials in the organic phase of TFN membrane is shown to have created higher peaks. Moreover, as discussed previously, the TFN membranes developed more apparent "leaf-like" folds in the SEM images, which are translated into the high peaks in the AFM images. The significant alterations of morphology attributable to the enhanced diffusion of diamine toward the IP zone are caused by the affinity of the MPD aqueous solution toward the hydrophilic nanomaterials. The diffusion rate of diamine into the organic solution of acid chlorides is increased by the unstable flow toward nanomaterials in the hexane phase. Therefore, the distribution of the reaction sites changes when the amide linkage is formed, causing higher peaks and lower valleys to form [2, 3, 8].

3.2 Membrane Hydrophilicity

The water contact angle is one of the most important factors applicable to the surface hydrophilicity of the membrane. In principle, it delivers information regarding the wettability of an ideal membrane surface because of the incorporating of the nanomaterials [5, 35]. In most cases, the intrinsic value of contact angle is perturbed by surface porosity and physical properties such as roughness and heterogeneity.

Based on the sessile drop method, water droplets with a volume of 3 μl were delivered onto the membrane surface, and the initial contact angle was measured after 3 s. A higher contact angle represents a more hydrophobic surface, and a lower contact angle reveals the higher surface energy, as well as the hydrophilic nature of membrane (see Fig. 8) [12]. As presented in Fig. 9, the contact angle data shows an interesting trend with the addition of graphene nanomaterial into the membrane matrix, tending to be more hydrophilic compared to pristine membrane. Yuan et al. [38] explained that the improved hydrophilicity of graphene is due to the functional groups that migrate spontaneously to the membrane surface to decrease the interface energy throughout the phase inversion process, which makes the nanocomposite membrane surface greater hydrophilic.

As suggested by [19] with increasing graphene concentrations into the polymer matrix, the nanomaterials move to the surface of the casting film to be exchanged with water, causing an increase in the contact angle, which means that more nanomaterials within the coagulation bath caused a membrane surface that was more hydrophobic

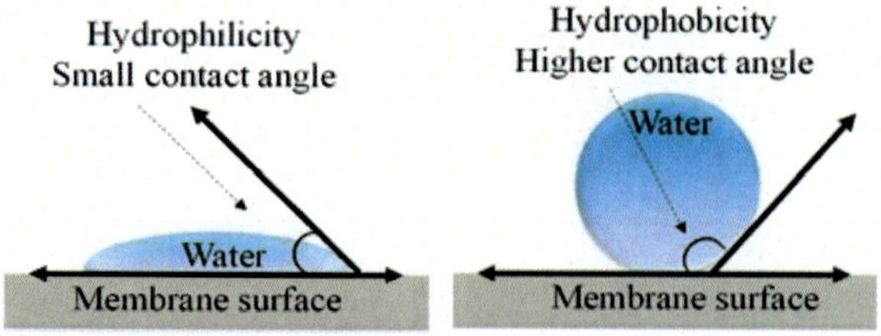

Fig. 8 Surface contact angle via sessile drop method

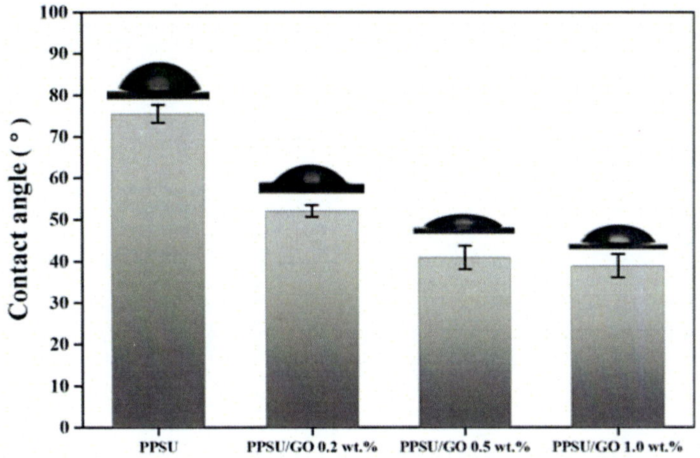

Fig. 9 Surface contact angle of pristine polymeric membrane and graphene-based nanocomposite membrane [27]

than that with low nanomaterial content. Surface hydrophilicity is a decisive parameter in determining the antifouling and water permeability characteristics of graphene-based membranes.

3.3 Surface Charge Determination of Nanocomposite Membrane

The surface zeta potential, which reveals the charge property of the graphene-based nanocomposite membrane samples, was considered by measuring the streaming potential using electrolyte solutions at different pH values. The identification of the surface charge as a function of pH is useful in helping to understand the acid–base properties of the membrane surface functional groups, along with anticipating the separation performance under different feed pH [18]. The surface charge of the nanocomposite membranes is negative in the higher pH range, and when it passes through an isoelectric point (IEP), it becomes positive in the lower pH range (Fig. 10). The graphene-based membrane gained a more effective negative charge than the pristine polymeric membrane. The alteration in the membrane surface charge with pH is attributed to the protonation (below IEP) and deprotonation (above IEP) of various GO functional group availability. Therefore, the nanocomposite membrane is more negatively charged under neutral and base conditions [14, 26].

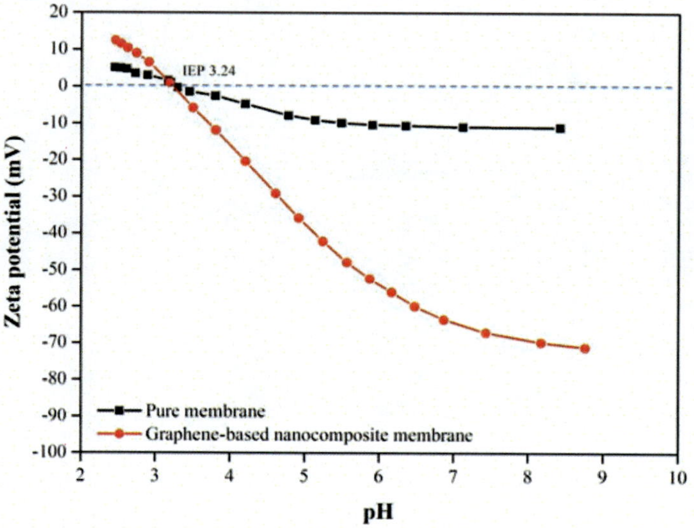

Fig. 10 Zeta potential of membranes as a function of pH [26]

3.4 Thermal Stability of Nanocomposite Membrane

The thermal stability of the graphene-based membrane samples was investigated by thermogravimetric analysis (TGA). The analysis depended on a high degree of precision in three measurements of the sample: weight, temperature, and weight loss with temperature in a given atmosphere. While thermal stability plots of nanocomposite materials can sometimes be difficult to interpret, this exploration is often used to determine their compositional analysis, using adequate pristine polymeric samples or reference plots [24]. A membrane weight loss curve can be used to define the point at which weight loss is most apparent. Furthermore, the feature of this thermo-analytical technique represents a useful tool to obtain information regarding thermal and oxidative stability, life expectancy, decomposition profile, moisture, and volatile content. As illustrated in Fig. 11a, the addition of GO nanomaterials to the polymer matrix improved the thermal stability of the membrane. As observed in the TGA curves, the decomposition of the membrane with temperature occurs in three steps. The first one corresponds to the evaporation of water embedded in the polymer chains (or absorbed moisture) and the evaporation of residual solvent. The second one corresponds to the degradation of the polymer chains, while for the nanocomposite membrane the degradation was more stable. The third one is attributed to the decomposition of the more stable oxygen functionalities, such as the carbonyls and phenols [23]. An increase in the thermal degradation temperature for the nanocomposite membranes is attributed to the obtainability of the polar functional groups of GO, leading to the strong interfacial bonding between the polymer matrix and the GO nanomaterials [27].

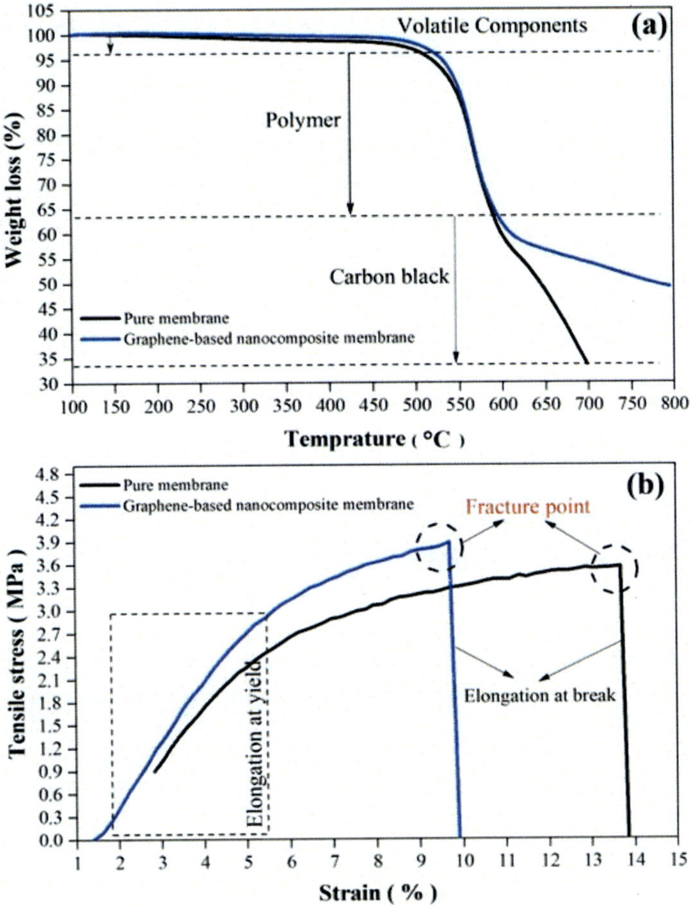

Fig. 11 **a** Thermal properties and **b** mechanical properties using tensile tests of graphene-based nanocomposite membrane [27]

3.5 Mechanical Properties

The presence of graphene in a polymer matrix can lead to significant strengthening, and for this purpose, a vast amount of research has been focused on the mechanical properties of graphene nanocomposite membrane. The evaluation of the mechanical strength in nanocomposite membrane is most commonly performed by the study of the stress–strain curves obtained during tensile testing [23, 27]. Figure 11b represents the comparison of mechanical properties of pristine membrane and nanocomposite membranes. Numerous parameters can affect the mechanical properties of nanocomposite membrane, including the structure of the graphene, preparation method, dispersion of the nanomaterials in the polymer matrix, nanomaterial–polymer matrix interactions, and the orientation of the nanomaterials [20, 39]. Sufficient mechanical strength is necessary to define the stability, end use, and processability of nanocomposite membranes. The interaction between the graphene nanomaterials entities and polymer components contributes a significant role in imparting mechanical properties to nanocomposite membranes. Furthermore, isotropy is an important property that is responsible for their mechanical properties. The good dispersion of graphene in a polymer matrix restricts chain movement and, thus, supports mechanical strength [32, 33]. However, the variation in mechanical properties depends on the size, uniformity, and volume fraction of the graphene nanomaterial entity, as demonstrated by [30]. Alam et al. [1] studied the effect of graphene with increasing content and observed that tensile strength along with toughness significantly increases, even with a very small amount of nanomaterials. The graphene-based nanocomposite membrane must possess good mechanical properties, e.g., toughness, to endure the pressure during water treatment application.

4 Nanocomposite Membrane Applications

Membrane separation has become an important technique to resolve water scarcity and environmental problems. Graphene-based nanocomposite membrane has nanopores that can be functionalized. They are used as a semipermeable barrier between two phases that retain contaminants, such as metal ions, organic–inorganic compounds, and microorganisms mainly from the aquatic environment. This superior property makes it ideal for higher permeate flux, higher selectivity, antifouling ability, and improved stability through controlled pore size and morphology [29]. A schematic of graphene-based nanocomposite membrane is shown in Fig. 12.

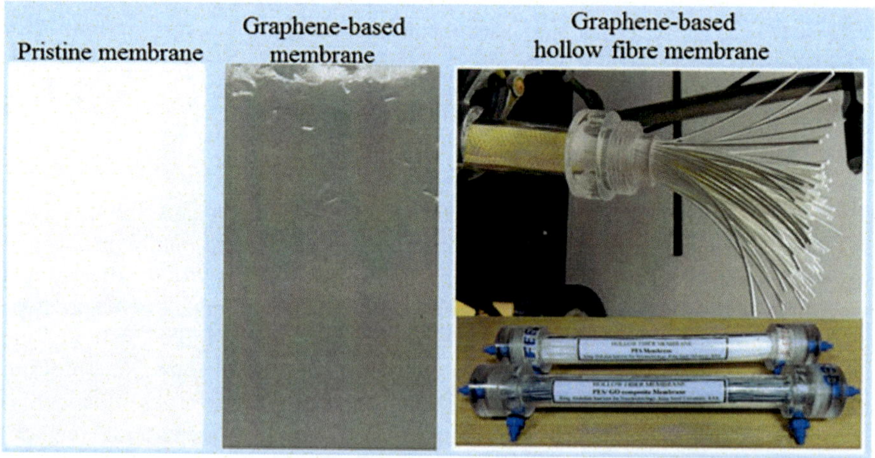

Fig. 12 Graphene-based flat sheet and hollow fiber nanocomposite membrane [1]

In recent years, graphene-based nanocomposite membranes have developed in rapid succession owing to its 2-D nanochannel between adjacent graphene layers and the adjustable nanoporous surface, including UF, NF, and RO. The application of these procedures has produced great economic and social benefits and has become one of the most important means of separation science.

4.1 Nanocomposite Ultrafiltration

Ultrafiltration is commonly used in the concentration, separation, and purification of assured solutions with low- and high-molecular-weight-soluble solutes and microorganisms. With the continual enhancement of membrane materials and process technology, UF in the field of water and wastewater treatment technology has been widely used. Incorporating graphene nanomaterials into the UF membrane is an area of great attention to many researchers. For example, [41] suggested the addition of GO nanomaterials to the polyethersulfone membranes applied in membrane bioreactor to treat milk-processing wastewater. The nanoparticles affect the pore formation process, and the addition of GO would also decrease the contact angle of the nanocomposite membrane significantly, thereby improving the hydrophilicity. Graphene-based nanocomposite membrane has less biofilm accumulation on the surface; therefore, the membrane bioreactor showed an increased capacity for the removal of organic matter, both in terms of COD and BOD of milk-processing wastewater. Furthermore, the nanocomposite membranes exhibited negative zeta potential, which induces an electrostatic repulsion between the microorganism and the membrane surface, hindering the surface attachment of the microorganism. Shukla et al. [27] used GO on polyphenylsulfone polymer matrix to create a novel nanocomposite UF membrane. It was found that with the incorporation of GO, the properties of membrane, including hydrophilicity, surface charge, and morphology were improved, preventing fouling as well as the adsorption of foulant on the surface (as shown in Fig. 13). Wang et al. [34] prepared novel polyvinylidene fluoride-doped with GO nanosheets via the immersion phase inversion process.

The GO-blended PVDF membrane always maintained a higher flux owing to the higher hydrophilicity of the prepared membranes, which intended that the antifouling performance of the membrane was significantly improved by the hydrophilic nature of the GO nanosheets. Yu et al. [37] modified GO by hyperbranched polyethylenimine (HPEI) and then blended it into a polyethersulfone casting solution to prepare nanocomposite membrane for antifouling and antibacterial application. When the HPEI-GO content in the membrane displayed a preferable antifouling and antibacterial performance against Escherichia coli (E. coli) and protein, the nanocomposite membrane provided a good antibacterial effect, as shown in Fig. 14. It is considered that irreversible damage can be induced on bacterial cells after direct contact with graphene nanomaterials because of membrane stress induced on bacterial cells resulting in the destruction of cell structures.

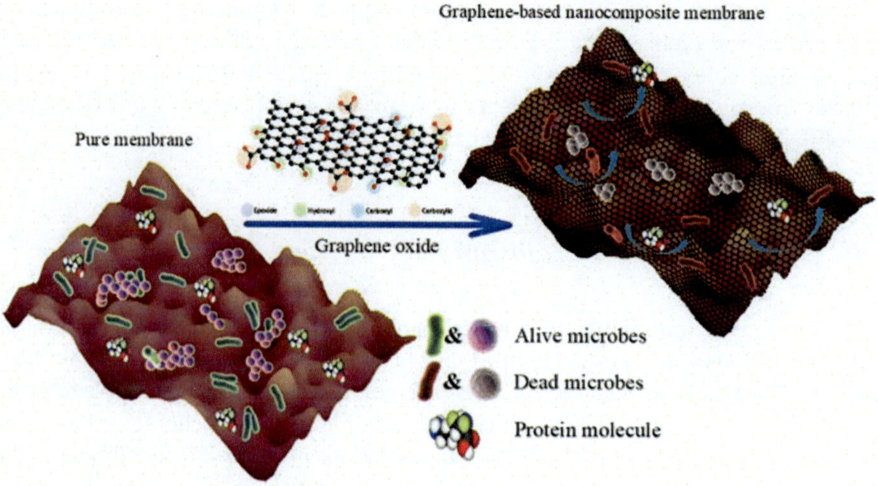

Fig. 13 Schematic model for antimicrobial and antifouling properties of graphene-based nanocomposite ultrafiltration membrane [28]

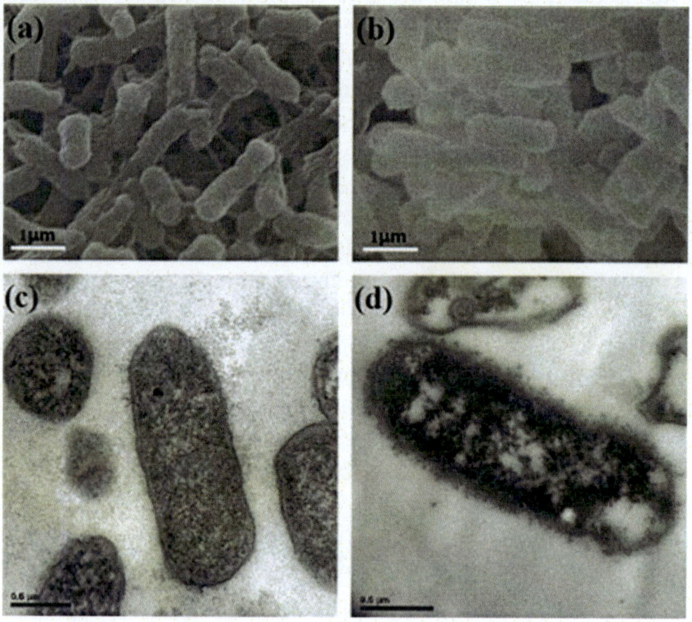

Fig. 14 SEM (**a**, **b**) and TEM (**c**, **d**) images of E. coli bacteria treated with membranes: (**a**), (**c**) pristine membrane; (**b**), (**d**) graphene-based nanocomposite membrane [37]. Copyright 2013 Elsevier

Ali et al. [2, 3] studied the combination of two nanomaterials, namely GO and silver (Ag). First, the silver-loaded graphene oxide (GO-Ag) nanocomposites were added at all concentrations as opposed to only the solitary polysulfone, resulting in a significant increase in the hydrophilicity, which means that the water flux has been influenced. The results also showed that adding GO-Ag nanomaterials into a polysulfone membrane inhibited the attachment, colonization, and biofilm formation against waterborne bacteria, such as E. coli and Staphylococcus aureus (S. aureus). The nanocomposite membrane showed that the cell wall and cell membrane were extensively damaged due to direct contact of bacterial cells with the membrane surface. In addition, the combination of GO-Ag nanomaterials exhibits the ability to enhance the oxidative stress of reactive oxygen species and lipid peroxidation, which play an important role in the killing of bacterial cells. Second, GO-Ag also displays efficient antifouling performance. The observation of graphene-based nanocomposite membrane reveals that the excellent properties of nanocomposite ultrafiltration membranes make them useful in a wide range of industrial and water separation applications.

4.2 Nanocomposite Nanofiltration

Nanofiltration is a molecular separation technique that occurs between UF and RO. Nanofiltration for water treatment is already on the market and currently seems to be the most mature method, although the current generation of water treatment devices can now capitalize on the new properties of nanomaterials and may prove to be of interest for both researchers and industries. It has a certain retention rate for inorganic salts and different rejection rates for different molecular weight particles. Graphene-based nanocomposite membrane can also be fabricated to be used for NF. The properties showed that nanocomposite membrane can produce potable water from the brackish groundwater [31].

Wang et al. [32, 33] demonstrated that nanoporous graphene membranes can successfully remove salts from water. Salt rejection may be attributed to physical sieving and Donnan exclusion in view that the hydrated radii of ions are slightly smaller than the radius of the carbon nanochannel. The rejection performance mainly depended on the charge base separation for nanocomposite NF membrane, following the Donnan exclusion mechanism. According to Donnan exclusion, the ionic concentrations at the membrane surface are not equivalent to those in the bulk solution. The counterion (ions with opposite charge of the nanocomposite membrane surface) concentration is higher at the membrane surface compared with the bulk solution. The co-ion (ions with the same charge of the nanocomposite membrane surface) concentration is lower at the membrane surface. When an external pressure is applied on the membrane surface, water can pass through the membrane, while co-ions are rejected owing to the Donnan potential. Simultaneously, counterions are also rejected owing to the requirements of electroneutrality. Han et al.

[9] fabricated the membrane by the synergistic assembling of graphene and multi-walled carbon nanotubes and applied organic dye, considering salt separation as well as antifouling performance. The incorporated graphene nanomaterials into the membrane, water flux, and rejection were significantly increased owing to the occurrence of graphene nanochannels. Therefore, more water molecules were able to pass through the membrane. Additionally, the higher salt rejection rate due to the nanomaterial nanochannel will also shrink at high ion strength because electrolytes screen the negatively charged carboxyl groups and suppress the electrostatic repulsion between graphene sheets. Shukla et al. [26] suggested that the addition of carboxylated–GO nanomaterials to the polyphenylsulfone matrix would affect the surface properties, thereby improving heavy metal ion (arsenic, chromium, cadmium, lead, and zinc) rejection at different concentrations. The rejection performance is governed by two main mechanisms—Donnan and dielectric exclusion. The rejection mechanisms briefly argue for ionic concentration effect, Donnan exclusion is the main influencing factor at lower concentrations, and the solvation energy barrier effect plays a secondary role in the rejection of ions. Alternatively, the solvation energy barrier effect plays a more dominant role, whereas the Donnan exclusion effect plays a secondary role at the higher ionic concentration. Figure 15 shows the heavy metal ion rejection performance mechanisms. Furthermore, [38] studied the salt rejection performance through carboxylated–GO nanocomposite membranes developed by a pressure-assisted self-assembly technique. The addition of nanomaterials reduced the contact angle and yielded a higher surface hydrophilicity. Therefore, the water permeability was significantly enhanced because the water molecules are first absorbed by the carboxyl groups on the membrane surface and penetrate into the empty space

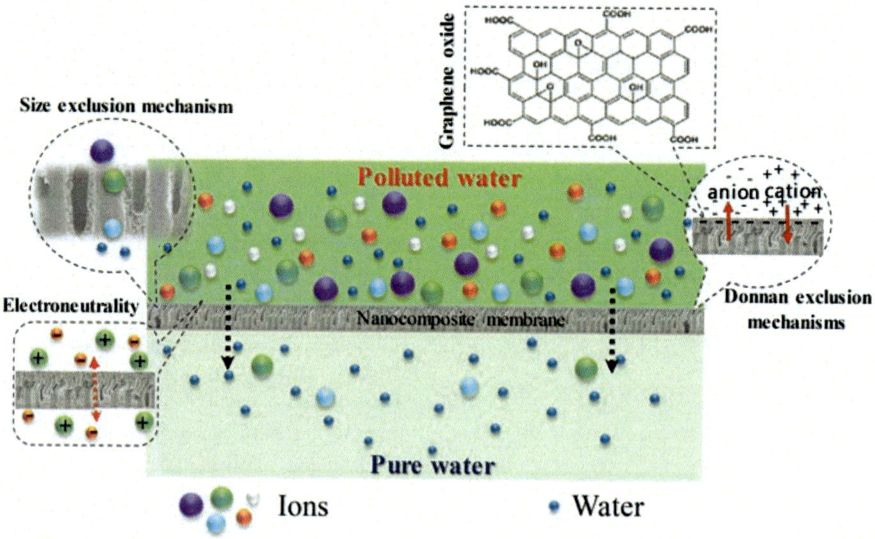

Fig. 15 Graphene-based nanocomposite membrane ions rejection mechanisms [26]

between carboxylated–GO layers as well as reduce the percentage of oxygen atoms, implying that the increasing unoxidized areas lead to faster water diffusion. The carboxylated–GO membrane demonstrated that the higher rejection of dianion and monovalent anion compared with the pristine GO membrane attributable to carboxyl groups would increase the negative charge distribution on the membrane surface, promoting stronger electrostatic repulsion between the anions and the membrane. Moreover, carboxylated–GO membranes have a similar trend with typical nanofiltration in rejecting the order of dianion over monovalent anion owing to the effect of the ionic mobility.

4.3 Nanocomposite Reverse Osmosis

Reverse osmosis was first introduced by Reid in the USA in 1953, who then developed a cellulose acetate-based polymeric asymmetric semipermeable membrane with high desalination rate and high permeability. In recent decades, a number of techniques have been used to prepare a new generation of polyamide nanocomposite membrane using a graphene nanomaterial via the interfacial polymerization of MPD and TMC, and they are still frequently employed [21]. The concept of these nanomaterials is to improve the surface properties and overcome the permeability–selectivity trade-off in graphene-based thin-film nanocomposite (TFN) membrane for water treatment applications. By incorporating a small quantity of graphene nanomaterials into the PA layer of the membrane, the water permeability of the resultant membrane could be improved by almost an order of magnitude without compromising salt rejection. This demonstrates the TFN membrane performance resulting from the unique characteristics that offer preferential flow paths for water molecules through its super-hydrophilic surface and structure [6]. Wang et al. [32, 33] have reported on the zeolitic imidazolate framework-8(ZIF-8)/graphene oxide TFN membrane via interfacial polymerization. The as-prepared (ZIF-8)/GO hybrids TFN membrane showed an inherent hydrophilicity of the polyamide layer, enhancing the water permeability as the ZIF-8/GO hybrid nanocomposite reduces the interaction of the polyamide chains and undermine the polymer chain packing to a certain degree. Furthermore, it is notable that the divalent salt rejection of the nanocomposite membrane increases significantly, up to 100%. This may be triggered by the improved negative charge density of nanocomposite membranes with the introduction of ZIF-8/GO nanocomposites. As stated by the electrostatic repulsive interaction mechanism, the negative charge of the membrane surface attracts a high-valent cation and repulses a high-valent anion. In addition, the nanocomposite membrane has an optimal antimicrobial activity with E.coli bacteria. He et al. [10] studied the impact of varying the graphene oxide content on poly(amide) layer and then applied it to desalinate seawater. They found that the graphene oxide nanomaterials disperse well across the composite membranes, leading to a lower membrane surface energy and enhanced hydrophilicity, increasing the pure water flux by up to 80% without significantly affecting the salt selectivity. The salt rejection of the nanocomposite membranes was enhanced up to 100%, attributed to physical and morphological factors simultaneously affecting diffusion and rejection. Owing to the high surface area of native graphene oxide, this will lead to a high surface of contact with the PA material within the hybrid nanocomposite membranes. The presence of different charges of salts induces charge effects that influence the water separation performance of the membrane, showing higher rejection. The nanocomposite membrane mainly has a negative surface charge due to the functional groups on the membrane surface; the membrane tends to repel negative anions. The molecular size and Donnan exclusion effects, caused by the acid groups attached to the polymer backbone in this hybrid graphene nanocomposite membrane, will determine the rejection rate of salts. Yin et al. [36] prepared a graphene-based TFN membrane by the in situ interfacial polymerization process for the removal of salts from aqueous solution. The results showed that the membrane had a very stable water flux and salt rejection during the extended filtration process, whereas an increased surface hydrophilicity will facilitate the interlayer space inside GO nanosheets, providing additional short paths for water permeation through the PA thin-film layer. In another study, [2, 3] successfully synthesized composite of Ag-doped graphene (GO-Ag) and investigated its antifouling and antibacterial activity of thin-film nanocomposite membrane. As a result, the GO-Ag nanomaterials revealed good dispersibility, and the membrane properties and structure, such as zeta potential, contact angle, and a smooth surface, were improved. Including a GO-Ag concentration of 80 ppm in the membrane increased the water flux recovery ratio (89%) and provided low irreversible resistance (10%) as well as eliminated 86% of viable E.coli cells in bacterial suspensions. Therefore, these graphene-based TFN membranes can potentially be used for water separation and purification applications.

5 Conclusion

Graphene is a novel nanomaterial, which includes many properties to develop a graphene-based nanocomposite membrane. The superior properties of nanoparticles including their thermal and mechanical stability, large specific surface area, surface properties, and composition, as well as the type of polymer, have strong influences on the desalination and water filtration performance of nanocomposite membranes; in addition, they are cost-effective. Therefore, this chapter provides systematic information regarding recently developed/modified methods, physiochemical properties, and filtration performance, such as bacterial and pollutants summarized for graphene-based nanocomposites membrane. However, graphene nanomaterials compatibility with polymer matrix and scalable manufacturing of graphene-based nanocomposite membranes will greatly enhance their applications. Graphene is a very appropriate nanomaterial to develop a more efficient and cost-effective membrane for filtration processes, aiding its further development to make them applicable in real water and wastewater treatment processes and different applications.

Acknowledgements The authors are thankful for the financial support from the King Abdullah Institute for Nanotechnology, Deanship of Scientific Research, King Saud University, Riyadh, Saudi Arabia. The authors thank the Deanship of Scientific Research and RSSU at King Saud University for their technical support.

References

1. Alam J, Shukla AK, Alhoshan M, Arockiasamy Dass L, Muthumareeswaran MR, Khan A, Ahmed Ali FA (2018) Graphene oxide, an effective nanoadditive for a development of hollow fiber nanocomposite membrane with antifouling properties. Adv Polym Technol 37:2597–2608
2. Ali AFA, Alam J, Shukla AK, Alhoshan M, Khaled JM, Al-masry WA, Alharbi NS, Alam M (2019) Graphene oxide-silver nanosheet-incorporated polyamide thin- film composite membranes for antifouling and antibacterial action against Escherichia coli and bovine serum albumin. J Ind Eng Chem 80:227–238
3. Ali FAA, Alam J, Shukla AK, Alhoshan M, Ansari MA, Al-Masry WA, Rehman S, Alam M (2019) Evaluation of antibacterial and antifouling properties of silver-loaded GO polysulfone nanocomposite membrane against Escherichia coli, Staphylococcus aureus, and BSA protein. React Funct Polym 140:136–147
4. An D, Yang L, Wang TJ, Liu B (2016) Separation performance of graphene oxide membrane in aqueous solution. Ind Eng Chem Res 55:4803–4810
5. Ayyaru S, Ahn YH (2017) Application of sulfonic acid group functionalized graphene oxide to improve hydrophilicity, permeability, and antifouling of PVDF nanocomposite ultrafiltration membranes. J Memb Sci 525:210–219
6. Bassyouni M, Abdel-Aziz MH, Zoromba MS, Abdel-Hamid SMS, Drioli E (2019) A review of polymeric nanocomposite membranes for water purification. J Ind Eng Chem 73:19–46
7. Calabrò V (2013) Engineering aspects of membrane bioreactors, handbook of membrane reactors. Woodhead Publishing. https://doi.org/10.1533/9780857097347.1.3
8. Ghanbari M, Emadzadeh D, Lau WJ, Matsuura T, Ismail AF (2015) Synthesis and characterization of novel thin fi lm nanocomposite reverse osmosis membranes with improved organic fouling properties for water desalination. RSC Adv 5:21268–21276
9. Han Y, Jiang Y, Gao C (2015) High-flux graphene oxide nanofiltration membrane intercalated by carbon nanotubes. ACS Appl Mater Interfaces 7:8147–8155
10. He L, Dumée LF, Feng C, Velleman L, Reis R, She F, Gao W, Kong L (2015) Promoted water transport across graphene oxide-poly(amide) thin film composite membranes and their antibacterial activity. Desalination 365:126–135
11. Hu Y, Lü Z, Wei C, Yu S, Liu M, Gao C (2018) Separation and antifouling properties of hydrolyzed PAN hybrid membranes prepared via in-situ sol-gel SiO_2 nanoparticles growth. J Memb Sci 545:250–258
12. Karkooti A, Yazdi AZ, Chen P, McGregor M, Nazemifard N, Sadrzadeh M (2018) Development of advanced nanocomposite membranes using graphene nanoribbons and nanosheets for water treatment. J Memb Sci 560:97–107
13. Kausar A, Ilyas H, Siddiq M (2018) Aptitude of graphene oxide-silver in advance polymer nanocomposite: a review. Polym Plast Technol Eng 57:283–301
14. Lai GS, Lau WJ, Goh PS, Ismail AF, Yusof N, Tan YH (2016) Graphene oxide incorporated thin film nanocomposite nanofiltration membrane for enhanced salt removal performance. Desalination 387:14–24
15. Li Y, Feng Z, Huang L, Essa K, Bilotti E, Zhang H, Peijs T, Hao L (2019) Additive manufacturing high performance graphene-based composites: a review. Compos Part A Appl Sci Manuf 124:105483

16. Lingamdinne LP, Koduru JR, Karri RR (2019) A comprehensive review of applications of magnetic graphene oxide based nanocomposites for sustainable water purification. J Environ Manage 231:622–634
17. Liu G, Jin W, Xu N (2015) Graphene-based membranes. Chem Soc Rev 44:5016–5030
18. Liu M, Zheng Y, Shuai S, Zhou Q, Yu S, Gao C (2012) Thin-film composite membrane formed by interfacial polymerization of polyvinylamine (PVAm) and trimesoyl chloride (TMC) for nanofiltration. Desalination 288:98–107
19. Mokkapati VRSS, Koseoglu-Imer DY, Yilmaz-Deveci N, Mijakovic I, Koyuncu I (2017) Membrane properties and anti-bacterial/anti-biofouling activity of polysulfone–graphene oxide composite membranes phase inversed in graphene oxide non-solvent. RSC Adv 7:4378–4386
20. Mukherjee R, Bhunia P, De S (2016) Impact of graphene oxide on removal of heavy metals using mixed matrix membrane. Chem Eng J 292:284–297
21. Nujić M, Habuda-Stanić M (2018) Toxic metal ions in drinking water and effective removal using graphene oxide nanocomposite. In: A new generation material graphene: applications in water technology, pp 1–471
22. Pal M, Mondal MK, Paine TK, Pal P (2018) Purifying arsenic and fluoride-contaminated water by a novel graphene-based nanocomposite membrane of enhanced selectivity and sustained flux. Environ Sci Pollut Res 25:16579–16589
23. Papageorgiou DG, Kinloch IA, Young RJ (2017) Mechanical properties of graphene and graphene-based nanocomposites. Prog Mater Sci 90:75–127
24. Rajisha KR, Deepa B, Pothan LA, Thomas S (2011) Thermomechanical and spectroscopic characterization of natural fibre composites. Interface Eng Nat Fibre Compos Maximum Perform 241–274
25. Scott K, Scott K (1995) Membrane materials, preparation and characterisation. Handb Ind Membr 187–269
26. Shukla AK, Alam J, Alhoshan M, Arockiasamy Dass L, Ali FAA, Mishra U, Ansari MA (2018) Removal of heavy metal ions using a carboxylated graphene oxide-incorporated polyphenylsulfone nanofiltration membrane. Environ Sci Water Res Technol 4:438–448
27. Shukla AK, Alam J, Alhoshan M, Dass LA, Muthumareeswaran MR (2017) Development of a nanocomposite ultrafiltration membrane based on polyphenylsulfone blended with graphene oxide. Sci Rep 7:41976–41987
28. Shukla AK, Alam J, Ansari MA, Alhoshan M, Ali FAA (2018) Antimicrobial and antifouling properties of versatile PPSU/carboxylated GO nanocomposite membrane against Gram-positive and Gram-negative bacteria and protein. Environ Sci Pollut Res 25:34103–34113
29. Song N, Gao X, Ma Z, Wang X, Wei Y, Gao C (2018) A review of graphene-based separation membrane: materials, characteristics, preparation and applications. Desalination 437:59–72
30. Suk JW, Piner RD, An J, Ruoff RS (2010) Mechanical properties of monolayer graphene oxide. ACS Nano 4:6557–6564
31. Surwade SP, Smirnov SN, Vlassiouk IV, Unocic RR, Veith GM, Dai S, Mahurin SM (2015) Water desalination using nanoporous single-layer graphene. Nat Nanotechnol 10:459–464
32. Wang Jing, Wang Y, Zhang Y, Uliana A, Zhu J, Liu J, Van Der Bruggen B (2016) Zeolitic imidazolate framework/graphene oxide hybrid nanosheets functionalized thin film nanocomposite membrane for enhanced antimicrobial performance. ACS Appl Mater Interfaces 8:25508–25519
33. Wang Jianqiang, Zhang P, Liang B, Liu Y, Xu T, Wang L, Cao B, Pan K (2016) Graphene oxide as an effective barrier on a porous nanofibrous membrane for water treatment. ACS Appl Mater Interfaces 8:6211–6218
34. Wang Z, Yu H, Xia J, Zhang F, Li F, Xia Y, Li Y (2012) Novel GO-blended PVDF ultrafiltration membranes. Desalination 299:50–54
35. Yin J, Deng B (2014) Polymer-matrix nanocomposite membranes for water treatment. J Memb Sci 479:256–275
36. Yin J, Zhu G, Deng B (2015) Graphene oxide (GO) enhanced polyamide (PA) thin-film nanocomposite (TFN) membrane for water purification. Desalination 379:93–101

37. Yu L, Zhang Y, Zhang B, Liu J, Zhang H, Song C (2013) Preparation and characterization of HPEI-GO/PES ultrafiltration membrane with antifouling and antibacterial properties. J Memb Sci 447:452–462
38. Yuan Y, Gao X, Wei Y, Wang X, Wang J, Zhang Y, Gao C (2017) Enhanced desalination performance of carboxyl functionalized graphene oxide nanofiltration membranes. Desalination 405:29–39
39. Zambare RS, Dhopte KB, Patwardhan AV, Nemade PR (2017) Polyamine functionalized graphene oxide polysulfone mixed matrix membranes with improved hydrophilicity and anti-fouling properties. Desalination 403:24–35
40. Zeng Z, Yu D, He Z, Liu J, Xiao FX, Zhang Y, Wang R, Bhattacharyya D, Tan TTY (2016) Graphene oxide quantum dots covalently functionalized PVDF membrane with significantly-enhanced bactericidal and antibiofouling performances. Sci Rep 6:20142–20152
41. Zinadini S, Vatanpour V, Zinatizadeh AA, Rahimi M, Rahimi Z, Kian M (2015) Preparation and characterization of antifouling graphene oxide/polyethersulfone ultrafiltration membrane: application in MBR for dairy wastewater treatment. J Water Process Eng 7:280–294

Carbon-Based Composite Hydrogels for Environmental Remediation

Omkar S. Nille, Akshay S. Patil, Govind B. Kolekar, and Anil H. Gore

Abstract This special chapter deals with the study of mechanistic and applicative properties of carbon-based composite hydrogels for the environmental remediation. The hydrogel types based on a synthetic and natural derived source of polymeric precursor as a backbone for the synthesis of hydrogel was also provided. The green, sustainable and biomass waste based carbonaceous materials like, activated carbon, graphene, graphene oxide, carbon nanotube, carbon dot etc. used in the designing of carbon-based composite hydrogel also discussed. The composite hydrogels are cheap, easy to use, sustainable, and can be easily synthesized. Carbon based materials which help in the crosslinking process through ionic interaction, electrostatic interaction, π-π interaction and weak Van der Waals forces. A unique property such as mechanical strength, porous nature, swelling ability, water insolubility, reusability and biodegradability of composite hydrogels helps in the adsorption and removal of environmental pollutants. We have briefly discussed the literature survey about different carbon-based composite hydrogels and their uses in removal and detection of variety of pollutants. In this chapter, further we have elucidated about the adsorption, removal and sensing of the heavy metals, organic, inorganic pollutants, dyes and pharmaceutical pollutants etc. These hydrogels are more efficient, reusable, biodegradable, ecofriendly, and hence these can be widely useful in the environmental remediation on large scale.

Keywords Carbon-based materials · Composite hydrogel · Sustainable and cheap · Biodegradable · Environmental remediation

---

O. S. Nille · A. S. Patil · G. B. Kolekar · A. H. Gore (✉)
Fluorescence Spectroscopy Research Laboratory, Department of Chemistry, Shivaji University, Kolhapur, MS 416004, India
e-mail: anilanachem@gmail.com; anil.gore@utu.ac.in; anilhgore@gmail.com

A. H. Gore
Department of Chemistry, UKA Tarsadia University, Maliba Campus, Tarsadi, GJ 394350, India

M. Jawaid et al. (eds.), *Environmental Remediation Through Carbon Based Nano Composites*, Green Energy and Technology,
https://doi.org/10.1007/978-981-15-6699-8_20

1 Introduction

Day by day increasing population and industrialisation due to which the consumption of environmental resources is increased by a human being, but the improper way of management causing more environmental pollution. Commonly untreated wastewater from various industries and drainage's, is the main reason for soil pollution and water pollution which is the major cause of many more environmental problems. These are directly or indirectly influencing diverse effect on human health, plants, animals and on many more environmental factors. There are lots of traditional ways have been followed for the treatment of wastewater like adsorption, ion exchange, ozonolysis, water membrane filtration, electrochemical treatment, chemical precipitation, ion exchange, etc. [31] which are time-consuming, major costs, and its process by-products are non-biodegradable. So, to overcome and fix the abovementioned problems, the use of hydrogel in environmental remediation [20] is one of the unique strategies. Hence, nowadays researchers have been focusing in the field of hydrogels and their composite materials for environmental remediation on a large scale.

The natural or synthetic hydrogel molecules are capable with inherent, unique physicochemical property such as cross-linked network, hydrophilicity, swellability in the solvent (water holding capacity), rapid gelation, good mechanical strength, a large amount of small porosity, biocompatibility and biodegradability, nontoxicity [20]. These charming properties of hydrogel increase its wide usage in water purification plant in the different industries. Also, hydrogels have well sorption capacity or the static exchange capacity is used as ion exchange in chromatographic technique.

Moreover, use of hydrogel in the agriculture is the best way, to avoid expenditure on the irrigation system, because in irrigation or direct application of water to crop its crisis as an evaporation process excess of water evaporated so we can prevent the wastage of water using hydrogel system. Controlled releasing property of hydrogel used in agriculture, bio-medical and drug delivery etc. As well as hydrogels are playing important role in the different areas like environmental remediation, tissue engineering, waste-water treatment, organic and inorganic pollutants, pharmaceutical, sensing, detection and removal of pollutants and heavy metals, etc. [20].

2 Hydrogels

2.1 What Is Hydrogel?

Hydrogels are the 3D cross-linked polymeric networked blends of natural or synthetic polymers. This is semi-solid; gel-like in nature having the capacity to hold a significant amount of water or fluids inside it due to the presence of high porous nature. The hydrogels were initially invented in china (1950s). Researchers Otto Wichterle and Drahoslav Lim are the pioneers of the hydrogels and he used these gels in tissue engineering and contact lenses [10].

The functional groups present on the surface of hydrogels like –OH, –COOH, and $-NH_2$, makes it unique and more interactive to form crosslinking and interact with functional groups present on the carbon composites. These functional groups make hydrogels stimuli-responsive (pH and temperature-sensitive). The cavities inside the hydrogels due to high porosity entrap the different types of pollutants, heavy metal ions, industrial dyes, pharmaceutical drugs, etc. Hydrophilic nature is one of the important phenomena for swelling capacity of the hydrogels.

2.2 Composite Hydrogels

Hydrogels make easily interact with Hydrogen bonding, ionic interactions, π-π interaction and weak Van der Waals forces with functional groups present on carbon composites to form composite hydrogels. When carbon-based nanomaterials, carbon composites, carbonaceous materials (activated carbon, carbon dots, graphene, graphene oxide, carbon nanotubes) are incorporated or easily blend with hydrogels to form carbon-based composite hydrogels [20].

3 Classification of Hydrogels

Hydrogels can be mainly classified on the basis of their sources from which it is synthesized. We can classify it generally depending on the polymeric material source, as naturally derived and manmade or synthetically derived. Mechanical, physical and reactive properties of hydrogels depend on these abovementioned sources. Bio-derived polymeric hydrogels have bio-compatibility, bio-degradability, nontoxic, cheaply and easily available but lack of mechanical properties like strength, reusability, etc. However, on the other side, the synthetically derived polymers have high mechanical strength, reusability, but they lack the bio-active properties (Fig. 1).

4 Synthesis of Carbon-Based Composite Hydrogel

Hydrogels are the gel-like, semisolid, 3D structure of different kinds of crosslinked polymers. Hydrogels can be synthesized by polymerisation of different types of naturally, synthetically derived or a mixture of both natural and synthetic polymers. In the synthesis process of a hydrogel, there are different steps involved, at the initiation polymerisation process different kinds of homogeneous or heterogeneous polymers are taken and mixed for the formation of blends of the polymers. After that, the next important part is the crosslinking process in which blends of polymers reacted with

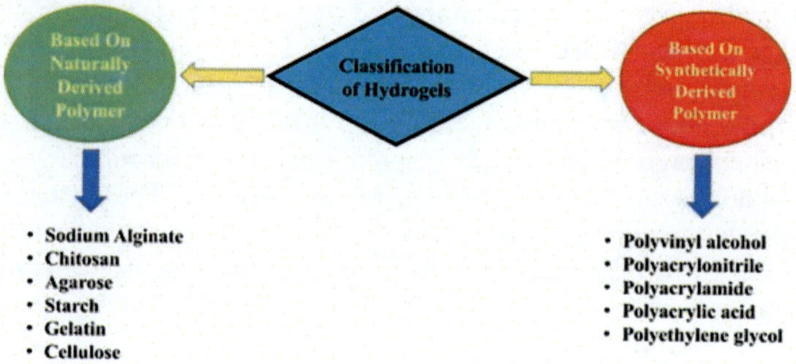

Fig. 1 Schematic representation for classification of hydrogels

the suitable crosslinkers for the formation of crosslinked gels. The crosslinkers and gelation process like freeze-thaw method gives the mechanical strength and stability to the hydrogel framework or structure of hydrogels. The crosslinked gels are placed for the setting of polymerisation that is nothing but the termination process. This hydrogel is placed for the desired time for the gelation. Composite hydrogels can be synthesized with the help of different kinds of carbon-based materials like activated carbon, bio-charcoal, graphene, graphene oxide, carbon dots, magnetic carbon, etc. incorporated with the polymeric network.

These carbonaceous materials contain different kinds of functionalities which help in the crosslinking process, through electrostatic interactions, ionic interactions, π-π interaction, weak Van der Waals forces like hydrogen bonding between polymers and carbon-based materials. The best way of synthesis of composite hydrogel is the process in which gelation of carbonaceous materials in the matrix of hydrogel framework. These carbonaceous materials can be incorporated in-situ that is while the blending of the polymers which forms the composite hydrogel. On the other hand, these carbonaceous materials can also be embedded after the gelation process that is ex-situ process. These carbon-based composite hydrogels give special properties to the hydrogels due to which it gives a wide range of applications in different fields (Fig. 2).

5 Crosslinking Methods

Crosslinking is one of the mandatory processes in the preparation of hydrogels, which plays an important role in the gelation process. There are two kinds of crosslinking processes one is physical crosslinking and on other hand chemical crosslinking. Hydrogels get excellent properties by the crosslinking of the polymers. The use of the said method influences the formation of singly networked or doubly networked 3D polymeric networked hydrogels, to get strength, active sites and porous nature.

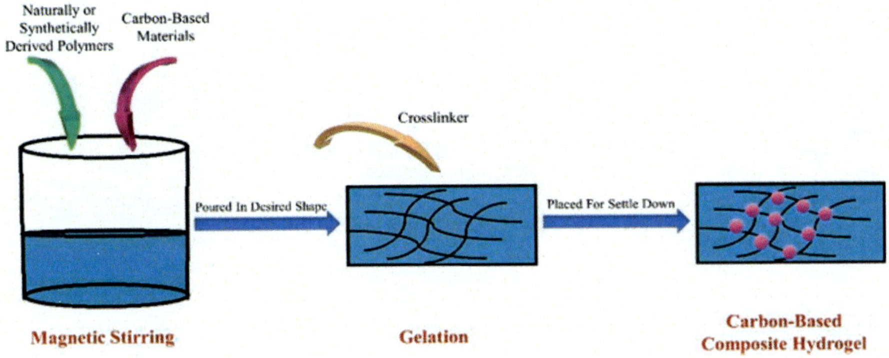

Fig. 2 Schematic representation of synthesis of carbon-based composite hydrogels

The physical method of crosslinking involves the freeze-thaw process. The polymers are blended which then poured in required shape followed by, freezing approximately below −30 to 40 °C for about 24 h and then these frozen hydrogels take out for thawing at room temperature for about 3–4 h and washed several times with deionised water. This process is repeated for 3 times, which gives greater stability, and mechanical strength to hydrogels. The chemical crosslinking process includes the process in which a mixture of polymeric material is placed in contact with 10% $CaCl_2$ which makes the crosslinking in polymers and helps for the gelation process. These chemically crosslinked hydrogels are washed with deionised water for several times to remove the excess of chloride ions and unreacted $CaCl_2$. The functional groups on the polymers and carbon-based materials also influence the process of crosslinking. The ionic functionalities such as carboxylic acid, hydroxyl, amine which makes crosslink through ionic interaction, hydrogen bonding, π-π interaction, and electrostatic interactions which gives more mechanical strength and stability to the carbon-based composite hydrogel (Fig. 3).

6 Types of Carbon-Based Composite Hydrogels

On the basis of different types of abovementioned carbonaceous materials, we can distinguish carbon-based composite hydrogel into the different type such as biochar, activated carbon, carbon nanotube, graphene/graphene oxide etc based composite hydrogels [20]. Different forms of carbon materials possess distinguishable properties like surface area, 0–3 dimensional framework, mechanical strength, presence of different functional groups on their surfaces such as hydroxyl, epoxide, carboxylic acid makes it more interactive, quantum dots are high fluorescent in nature having high quantum yield can be easily incorporated into the hydrogel matrix [2], hence due to these outstanding properties scientists and researchers all over the globe attracted in the field of carbon-based composite hydrogels. In the recent year (2017) the

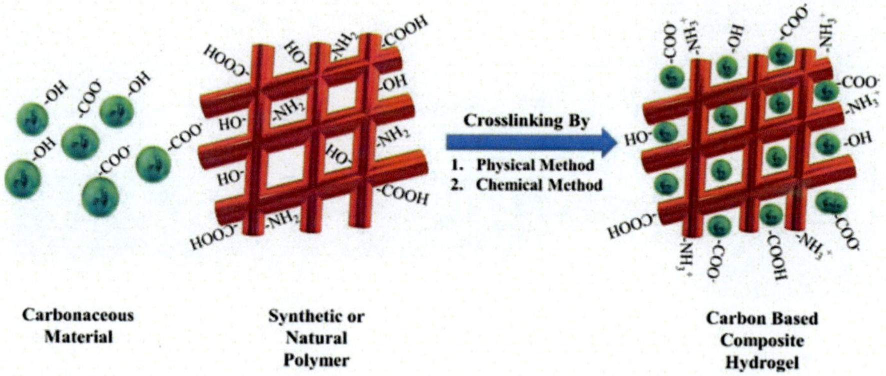

Fig. 3 Schematic representation of crosslinking between carbonaceous materials and polymers to form composite hydrogel

researcher Wu et al. reported on the sodium alginate immobilized with β cyclodextrin and graphene oxide which is used for the removal of methylene blue dye from solution [23]. Recently, Yang et al. synthesized double network hydrogel beads by directly adding a mixture of graphene oxide and sodium alginate solution. These prepared hydrogel beads showed maximum adsorption capacity to cationic metals like Mn^{2+} about 56.49 mg/g [24]. Zhuang et al. worked on the TiO_2 Nanotube/graphene oxide hydrogel for the removal and separation of the ciprofloxacin [30]. The use of carbon-based magnetic hydrogel has the special property that is we can easily separate it from solution by using an external magnetic field. The different types of reported carbon-based composite hydrogels and their environmental remediation applications are given in Table 1.

6.1 *Activated Carbon Composite Hydrogels*

Activated carbon possess large surface area, which helps for the adsorption of the pollutants on its surface. Activated carbon is used for the application in the wastewater treatment, removal of dyes etc. The major problem of the use of activated carbon is that, the dispersion of the activated carbon, which is not so easy to separate. Lin et al. has designed alginate gel beads entrapped with activated carbon [13]. They have used activated carbon composite beads for the study of selectivity of these beads on cationic, anionic and neutral organic compounds. The concerned study shows that the prepared beads have a negative charge and adsorbs positively charged and neutral compounds. Also, these beads have the potential to remove agrochemicals.

Table 1 Carbon-based composite hydrogels and their applications for environmental remediations

S. No.	Types of carbon-based composite hydrogels	Materials used	Applications (detection/removal)	Maximum adsorption efficiency (mg/g)/removal (%)	References
1	Ball milled biochar sodium alginate (beads)	Biochar, sodium alginate	Cd (II)	227.1 mg/g	[22]
2	Sodium alginate beads entrapped with activated carbon (AG-AC)	Activated carbon, sodium alginate	Humic acid, gallic acid, methylene blue, p-Chlorophenol	–	[13]
3	Graphene/agarose based composite hydrogels	Graphene, agarose	Malachite green	90%	[14]
4	TiO_2/graphene hydrogel	TiO_2, graphene	Ciprofloxacin	181.8 mg/g	[30]
5	Alginate/graphene double network nanocomposite hydrogel	Sodium alginate, reduced graphene oxide	Cu^{2+}, $Cr_2O_7{}^{2-}$	169.5 mg/g 72.5 mg/g	[29]
6	Graphene oxide/calcium alginate (GO/CA)	Graphene oxide, Sodium alginate	Tetracycline	131.6 mg/g	[28]
7	Graphene oxide/alginate beads (go/alginate)	Graphene oxide, sodium alginate	Methylene blue	4.25 mmol/g	[17]
8	Sodium alginate/graphene oxide double network composite hydrogel	Sodium alginate, graphene oxide	Mn^{2+}	56.49 mg/g	[24]
9	Calcium alginate/graphene oxide aerogel (mp-CA/GO)	Graphene oxide, sodium alginate	Pb^{2+}, Cu^{2+}, and Cd^{2+}	368.2, 98.1 and 183.6 mg/g	[16]
10	Graphene oxide/sodium alginate/polyacrylamide (GO/SA/PAM)	Graphene oxide, sodium alginate, polyacrylamide	Rhoda mine 6 G, methylene blue, methyl orange	–	[14]

(continued)

Table 1 (continued)

S. No.	Types of carbon-based composite hydrogels	Materials used	Applications (detection/removal)	Maximum adsorption efficiency (mg/g)/removal (%)	References
11	Graphene oxide/chitosan (GO/CS)	Graphene oxide, chitosan	Cu(II) Pb(II) Methylene blue Eosin Y	70 mg/g 90 mg/g >300 mg/g >300 mg/g	[5]
12	Graphene oxide/cellulose	Graphene oxide, cellulose	Cu^{2+} ions in aqueous solution	94.34 mg/g	[4]
13	Sodium alginate immobilized with β-cyclodextrin and graphene oxide (SCGG)	Sodium alginate, β-cyclodextrin and graphene oxide	Methylene blue	133.24 mg/g	[23]
14	Graphene oxide/polyethyleneimine GO/PEI	Graphene oxide, polyethyleneimine	Methylene blue, rhodamine B	–	[7]
15	Graphene oxide/polyvinyl glycol/ polyvinyl alcohol (GO/PEG/PVA)	Graphene oxide, polyvinyl glycol, polyvinyl alcohol	Cu^{2+} ions	917 mg/g	[18]
16	Graphene oxide/ poly (acrylic acid) (PAA) hydrogel	Graphene oxide, Poly(acrylic acid)	Cu^{2+} ions	209 mg/g	[21]
17	Graphene oxide/ fungus hyphae hydrogel films	Graphene oxide, fungal hyphae	Orange G dye	15 mg/L (up to 10 cycles)	[26]
18	Graphene oxide quantum dot/PVA (GOQDs/PVA)	Graphene oxide quantum dot, PVA	Fe^{2+}, Co^{2+}, Cu^{2+} ions	1×10^{-7} M (minimum detection level)	[3]
19	Carbon dots rooted Agarose (Agr/CD)	Agarose, carbon dots	Cr^{6+}, Cu^{2+}, Fe^{3+}, Pb^{2+}, and Mn^{2+} ion	27.75%, 54.85%, 38.48%, 83.97% and 35.41%	[6]

(continued)

Table 1 (continued)

S. No.	Types of carbon-based composite hydrogels	Materials used	Applications (detection/removal)	Maximum adsorption efficiency (mg/g)/removal (%)	References
20	Sulphonated carbon dot/chitosan (Ch-SO_3-CDs)	Sulphonated carbon dot, chitosan	Ca^{2+}, Mg^{2+}	68.01%, 56.35%	[2]
21	Carbon dots/PEG diacrylate (CDs/PEGDA)	Carbon dots, PEG diacrylate	Hg^{2+}	–	[8]
22	EDTA modified Fe_3O_4/sawdust carbon EDTA@ Fe_3O_4/SC	EDTA modified Fe_3O_4, sawdust carbon	Methylene blue, and brilliant green	227.3 mg/g and 285.7 mg/g	[9]
23	Magnetic graphene nanocomposite (PAA/GO/Fe_3O_4)	Magnetic graphene oxide, PAA	Pb^{2+}, Cu^{2+}, Cd^{2+}	85% (up to 5 cycles)	[14, 27]
24	Magnetic attapugite/fly ash/poly(acrylic acid) (ATP/FA/PAA)	Magnetic attapugite, fly ash, poly(acrylic acid)	Pb^{2+}-ions	38 mg/gm	[21]
25	Magnetic graphene oxide/poly(N-vinylimidazole-co-acrylic acid) hydrogel (MGO/PNA)	Magnetic graphene oxide, Poly(N-vinylimidazole-co-acrylic acid)	methyl violet, methylene blue, tartrazine, amaranth	609.8 mg/g 625.0 mg/g 613.5 mg/g 609.8 mg/g	[25]

6.2 Graphene Composite Hydrogels

Graphene is one of the allotropic forms of carbon, having two-dimensional (2D) layered structure. In graphene, the strong π-π interaction makes it hydrophobic and easy to reunite. Huang et al. have reported graphene-based hydrogels for water treatment [14]. They have used agarose as a stabiliser and reducing agent to form Graphene-Agarose hydrogel. Surface functional groups, large specific surface area strong hydrogen bonding nature of graphene and the porous structure of hydrogel helps for the adsorption of dyes. They have used malachite green as a model pollutant dye. The results showed excellent dye removal property giving 50% dye removal efficiency within 12 h and 90% dye removal after 7 days [14].

6.3 Graphene Oxide Composite Hydrogels

As compared to single networked hydrogels double networked hydrogels have much more excellent mechanical properties. Zhuang et al. have comparatively studied mechanical and adsorption properties of the single and double networked hydrogel [29]. There is the formation of a single network between Graphene oxide and alginate. The hydrothermal reduction of single networked hydrogels is carried out in which, graphene oxide is reduced to form a double networked hydrogel. They have carried out adsorption study on model pollutants $Cr_2O_7{}^{2-}$ and Cu^{2+} toxic anionic and cationic heavy metal ion. The results showed that the double networked hydrogels have higher adsorption capacity as compared to a single networked hydrogel. The reason behind this is that there is an increase in mechanical strength, lower swelling ratio and functionalities present on the graphene. The double networked hydrogels show improved adsorption capacity as well as good reusability than single networked hydrogels [29].

6.4 Magnetic Carbon Composite Hydrogels

The magnetic separation method is very useful to separate out magnetic carbonaceous materials from waste-water after adsorption by using an external magnetic field. The material is having properties like high adsorption efficiency, recyclability, and easy separation. Lei et al. have reported sodium alginate derived magnetic carbon material which used for the removal of toxic Cr(VI) from water [12]. The iron alginate [SA-Fe(III)] hydrogel beads are prepared by crosslinking of sodium alginate and $FeCl_3$ as a carbon precursor and magnetic precursor. The dried beads are heated and grinded and again heated at different temperatures in N_2 atmosphere for the carbonization. The resultant material is used for the adsorption of Cr(VI) from water. Yao et al. have prepared magnetic hydrogels by co-polymerisation of N-vinylimidazole (NVI) and acrylic acid (AA) embedded with magnetic graphene oxide modified with ethylenediamine [25]. This magnetic hydrogel is versatile material capable of adsorption for cationic and anionic dyes. The concerned study shows maximum adsorption for the cationic dye's methyl violet, and methylene blue having maximum adsorption capacity 609.8 mg/g and 625.0 mg/g respectively as well as anionic dyes tartrazine and amaranth having maximum adsorption capacity 613.5 mg/g and 609.8 mg/g respectively.

6.5 Carbon Nanotube Composite Hydrogels

Carbon nanotubes (CNTs) having one dimensional (1D) structure. CNTs having excellent properties like high specific surface area, high mechanical strength, high adsorption properties. The hydrogels embedded with CNT shows high mechanical strength, toughness and greater adsorption properties. Gu et al. has reported Graphene oxide (GO)/CNT hybrid hydrogel. The GO/CNT hybrid hydrogel is a pH-dependent and this helps for the adsorption of Uranium (VI) and can be used for nuclear industrial effluents and water treatment [1].

6.6 Carbon Dots Composite Hydrogels

Carbon dots are the class of zero dimensional (0D) materials having a size below 10 nm. The distinguishable properties are observed at the level below 10 nm having properties like low toxicity, tunable fluorescence, biocompatible and chemically inert. The carbon dots can be easily incorporated in the hydrogel matrix and easily helps for sensing, detection and removal of heavy metals, dyes, and other pollutants. The Baruah et al. has synthesized sulphur doped carbon dots incorporated with chitosan hydrogel. The synthesized hydrogel films successfully separate and remove Ca^{2+} and Mg^{2+} from solution [2].

7 Applications of Carbon-Based Composite Hydrogel

Improper management of wastewater treatment plants, effluent treatment plants (ETP), untreated waste water from the mines, foundries, metallurgies, pharmaceutical pollutants, dye industries, textile industries, chemical industries, etc. across the world is mixed through the drainage in potable water sources. This polluted water contains heavy metal ions like Pb, Fe, Cu, Mn, Cr, Sn, Cd, Hg, etc. [3, 2, 18, 8] which cannot be removed with high efficiency, which we consume in trace amounts that effect on our health and causes severe health damage. Fertilizers and pesticides

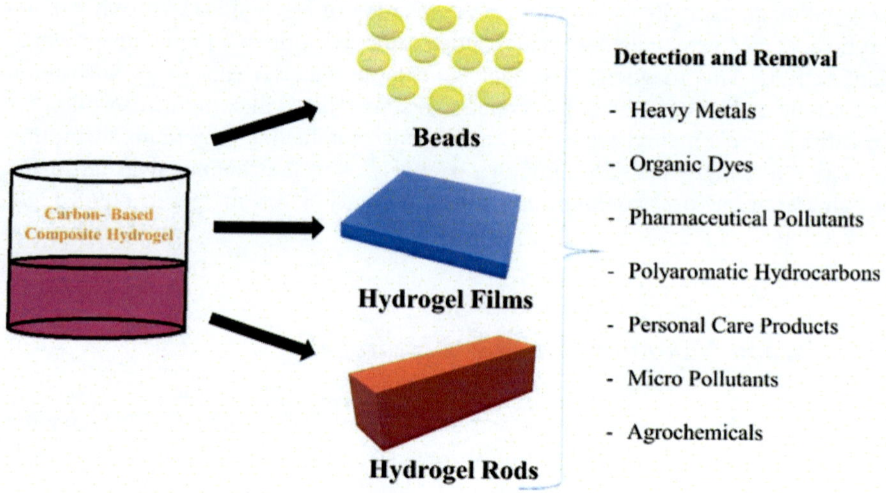

Fig. 4 Schematic representation of different forms of composite hydrogels and its applications

from the soil are mixed in rivers through rain-water, which also effects on water pollution. So, to overcome these diverse issues carbon-based composite hydrogels are more effective, sustainable, fast, low cost, etc. Over conventional methods of environmental remediation, composite hydrogel strips, beads, and films are playing an important role in the detection and removal of the pollutants. Hydrogel beads can help to easily remove water impurities, pollutants, sludges and can be easily separated by settling down. Organic and inorganic compounds, heavy metals, dyes, pharmaceuticals, etc. can be removed [8, 13, 19] (Fig. 4).

Swelling property, reusability and biodegradability are the unique and novel characteristics of the hydrogels [20]. Carbon-based composite hydrogel strips can sense the heavy metals and other pollutants from the water and soil. With due, all these unique and distinguishable properties broad area is generated for researchers and scientists to work in the field of environmental remediation.

7.1 Water and Wastewater Treatment

The need for pure, potable water is increased tremendously all over the world. Increasing water pollution has made crisis and health hazards [2]. The traditional ways of water treatment are time consuming, costly and tedious. So as to overcome the problem of water pollution, scientists and researchers have been working for new simple, low-cost methods. The new methods are developed in which carbon-based nanomaterials are embedded in polymeric network hydrogels for the wastewater treatment.

Jiang and Liu et al. synthesized a novel ternary magnetic hydrogel beads from magnetic attapugite/fly ash/poly(acrylic acid) (ATP/FA/PAA) used for the removal of Pb^{2+} ion [21]. Fly ash contains the amount of carbon content and helps for the adsorption of Pb^{2+} ion. These novel ternary hydrogel beads system has a maximum adsorption capacity of about 38 mg/gm of Pb^{2+} ion in 100 mg/L at 5 pH. The resolution of the hardness of water is quite difficult due to the presence of Ca^{2+} and Mg^{2+} ions in water.

In 2016 Baruah et.al. have prepared chitosan-based embedded with sulphonated carbon dot hydrogel films for the removal of Ca^{2+} and Mg^{2+} ions from pond water with a high removal efficiency of 68.01% and 56.35% respectively [2]. The research group of Zhang et al. have enthusiastically novel worked on graphene oxide/ fungus hyphae hydrogel films for the removal of Orange G dye [26]. It shows great dye removal efficiency through syringe filtration process of 15 mg/L of 10 mL dye solution at 2 pH of Orange G dye up to 10 cycles.

Yu et al. have synthesized a novel cost-effective and environmentally friendly nanocomposite hydrogel for the water treatment [21]. They have successfully prepared graphene oxide/poly(acrylic acid) (PAA) hyperbranched polymer hydrogel. The synthesized nanocomposite hydrogel shows high adsorption capacity of 209 mg/g of Cu^{2+} ions from solution at pH 5.

7.2 *Removal and Separation of Pollutants*

There is a tremendous amount of hazardous pollutants are contaminated through industries, pharmaceuticals, foundries, agrochemicals etc. in an environment which affects living organisms. There are different carbonaceous materials available for the removal of pollutants, but these martials disperse in the solution, hence immediate separation of these materials after adsorption is very difficult. So, there is now a challenge to remove and easy separation of pollutant simultaneously. The carbon-based composite hydrogels can promisingly help for the environmental remediation.

Zhu et al. have proposed significant study in the adsorption and separation of tetracycline as a model pharmaceutical pollutant [28]. They have synthesized graphene oxide/calcium alginate hydrogel through a freeze-drying method, which is having the potential of more adsorption capacity towards the tetracycline in aqueous solution. The maximum adsorption capacity of synthesized hydrogel for tetracycline is 131.6 mg/g at pH 6 by Langmuir Model.

There is an urgent need to develop simple advanced system having the potential to remove both cationic and anionic dyes as well as heavy metals and water contaminants in less time with low expenditure. By considering all above problems, group of Yunqiang Chen has successfully developed a hydrogel-based column for continuous water filtration [5]. Graphene oxide/chitosan hydrogel is used as a column packing material/membrane for rapid and continuous removal of Methylene Blue (cationic) and Eosin Y (anionic) dyes as well as Cu(II) and Pb(II) heavy metal ions by column packing from the contaminated water. The GO/Chitosan hydrogel column packing system shows the more adsorption capacity which is greater than 300 mg/g of both cationic and anionic type dyes. It also shows the removal of Cu(II) and Pb(II) ion with an adsorption capacity of 70 mg/g and 90 mg/g respectively. This column-based water filtration system is very useful for continuous and rapid water filtration.

7.3 Detection and Sensing of Pollutants

The detection and sensing of different kinds of pollutants at trace level from aqueous solution, water is a quite difficult task. The carbon-based composite hydrogel is promising material can be used as a detector or sensor for different pollutants. The detection and sensing of pollutants can be studied by using visual and photophysical methods. There are so many methods are available for detection and sensing like fluorescent carbon materials, quantum dots, different spectroscopic methods which lacks the sensitivity, tedious methods and time-consuming. Instead of that, by using carbon-based composite hydrogels we can easily detect and sense the pollutants by the naked eye and/or by photophysical study [6, 11].

Upama Baruah and Devasish Chowdhury have synthesized functionalised graphene oxide quantum dots/ PVA hybrid hydrogels used for the sensing of Fe^{2+}, Co^{2+} and Cu^{2+} ions in aqueous solution by colorimetric analysis [3]. When the synthesized hydrogel is put into the solution of Fe^{2+}, Co^{2+} and Cu^{2+} ions it shows brown, orange and blue colour respectively indicating that Fe^{2+}, Co^{2+} and Cu^{2+} ions are present in the solution. The minimum detection limit of the abovementioned ion is 1×10^{-7} M is studied by UV-Visible spectroscopy. This method shows easy, rapid and sensitive composite hydrogel for detection and sensing of Fe^{2+}, Co^{2+} and Cu^{2+} ions present in the solution.

Gogoi et al. have synthesized Carbon Dots Rooted Agarose Hydrogel film for the naked eye colour detection and separation of a heavy metal ion [6]. The prepared hydrogel strip when dipped into the heavy metal ion solution of Cr^{6+}, Cu^{2+}, Fe^{3+}, Pb^{2+}, and Mn^{2+} ion, then instantly hydrogel strips changes its original colour and shows yellow, blue, brown, white, and tan orange colour respectively. The maximum adsorption capacity of the above metals ions is 27.75, 54.85, 38.48, 83.97, and, 35.41 respectively. The synthesized hydrogel can be used as an efficient filtration membrane for the detection and separation of a heavy metal ion in aqueous solution. More recently, Vaibhav Naik from our research group designed a N-CDs/Agarose smart hydrogel hybrid strips and successfully implemented for highly sensitive naked eye colorimetric as well as fluorometric sensing of Dopamine selectively over other dopamine like organic moiety [15]. The aforementioned carbon-based composite hydrogels and its application with efficiency are tabulated in Table 1.

8 Conclusion and Future Perspective

These carbon-based composite hydrogels are facile, non-toxic, more effective and biodegradable; hence it can be used in environmental remediation. The designing of the composite hydrogels is an easy method, in which carbonaceous materials can be crosslinked with the naturally or synthetically derived polymers. These can be the replaceable materials for the traditional methods for the removal of pollutants in the environment. The main and important fact is that large scale production of a composite hydrogel is quite possible by utilizing sustainable, cheap and easily available sources. Hence in the future, the carbon-based composite hydrogels will be a complementary solution to the environmental remediation and enhance the quality of water which is very important for healthy growth of every living thing on this earth planet.

References

1. Adewunmi AA, Ismail S, Sultan AS (2016) Carbon nanotubes (CNTs) nanocomposite hydrogels developed for various applications: a critical review. J Inorg Organomet Polym Mater 26:717–737. https://doi.org/10.1007/s10904-016-0379-6
2. Baruah U, Konwar A, Chowdhury D (2016) A sulphonated carbon dot-chitosan hybrid hydrogel nanocomposite as an efficient ion-exchange film for Ca^{2+} and Mg^{2+} removal. Nanoscale 8:8542–8546. https://doi.org/10.1039/c6nr01129b
3. Baruah U, Chowdhury D (2016) Functionalized graphene oxide quantum dot-PVA hydrogel: a colorimetric sensor for Fe^{2+}, Co^{2+} and Cu^{2+} ions. Nanotechnology 27:0. https://doi.org/10.1088/0957-4484/27/14/145501
4. Chen X, Zhou S, Zhang L, You T, Xu F (2016) Adsorption of heavy metals by graphene oxide/cellulose hydrogel prepared from NaOH/urea aqueous solution. Mater (Basel) 9. https://doi.org/10.3390/MA9070582
5. Chen Y, Chen L, Bai H, Li L (2013) Graphene oxide-chitosan composite hydrogels as broad-spectrum adsorbents for water purification. J Mater Chem A 1:1992–2001. https://doi.org/10.1039/c2ta00406b
6. Gogoi N, Barooah M, Majumdar G, Chowdhury D (2015) Carbon dots rooted agarose hydrogel hybrid platform for optical detection and separation of heavy metal ions. ACS Appl Mater Interfaces 7:3058–3067. https://doi.org/10.1021/am506558d
7. Guo H, Jiao T, Zhang Q, Guo W, Peng Q, Yan X (2015) Preparation of graphene oxide-based hydrogels as efficient dye adsorbents for wastewater treatment. Nanoscale Res Lett 10:0–9. https://doi.org/10.1186/s11671-015-0931-2
8. Guo J, Zhou M, Yang C (2017) Fluorescent hydrogel waveguide for on-site detection of heavy metal ions. Sci Rep 7:1–8. https://doi.org/10.1038/s41598-017-08353-8
9. Kataria N, Garg VK (2019) Application of EDTA modified Fe_3O_4/sawdust carbon nanocomposites to ameliorate methylene blue and brilliant green dye laden water. Environ Res 172:43–54. https://doi.org/10.1016/j.envres.2019.02.002
10. Kopecek J (2009) Hydrogels: from soft contact lenses and implants to self-assembled nanomaterials. J Polym Sci Part A: Polym Chem 47:5929–5946. https://doi.org/10.1002/pola.23607
11. Kundu A, Layek RK, Kuila A, Nandi AK (2012) Highly fluorescent graphene oxide-poly(vinyl alcohol) hybrid: An effective material for specific Au^{3+} ion sensors. ACS Appl Mater Interfaces 4:5576–5582. https://doi.org/10.1021/am301467z

12. Lei Z, Zhai S, Lv J, Fan Y, An Q, Xiao Z (2015) Sodium alginate-based magnetic carbonaceous biosorbents for highly efficient Cr(VI) removal from water. RSC Adv 5:77932–77941. https://doi.org/10.1039/c5ra13226f
13. Lin YB, Fugetsu B, Terui N, Tanaka S (2005) Removal of organic compounds by alginate gel beads with entrapped activated carbon. J Hazard Mater 120:237–241. https://doi.org/10.1016/j.jhazmat.2005.01.010
14. Lu H, Zhang S, Guo L, Li W (2017) Applications of graphene-based composite hydrogels: a review. RSC Adv 7:51008–51020. https://doi.org/10.1039/c7ra09634h
15. Naik V, Zantye P, Gunjal D, Gore A, Anbhule P, Kowshik M, Bhosale SV, Kolekar G (2019) Nitrogen-doped carbon dots via hydrothermal synthesis: naked eye fluorescent sensor for dopamine and used for multicolor cell imaging. ACS Appl Bio Mater 2:2069–2077. https://doi.org/10.1021/acsabm.9b00101
16. Pan L, Wang Z, Yang Q, Huang R (2018) Efficient removal of lead, copper and cadmium ions from water by a porous calcium alginate/graphene oxide composite aerogel. Nanomaterials 8. https://doi.org/10.3390/nano8110957
17. Platero E, Fernandez ME, Bonelli PR, Cukierman AL (2017) Graphene oxide/alginate beads as adsorbents: influence of the load and the drying method on their physicochemical-mechanical properties and adsorptive performance. J Colloid Interface Sci 491:1–12. https://doi.org/10.1016/j.jcis.2016.12.014
18. Serag E, El Nemr A, El-Maghraby A (2017) Synthesis of highly effective novel graphene oxide-polyethylene glycol-polyvinyl alcohol nanocomposite hydrogel for copper removal. J Water Environ Nanotechnol 2:223–234. https://doi.org/10.22090/jwent.2017.04.001
19. Tang S, Chen Y, Xie R, Jiang W, Jiang Y (2016) Preparation of activated carbon from corn cob and its adsorption behavior on Cr(VI) removal. Water Sci Technol 73:2654–2661. https://doi.org/10.2166/wst.2016.120
20. Thakur S, Sharma B, Verma A, Chaudhary J, Tamulevicius S, Thakur VK (2018) Recent progress in sodium alginate based sustainable hydrogels for environmental applications. J Clean Prod 198:143–159. https://doi.org/10.1016/j.jclepro.2018.06.259
21. Van Tran V, Park D, Lee YC (2018) Hydrogel applications for adsorption of contaminants in water and wastewater treatment. Environ Sci Pollut Res 25:24569–24599. https://doi.org/10.1007/s11356-018-2605-y
22. Wang B, Gao B, Wan Y (2018) Entrapment of ball-milled biochar in Ca-alginate beads for the removal of aqueous Cd(II). J Ind Eng Chem 61:161–168. https://doi.org/10.1016/j.jiec.2017.12.013
23. Wu Y, Qi H, Shi C, Ma R, Liu S, Huang Z (2017) Preparation and adsorption behaviors of sodium alginate-based adsorbent-immobilized β-cyclodextrin and graphene oxide. RSC Adv 7:31549–31557. https://doi.org/10.1039/c7ra02313h
24. Yang X, Zhou T, Ren B, Hursthouse A, Zhang Y (2018) Removal of Mn (II) by sodium alginate/graphene oxide composite double-network hydrogel beads from aqueous solutions. Sci Rep 8:1–16. https://doi.org/10.1038/s41598-018-29133-y
25. Yao G, Bi W, Liu H (2020) pH-responsive magnetic graphene oxide / poly (NVI-co-AA) hydrogel as an easily recyclable adsorbent for cationic and anionic dyes. Colloids Surfaces A 588:124393. https://doi.org/10.1016/j.colsurfa.2019.124393
26. Zhang L, Li X, Wang M, He Y, Chai L, Huang J, Wang H, Wu X, Lai Y (2016) Highly flexible and porous nanoparticle-loaded films for dye removal by graphene oxide-fungus interaction. ACS Appl Mater Interfaces 8:34638–34647. https://doi.org/10.1021/acsami.6b10920
27. Zhang W, Shi X, Zhang Y, Gu W, Li B, Xian Y (2013) Synthesis of water-soluble magnetic graphene nanocomposites for recyclable removal of heavy metal ions. J Mater Chem A 1:1745–1753. https://doi.org/10.1039/c2ta00294a
28. Zhu H, Chen T, Liu J, Li D (2018) Adsorption of tetracycline antibiotics from an aqueous solution onto graphene oxide/calcium alginate composite fibers. RSC Adv 8:2616–2621. https://doi.org/10.1039/c7ra11964j
29. Zhuang Y, Yu F, Chen H, Zheng J, Ma J, Chen J (2016) Alginate/graphene double-network nanocomposite hydrogel beads with low-swelling, enhanced mechanical properties, and

enhanced adsorption capacity. J Mater Chem A 4:10885–10892. https://doi.org/10.1039/c6ta02738e

30. Zhuang Y, Yu F, Ma J (2015) Enhanced adsorption and removal of ciprofloxacin on regenerable long TiO_2 nanotube/graphene oxide hydrogel adsorbents. J Nanomater. https://doi.org/10.1155/2015/675862
31. Zinicovscaia I, Cepoi L (2016) Cyanobacteria for bioremediation of wastewaters. In: Cyanobacteria for bioremediation of wastewaters, pp 1–124. https://doi.org/10.1007/978-3-319-26751-7